7th International Materials Specialty Conference 2018

Held as Part of the Canadian Society for Civil Engineering Annual Conference 2018

Fredericton Canada
13-16 June 2018

Editors:

Xiomara Sanchez
Shahria Alam

ISBN: 978-1-5108-8975-0

Printed from e-media with permission by:

Curran Associates, Inc.
57 Morehouse Lane
Red Hook, NY 12571

Printed by Curran Associates, Inc. (2019)

For permission requests, please contact Canadian Society for Civil Engineering
at the address below.

Canadian Society for Civil Engineering
521-300 rue St-Sacrement
Montréal, QC
H2Y 1x4

Phone: (514) 933-2634
Fax: (514) 933-3504

info@csce.ca

Additional copies of this publication are available from:

Curran Associates, Inc.
57 Morehouse Lane
Red Hook, NY 12571 USA
Phone: 845-758-0400
Fax: 845-758-2633
Email: curran@proceedings.com
Web: www.proceedings.com

TABLE OF CONTENTS

Building Tomorrow's Society

Bâtir la Société de Demain

Fredericton, Canada

June 13 to 16, 2018 / *13 au 16 juin, 2018*

EFFECTS OF STICKY RICE ON COMPRESSIVE STRENGTH OF CEMENT STUCCO CONTAINING LIME AND SLAG

Zhang, Zhe[1,3], Miao, Buquan[1], Poitras, Gérard J.[1] and Xun, Yong[2]

[1] Département de génie civil, Université de Moncton, Moncton, NB, Canada

[2] School of Civil Engineering, Yancheng Institute of Technology, Yancheng, Jiangsu, China

[3] ezz6686@umoncton.ca

Abstract: As the concept of green building becomes increasingly popular, it is beneficial to find an eco-friendly stucco with better mechanical property and durability. As a traditional organic admixture, sticky rice has been widely used in the construction of ancient Chinese architectures, and it has been demonstrated that adding small quantity (up to 1%) of sticky rice can significantly improve the compressive strength of lime mortar. This paper studies the 56-day compressive strength of cement-based stucco as affected by sticky rice slurry, lime and slag contents. The results show that the 56-day compressive strength of stucco decreased with lime and sticky rice (higher than 1%). The highest strength was achieved with 1% sticky rice, 10% lime and 30% slag.

1 INTRODUCTION

Sticky rice mortar is one of the most widely used building materials in ancient times. It was first found applied to the brick tombs of Deng County, Henan Province, during the Northern and Southern Dynasties, Yang *et al.* (2010). Archaeologists detected sticky starchy substances in the cementation material of the tomb. In the third century, the application of sticky rice mortar has already become a relatively mature technology, Peng and Jackson (2015). From the archaeological findings, sticky rice mortar is mainly used for tombs, urban construction and water conservancy projects, Qiu (1980). Today, many ancient buildings using sticky rice mortar, such as the ancient city walls in Nanjing and a Qing Dynasty stone pagoda located in Baohe Town of Chongqing, which was built in the 16th century with a height of 10 meters and an inclination of more than 45°, still have high compressive strength, despite the erosion of wind and rain, Song (2008). It has been found that the sticky rice slurry acts as a biomineralization templating agent in the hydration reaction of the cementitious material, which guides the size, shape and structure of the crystalline particles, Wei *et al.* (2012). In this process, the particles of the sticky rice paste and the cementitious material distribute evenly, and wrap each other, making the structure of the hydration products denser. This dense microstructure contributes to higher sticky rice mortar compressive strength and surface hardness of the micro-based foundation, Zeng *et al.* (2008). Similar typical organic / inorganic composite structures are found in bones and shells of animals as well. It was also found that the alkaline environment formed by the alkaline materials can inhibit the growth of bacteria and has a preservative effect on the sticky rice paste, which can improve the durability of the material, Yang *et al.* (2010). The study of contemporary applications of sticky rice mortar is of significance in improving the strength of the material and the durability of the structures where this material is used in small quantity (up to 1%). Furthermore, studies have found that after the slag was added to the sticky rice mortar, the process of the biomineralization reaction was accelerated, and the microstructure of the mortar became much denser, Qiu (2014). As an industrial waste, slag can chemically react with cement and lime, and its utilization can also reduce environmental pollution.

As a durable and inexpensive building material for inner and exterior walls, cement-based stucco containing lime and other cementitious material such as slag and fly ash is gaining popularity during the last decade. However, few studies have been done to optimize the mix proportion. Moreover, incorporating sticky rice might be as beneficial to stucco as it is to mortar. In this paper, the effect of sticky rice, lime, and slag on cement-based stucco will be studied from the perspective of improving the compressive strength and durability of stucco.

2 MATERIAL AND EXPERIMENTAL METHODS

2.1 Material

The materials used for the stucco mixture were the following:

- Lime: High calcium hydrated lime with a calcium content of 48% and a neutralizing value ($CaCO_3$) of 130%;
- Slag: Ground granulated blast-furnace slag which meets the requirements of CSA A3001;
- Cement: General Use Portland Cement which meets the requirements of CSA A3001;
- Sticky rice slurry: The sticky rice flour comes from Golden Panda Inc. To make the sticky rice slurry, sticky rice flour and water were mixed together with a sticky rice/water ratio of 0.1. The mixture was stirred and poured into a rice cooker and boiled for an hour. Water was added regularly during this process to keep the water level unchanged.
- Plasticizer: superplasticizer which meets the requirements of ASTM C494;
- Water: Tap water.

2.2 Mix proportion and nomenclature

In this study, the water to cementitious material ratio was set at 0.47, and the cementitious material to sand ratio was set at 0.5. Based on the total mass of cementitious materials, three different ratios of sticky rice (1%, 2% and 3%), slag (15%, 30% and 50%), and lime (10%, 20% and 30%) were chosen for this study. The rest of the cementitious material was Type GU Portland cement.

2.3 Preparation of specimens

Workability is an important parameter in stucco applications. If the mix is too dry and harsh, it would be difficult to place; if the mix is too fluid, segregation can occur. The flow diameter, as measured following the procedure of ASTM C1437, was controlled to 260 ± 10 mm by adding appropriate amounts of superplasticizer for all the mixes in this study.

For each mix, three cubes of 50x50x50 mm were prepared following the procedure of ASTM C109. After 24 hours, the moulds were removed and the specimens were wrapped with plastic sheet and placed in a curing chamber at 23 ± 1°C. The test age was set to 56 days as the mixes contain large amount of cementitious materials.

2.4 Compressive strength test

The procedure of ASTM C109 was followed to perform compression tests using an ELE compression machine with a capacity of 2000 ±0.05 kN. The load was then applied at a rate of 0.02 MPa/s.

3 EXPERIMENTAL RESULTS AND DISCUSSIONS

3.1 Results of compressive strength tests

The compressive strength of each specimen at 56 days is shown in Table 1. The average compressive strength X, the standard deviation S and the coefficient of variation CV (= S/X) are also presented. It can be seen that the results are very consistent for all the mixes with CV values under 2% in most cases.

Table 1: Compressive strength (MPa) at 56 days

Mix	Compressive strength X_1	X_2	X_3	Average X	S. dev. S	CV = S/X (%)
R1L10S15	50.1	50.8	51.4	50.7	0.62	1.2
R1L10S30	58.3	57.2	57.6	57.7	0.57	1.0
R1L10S50	48.9	48.1	48.6	48.5	0.42	0.9
R1L20S15	44.3	45.4	45.2	45.0	0.61	1.3
R1L20S30	44.3	44.7	45.1	44.7	0.45	1.0
R1L20S50	43.6	43.1	44.3	43.7	0.56	1.3
R1L30S15	42.1	42.9	42.0	42.3	0.48	1.1
R1L30S30	42.1	41.5	40.6	41.4	0.75	1.8
R1L30S50	33.3	33.6	32.9	33.2	0.34	1.0
R2L10S15	44.3	44.9	45.4	44.9	0.57	1.3
R2L10S30	49.9	50.2	51.4	50.5	0.80	1.6
R2L10S50	47.6	47.7	48.8	48.0	0.65	1.4
R2L20S15	41.7	42.0	43.4	42.4	0.94	2.2
R2L20S30	40.7	39.7	40.2	40.2	0.48	1.2
R2L20S50	43.9	43.9	42.1	43.3	1.06	2.4
R2L30S15	42.4	42.8	42.2	42.5	0.32	0.8
R2L30S30	39.9	40.1	39.4	39.8	0.36	0.9
R2L30S50	34.4	35.2	34.2	34.6	0.52	1.5
R3L10S15	43.1	42.6	41.8	42.5	0.63	1.5
R3L10S30	44.7	42.9	43.3	43.6	0.99	2.3
R3L10S50	41.9	40.9	41.1	41.3	0.56	1.3
R3L20S15	42.4	41.8	42.3	42.2	0.36	0.9
R3L20S30	45.3	47.2	46.2	46.2	0.96	2.1
R3L20S50	47.7	45.8	46.2	46.6	0.99	2.1
R3L30S15	44.1	43. 9	43.8	43.9	0.19	0.4
R3L30S30	43.5	44.6	41.6	43.2	1.51	3.5
R3L30S50	37.2	39.8	39.3	38.7	1.36	3.5
R0L10S30	52.7	53.8	53.1	53.2	0.54	1.4

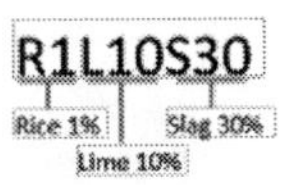

Nomenclature used: letters R, L and S are short for sticky rice, lime and slag respectively. The number following each letter represents the ratio in percentage.

3.2 Effect of lime on compressive strength

To illustrate the effects of lime, the stucco's 56 day compressive strengths containing 1% sticky rice are shown in Figure 1. It can be observed that the strength decreased as the lime proportion increased for all the three groups (S15, S30 and S50). The stuccos achieved the highest strength with 10% lime. The same trends were observed for nearly all groups with other sticky rice proportions (2% and 3%). Similar results were reported by Pandey and Sharma (2000) for mortars. It can be concluded that increasing lime proportion is not beneficial in terms of 56 day compressive strength of stucco.

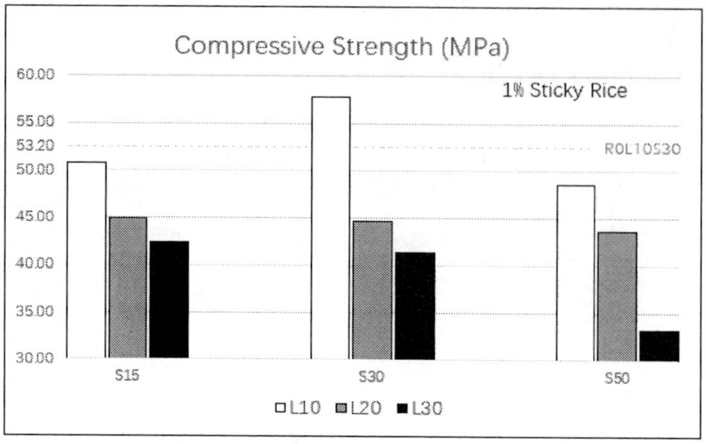

Figure 1: Effect of lime on 56-day compressive strength

3.3 Effect of slag on Compressive strength

Figure 2 shows the stucco's 56-day compressive strength of mixes containing 10% lime. Within each of the three groups (R1, R2 and R3), the stucco's compressive strengths increased when the slag proportion increased from 15% to 30%, but decreased when the slag proportion increased from 30% to 50%. The highest compressive strength were achieved with 30% slag. This result is consistent with the findings of Liu (2015), where he reported that when replacing 30% of cementitious material with slag, the compressive strength increased at first, and then declined when the replacement ratio reached 45%. Furthermore, the highest compressive strength of each group dropped with higher proportions of sticky rice.

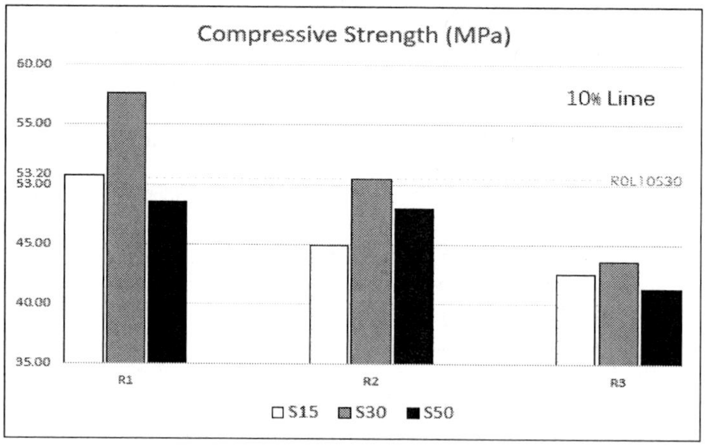

Figure 2: Effect of slag on 56-day compressive strength

3.4 Effect of sticky rice on compressive strength

Figure 3 shows the stucco's 56 day compressive strength of specimens, as affected by sticky rice proportion, for mixes containing 10% lime. As shown in the figure, the compressive strength gradually decreased as the sticky rice proportion increased from 1% to 3%, for all the three groups (S15, S30 and S50). The stucco achieved the highest compressive strength with 1% sticky rice. Similar results were also observed for the other groups containing 20% and 30% lime. This is in accordance with previous findings reported by Peng and Jackson (2015) which concluded that mortar with 1% rice had the highest compressive strength at 28 days, and they also concluded that sticky rice contributed to refining the calcite grain size and microstructures. To confirm this finding, one more mix (R0L10S30) were tested and the results are shown in Table 1 and Figure 3. As compared to the mix R1L10S30, adding 1% sticky rice improved the 56-day compressive strength by about 8%.

It can also be seen that the mixes containing 30% slag achieved the highest strength among the three groups.

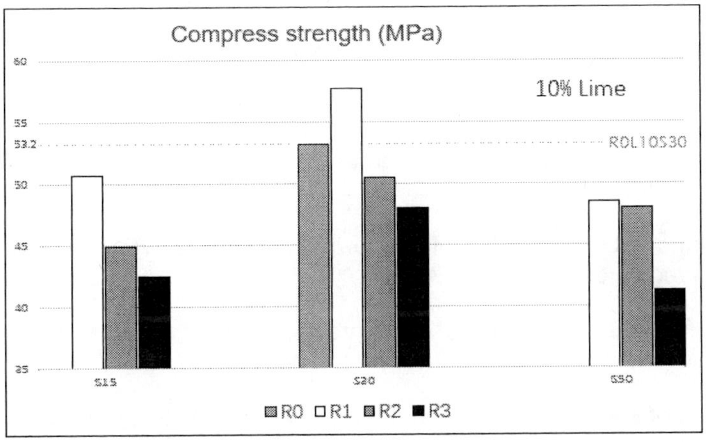

Figure 3: Effect of sticky rice on 56-day compressive strength

From Table 1, it can be observed that, among all the mixes studied, the highest compressive strength was achieved with the mix R1L10S30 which contains 1% sticky rice, 10% lime and 30% slag. This mix is considered as the optimal for this study and will be used for field tests in future studies.

4 CONCLUSIONS

From the findings of this study, the following conclusions can be drawn:

1. Cement-based stucco mixes with reasonable workability can be formulated with lime, slag and sticky rice slurry.
2. The 56-day compressive strength decreased as the lime proportion increased.
3. The 56-day compressive strength increased when the slag proportion increased from 15% to 30%, but decreased when the slag proportion increased form 30% to 50%. The highest compressive strength was achieved with 30% slag.
4. The 56-day compressive strength gradually decreased as the sticky rice proportion increases from 1% to 3%. It is therefore recommended to not use sticky rice with a dosage higher than 1%.
5. The highest 56-day compressive strength was achieved with the mix containing 1% sticky rice, 10% lime and 30% slag.

Further investigations need to be done for the long-term durability of such mixes.

Acknowledgements

This paper is part of a joint research program between the Université de Moncton (Canada) and Yancheng Institute of Technology (China). The authors would like to thank the support of the Université de Moncton for providing research funds, facilities as well as technical supports to conduct this study.

References

ASTM C109 2016. Standard Test Method for Compressive Strength of Hydraulic Cement Mortars (Using 2-in. or [50-mm] Cube Specimens). ASTM International.

ASTM C494 2017. Standard Specification for Chemical Admixtures for Concrete. ASTM International.

ASTM C1437 2015. Standard Test Method for Flow of Hydraulic Cement Mortar. ASTM International.

CAN/CSA-A3001 2013. Cementitious Materials for Use in Concrete. Canadian Standards Association.

Liu, P. 2015. Effect of different particles size of the steel slag on the compressive strength of concrete. *Fly Ash Compr.* Util., **2**: 34–36.

Pandey, S.P. and Sharma R.L. 2000. The influence of mineral additives on the strength and porosity of OPC blinder. *Cem. Concr. Res.*, **30**(1): 19-23.

Peng, Z. and Jackson, M.D. 2015. Material characteristics of ancient Chinese lime binders and experimental reproductions with organic admixtures. *Constructions and Building Material,* **84**: 477-488.

Qiu, X. 2014. Application of Bao steel's BSSF slag in C30 concrete for pavement. *Coal Ash,* **4**: 31–34.

Qiu, F. 1980. When did the lime production begin? The 14th century isotopic geochemical method can answer the question. *Archeol Cul Rel*: 126-30.

Song, Y. 2008. Exploitation of the Works of Nature. Shanghai: Rarebooks Publishing House.

Wei, G.F., Fang, S.Q., Zhang, B.J., Wang, X.Q. and Li, Z.G. 2012. Study on the Formation Mechanism of Liesegang Ring in Carbonization of Traditional Glue Rice Mortar. *Spectroscopy and Spectral Analysis,* 32 (8): 2181-2184.

Yang, F., Zhang, B.J. and Ma, Q. 2010. Study of sticky rice-lime mortar technology for restoration of historical masonry construction. *Accounts of Chemical Research,* **43**(6): 936-944.

Zeng, Y., Zhang, B.J. and Liang, X. 2008. A case study and mechanism investigation of typical of mortars used on ancient architecture in China. *Thermochimica Acta,* **473**(1/2): 1-6.

Zhao, P., Jackson, M.D., Zhang, Y., Li, G., Monteiro, P.J.M. and Yang, L. 2015. Material characteristics of ancient Chinese lime binder and experimental reproductions with organic admixtures. *Construction and Buildings Materials,* **84**: 477-488.

Building Tomorrow's Society
Bâtir La Société De Demain

Fredericton, Canada
June 13 to 16, 2018 / *13 au 16 juin, 2018*

ALKALI-ACTIVATED CONCRETE

Amer, Salma [1], Bakhoum, Michel [1,2], Yassa, Andrew [1], El Mahallaoui, Amr [1], Fahem, George [1], El Nahas, Eman [1], Hussam, Ahmed [1], and Abou-Zeid, Mohamed N.[1]
[1]The Department of Construction Engineering, the American University in Cairo, Egypt
[2]michelbakhoum@aucegypt.edu

Abstract: Concrete is the second most used commodity in the world after only water. It is also one of the most polluting materials to the environment. The production process of one ton of Ordinary Portland Cement (OPC) releases approximately 0.84 ton of carbon dioxide. Alkali-Activation has been developed as a technique of combining various latently hydraulic cementitious matrices with an alkali, to produce 'clinker-free' binders in concrete mixes, therefore inhibiting the need for OPC. This study aims at using and comparing results from three different types of cementitious matrices, namely Fly Ash, water quenched Ground Granulated Blast Furnace Slag (GGBFS) and air-cooled Blast Furnace Slag (BFS) with three different concentrations of the alkali-activator, a mix between Sodium Silicate solution and Sodium Hydroxide flakes. A conventional average OPC mix has also been casted to be used as a control mix. It is important to note that all cementitious matrices used are found as byproducts of different industries; both BFS and GGBFS are byproducts of iron production and Fly Ash is produced as a byproduct from burning coal in electric generation power plants. Conventional concrete tests have been performed, as well as the accelerated mortar bar test and accelerated corrosion test. Other tests have also been performed for the cementitious matrices including elemental composition, unit weight and fineness of the granules. Results reveal that Fly Ash activation is the most promising in terms of compressive and flexural strengths. It, however, requires heat curing which may limit its applications to precast, repair works and water plugging.

1 INTRODUCTION

Ordinary Portland Cement Concrete has been widely used since the 1800's. Research estimates that each ton of Cement produced, releases into the environment about 0.84 ton of Carbon Dioxide; directly during decomposition and indirectly because of the energy consumed during production. The total contribution of Cement to global Carbon Dioxide emissions is estimated to be between 5 and 8 percent. Today around 4.2 billion tons of Cement are produced annually with an expected growth rate of 2.5% per annum. Simultaneously, latently hydraulic cementitious matrices such as Fly Ash, Silica Fumes and Slag which are byproducts from various industries have been used as admixtures to replace some of the cement content. This leads to an improvement in concrete properties such as strength and permeability. The presence of cement, however, is required for the cementitious matrices' reaction to occur and produce the binding material in concrete. The idea behind alkali-activation is inhibiting the need for the cement's reaction and 'activating' the unaccompanied cementitious matrices.

1.1 Alkali-Activation

Alkali-Activation has been introduced as a method of initiating the reaction of the cementitious matrices without the need for cement. The process of Alkali-Activation is best demonstrated by the reaction of concrete containing both cement and a cementitious matrix, Fly Ash for example. Two chemical reactions take place. The first is the traditional cement reaction, in which the active components of cement, namely Tricalcium Silicate (C_3S) and Dicalcium Silicate (C_2S) react with water to produce Calcium Silicate (CSH), which is the binding component and Calcium Hydroxide (CH). In the following reaction, Calcium Hydroxide from the first reaction reacts with the Silicon Dioxide (S) present in Fly Ash to produce more Calcium Silicate (CSH). This reaction is represented as follows:

[1] $C_3S/C_2S + H \rightarrow C\text{-}S\text{-}H + CH$

[2] $S + CH \rightarrow C\text{-}S\text{-}H$

The idea behind Alkali-Activation is to externally introduce the Hydroxide ion to replace the Calcium Hydroxide from the first reaction through an alkali activator(s). By doing so, the need for cement in concrete is inhibited. This results in the formation of a 'clinker-free' binder, which both reduces requirement for cement and utilizes byproducts from other industries. The exact details of the chemistry and the exact chemical reactions that occur in Alkali-Activated Concrete (AAC) are not yet fully comprehended. However, the final components of the reaction and the factors affecting the reaction are commonly known. Factors include the chemical composition of the cementitious matrix used, the activator(s) used and their concentrations, as well as external factors such as the curing temperature, fineness of the cementitious matrix granules and mixing time. Commonly activators used include Sodium Silicates, Sodium Hydroxide, Sodium Carbonates, Sodium Alumiosilicates glass, Sodium Sulfate and Potassium Hydroxide. These can be used separately or combined.

2 LITERATURE REVIEW

1) <u>Self-Cured Alkali Activated Slag Concrete Mixes an Experimental Study:</u>

In this paper, the Alkali-Activated slag concrete was made using Ground Granulated Blast Furnace Slag (GGBFS) through activation by sodium silicate and sodium hydroxide solution. The effects of Sodium Oxide concentration and the ratio of Silicon Dioxide to Sodium Oxide in the activator on the strength and workability were studied. The results showed that as sodium oxide concentration increases, the compressive and tensile strength increases. However, as the silicon dioxide/sodium oxide ratio increases, the workability decreases. In addition, the effect of using portable or de-ionized water was not significant on the strength of the alkali-activated slag concrete. (Mithun et al, 2014)

2) <u>Performance of Alkali-Activated Slag with Various Alkali Activators:</u>

In this paper, various activators were used to activate the Ground Granulated Blast Furnace Slag and the performance of each mix was studied. The activators used were Sodium Silicate, Sodium Hydroxide and Sodium Carbonate. The workability in addition to the strengths were measured and the results of the mixes were compared. The results showed that the mixes with sodium hydroxide and sodium silicate set quickly. The Sodium Silicate activator yielded the highest compressive and flexural strengths, followed by the Sodium Carbonate and finally the Sodium Hydroxide. (Rajesh et al, 2013)

3) <u>Alkali-Activated Materials, State of the Art RILEM</u>

This book by RILEM (International Union of Laboratories and Experts in Construction, Materials, Systems and Structures) provides a comprehensive study of Alkali-Activated Concrete. It assesses various aspects regarding Alkali-Activation, such as its history, binder chemistry, mix designs standards, testing and potential applications. It refers to several other literature as it discusses the topics of Alkali-Activated Concrete. An important issue discussed in the Fly Ash section is that for Fly Ash to achieve proper strength gain oven curing is required.

4) <u>Handbook of Alkali Activated Cements, Mortars and Concretes</u>

This book gathers various literature on the topic of Alkali-Activated materials in one place, summarizing them under various headings. Topics discussed include chemistry and mix design, properties of Alkali-Activated Concreters, Durability, applications and lifecycle assessment.

3 OBJECTIVES

The main objective of this research is to study Alkali-Activated Concrete as a potential alternative to the ordinary Portland Cement Concrete. This will be done by examining different cementitious matrices with different activator concentrations to possibly highlight potential benefits of Alkali-Activated Concrete.

4 EXPERIMENTAL WORK

4.1 Mix Proportions

Since no specifications or mix design procedures are available, the following criteria were set in developing the mix proportions. These are as follows; the weights of cementitious matrix, coarse aggregates and fine aggregates were all fixed per 1m³. The alkali to cementitious matrix ratio was kept almost constant and within the effective range for each type of cementitious matrix. The silicate/hydroxide ratio was changed between the subsequent mixes. It is important to note that the following naming system has been used for the mixes; the letters indicate the material as follows, GGBFS shortened to GG, BFS kept as is and Fly Ash abbreviated to FA. The number following the letters indicates the Silicate to Hydroxide ratio moving from highest to lowest.

4.2 Material Properties

The cementitious matrices used were Type F Fly Ash, quenched (water-cooled) Ground Granulated Blast Furnace Slag (GGBFS), Egyptian air-cooled Blast Furnace Slag (BFS) and OPC. As for the activators, a combination of two activators was used; firstly Sodium Hydroxide Caustic Soda flakes with a purity of 99% and secondly Sodium Silicate solution with Sodium Oxide (Na_2O) content 12%, Silicon Dioxide (SiO_2) content of 32% and a molar ratio of 2.6. Fine aggregates used were well-graded, washed sand. The coarse aggregates used were well-graded washed dolomite.

4.3 Mix Proportions

Table 1: Mix Proportions per 1m³

Mix	Cementitious (kg/m³)	Coarse (kg/m³)	Fine (kg/m³)	Sodium Hydroxide (kg/m³)	Sodium Silicate (kg/m³)	Water (kg/m³)	AL/ cement	Silicate/ Hydroxide ratio
OPC	400	1150	620	-	-	192	-	-
BFS 1	350	1150	620	0	65	158	-	-
BFS 2	400	1150	620	16	110	105	-	-
BFS 3	400	1150	620	25	90	120	-	-
GG 1	400	1150	620	15	115	110	0.33	7.667
GG 2	400	1150	620	16	110	105	0.32	6.88
GG 3	400	1150	620	18	120	100	0.35	6.78
GG 4	400	1150	620	30	100	85	0.33	3.33
FA 1	400	1150	620	30	110	45	0.35	3.67
FA 2	400	1150	620	50	130	40	0.45	2.60
FA 3	400	1150	620	53	128	45	0.45	2.42
FA 3'	400	1150	620	53	128	45	0.45	2.42
FA 4	400	1150	620	55	125	47	0.45	2.27

4.4 Specimen Preparation

The reaction between the activators, namely sodium silicate and sodium hydroxide produces large amounts of heat. The activators were, therefore, separately mixed together first according to the required proportions and left to cool down for 15 minutes before being added to the remaining components of the mix in the mixer (cementitious matrix, coarse and fine aggregates, and water). Alkali-activated mixes were used to cast 15x15x15 cm cubes and 15x15x75 cm beams. The GGBFS and BFS mixes were kept in the molds in room temperature for 48 hours, then were demolded and cured in air in room temperature until testing. The Fly Ash mixes were kept in the molds at room temperature for 72 hours, then demolded, wrapped with plastic and cured in the oven at 90 ^{0}C for 24 hours. Afterwards, the specimens were left in room temperature until testing. In order to examine the importance of the oven curing for Fly Ash, the specimens from one of the Fly Ash mixes (FA 3'), which had the same mix proportions as mix (FA 3), was cured in air in normal room temperature.

4.5 Methodology

As aforementioned, tests were performed on the cementitious matrices, on fresh concrete and on hardened concrete.

4.5.1 Material Properties

- Elemental Composition (EDX): The Energy-dispersive X-ray Spectroscopy test, is an analytical test that identifies the elemental composition and percentages of elements present in materials.
- Density
- Blaine Fineness: the Blaine Fineness test was performed using the Blaine Permeability Apparatus and is used to determine the material fineness.

4.5.2 Fresh Concrete

- Slump
- Air content
- Unit Weight
- Temperature

4.5.3 Hardened Concrete

- Compressive strength (7, 28 Days)
- Three Point Flexural strength for reinforced beams (7, 28 Days): performed on beams with a steel cage of 4 steel bars (10mm diameter each) and 4 stirrups (8 mm diameter each).
- Accelerated Corrosion test: to examine the effect of alkali-activation on reinforcement corrosion.
- Accelerated Mortar Bar Test: to test for Alkali-Aggregate Reaction in case of alkali-activation.

5 RESULTS

5.1 Material Properties

5.1.1 EDX

The EDX results show a difference in the chemical composition of the three materials. Besides Oxygen, Fly Ash is rich in Silicon and Aluminum and this classifies it as low Calcium content cementitious matrix. On the other hand, GGBFS and BFS are classified as high Calcium cementitious matrix because of their high Calcium content.

Table 2: Chemical Composition of the Cementitious Matrices

Element / Percentage (%)	BFS	GGBFS	Fly Ash
O	32.9	46.1	51.3
Si	4.9	8.5	26.0
Ca	32.6	40.3	2.5
Fe	23.2	2.2	5.4
Al	1.1	2.2	9.9

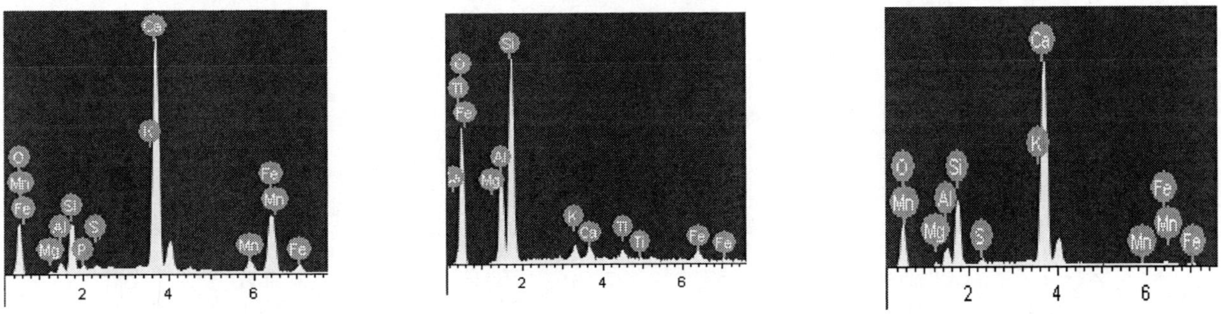

Figure 1: Results from EDX, from left to right BFS, GGBFS, Fly Ash

5.1.2 Density & Blaine Fineness

Density results show that the Fly Ash and GGBFS had considerably lower densities than OPC. OPC and Fly Ash, which were both supplied in powder form, had the highest fineness. GGBFS and BFS were supplied as large granules and lumps respectively and had to be grinded to achieve a lower fineness. Fineness was performed to ensure that the grinded material has a roughly uniform fineness after grinding.

Table 3: Blaine fineness and Density of the Cementitious Matrices

Cementitious Matrix	Blaine Fineness (cm^2 / g)	Density (g / cm^3)
OPC	3508	3.22
GGBFS	3171	2.81
BFS	2912	3.18
Fly Ash	3752	2.38

5.2 Ordinary Portland Cement Concrete Fresh and Hardened Results

Table 4: Fresh Concrete Results

Mix	Slump (mm)	Air Content (%)	Temperature (°C)	Unit Weight (kg/m^3)
OPC	20	1.3	31	2447

Table 5: Hardened Concrete Results

	7 Days		28 Days	
Mix	Compressive (MPa)	Flexural (MPa)	Compressive (MPa)	Flexural (MPa)
OPC	28.8	17.6	39.1	24.6

5.3 Air Cooled BFS Fresh and Hardened Concrete Results

BFS yielded very poor results for both fresh and hardened tests. The mixes did not develop any cohesion with the paste surrounding the aggregates only. The setting behavior of the concrete was also abnormal, with the mix suddenly in one cycle of the mixer becoming very unworkable. Hardened concrete achieved very low strengths. The mixes were therefore excluded from further analysis and comparison.

5.4 Water Quenched GGBFS Fresh and Hardened Concrete Results

Table 6: GGBFS Fresh Concrete Results

Mix	Slump (mm)	Air Content (%)	Temperature (°C)	Unit Weight (kg/m^3))
GG 1	20	5.5	23	2091
GG 2	Collapse	2.5	25	2267
GG 3	30	4.0	24	2360
GG 4	10	5.0	26	2298

All the GGBFS mixes were unworkable except mix GG2, which was highly workable. Also, all the GGBFS mixes had lower temperatures than the conventional OPC concrete mix performed.

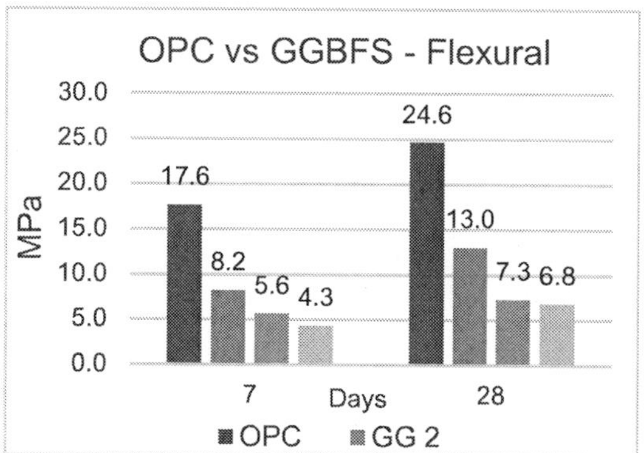

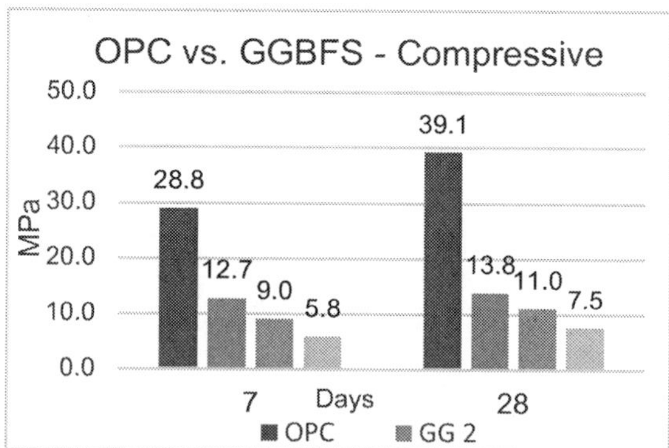

Figure 2: OPC vs GGBFS Compressive and Flexural Strengths

Mix GG1 showed extremely low compressive and flexural strength compared to the other mixes, most likely because of an error during mixing. Its results are excluded from further analysis. All GGBFS mixes showed lower 7 and 28 days compressive and flexural strength compared to the conventional OPC mix. Compressive and flexural strength of the GGBFS concrete decreases as the ratio Silicate/Hydroxide decreases. For example, mix GG4 which has the lowest Silicate / Hydroxide content, yielded the lowest compressive and flexural strength. Green formations with a distinctive smell were noticed over the surfaces and inside the GGBFS casted beams and cubes. Samples of these formations were biologically tested four times on four different media with no conclusion reached.

5.5 Fly Ash Fresh and Hardened Concrete Results

Table 7: Fresh Concrete Results

Mix	Slump (cm)	Air Content (%)	Temperature (°C)	Unit Weight (kg/m^3)
FA 1	0	1.8	40	2380
FA 2	Poor fresh concrete status			
FA 3	0	1.5	42	2312
FA 3'	0	1.5	42	2312
FA 4	Collapse	1.2	44	2360

All the Fly Ash mixes showed very low workability except mix FA 4, which was highly workable. Mix FA2 was extremely unworkable and very sticky to the extent that it was impossible to perform any of the fresh concrete tests for it and no beams were cast for this mix. All the Fly Ash mixes had higher temperatures than the conventional OPC concrete mix.

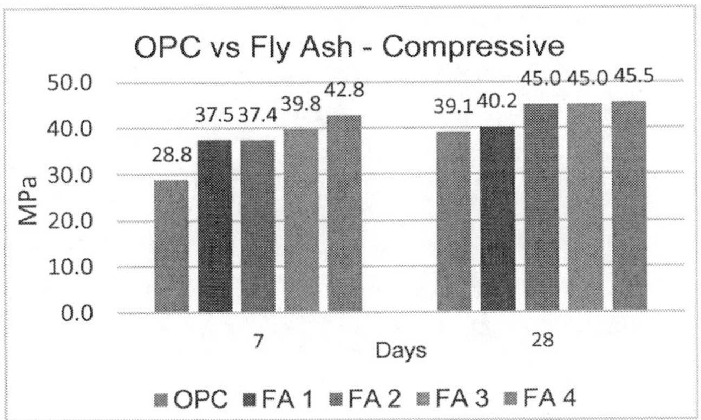

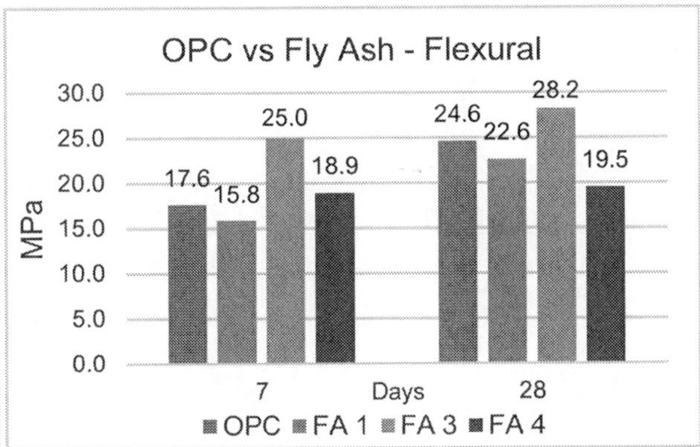

Figure 3: OPC vs Fly Ash Compressive and Flexural Strengths

The Fly Ash mixes showed higher 7 and 28 days compressive strengths compared to the conventional OPC mix. They also showed similar 7 and 28 days flexural strength compared to OPC. As the ratio of Silicate/Hydroxide decreased, the 7 days compressive strength of the Fly Ash concrete increased. For example, mix FA4 which had the lowest Silicate/Hydroxide content, had the highest 7 days compressive strength. At 28 days, the effect of this Silicate/Hydroxide became negligible; 28 days strength was approximately 45 MPa for most mixes. Moreover, Fly Ash mixes gained a much higher percentage of their 28-days strength by day 7 compared to the conventional mix.

5.5.1 Oven-Curing versus Air-Curing for Fly Ash Activated Concrete

Mixes FA 3 and FA 3' have the exact mix proportions but mix FA 3 was oven cured, while FA 3' was air cured. As the results show, the compressive and flexural strengths of the air cured mix were extremely lower than those of the oven cured mix. It can therefore be concluded that oven curing is essential for Fly Ash's strength gain.

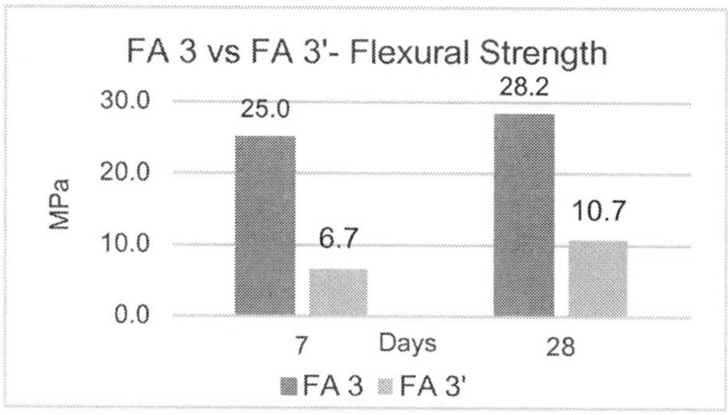

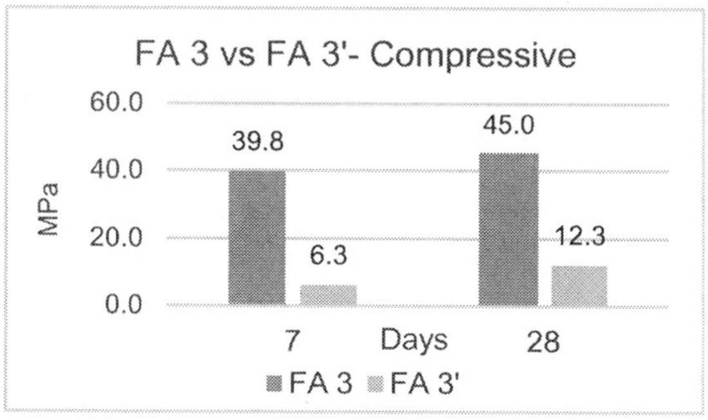

Figure 4: FA 3 (Oven cured) vs FA 3' (Air-cured) Compressive and Flexural Strength

5.6 Accelerated Corrosion Test

The Accelerated Corrosion Test was conducted only on OPC and Fly Ash. Special beams were cast with 3 steel bars extruding from the sides. A Sodium Chloride solution with 20% concentration was used to accelerate the corrosion rate. Current was measured once every week for 3 weeks between the top and bottom bars. As per test standards, the lower the current, the better the resistance to corrosion. The results showed that Fly Ash has better resistance to corrosion than the conventional concrete since a lower current was measured from the Fly Ash beam compared to that measured from the OPC beam.

Table 8: Accelerated corrosion Results

	Current (mA)		
	week 1	week 2	week 3
OPC	0.036	0.045	0.050
Fly Ash	0.026	0.029	0.032

Figure 5: Corrosion beams with extruding reinforcement

1.2 Accelerated Mortar Bar Test

Special OPC and Fly Ash bars were casted (25mmx25mmx285mm) with the required gradation and testing procedures according to ASTM C1260. The results show that the specimens experienced minor shrinkage whereas the standards anticipate expansion. It can be concluded that since no expansion occurred, the threat of alkali-aggregate is not expected to occur.

Table 9: Accelerated mortar bar results

Material	% Change in length
OPC	0.02
FA	0.04

Figure 6: Mortar bar specimens

6 ANALYSIS

6.1 6.1 7/28 Days Compressive Strength Ratio

The compressive strength 7/28 days ratios for the water quenched GGBFS yielded an average of 84% and the oven-cured Fly Ash an average of 90%. These are considerably higher than the 74% of the conventional concrete mix, showing that GGBFS and Fly Ash had higher initial strength gain compared to the conventional mix. Some oven-cured Fly Ash mixes almost gained maximum compressive strength after 7 days. After 7 days, the air-cured Fly Ash mix gained only 51% of its maximum compressive strength which is very low. This further proves that heat curing is a must for the Fly Ash.

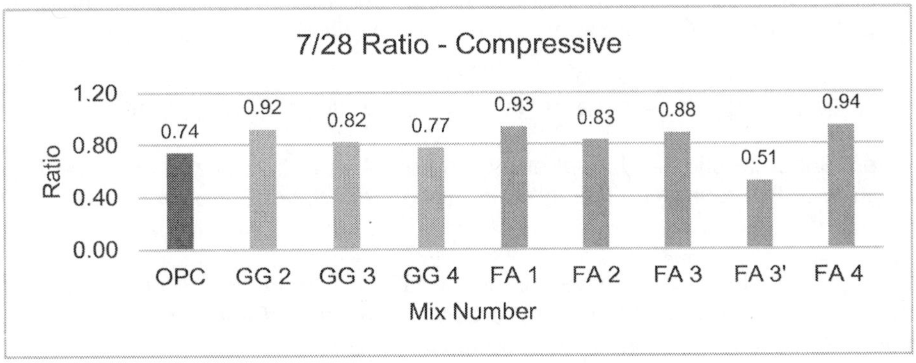

Figure 8: Compressive strength 7/28 ratio

6.2 Compressive Strength and Silicate/Hydroxide Ratio

The compressive strength versus the Silicate to Hydroxide ratio was plotted for both the GGBFS and Fly Ash mixes. The trends show that for GGBFS, as the Sodium Silicate to Sodium Hydroxide ratio decreases, the compressive strength decreases. For the Fly Ash, however, as this ratio decreases, the compressive strength of the Fly Ash increases.

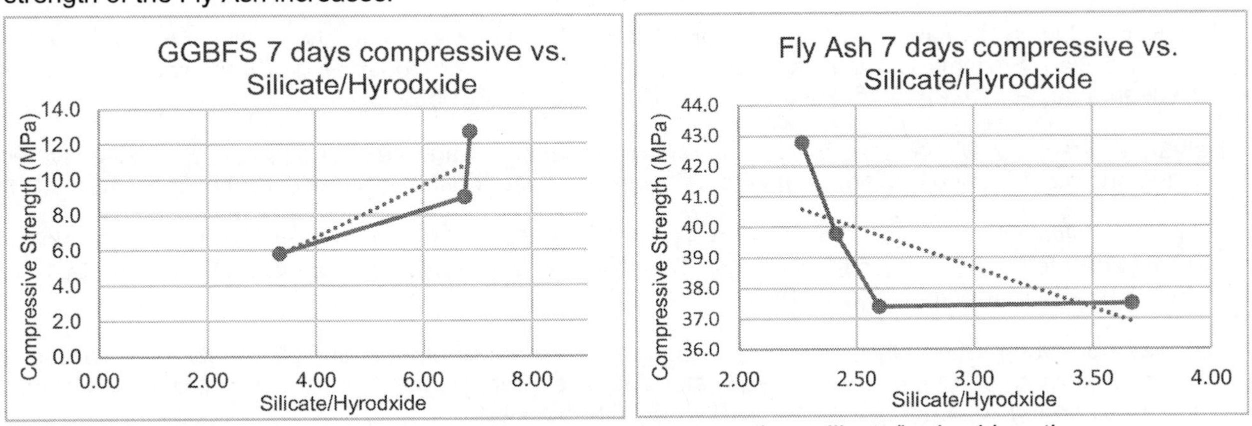

Figure 9: GGBFS and Fly Ash compressive strength vs silicate/hydroxide ratio

7 CONCLUSIONS

This study provides a comparison between three different cementitious matrices namely, BFS, GGBFS and Fly Ash. These were tested with an activator consisting of a combination of sodium silicate and sodium hydroxide. In light of the materials, methodology and results, the following conclusions can be stated:

- The poor fresh and hardened results obtained from air-cooled BFS may be an indication of the inert nature of the crystalline structures formed during the air-cooling process.
- Workability represents a major challenge for Alkali-Activated Concrete, as most mixes did not yield acceptable workability.
- Water Quenched GGBFS yielded lower compressive and flexural strengths after 7 and 28 days compared to conventional concrete. This could be attributed to the poor available quality of GGBFS in Egypt.
- Egyptian air-cooled BFS results show extremely poor results in compressive and flexural strengths.
- Fly Ash achieved higher 7 and 28 days strength than conventional concrete and similar ranges of flexural strengths. It exhibited early strength gain properties, however, heat curing is essential for the strength gain to take place.
- Results from the accelerated corrosion and mortar bar tests most likely eliminated the threat of alkali-aggregate reaction and reinforcement corrosion respectively.

8 RECOMMENDATIONS FOR FURTHER RESEARCH AND APPLICATION

Alkali-Activated Concrete still needs further experimental work in order to both, verify as well as to build upon the findings of this research. The following recommendations are proposed for future experimental work:

- Experiment with a wider set of activators, concentrations and activation techniques.
- Experiment with admixtures and mix proportion modifications for workability.
- Experiment with different activation techniques for BFS.
- Perform compressive and flexural strength tests beyond 28 days.
- Conduct the Mortar Bar Test at further ages.
- Conduct the Chemical Durability Test and permeability test.
- Investigate different curing temperatures and durations for Fly Ash.
- Explore potential use of Fly Ash mixes in steam-cured precast concrete, water and fluid plugging and repair works.

References

Arbi, Kamel, Marija Nedeljković, Yibing Zuo, and Guang Ye. "A Review on the Durability of Alkali-Activated Fly Ash/Slag Systems: Advances, Issues, and Perspectives." *Industrial & Engineering Chemistry Research* 55, no. 19 (2016): 5439-453. doi:10.1021/acs.iecr.6b00559.

Behfarnia, Kiachehr, Mohammad Shojaei, and Reza Mohebi. "H and flexural strength of alkali-activated slag concrete designed by Taguchi method." *Asian Journal of Civil Engineering* 16, no. 505-513 (2015).

Junaid, M. Talha, Obada Kayali, Amar Khennane, and Jarvis Black. "A mix design procedure for low calcium alkali activated fly ash-based concretes." *Construction and Building Materials* 79 (2015): 301-10. doi:10.1016/j.conbuildmat.2015.01.048.

Lee, Kwang-Myong, Sung Choi, Jinkyo F. Choo, Young-Cheol Choi, and Sung-Won Yoo. "Flexural and Shear Behaviors of Reinforced Alkali-Activated Slag Concrete Beams." *Advances in Materials Science and Engineering* 2017 (2017): 1-12. doi:10.1155/2017/5294290.

M, Mithun B., and Mattur C. Narasimhan. "Self-Cured Alkali Activated Slag Concrete Mixes An Experimental Study." *International Journal of Civil and Environmental Engineering*8, no. 4 (2014).

Pacheco-Torgal, Fernando. *Handbook of Alkali-Activated Cements, Mortars and Concretes.* S.l.: WOODHEAD, 2014.

Pacheco-Torgal, Fernando, Joa˜o Castro-Gomes, and Said Jalali. "Alkali-Activated Binders: A Review Part 1. Historical Background, Terminology, Reaction Mechanisms and Hydration Products." *Construction and Building Materials*22, no. 7 (July 2008).

Provis, John. *Alkali Activated Materials : State-of-the-Art Report, RILEM TC 224-AAM.* Springer., 2014.

Rajesh, D.V.S.P., A. Narender Reddy, U. Venkata Tilak, and M. Raghavendra. "Performance of Alkali-Activated Slag with Various Alkali Activators." *International Journal of Innovative Research in Science, Engineering and Technology* 2, no. 2 (February 2013).

Building Tomorrow's Society
Bâtir la Société de Demain

Fredericton, Canada
June 13 to 16, 2018 / *13 au 16 juin, 2018*

AN OVERVIEW OF THE EFFECT OF FLY ASH AND SILICA FUME ON THE CEMENT HEAT OF HYDRATION

Abdel-Raheem, Mohamed[1,2], Cortina, Jennifer[1], Quintana, Ofsman[1], Morales, Melissa[1], Marroquin, Yahaira[1,] Ramos, Dalinda[1] and Hernandez, Sylvia[1]
[1] Department of Civil Engineering, University of Texas Rio Grande Valley, USA
[2] mohamed.abdelraheem@utrgv.edu

Abstract: Over years, the incorporation of pozzolanic supplementary cementitious materials (PSCM) in concrete have been increasing tremendously. Two of the most notable PSCM are fly ash (FA) and silica fume (SF), which have been proven to enhance some of the properties of concrete. Previous research shows that the use of PSCM in concrete serves many purposes. Several studies focused on the assessment of the ability of PSCM to reduce the heat released during the hydration process of cement when incorporated in the concrete mix. It is very desirable to minimize the cement HH especially in mass concreting as it may induce thermal stresses that may lead to the failure of the structure. This study focuses on the performance of FA and SF with respect to their effect on the cement heat of hydration (HH) as reported in the literature. The paper summarizes the most important findings of previous studies conducted on the effect of FA and SF on the HH, and shed the light on the conformity and discrepancies between different studies results. The paper should prove useful for future researchers and practitioners interested in studying the effect of FA and SF on the HH of cement.

1 INTRODUCTION

The hydration of cement is an exothermic reaction that can release a tremendous amount of heat. This heat of hydration (HH) can cause a rise in the concrete temperature. Due to environmental effects, the exterior of concrete tends to cool at a faster rate than its interior; this creates a temperature gradient between the inner and out concrete. As such, the rates of expansion and contraction of the concrete core and its surface differ due to this temperature gradient, which can create thermal stresses. If the thermal stresses developed exceeds the tensile strength of concrete, thermal cracks start to occur. Thermal cracks usually occur in random pattern on the surface of concrete; they occur few days after stripping the forms perpendicular to the longest axis of the concrete member (NRMCA 2009). This phenomenon becomes very critical in mass concrete members due to the huge amount of HH released and the substantial difference in temperature between the core and the surface of the mass concrete member. As such, it is recommended that the temperature difference between the different layers of the concrete member should not exceed 20°C (Portland Cement Assoc. 1997).

Mass concrete members are recognized by their large dimensions. The American Concrete Institute (ACI) considers any concrete member that has a minimum dimension of 4 feet (1.3m) as a mass concrete member (ACI 2007). Other factors of considerations that would classify the structural element as a mass concrete member is the volume to surface ratio, and the inclusion of high heat-generating admixtures in the concrete mix (NRMCA 2009). Examples of mass concrete structures are mat foundations, dams, and thick walls. Special precautions should be taken to control the HH when dealing with mass concrete structure.

There are several methods that can be used to control the heat gradient in mass concrete structures. Controlling the HH can be achieved by altering the concrete mix design by using low-heat cement or incorporating PSCM that are known to reduce the HH, such as slag and FA (Briley 2004, NRMCA 2009, Patil 2015). Also, the precooling of concrete aggregates contributes significantly to reducing the peak temperature of the concrete mix after placement. Several construction methods are known to be efficient in reducing the HH; one of them is post cooling of concrete using cooling pipes (Zhu 1999, Kim et al. 2001, yang et al. 2012). Proper insulation of the concrete members and pouring concrete in thin lifts can reduce the heat gradient significantly (Gajda and John 2016). This study focuses on the effect of two types of PCSM on the cement HH, which are FA and SF. The study presents the results documented in the literature about the effect of FA and SF on the cement HH, and highlights the consistency and discrepancies of the results reported.

2 PROBLEM STATEMENT AND OBJECTIVE

The effects of FA and SA on different properties on concrete have been investigated in a multitude of studies. A particular interest was given to the effect of those materials on the HH of cement. There are several studies that focused on this aspect. However, there is not a single a study that presents a thorough review of the work that has been done before. This paper aims to provide an organized and categorized concise review and analysis of findings of previous studies from the literature that addressed the effect of FA and SF on the cement HH.

3 METHODOLOGY

The methodology adopted on this research focused on reviewing the studies investigated the effect of FA and SF on the HH of cement. The literature included all the literature between 1980 and 2016. The identified studies were grouped and categorized by material type, as shown in Figure 1. The main points of each study were summarized with a main emphasis on the findings. The discrepancy between the results obtained from different studies were highlighted to identify existent gaps and inconclusive findings in the literature.

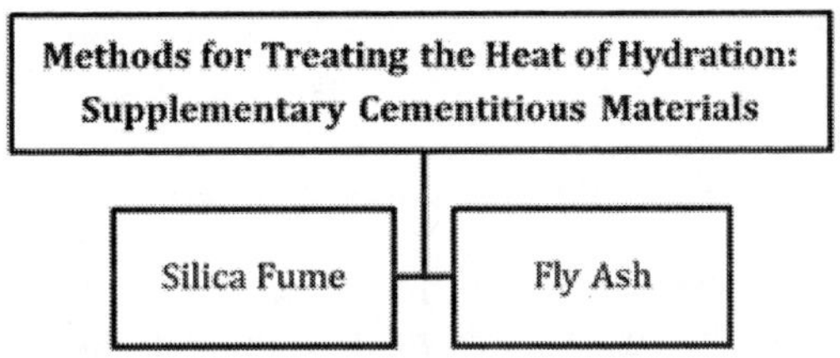

Figure 1: Categorization of previous studies focused on the utilization of supplementary materials to treat the HH

4 SUPPLEMENTARY MATERIALS USED FOR REDUCING THE HEAT OF HYDRATION

The effects of many PSCMs on the HH of cement have been considered on a multitude of studies. However, the majority of research focused primarily on FA and secondarily on SF. The following section will discuss the details of these studies and the most important findings reported.

4.1 Fly Ash

Fly ash is a common PSCM that was investigated in many papers for its effect on the HH of cement. A conduction calorimeter was used to test FA as a 10% and 20% replacement of PC. Also, the study investigated the effect of FA when mixed with a plasticizer. The results showed that FA was able to reduce HH; as the quantity of FA increased the HH decreased. The tests done on the mixes with the plasticizers showed that there was more heat release compared to the samples made with FA and PC only; however, it was still less than the HH of the control sample (Meland 1983).

Three different types of Turkish fly ash were tested using standard ASTM C186 in percent replacements of PC of 10%, 20%, and 40% (Tokyay 1988). One of the fly ashes was a high-calcium, while the other two were low-calcium. The results showed that the low-calcium fly ashes were more effective in decreasing the HH when compared to the high-calcium FA. Of all mixes that contained the three FAs used at different percentages, the mix that had a higher HH than the control mix was the one that contained the high-calcium FA at a 10% replacement. This occurred due to the high heat evolving lime and anhydrite in the high-calcium FA (Tokyay 1988).

Different mixtures were tested with FA replacing cement to up to 70% fly ash (De Rojas et al. 1993). The four mixtures that were tested contained 50%, 60%, and 70% FA replacements of PC in addition to a control mix made with 100% PC for comparison purposes. The results show as the percentage of FA increases in the mix, the peak temperature decreases, as shown in Figure 2 (De Rojas et al. 1993).

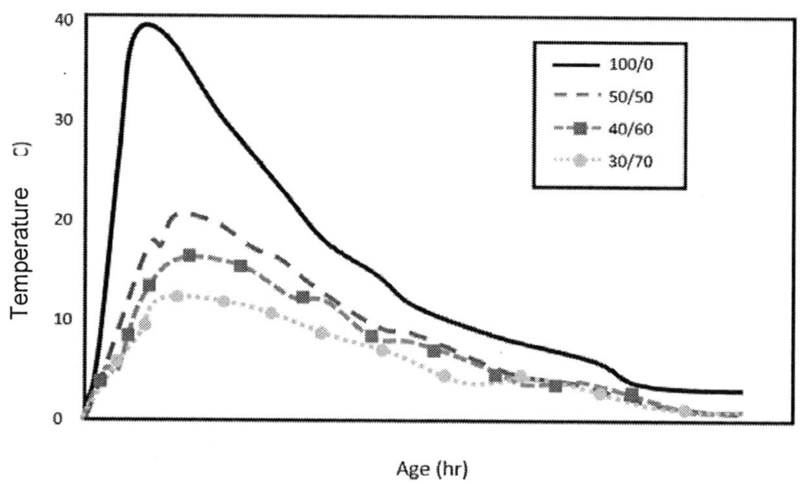

Figure 2: The heat of mixes at different fly ash percentages (De Rojas et al. 1993)

The testing of the effect of FA on the HH was done in different room temperatures using mixtures of 70% PC and 30% FA (Alshamshi 1994). The room temperatures used in the study were 20°C, 35°C, and 45°C. The peak temperatures of the mixes were compared to each other. The results showed that the increase of the room temperature caused the heat peak to increase. While the HH did not decrease in hotter temperatures, it was still concluded that using FA as a partial replacement could provide an advantage in reducing the effects caused by the HH such as thermal cracking (Alshamshi 1994).

A semi-adiabatic calorimeter was used to compare the results of a FA mix to another made with SF (De Rojas and Frias 1995). The mixtures were done with 30% replacement of PC each and the results, see Figure 3, shows that FA can decrease the HH. This is due to the fact that FA has a slower pozzolanic activity compared to SF (De Rojas and Frias 1995).

In another study, FA was used as a replacement of 10% and 30% of PC pastes and mortars (Frias et al. 2002). The findings of this study verified that the increase in the amount of FA used in the mix has a significant effect on decreasing the HH. The tests on the mortars further also showed a significant reduction in the HH with a difference of up to 80 J/g of the HH of the control mix (Frias et al. 2002).

Atis (2002) used an adiabatic system to test the HH associated with using high volumes of FA in the amount of 50% and 70% replacements of PC. Separate samples that contained plasticizer were tested along with a mix of 100% PC to be used as the control. The results showed that the samples that contained FA had lower peak temperature compared to the control mix; the addition of the plasticizer increased the peak temperature. The study concluded that as the quantity of the FA increased, the peak temperature decreased (Atis, 2002).

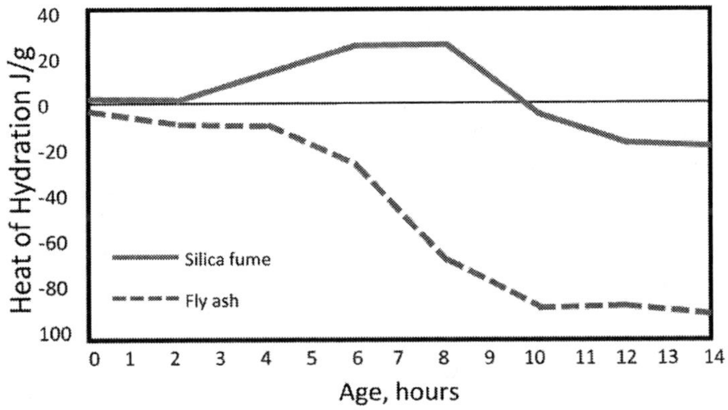

Figure 3: Comparison of the HH between SF and FA (De Rojas and Frias 1995)

An isothermal calorimeter was used to test class C fly ash with different w/c ratios (Langan et al. 2002). The samples contained w/c ratios of 0.35, 0.40, and 0.50 with replacement of 20% of PC using FA and were tested for 72 hours. The results found in this paper showed that higher w/c ratios delay the peak temperature. The heat evolution after one day was lowered as the w/c ratio increased. FA and SF were also mixed together in a sample that contained 20% FA, 10% SF, and 70% cement. This mixture was able to reduce the heat slightly and delay the peak temperature. (Langan et al. 2002).

Class C and Class F FA were both tested and compared to each other in mixtures that contained replacements of 15%, 25%, and 35% of PC (Schindler and Folliard 2003). Both Class C and Class F fly ashes with the highest replacement amount (35%) had the lowest adiabatic temperature rise when compared to the control sample (100% cement). However, after 40-50 hours, the heat release of the samples made with Class C FA showed an increase exceeding the HH released by the control sample. It was concluded that Class F FA is a better option for reducing the HH than FA Class C.

A separate study tested Class F FA at different w/b ratios as a replacement of 25%of PC (Pane and Hansen 2005). The test from the w/b ratios of 0.35 and 0.45 showed that decreasing the w/b ratio lowers the heat released (Pane and Hansen 2005).

More FA tests were conducted as well as a combination of FA with metakaolin (Snelson et al. 2008). The experiments on FA consisted of 10%, 20%, 30%, and 40% replacements of PC. The results confirmed previous findings with respect to decreasing the HH by increasing the percentage of FA added. The addition of metakaolin to the mixture proved to not be as effective as fly ash alone. A mixture of 15% metakaolin and 5% fly ash caused the HH to be greater than the control sample due to the large amount of metakaolin added (Snelson et al. 2008).

Additional studies were conducted on samples that contained 20%, 40%, and 60% FA as a replacement of PC (Maia et al. 2011). The results -see Figure 4- shows that increasing the amount of FA in the mix leads to a significant decrease in the HH released (Maia et al. 2011).

Different fly ash and slag combinations were tested using adiabatic calorimeter to ass their effect on the HH (Lee et al. 2014). The two samples tested included a mix of 30% cement, 50% slag, and 20% fly ash and another one of 40% cement, 40% slag, and 20% fly ash. In both mixtures, the fly ash content stayed the same, but the tests showed a reduction in the peak temperature of both mixes. The peak temperature of the control mix was 37.5°C while the 50% slag mix had a peak of 24.5°C and the 40% slag one peaked at 30°C. There was no information provided in the study regarding the cause of the decrease in the peak temperature and it was not stated if it was due to the decrease in cement or effect of slag (Lee et al. 2014).

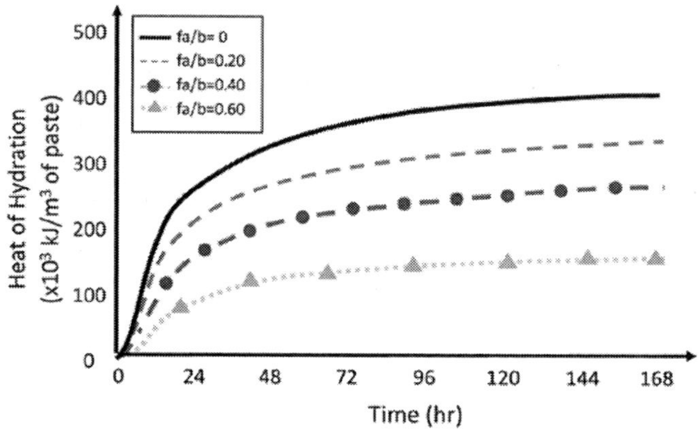

Figure 4: The heat of hydration at different fly ash ratios with w/b=0.45 (Maia et al. 2011)

The most recent study on FA used it as replacement of 25% of PC in comparison to a 100% PC control sample (Alhozaimy et al. 2015). The results from that test showed that the HH for the control sample was 252.76 J/g, while the FA sample had a HH of 206.97 J/g. A total decrease of 19.1% in the HH was observed when the samples were compared (Alhozaimy et al. 2015).

4.2 Silica Fume

The impact of SF on the HH of concrete have been discussed in multiple studies. An early study from 1983 documented different tests conducted of SF blended together with slag to reduce the heat released from the hydration of cement (Lessard et al. 1983). The mixture used in the study consisted of one mixture made with 35% slag and 15% condensed silica fume as a replacement of Portland Cement (PC). When it was compared to the control sample that is made of 100% PC, it was found that mixture of SF and slag led to a substantial reduction in the heat released. However, the study did not specify whether this decrease was due to the effect of slag, silica fume, or the lower cement content. The results from the study also showed a reduction of 30% and 22% of the HH, when isothermal temperature of 20°C and 40°C were used respectively (Lessard et al. 1983).

Another study by Meland (1983) provided more results on the performance of SF replacement of PC with respect to the HH released. Samples made with SF as 10% and 20% replacement of PC as well as another sample composed of PC, SF and lignosulphonate (plasticizer) were prepared and tested in a conduction calorimeter. The results show that the HH decreased slightly with the addition of S. It was noticed that by increasing the percentage of PC replacement with SF, the HH continued to decrease. On the other hand, when the lignosulphonate was incorporated in the mix with the silica fume, a larger amount of the HH was released compared to the control mix. This increase in the heat was explained by the possible absorption of the lignosulphonate by SF, which increased the amount of heat released (Meland 1983).

Cheng-yi and Feldman (1985) presented a study to test samples made with SF replacements of PC of 10%, 20%, and 30% using a conduction calorimeter. After testing the samples for 72 hours, the results showed that increasing the amount of SF in the mix yields a greater heat release. This was attributed to the early pozzolanic reaction of SF or the acceleration of cement hydration (Cheng-yi and Feldman 1985).

The Langavant Calorimeter method, a semi-adiabatic method, was used to test SF at 20°C (De Rojas and Frias 1995). Percentage replacements of PC were done using 5%, 10%, 15%, and 30% of SF. Also, samples containing 6 different types of SF were also tested at 10% replacements of PC. The results showed an increase in the HH when the SF was used in relatively small quantities of 5%, 10%, and 15% as a replacement of cement compared to control samples made with plain cement. However, when the quantity of SF was increased to a replacement of 30% of PC, the results showed a small decrease in HH. It was then concluded that the pozzolanic activity in the silica fume caused the increase in heat. However, the large replacement of PC in the 30% samples decreased the HH released from the PC hydration, which counteracted the heat released due to the pozzolanic reaction of SF. The tests conducted on the different samples representing the 6 types of SF showed an increase in the heat released compared to the control samples. Some of the experiments results are shown Figure 5 (De Rojas and Frias 1995).

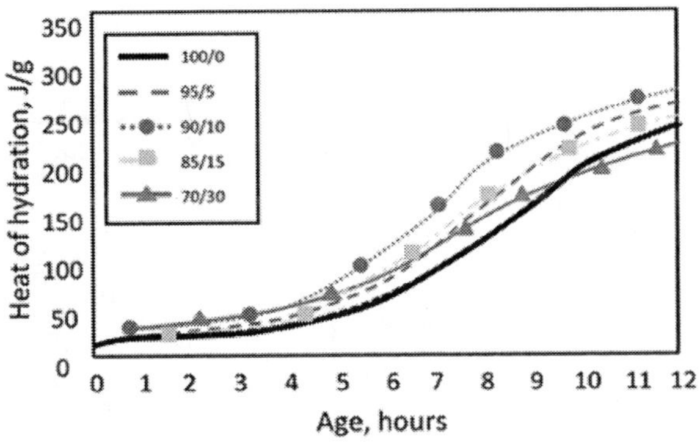

Figure 5: HH at different SF percentages (De Rojas and Frias 1995)

De Rojas and Frias (1996) continued their previous work where they further proved that only a large amount of SF could slightly decrease the HH. It was explained that this due to the great reduction in the amount cement and consequently a reduction in the HH associated with the cement hydration. However, SF still undergoes a pozzolanic reaction, which contributes to the heat release but at a lower rate (De Rojas and Frias 1996).

Semi-adiabatic experiments were done for 22 hours on sample made with 10% SF replacement of PC using different water-cement (w/c) ratios (Persson 1997). The results show that as the w/c ratio increased, the HH increased as well. Additional tests with silica fume and superplasticizers were done and the results showed that the mixture with 10% silica fume, 1.63% superplasticizer, and a w/c ratio of 0.24 had the lowest heat release (Persson 1997).

Mortar mixtures that contained SF were tested in another study in PC replacements of 10% and 30%. The results found were similar to ones in previous studies where the 10% mixture released a higher amount of heat compared to the control samples. However, the 30% samples released less heat compared to the control samples (Frias et al. 2000).

In another study, samples made with 10% SF replacements of PC were tested with different w/c ratios of 0.35, 0.40, and 0.50 (Langan et al. 2002). An isothermal calorimeter was used at 25°C to test the samples. It was found that samples contained SF released higher amount of heat compared to the control samples. As for the samples made with the different w/c ratios, the conclusion was that the increase in the w/c ratio caused the SF samples to liberate the HH at a higher rate compared to the control sample (Langan et al. 2002).

More tests using different water to binder (w/b) ratios were done using an isothermal calorimeter (Pane and Hansen 2005). The tests performed consisted of a 10% SF replacement of PC with w/b ratios of 0.45 and 0.35. The results showed that the SF samples released a low amount of heat at the beginning, but it quickly increased. The mixture made with a w/b ratio of 0.35 released less heat than the 0.45 mixture, but both mixtures still produced more heat than the 100% cement sample (Pane and Hansen 2005).

Silica fume mixtures were tested at percentages of 10%, 20% and 30% PC replacements (Mostafa and Brown 2005). Similar to previous results, low heat release was detected at the beginning of the tests, but after 25 hours, the mixture made with 10% SF replacement had a higher HH than the rest of the samples. It further proved that the decrease in HH for the samples made with larger quantities of SF is related to the reduction in the amount of PC in the mix rather than the effect of SF (Mostafa and Brown 2005).

A different study used semi-adiabatic calorimetry for 10 days to test 16 different samples (Kadri and Duval 2009). The samples had different percentages of SF replacements of PC of 0%, 10%, 20%, 30%, and 40% as well as different w/b ratios of 0.25, 0.30, 0.35, and 0.45. The results of some of the tests discussed in this study is shown in Figure 6. The graphs show that during the first 24 hours, the HH of the SF samples is greater than the HH of the control mix. The findings of this study also show that as the w/b ratio increases, the heat released increases (Kadri and Duval 2009).

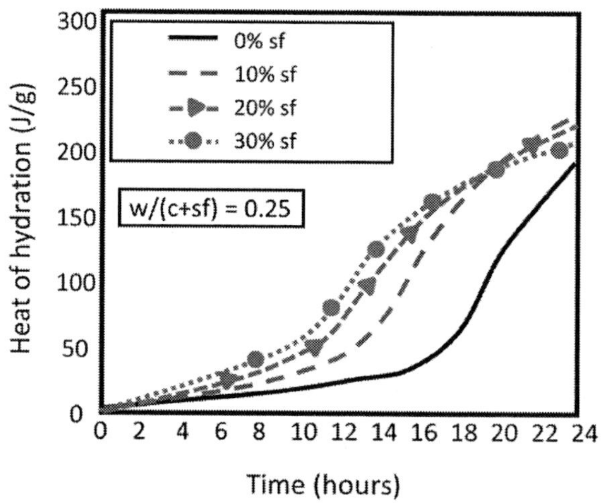

Figure 6: Heat of hydration of different SF percentages at two w/b ratios (Kadri and Duval 2009)

Additional tests were done using samples that contained SF in the percentages of 5%, 10%, and 15% as PC replacement to compare the amount of HH released (Maia et al. 2011). The results provided in this study were different than the ones in different papers. In this study, it was reported that SF decreased the HH slightly compared to a control mixture, as shown in Figure 7. The results show that as the percentage of SF increases, the HH released decreases (Maia et al. 2011). The findings of this study show discrepancy in the results compared to other studies.

Kadri et al. (2011) conducted a test using only one sample that contained SF as a 10% replacement of PC. It was compared to a control mix that contained 100% PC. The results concurred what was stated in previous studies; the addition of SF increases the HH (Kadri et al. 2011).

A recent study provided results for experiments that were conducted for 72 hours on mixtures that contained ground SF as a 25% replacement of PC (Alhozaimy et al. 2015). The results for the SF mixture showed that there was 15% reduction in the HH when compared to a control sample. The results from the concrete samples that contained ground SF also showed that the total HH was reduced along with its peak heat. The results obtained using ground silica fume are different from the results of the SF samples. This difference in behavior was attributed to the to the high crystallinity of the ground silica fume, which diminishes its chemical reactivity.

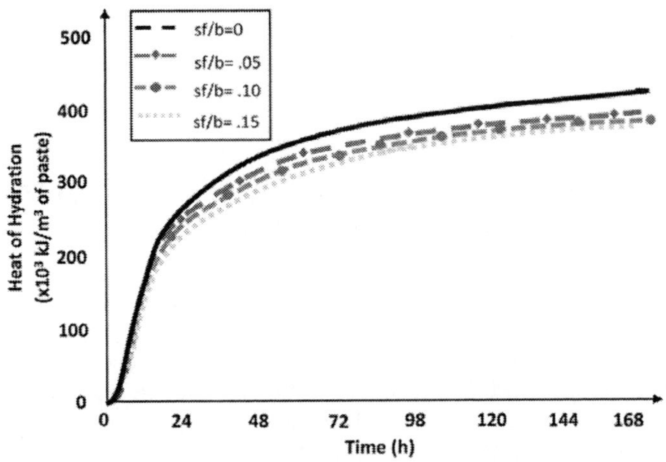

Figure 7: Influence of SF content on the HH (Maia et al. 2011)

5 FINDINGS

This paper presents a review of 14 previous studies that were conducted to assess the effect of FA the HH of a cement mix. The results reported in these studies seem to be consistent regarding the ability of FA to reduce the HH. The observed behavior was that increasing the amount of FA in the mix leads to a greater decrease in the amount of heat released and the peak temperature. However, special attention should be given to the calcium and anhydrite contents in the type of FA used as higher contents of calcium and anhydrite may increase the HH released (Tokyay 1988, Schindler and Folliard 2003). When FA was combined with other PSCMs and admixtures, the results were contradictory. The results show that FA is compatible with slag as they positively reinforce the tendency to lower the HH in a cement mix when combined together. On the other hand, Metakaolin and plasticizers seem to counteract the capability of FA to lower the HH of the mix. The most important results of previous studies conducted on the effect of FF on the HH is given in Table 1.

Table 1: Summary of the most important findings about the effect of FA on the HH

Material	Replacement (%)	HH (J/g)	Reference
Fly Ash	10%	185	(Meland, 1983)
(unclassified)		250	(Snelson et al., 2008)
		260	(Frias et al., 2000)
	20%	170	(Meland, 1983)
		235	(Snelson et al., 2008)
		275	(Maia et al., 2011)
	25%	206	(Alhozaimy et al. 2015)
		220	(Pane and Hansen, 2005)
		250	(Pane and Hansen, 2005)
		215	(Pane and Hansen, 2005)
		225	(Pane and Hansen, 2005)
	30%	275	(De Rojas et al., 1993)
		330	(Cheng-yi and Feldman, 1985)
		200	(Frias et al., 2000)
		200	(Snelson et al., 2008)
		275	(De Rojas et al., 1996)
	40%	285	(Snelson et al., 2008)
		205	(Maia et al., 2011)
	60%	120	(Maia et al., 2011)

The results of the SF as reported in the literature are quite contradictory and inconclusive. Most studies showed that the addition of SF leads to an increase in the HH of the mix. However, increasing the amount of SF beyond a certain limit leads to a decrease in the heat released. The explanation of this observation is quite inconclusive as it is not clear whether the reduction of the amount of PC used is the main factor for the decrease in the HH or not. The threshold for the percentage of SF replacement needed to start noticing a decrease in the HH is undetermined. A study reported contradictory results regarding the effect of SF on the HH of a cement mix. The study indicates that increasing the percentage of SF replacements of PC will lead to a decrease in the HH generated (Maia et al. 2011).

6 CONCLUSIONS

This paper presents a comprehensive review for 28 studies conducted on the effect of FA and SF on the HH of a cement mix. For the most part, the results on the behavior of FA are conclusive and consistent. The addition of FA leads to a decrease in the HH of the mix. However, careful consideration should be given the calcium and anhydrite contents in FA, as they may lead to different results. The majority of studies conducted on SF reported an inefficient behavior regarding the ability of SF to lower the HH of a cement mix. However, the results are contradictory given the fact that a recent study reported opposite results. As such, new methods are needed to study the behavior of PSCM with respect to their effect on the HH. Also, further classification of SF types needs to be established based on the chemical composition of the type used.

REFERENCES

ACI 2005. "207.1 R-05 Guide to Mass Concrete." In American Concrete Institute.

ACI 207.2R 2007. Report on Thermal and Volume Change Effects on Cracking of Mass Concrete. American Concrete Institute. www.concrete.org.

Alhozaimy, A., G. Fares, O. A. Alawad, and A. Al-Negheimish 2015. "Heat of hydration of concrete containing powdered scoria rock as a natural pozzolanic material." Construction and Building Materials 81: 113-119.

Alshamshi, A. M 1994. "Temperature rise inside pastes during hydration in hot climates." Cement and concrete research24, no. 2: 353-360.

Atiş, Cengiz Duran 2002. "Heat evolution of high-volume fly ash concrete." Cement and Concrete Research 32, no. 5: 751-756.

Awal, ASM Abdul, and M. Warid Hussin 2011. "Effect of palm oil fuel ash in controlling heat of hydration of concrete." Procedia Engineering 14: 2650-2657.

Bai, J., and S. Wild 200. "Investigation of the temperature change and heat evolution of mortar incorporating PFA and metakaolin." Cement and Concrete Composites 24, no. 2: 201-209.

Briley, G. C. 2004. Cooling for Dams. ASHRAE Journal, 46(3), 66.

Cheng-Yi, Huang, and Rolf F. Feldman 1985. "Hydration reactions in Portland cement-silica fume blends." Cement and Concrete Research 15, no. 4: 585-592.

De Lomas, M. García, MI Sánchez De Rojas, and M. Frías 2007. "Pozzolanic reaction of a spent fluid catalytic cracking catalyst in FCC-cement mortars." Journal of Thermal Analysis and Calorimetry 90, no. 2: 443-447.

De Rojas, MI Sanchez, Ma P. Luxan, M. Frías, and N. Garcia 1993. "The influence of different additions on portland cement hydration heat." Cement and Concrete Research 23, no. 1: 46-54.

De Rojas, MI Sanchez, and M. Frias 199. "The influence of silica fume on the heat of hydration of Portland cement." Special Publication 153: 829-844.

De Rojas, MI Sanchez, and M. Frías 1996. "The pozzolanic activity of different materials, its influence on the hydration heat in mortars." Cement and Concrete Research 26, no. 2: 203-213.

Frıas, M., MI Sanchez De Rojas, and J. Cabrera 2000. "The effect that the pozzolanic reaction of metakaolin has on the heat evolution in metakaolin-cement mortars." Cement and Concrete Research 30, no. 2: 209-216.

Ismail, Mohammad, Ainul Haezah Noruzman, Muhammad Aamer Rafique Bhutta, Taliat Ola Yusuf, and Ibrahim Hassan Ogiri 2016. "Effect of vinyl acetate effluent in reducing heat of hydration of concrete." KSCE Journal of Civil Engineering20, no. 1: 145-151.

Kadri, El-Hadj, and Roger Duval 2009. "Hydration heat kinetics of concrete with silica fume." Construction and Building Materials 23, no. 11: 3388-3392.

Kadri, El-Hadj, Said Kenai, Karim Ezziane, Rafat Siddique, and Geert De Schutter 2011. "Influence of metakaolin and silica fume on the heat of hydration and compressive strength development of mortar." Applied Clay Science 53, no. 4: 704-708.

Kim, J. K., Kim, K. H., & Yang, J. K. 2001. Thermal analysis of hydration heat in concrete structures with pipe-cooling system. Computers & Structures, 79(2), 163-171.

Korpa, A., T. Kowald, and R. Trettin 2008. "Hydration behaviour, structure and morphology of hydration phases in advanced cement-based systems containing micro and nanoscale pozzolanic additives." Cement and Concrete Research 38, no. 7: 955-962.

Langan, B. W., K. Weng, and M. A. Ward 2002. "Effect of silica fume and fly ash on heat of hydration of Portland cement." Cement and Concrete research 32, no. 7: 1045-1051.

Lee, M. H., B. S. Khil, and H. D. Yun 2014. "Influence of cement type on heat of hydration and temperature rise of the mass concrete.

Lessard, S., P. C. Aitcin, and Micheline Regourd 1983. "Development of a low heat of Hydration blended cement." Special Publication 79: 747-764.

Maia, Lino, Miguel Azenha, Rui Faria, and Joaquim Figueiras 2011. "Influence of the cementitious paste composition on the E-modulus and heat of hydration evolutions." Cement and Concrete Research 41, no. 8: 799-807.

Massazza, Franco 1993. "Pozzolanic cements." Cement and Concrete composites 15, no. 4: 185-214.

Meland, I 1983. "Influence of condensed silica fume and fly ash on the heat evolution in cement pastes." Special Publication 79: 665-676.

Mostafa, N. Y., and P. W. Brown 2005. "Heat of hydration of high reactive pozzolans in blended cements: isothermal conduction calorimetry." Thermochimica acta 435, no. 2: 162-167.

National Ready Mix Concrete Association (NRMCA) 2009. "Thermal Cracking of Concrete" Concrete in Practice. https://www.nrmca.org/aboutconcrete/cips/42p.pdf.

Pane, Ivindra, and Will Hansen 2005. "Investigation of blended cement hydration by isothermal calorimetry and thermal analysis." Cement and concrete research 35, no. 6 1155-1164.

Patil, A 2015. Heat of Hydration in the Placement of Mass Concrete. International Journal of Engineering and Advanced Technology (IJEAT) ISSN: 2249 – 8958, Volume-4 Issue-3.

Persson, Bertil 1997. "The effect of silica fume on the principal properties of concrete." In Symposium on Advanced Design of Concrete Structures, Gothenburg, pp. 161-168.

Portland Cement, Concrete, and Heat of Hydration. (1997, July). Concrete Technology Today, Portland Cement Association (PCA). 18(2), 1-4. Retrieved July, 2014, from http://cement.org/tech/pdfs/pl972.pdf

Schindler, Anton K., and Kevin J. Folliard 2003. "Influence of supplementary cementing materials on the heat of hydration of concrete." In Advances in Cement and Concrete IX Conference, Copper Mountain Conference Resort in Colorado.

Snelson, D. G., Wild, S., & O'Farrell, M. 2008. Heat of hydration of Portland Cement–Metakaolin–Fly ash (PC–MK–PFA) blends. Cement and Concrete Research, 38(6), 832-840.

Su, Lei, Baoguo Ma, Shouwei Jian, Zhiguang Zhao, and Min Liu 2013. "Hydration heat effect of cement pastes modified with hydroxypropyl methyl cellulose ether and expanded perlite." Journal of Wuhan University of Technology-Mater. Sci. Ed.28, no. 1 122-126.

Tokyay, Mustafa 1988. "Effects of three Turkish fly ashes on the heat of hydration of PC-FA pastes." Cement and Concrete Research 18, no. 6: 957-960.

Yang, J., Hu, Y., Zuo, Z., Jin, F., & Li, Q. 2012. Thermal analysis of mass concrete embedded with double-layer staggered heterogeneous cooling water pipes. Applied Thermal Engineering, 35, 145-156.

Zhu, B. 1999. Effect of Cooling by Water Flowing in Nonmetal Pipes Embedded in Mass Concrete. J. Constr. Eng. Manage., 125(1), 61-68.

Building Tomorrow's Society
Bâtir la Société de Demain

Fredericton, Canada
June 13 to 16, 2018 / *13 au 16 juin, 2018*

AN OVERVIEW OF THE RECYCLED WASTE MATERIAL USED AS REPLACEMENTS OF AGGREGATES IN CONCRETE

Abdel-Raheem, Mohamed[1,2], Rodriguez, Arlynn [1] and Allam, Mohamed[1], Beas, Moises[1], Gonzalez, Juan[1], Tobias, Cesar[1]
[1] Department of Civil Engineering, University of Texas Rio Grande Valley, USA
[2] mohamed.abdelraheem@utrgv.edu

Abstract: Solid waste material has been a major concern around the world due to its negative impact on the environment. Over years, there has been a dramatic increase in the amount of solid wastes stored in disposal sites. In an attempt of minimizing the negative impacts of industrial wastes on the environment, many research have been conducted on the effect of these wastes when incorporated in concrete. The majority of the research in this area focused on the partial replacement of cement with these wastes. However, other studies investigated the replacement of fine and coarse aggregates with these wastes. This Paper presents a comprehensive summary for previous studies that focused on the partial replacement of aggregates in concrete with different industrial waste. The paper addresses the findings of these studies in an attempt of assisting future researchers identify useful alternative substitutes for concrete aggregates that enhance its characteristics and contribute to preserving the environment.

1 INTRODUCTION

The construction industry is one of the most demanding and most rapidly expanding industries in the world, as result of the urban expansion, rapid population growth, industrialization and the need to improve the standards of living. This leads to the consumption of humongous amounts of materials, either to create new materials or directly use them for construction. Concrete due its strength to bear large compression loads was determined to be the most effective building material. In the last decade, concrete production has increased from 3,310 million tons to 4,200 million tons annually; and this number is subject to an annual increase. The manufacturing of Portland Cement, which is the main hydraulic binder used for concrete is responsible for huge emission of CO_2 gas that is a main cause of the greenhouse effect. Currently, there are several approaches that can be used to reduce cement consumption. One method is to minimize the concrete cross sections of the load carrying members in the structural designs. Another approach is to substitute portions of the cements needed for manufacturing of concrete with waste materials that have comparable properties to that of cement, such as fly ash. Using recycled wastes as a substitute or replacement for a specific components in concrete has become a great alternative to aid in overcoming pollution and decrease waste. Recycled wastes are analyzed and those that have similar properties as cement, fine aggregates, or coarse aggregate are chosen as replacements for traditional components of concrete. In turn, this reduces the overall cost and pollution in order to produce an efficient material which could be a suitable substitution. This paper presents a comprehensive review of all the studies that focused on the utilization of waste materials as a substitute for portions of the concrete aggregates. done on materials suitable for cement replacement.

2 PROBLEM STATEMENT AND OBJECTIVES

Several studies have been conducted to assess the effect of using waste materials on the properties of concrete when used as partial replacements of aggregates. However, there is not a single paper that provides a comprehensive review of all the work that has been performed. As such, there is a need for a study that succinctly summarize previous studies that investigated effects of utilizing waste material in concrete as a partial replacement of aggregates. This paper sheds the light on previous research attempts in this area. This study presents a concise description of each of the previous studies and highlights the most important findings.

3 METHODOLOGY

The methodology used in this research consists of two main steps. The work started by collecting scholarly articles that addressed previous experiments conducted on the utilization of waste materials in lieu of portions of the aggregates in the concrete mix. The collected papers were classified based on the aggregate type that was replaced in the study. As such, previous studies have been grouped into two categories: fine aggregates and coarse aggregates.

4 WASTE MATERIALS IN CONCRETE

The main objective of all the studies compiled was to explore alternative venues to dispose industrial wastes by utilizing them in concrete in the hope of enhancing or at least maintaining the normal level of the concrete quality. Many studies revolved around substituting either the coarse aggregate or fine aggregate in concrete. The different sizes, shapes and chemical composition of the aggregate have direct impacts not only on the strength and durability or the concrete which may be an engineers' main concern but also on the workability and weight of the concrete. The different aggregates chosen to use in a specific mix also play a huge role in the appearance of the surface being poured.

The materials that were used to substitute fine aggregates in concrete were: waste foundry sand, bottom ash, plastics, crushed glass, waste glass, waste marble aggregates, waste iron, Class F fly ash, scale and steel chips, concrete metallic chips, copper slag, fine crushed bricks, waste tire steel beads, and steel slag. For the replacement of coarse aggregate, the materials used were, coarse rigid polyurethane foam waste, rubber tire particles, crushed concrete, rubber tire waste, oil palm shells, ground coarse fly ash, recycled concrete, and ceramic industry waste.

The majority of the studies focused mainly on the effect of incorporating these wastes in concrete; however, some studies considered other properties of concrete such as tensile and flexural strengths as well as the ability for the concrete mixes to shield radiation.

5 FINE AGGREGATES

The review of the literature shows that there are seven studies that have been conducted to evaluate the quality of concretes made with waste material used as partial replacement of fine aggregates. The main measure of quality used in these studies was the concrete compressive strength.

Siddique (2003) used Class-C fly ash to partially replace fine aggregate in concrete. Class-C fly ash was used in the increments of 10%, 20%, 30%, 40% and 50% as a replacement of sand in the concrete mix. Cube specimens (150 mm) were casted for compressive strength testing in this study. The concrete was cured for 7, 28, 91, and 365 days. The compressive strength of concrete increased with the increasing the percentage of the Class-C fly used as a replacement of the fine aggregate in the concrete mix (Siddique, 2003).

Aggarwal el al. (2007) investigated the effect of use of bottom ash and waste foundry sand in equal quantities as partial replacement for fine aggregates in various percentages from 0% to 60%. Bottom ash is a substance produced when coal is burned off; it is heavier than fly ash that is why it settles at the bottom of the boiler. When replacing fine aggregate with waste foundry ash and bottom ash the mechanical behavior of concrete showed strengths comparable to that of conventional concrete. Furthermore, it was observed that the greatest increase in compressive, splitting tensile strength and flexural strength was achieved by substituting 30% of the natural fine aggregate (Aggarwal et al., 2007). The maximum acceptable replacement was at 50% as increasing the percentage of fine aggregate replacement beyond this point would result in a decrease in its compressive strength (Aggarwal et al., 2007).

In 2007, Batayneh et al. studied the properties of concrete made with recycled construction material from constructions sites. The recycled materials used composed of plastics and crushed glass and were used as a substitute for the sand in the concrete mix. Different mixes incorporating the waste materials were made. Waste plastics were reused by grinding them into small particles. A concrete mix was made using the fine plastic particles as 20% replacement of sand (Batayneh et al., 2007). The crushed glass was manually crushed and sieved replacing from 0% to 20% of the sand in the concrete mix. After 28 days, the results show that the compressive strength of concrete mixes was improved by the partial replacement of the fine aggregates with crushed glass; however, when replacing the fine aggregates with 20% of plastics with crushed concrete, the compressive strength of concrete exhibited lower values than the compressive strengths of the control mix (Batayneh et al., 2007).

Debieb and Kenai (2008) also conducted a study in which fine aggregate was replaced in concrete using fine crushed bricks. Cubes of 100 mm were casted and tested for 3, 7, 28, and 90 days (Debieb & Kenai, 2008). The finer crushed aggregates were used, the more decrease in the compressive strength of concrete (Debieb & Kenai, 2008).

Ismail and Al-Hashmi (2008) completed a second study by replacing fine aggregate in concrete with waste iron. The procedure for the study was executed by comparing the compressive strength of a control concrete mix to different concrete mixes that contained waste iron in the increments of 10%, 15%, and 20% as replacement of fine aggregates. Both the control and incremental replacement mixes were cured for 3, 7, 14, and 28 days. A total of 115 cubes of concrete were used for compressive strength testing and 87 prisms for flexural strength (Ismail and Al-Hashmi, 2008). The results show that the compressive and tensile strength improved with the addition of the waste iron.

Ismail and Al-Hashmi (2009) conducted a second study where they replaced fine aggregates in concrete with waste glass. Waste glass can be produced in many such as packaging or container glass, bulb glass, flat glass, and cathode ray tube glass. A control mix was compared to several concrete mixes that contained waste glass as 10%, 15%, and 20% replacements of fine aggregates. The results show a decrease in the slump by increasing the percentage of the waste glass in the mix. However, the compressive strength increased by 4.23% when replacing 20% of fine aggregates with waste glass (Ismail & Al-Hashmi, 2009).

Alwaeli and Nadziakiewicz (2012) also replaced fine aggregate in concrete using scale and steel chips. In this study, two distinct types of concrete were made with steel chips scale (ScC) and (SchC). The sand was replaced in proportions of 25%, 50%, 75%, 100%. The concretes were produced in accordance with the Polish standards. Cubes of 100 mm length were used for gamma radiation testing, 84 slabs were casted with measurements of 250 x 250 x 26 mm. The results were the following: the compressive strength of concrete containing steel chips is better than ordinary concrete when used as fine aggregate replacement of up to 25% (Alwaeli & Nadziakiewicz, 2012). At the same time, addition of scale and steel chips enhances the shielding of gamma radiation (Alwaeli & Nadziakiewicz, 2012).

A summary of different waste materials used as a replacement of fine aggregates in concrete and their effect of on the compressive strength is shown in Figure 1.

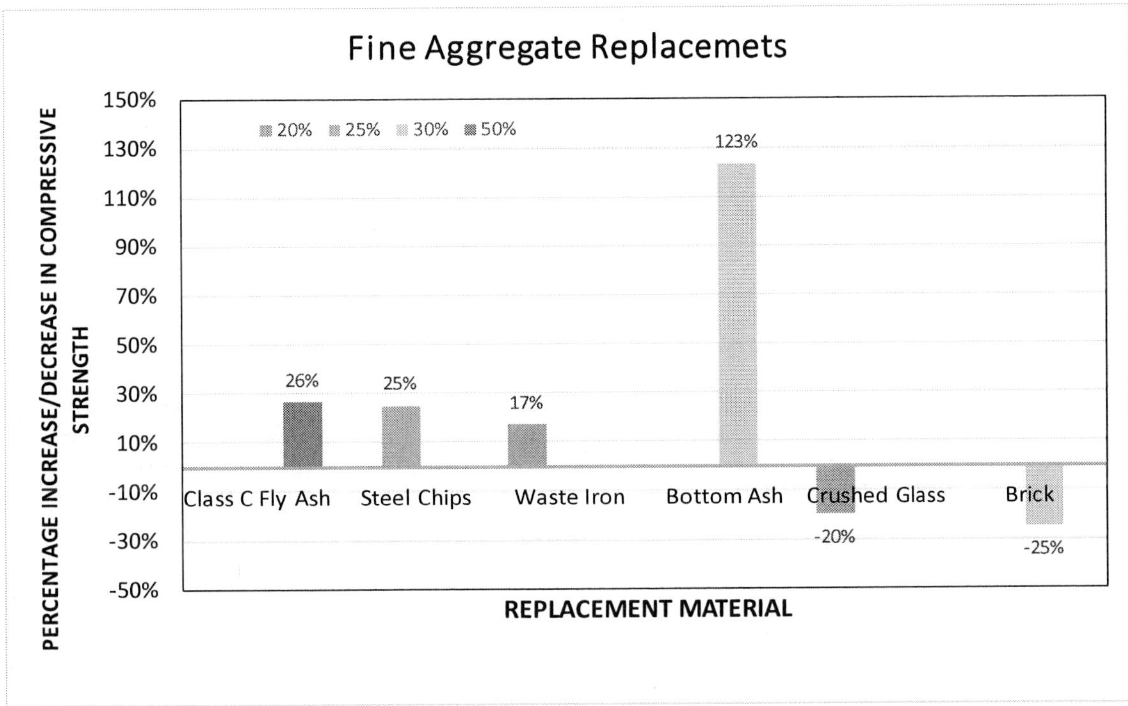

Figure 1: Comparison of the compressive strength of with different fine aggregate replacements

6 COARSE AGGREGATES

The literature included only four studied that explored the utilization of waste materials as partial replacements of coarse aggregates. The waste materials used in these studies included: ceramic industry waste, recycled concrete, rubber chips, and oil palm shells.

Basri et al. (1999) conducted a study to explore the possibility of using oil palm shells as a partial replacement of the coarse aggregate in concrete. The objective of using oil palm shells is to produce lightweight concrete. The natural thickness of this material ranges from 1.5 – 2.5 mm. Specimens were made using oil palm shells as a partial replacement of coarse aggregates. The concrete was tested for the slump and compressive strength and the results showed that the compressive strength of the oil palm shell specimens was about 41%-51% lower than the regular concrete; however, it was still within the range for structural lightweight concrete (Basri et al, 1998).

In another study, recycled concrete was used as a replacement of the coarse aggregates. The waste was obtained from crushed concrete that had been removed from a demolished structure. The results of experiments conducted showed that the quality of the recycled concrete used as a replacement of coarse aggregate is lower than that of natural aggregate. This is mainly because of the effect of mortar along with the remnants of other reactions and materials attached to the natural aggregates that were used in the recycled concrete. The replacement of the coarse aggregate with recycled concrete showed a reduction in compressive, flexural strength and splitting tensile strength compared to the control concrete mix.

A third study investigated the effect of utilizing waste tire rubber as partial replacement of coarse aggregates in concrete. In preparing rubberized concrete samples, 15% of the volume of gravel was replaced with rubber chips. With this replacement, the compressive strength was reduced by 45% and the indirect tensile strength was reduced by 23%. It was found that the rubber chips' size had a direct impact on the compressive strength. Reducing the size of the rubber chips should result in an increase in its strength. (Huang et al, 2004)

A study investigated the effects of using ceramic industry waste on concrete mix when used as a replacement of the coarse aggregate. This ceramic waste material was broken in 20 mm size to be used as coarse aggregate. Different mixes were made using different water cement (w/c) ratios of 0.35 ,0.40 ,0.45, 0.50, 0.55 and 0.60. The ceramic waste was used as a 100% replacement of the coarse aggregate, but the sand and cement varied depending on the w/c ratio. The specimens made from different batches were tested for the compressive and flexural strengths and were compared to the control concrete mix. The results show the utilization of ceramic waste as coarse aggregates enhanced the workability of concrete and increased the compressive and flexural strengths as well as the modulus of elasticity of the concrete. (Gobinath et al, 2011)

7 FINDINGS

From previous research, some interesting findings can be found about the impact of using waste materials as replacements of aggregates on concrete. The change in concrete properties depend on the type of material used as an aggregate replacement, and the percentage of replacement. A complete summary of the most important findings for each study is summarized in Table 1.

Table 1: Findings of Coarse and Fine Aggregate Replacements

Type of Aggregate	Material	Findings	Source	Speci men Type	Speci men Size
Coarse	oil palm shells	OPS has better workability, lower compressive strength, light weight concrete	Concrete using oil palm shells as aggregate Basri et al. (1999)	Cube Molds	100x10 0x100 mm3
Fine	copper slag	increase in density, up to 50 percent replacement had a comparable compressive strength and then it goes down,	Copper slag as sand replacement for high performance concrete , Khalifa S. Al-Jabri ,a, Makoto Hisada, Salem K. Al-Oraimi , Abdullah H. Al-Saidy	Cube Molds	150x15 0x150 mm
Fine	bottom ash	density and workability decreased, compressive strength was lower than control	EFFECT OF BOTTOM ASH AS REPLACEMENT OF FINE AGGREGATES IN CONCRETE, P. Aggarwal , Y. Aggarwal, S.M. Gupta		
Fine	class F fly ash	compressive strength increased	Effect of fine aggregate replacement with Class F fly ash on the mechanical properties of concrete. Rafat Siddique	Cylinder s	150x30 0 mm
Fine	glass aggregate		"Glasscrete" concrete with glass aggregateby Weihua Jin, Christian Meyer and stephen Baxter		
Coarse	rubber tire particles	reduction of both compressive and flexural strength	The Use of Rubber Tire Particles in Concrete to replace mineral aggregate , H. A. Toutanji	Cylinder s	100x20 0
Fine	glass aggregate	optimum flexural strength was at 20 percent replacement, slump decreased	Recycling of waste glass as a partial replacement for fine aggregate in concrete, Zainab Z. Ismail , Enas A. AL-Hashmi		
Fine	scale and steel chips	better compressive strength until 25 percent, better absorption of gamma radiation	Recycling of scale and steel chips waste as a partial replacement of sand in concrete, Mohamed Alwaeli , Jan Nadziakiewicz	Cube Molds	100x10 0x100 mm3

Type of Aggregate	Material	Findings	Source	Specimen Type	Specimen Size
Fine	iron	flexural strength increase, slump decrease	Reuse of waste iron as a partial replacement of sand in concrete, Zainab Z. Ismail , Enas A. AL-Hashmi		
Fine & Coarse	scrap tire rubber	compressive strength, modulus of elasticity tensile and flexural strength decreased	Scrap-tire-rubber replacement for aggregate and filler in concrete, Eshmaiel Ganjian a, Morteza Khorami b, Ali Akbar Maghsoudi c		
Coarse	metallic chips	better tensile strength , better compression with small percentage	Strengthening of Concrete Metallic Chips, Y. Bouafia, S. Djebali, D. Atlaoui, and M. S. Kachi	Cube Molds	90x90x 100 mm
Fine & Coarse	crushed bricks	lower density, higher water absorption, decrease in compressive strength	The use of coarse and fine crushed bricks as aggregate in concrete, Farid Debieb , Said Kenai	Cube Molds	100x100x100 mm & 70x70x 280
Fine	fly ash	higher strength and environmental friendly	Use of ground coarse fly ash as a replacement of condensed silica fume in producing high-strength concrete, Chai Jaturapitakkul, Kraiwood Kiattikomol, Vanchai Sata, Theerarach Leekeeratikul		
Fine	solid waste incineration fly ash	the physical properties were not deteriorated, it maintained close to the control	Use of municipal solid waste incineration fly ash in concrete, J.E. Aubert, B. Husson, A. Vaquier	Cylinders	15x20 cm
	recycled plastic	density reduced, compressive strength decreased, decrease in tensile strength,	Use of recycled plastic in concrete: A review, Rafat Siddique a, Jamal Khatib b, Inderpreet Kaur		
	steel fibers	slightly increased compressive strength, lower workability	Use of steel fibers recovered from waste tires as reinforcement in concrete: Pull-out behavior, compressive and flexural strength, M.A. Aiello, F. Leuzzi, G. Centonze, A. Maffezzoli	Cube Molds	150x150x150
	steel beads	decreased workability , reduction in compressive strength, increased tension	Use of waste tire steel beads in Portland cement concrete, Christos G. Papakonstantinou, Matthew J. Tobolski	Cylinders	4x8 inch
Coarse	coarse rigid polyurethane foam waste	good workability, higher porosity, decrease in compressive strength and dynamic modulus of elasticity	Valorization of coarse rigid polyurethane foam waste in lightweight aggregate concrete, Amor Ben Fraj, Mohamed Kismi, Pierre Mounanga	Cube Molds	50x50x 50 mm

8 CONCLUSIONS

The increment of waste materials and the awareness of waste management and environmental issues had led to a progress in the utilization of various waste materials in concrete cease to be toxic. Concrete mixtures can be made containing a volume fraction of waste materials as a replacement of natural aggregates. This serves as an alternative venue for disposing waste material sustainably. The results show that some materials can have a positive impact on some properties of concrete, such as the compressive strength and workability. While the majority of waste material have negative effect on the compressive strength of concrete, a life cycle cost analysis study is needed to determine the feasibility of this approach.

References

Aggarwal, P., Aggarwal, Y., & Gupta, S. M. (2007). Effect of bottom ash as replacement of fine aggregates in concrete. Asian journal of civil engineering (building and housing), 8(1), 49-62.

Alwaeli, M., & Nadziakiewicz, J. (2012). Recycling of scale and steel chips waste as a partial replacement of sand in concrete. Construction and Building Materials, 28(1), 157-163.

Basri, H. B., Mannan, M. A., & Zain, M. F. M. (1999). Concrete using waste oil palm shells as aggregate. Cement and concrete Research, 29(4), 619-622.

Batayneh, M., Marie, I., & Asi, I. (2007). Use of selected waste materials in concrete mixes. Waste management, 27(12), 1870-1876.

Debieb, F., & Kenai, S. (2008). The use of coarse and fine crushed bricks as aggregate in concrete. Construction and building materials, 22(5), 886-893.

Hebhoub, H., Aoun, H., Belachia, M., Houari, H., & Ghorbel, E. (2011). Use of waste marble aggregates in concrete. Construction and Building Materials, 25(3), 1167-1171.

Huang, B., Li, G., Pang, S. S., & Eggers, J. (2004). Investigation into waste tire rubber-filled concrete. Journal of Materials in Civil Engineering, 16(3), 187-194.

Ismail, Z. Z., & AL-Hashmi, E. A. (2008). Reuse of waste iron as a partial replacement of sand in concrete. Waste Management, 28(11), 2048-2053.

Ismail, Z. Z., & Al-Hashmi, E. A. (2009). Recycling of waste glass as a partial replacement for fine aggregate in concrete. Waste management, 29(2), 655-659.

Senthamarai, R. M., Manoharan, P. D., & Gobinath, D. (2011). Concrete made from ceramic industry waste: durability properties. Construction and Building Materials, 25(5), 2413-2419.

Siddique, R. (2003). Effect of fine aggregate replacement with Class F fly ash on the mechanical properties of concrete. Cement and Concrete research, 33(4), 539-547.

Sri Ravindrarajah, R., & Tam, C. T. (1985). Properties of concrete made with crushed concrete as coarse aggregate. Magazine of concrete research, 37(130), 29-38.

Toutanji, H. A. (1996). The use of rubber tire particles in concrete to replace mineral aggregates. Cement and Concrete Composites, 18(2), 135-139.

Building Tomorrow's Society

Bâtir la Société de Demain

Fredericton, Canada

June 13 to 16, 2018 / *13 au 16 juin, 2018*

ANALYSIS OF PLASTIC DEFORMATION AND FATIGUE OF RECYCLED MATERIALS FOR ITS IMPLEMETATION ON RURAL ROADS

Figueroa-Infante, Ana[1,2] ,
[1] De La Salle University, Colombia
[2 4] afigueroa@unisalle.edu.co

Abstract: This research shows the results on the rutting and fatigue for the materials derived from maintenance of flexible pavements and the accumulation of tires that have become a threat to the population in terms of health and invasion of the public space. The two types of materials are waste that comes from different processes: the first one comes from recycling asphalt pavement and the second one from tire wear unused for different vehicles. The studies include analysis of the materials in the original condition and a proposal for improving the material, considering the reduced production of greenhouse gases. The product obtained in this project intended to be used in secondary and tertiary roads (rural roads) from Colombia, to improve the connection between agricultural production areas and urban centers in such a way that it contributes to the improvement and development of the competitiveness of these areas.

Key words: Recycled Materials, Rural Roads, Rubber-GCR, Trapezoidal Fatigue, Rutting, RAP

1 INTRODUCTION

Pavement recycling is a well-known practice that requires rejuvenation for its reuse [1]. Recent research has especially focused on the optimization of percentages of Reclaimed Asphalt Pavement-RAP permitted in the fabrication of new mixtures [2]. Studies on the utilization of RAP date back to the 80's with support from the Illinois Department of Transportation and the University of Illinois who implemented a new design methodology and experimented with the addition of residual asphalt to existing mixtures [3]. In successive years, new research began to focus on maximizing the amount of RAP used in new pavement mixtures without compromising performance appreciably; typically this turned out to be approximately 30% RAP, and 70% raw materials. However, there are countries, such as Holland, with experience using greater amounts of RAP. In Colombia, studies specifically focused on the use of RAP have taken place as part of the country's commitment to environmental sustainability. Likewise, in 2009 La Universidad de la Salle [4], with support from the Engativá city government, began an investigation to optimize manufacturing processes for the utilization of RAP acquired from the city's already existing stockpiles of RAP for which there was currently no possibility of use. This investigation laid the foundation for much more research on the use of RAP in Colombia. Since then, research on this topic has continued through some universities and governmental institutions [5].

Just as the analysis of adhesion and cohesion of RAP is important, so is that of fatigue and footprint in the study of new pavements. All of these factors, together, are essential for the categorization of its performance under vehicular loads: now an especially important quality control check given that the original composition of the pavement has been altered and, as such, requires additional analyzation. [6][7][8]. In 2011, the Federal Highway Administration of the United States published an official document on the state of recycling practices in the United States and methodologies for their realization. This document identified cost effectiveness, environmental responsibility, and performance as the most essential aspects for a well-designed recycling process. Additionally, it indicates that it is necessary to promote the use of the highest possible levels of RAP, while still conserving performance when compared with asphaltic mixtures of 100% raw materials. Table 1 shows the quantities of RAP used for various projects in the United States.

Table 1: Field projects using high levels of RAP

State	% RAP	Date of Construction
North Carolina	40%	Sep-07
South Carolina	30 and 50 %	Oct-07
Wisconsin	25%	Nov-07
Florida	45%	Dec-07
Kansas	30 to 40 %	May-08
Delaware	35%	Summer 2008
Minnesota (MnROAD)	30%	2008
Illinois	10 to 50 % allowed	2008

Source: http://www.fhwa.dot.gov/pavement/recycling/RAP/

2 THEORETICAL FRAMEWORK

2.1 Analysis of fatigue in asphaltic materials

Study of the fatigue of asphalt mixtures allows for the determination of the maximum deformation before failure after a certain number of load cycles. In this way, the useful life of such mixtures can be predicted. Damage occurs in the flexible layers of the pavement and presents itself when high levels of deformation are generated due to traction in the lower area of the asphalt layer; a product of repetitive traffic loads, as in [9][10]. According to the literature, the phenomenon of fatigue in asphaltic mixtures is defined as a permanent and progressive change that occurs at a point in the material due to varying forces and deformations. Such fatigue will eventually result in the appearance of fissures which can then lead to a total fracture of the asphaltic mixture due to the application of a certain number of traffic load cycles [11]. In [12], two phases of damage are identified: the first, known as the initiation phase, is when micro-fissures appear, generating a reduction of structural rigidity; and the second, known as propagation, is when union of micro-fissures begins to occur, causing a propagating effect of micro-fissure growth. According to that same author, the key factors that influence fatigue are: method of sample compression, method of loading and type of performance test applied, design parameters, and environmental variables.

In accordance with [13], performance testing methods used to measure the behavior of fatigue can be grouped into the following categories: simple flexibility tests, supported flexibility tests, direct axial tests, diametric tests, triaxial tests, and road tests. Laboratory tests have verified that the relationship between deformation, ε (produced by loading), and duration or the life of the fatigue of the material, N (represented by the number of times supporting a load before failure due to fatigue), can be described using the equation:

$$N = K_1 * \varepsilon^{K_2} \quad (1)$$

Where:

N: Number of load cycles until material fatigue.

ε: Uniform deformation of traction (in micro-deformations $^{\mu m}/_{m}$).

K_1 and K_2: Constants that describe the behavior of the fatigue of the material.

It has been confirmed that K_1 and K_2 are principally affected by: the modulus of the mixture, the contents of the asphalt, the viscosity of the asphalt, granulometry and natural state of the aggregates, air content, and pavement temperature.

Understanding the behavior of bituminous mixtures under fatigue is fundamental for dimensioning pavements using analytical methods. Failure due to fatigue was first observed in metals when they were subjected to a lower tension and fissured after repeated applications. From the very first studies of this type of fissure due to fatigue a logarithmic relationship between tension (or imposed deformation) and the number of load applications was deduced as:

$$log\ N\varepsilon = a - (b * log\ \varepsilon) \quad (2)$$

$$log\ N\sigma = c - (d * log\ \sigma) \quad (3)$$

Where:

$N\varepsilon$, $N\sigma$: Number of load applications in a test of fatigue due to deformation or controlled tension.

ε: Imposed deformation in a test of fatigue due to controlled deformation

σ: Imposed tension in a test of fatigue due to controlled tension.

$a, b, c\ y\ d$: Characteristics of the material under test.

3 MATERIALS AND METHODS

3.1 RAP Material Used

The selection of RAP samples was obtained from el Instituto de Desarrollo Urbano (IDU) which has stockpiles from various city projects in which road improvement projects are being completed through tearing up and replacing layers of asphalt. Record of stored asphalt selected for this phase of the investigation. Analysis for these samples is performed in accordance with the Instituto Nacional de Vias-INVIAS specification, from Colombia. See Figure 1.

Figure 1: Samples of RAP from different stockpiles

4 ANALYSIS OF RESULTS

4.1 Granulometry

Granulometry was performed on the RAP material (see Table 2) to determine the distribution of the aggregates and how they corresponded to normal Colombian rutting mixtures.

Table 2: Granulometry of original RAP

Sieve	Aperature (mm)	weight (gr)	% Filtered	% Passed	Lower Limit	Upper Limit
1	25.00	0.00	0.00	100.00	100	100
3/4	19.00	600.90	6.49	93.51	80	95
1/2	12.50	1273.60	13.75	79.77	0	0
3/8	9.50	678.80	7.33	72.44	69	75
No. 4	4.75	1782.50	19.24	53.20	47	62
No. 8	2.26	1661.90	17.94	35.26	35	50
No. 50	0.30	2348.20	25.35	9.91	13	23
No. 200	0.08	853.70	9.22	0.70	3	8
Fondo		64.40	0.70	0.00		

The graphic distribution of the aggregates is shown in Figure 2.

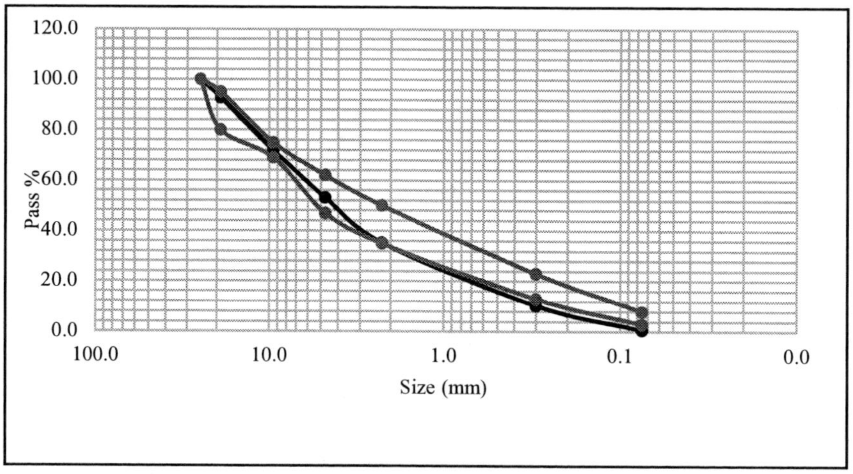

Figure 2: Granulometry curve of original RAP material

From the graph, it is observed that the granulometry does not comply with the lower limit imposed. For this reason, a granulometry fix was applied to the RAP for such compliance. See Table 3 and Figure 3.

Table 3 Granulometric fix of the RAP

Sieve	Aperature (mm)	weight (gr)	% Filtered	% Passed	Lower Limit	Upper Limit
1	25	0.00	0.00	100.00	100	100
3/4	19	87.50	12.50	87.50	80	95
1/2	12.5	54.21	7.74	79.75	0	0
3/8	9.5	51.31	7.33	72.42	69	75
No. 4	4.75	134.76	19.25	53.17	47	62
No. 8	2.26	74.69	10.67	42.50	35	50
No. 50	0.3	171.50	24.50	18.00	13	23
No. 200	0.075	88.20	12.60	5.40	3	8
Fondo		37.80	5.40	0.00		
Total		699.97				

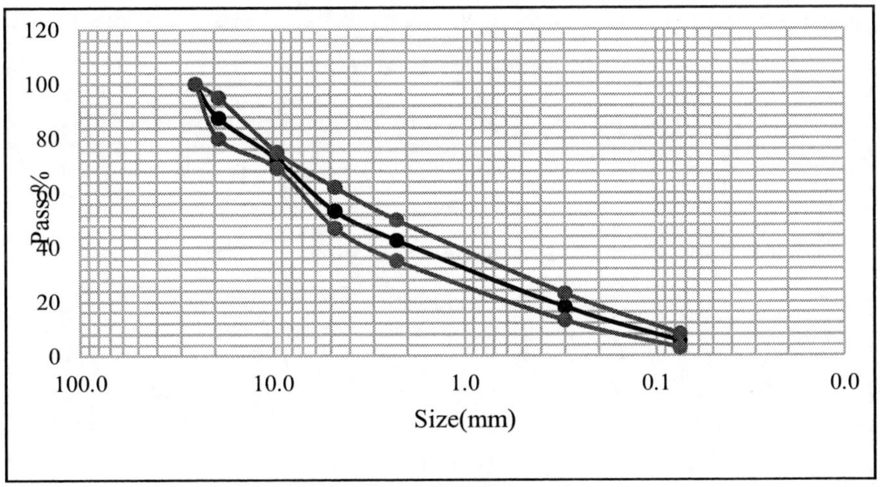

Figure 3: Granulometry curve of RAP material fixed with raw aggregate

4.2 Trapezoidal fatigue

In this investigation, tests were performed on trapezoidal samples whose fatigue behavior was studied using flexion tests at 25°C and 10 Hz. The first part of the test involved subjecting a sample to sinusoidal displacement with constant amplitude. This was done by affixing the bottom base and applying a continuous sinusoidal wave to the upper base at a set amplitude. Rupture is defined as the moment when the force applied to the end of the sample is reduced to half its initial value for a given displacement. It is necessary to evaluate various amplitudes of displacement in order to determine the slope of the law of fatigue, b, and the value of admissible deformation, $\varepsilon 6$, for one million cycles. The samples are trapezoidal with thickness of 25mm, length of 250mm, and long and short base widths of 70mm and 25mm, respectively. Samples were compressed using the universal set to guarantee that the density of the volumetric design of the mixture was obtained. See Figure 4.

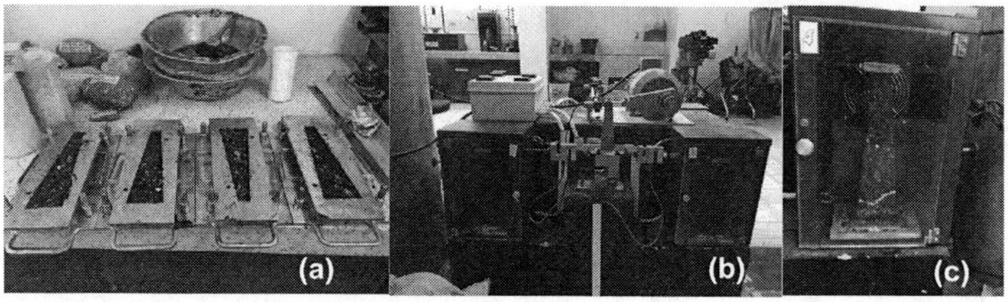

Figure 4: (a) Trapezoidal briquette test samples, (b) Experimental set up, (c) failure of original RAP briquette.

The law of fatigue of original RAP material for a type MDF-25 mixture was observed with basis in the French standard NFP98 at room temperature (20°C), taking into account three controlled deformations of 180, 300, and 440 μm in each of the chambers of the fatigue bank. See Figure 5.

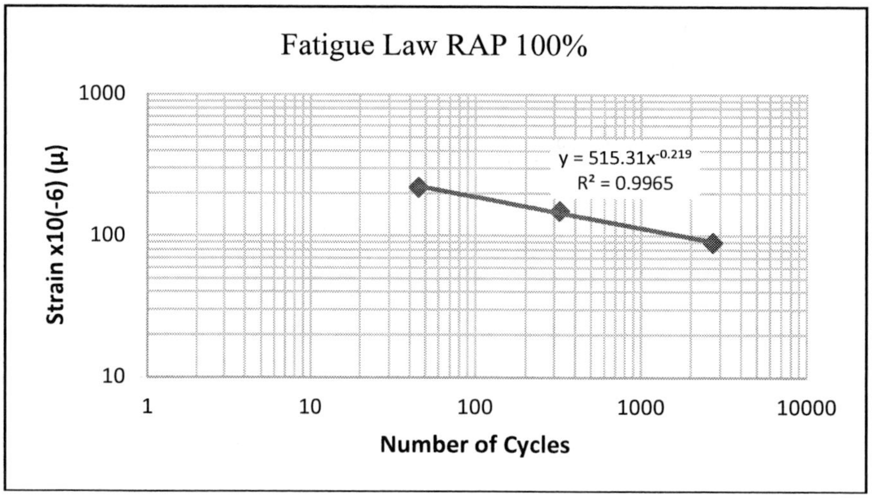

Figure 5: Law of fatigue for original RAP material

In order to analyze RAP fatigue in contrast with raw aggregate the same process is performed, obtaining the law of fatigue for a new type MDF-25 mixture with 12% asphaltic emulsion CRL-1. See Figure 6.

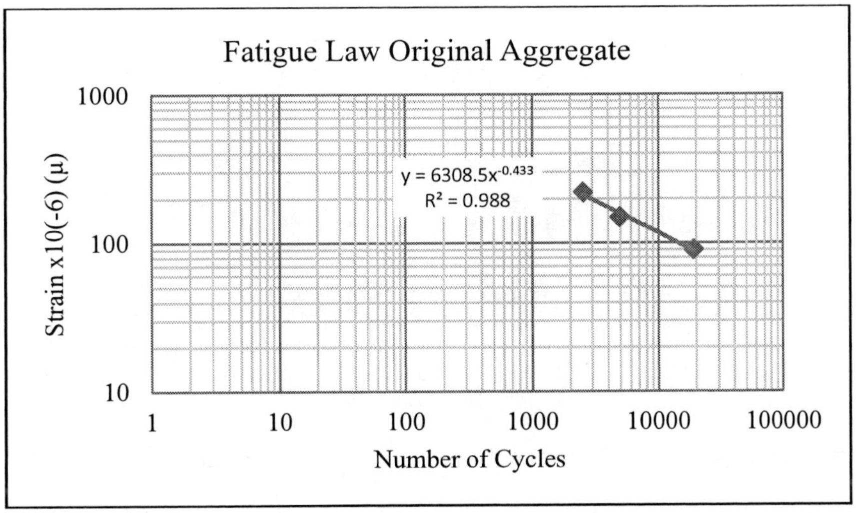

Figure 6: Law of fatigue for a type MDF-25 mixture with raw aggregate

Comparison of the two mixtures (100% RAP and raw aggregate) reveals a large difference in slope between the two curves such that for one million cycles (the design reference value) the RAP mixture exhibits more deformation. This implies that it is necessary to enhance this mixture by means of a rejuvenator or additive that would help to reduce the magnitude of deformations per load cycle. In order to know improve the fatigue behavior of this material was added certain amount[1] of rubber-CGR as a part of aggregate (dry condition). See Figure 7.

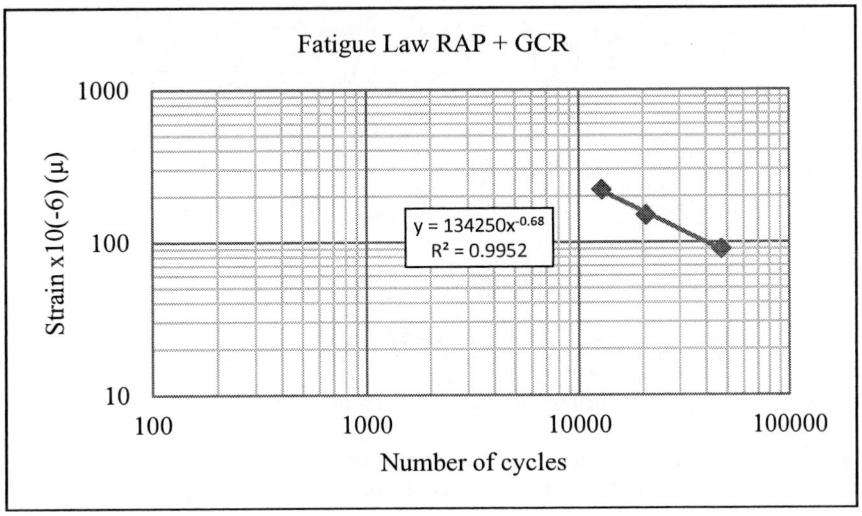

Figure 7: Law of fatigue for a type MDF-25 mixture RAP+ Rubber-CGR

4.3 Rutting

In order to determine rutting an experimental set up was constructed as per the INVIAS standard 756-13, and in accordance with the same standard, briquettes were made. See Figure 8.

---

[1] This value is protected as a part of this research

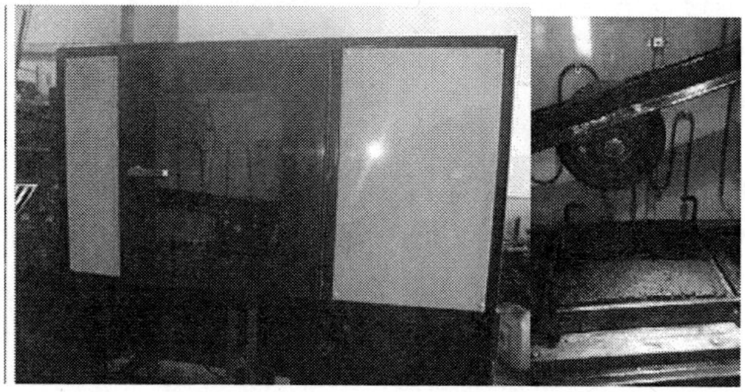

Figure 7: Experimental set up constructed as per the INVIAS standard for the determination of rutting in asphaltic mixtures[2].

4.4 Determination of plastic deformation

In order to determine performance under expected conditions, the experimental set up of footprint was designed and built according to the I.N.V.E. 756-13 specification. In order to determine its response, the material was subjected to extreme conditions; this was considered heavy traffic. See Figure 9.

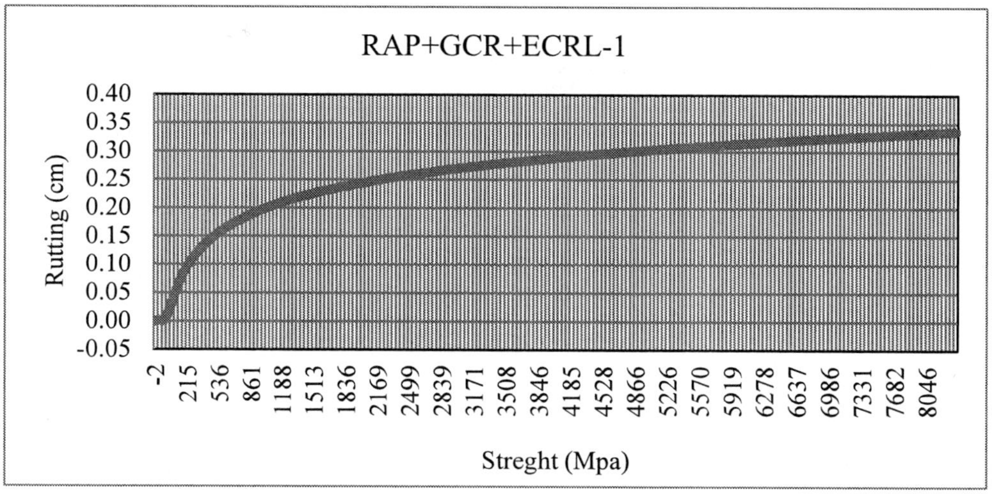

Figure 8: Plastic deformation of RAP mixture modified with GCR and ECRL-1

This first set of briquettes produced a maximum plastic deformation of 0.30 cm, however, this data should be contrasted with other test samples and different contents of residual asphalt.

5 CONCLUSIONS

One of the most influential variables during RAP acquisition was the content of binder present in the aged mixture. This parameter is essential to guarantee the homogeneity of the selected RAP.

Due to the diverse granulometries that are obtained during the recycling process in situ, the selection of RAP should comply with a desired granulometry for reuse. This implies a reliance on the addition of material

---

[2] Experimental set up financed by Vicerrectoría de Investigación, Universidad de la Salle,2015.

to sieves which don't comply with granulometry: in this case, sand and other fine material to keep on the frame of the newly designed type MDF-25 mixture.

Given that adhesion is lost completely due to aging processes, fatigue tests for virgin RAP without binder were not realized. For this reason, an adjustment with CRL-1 emulsion was performed.

The mixture of 100% RAP was compared with the mixture of raw aggregate using the law of fatigue and produced evidence of the necessity of the application of some type of rejuvenator to the RAP mixture to reduce deformations after one million cycles (the design reference value). However, for roads with lighter traffic loads and shorter lifespans, this mixture is viable in terms of mobility.

The plastic deformation of the RAP mixture modified with GCR and ECRL-1 was determined to be 0.30 cm during the first phase. However, this value should be contrasted with other test samples and different contents of residual asphalt.

Acknowledgements

- La Universidad de la Salle for their financial support in this investigation.
- The company Incoasfaltos, Concrescol S.A., and el Instituto de Desarrollo Urbano-IDU for their contribution of materials.

References

[1] Sondag, M. S., Chadbourn, B. A., & Drescher, A. (2002). Investigation of recycled asphalt pavement (RAP) mixtures.

[2] Fernández del Campo, J.A. (2003) Recycling in Road Pavements, MAIREPAV03—Third International Symposium on Maintenance and Rehabilitation of Pavements and Technological Control (University of Minho, Guimaraes).

[3] Civil Engineering Studies, Illinois Center for Transportation Series No. 07-001. UILU-ENG-2007-2014.ISSN: 0197-9191.

[4] Espinosa Rodríguez, Diego Armando. Guía de mantenimiento para pavimentos asfálticos de vías locales en Bogotá. Monografía de grado dirigida por Ana S. Figueroa, Programa de Ingeniería Civil, Universidad de la Salle. 2009.

[5] Ostos Ascencio, Liliana Duarte Sanmiguel, Juan M.Vargas Fonseca, German L Camacho Tauta, Javier Reyes-Ortiz, Oscar Javier. Comportamiento de mezclas asfálticas del Instituto de Desarrollo Urbano con adición de pavimento reciclado. VI Jornadas de Pavimentos y Mantenimiento Vial. Bogotá D.C. Colombia. 2012.

[6] Figueroa Infante Ana Sofía, Velásquez Raúl, Reyes Lizcano Freddy Alberto y Bahia Hussain. The Effect of Water Conditioning for Extended Periods on the Properties of Asphaflt Binders. Transportation Research Record: Journal of the Transportation Research Board. Volume 2372 / Asphalt Materials and Mixtures 2013, Vol. 3. Pages 34-45. Transportation Research Board of the National Academies, ISNN 0361-1981. 2013.

[7] Herrera Olga, Daza Carlos E., Figueroa Infante Ana Sofía, Fernández Wilmar y Reyes Lizcano Freddy Alberto. Análisis de la alteración del asfalto colombiano en inmersión en agua. Ingeniería. Universidad. Bogotá, Colombia, 17(2), p 427-442. Julio -diciembre 2013.

[8] Qin Q et al. Field aging effect on chemistry and rheology of asphalt binders and rheological predictions for field aging. Fuel 121 (2014) 86-94.

[9] Hsu, T., & Tseng, K. (1996). Effect of rest periods on fatigue response of asphalt concrete mixtures. Journal of Transportation engineering, 122 (4), 316–322.

[10] Masad, E., Muhunthan, B., Shashidhar, N., & Harman, T. (1998). Aggregate Orientation and Segregation in Asphalt Concrete. ASCE Getotechnical Special Publication, 85, 69–80.

[11] Rondón, H., Reyes, F., Gónzalez, L., & Vásquez, S. (2012). Ahuellamiento y fatiga en mezclas asfalticas (Colección tierra y vida.). Bogota D.C, Colombia: Ecoe ediciones Ltda.

[12] Lundstrom, R., Di Benedetto, H., & Isacsson, U. (2004). Influence of asphalt mixture stiffness on fatigue failure. Journal of materials in civil engineering, 16(6), 516-525.

[13] Rao Tangella, R., Craus, J., Deacon, J. A., & Monismith, C. L. (1990). Summary report on fatigue response of asphalt mixtures. (Rep. to strategic Highway Research Program). Washington D. C.

Building Tomorrow's Society

Bâtir la Société de Demain

Fredericton, Canada

June 13 to 16, 2018 / *13 au 16 juin, 2018*

CHARACTERIZATION OF LIGHTWEIGHT CELLULAR CONCRETE

Ni, Frank Mi-Way[1, 5], Averyanov, Sergey[2], Melese, Eskedil[3], and Tighe, Susan[4]
[1] University of Waterloo, Department of Civil and Environmental Engineering, Canada
[2] University of Waterloo, Department of Civil and Environmental Engineering, Canada
[3] University of Waterloo, Department of Civil and Environmental Engineering, Canada
[4] University of Waterloo, Department of Civil and Environmental Engineering, Canada
[5] fmni@uwaterloo.ca

Abstract: Lightweight cellular concrete (LCC) is an alternative construction material with growing applications over the past three decades. It is mainly composed of hydraulic cement, water, and pre-formed foam. Its lightweight, ease of placement, and insulating property make it a suitable material for structural backfill, utility insulation, and road construction. Being an innovative material with diverse potential applications in the construction industry, it is worth knowing typical engineering characteristics of this material. So far, various laboratory tests were conducted on LCCs at different Canadian laboratories. These include tests for resilient modulus, elasticity modulus, shear strength, tensile strength, hydraulic conductivity, thermal conductivity, and freeze-thaw resistance. The laboratory tests were conducted on LCCs with densities ranging from 400kg/m^3 to 600kg/m^3. These are typical densities of the LCCs that have been so far used in Canada for different applications. According to the test results, the resilient modulus and modulus of elasticity of the LCC vary over the ranges of 0.12 – 0.35 MPa and 850 MPa – 1,100 Mpa respectively, while the Poisson's ratio range is 0.19 – 0.26. The freeze-thaw test results, on the other hand, indicates the relative dynamic modulus of the LCCs exceeds 80% after 350 freeze-thaw cycles. These show that LCC has excellent resistance to freeze-thaw even though their stiffness is low compared to chemically stabilized pavement materials. In this paper, the mechanical and thermal properties of the lightweight cellular concrete that have been determined in various Canadian laboratories using standard and non-standard test methods will be compiled and presented. Besides, comparison of the test results with the results of similar tests that have been performed outside Canada will be incorporated. Furthermore, results of some mechanical testing based on different densities will also be shown and analyzed.

1 INTRODUCTION

Lightweight Cellular Concrete (LCC) or Foamed Concrete is a cementitious material with a typical plastic density ranging from 375 to 1600 kg/m^3 (Ozlutas 2015) that contains a homogeneous air bubble structure in the mix. It was first patented in 1923 (Valore 1954) and been using as a void filler material in the 1970s. The application of LCC in construction provides several benefits such as reducing earth pressure, resistance to freezing and thawing, mitigate settlement, good thermal insulation (Tiwari 2017, Maruyama and Camarini 2015). LCC is often used in construction projects which required light material and excellent workability. For instance, construction project such as bulk filling, trench reinstatements, floor screeds, thermally insulating foundations, and stabilizing soils (Mydin 2010, Ozlutas 2015).

LCC has been a useful backfilling material in the world. However, researchers have found that LCC could be a viable option as structural materials (Jones and McCarthy 2005). Studies have proven that LCC could be a feasible choice for structural applications such as stabilization of weak soil (Drusa et al. 2011, Lee et al. 2009), replacement of weak soil in sandwich solution for foundation slabs (Hulimka et al. 2013), industrial concrete floor (Kadela and Kozlowski 2016).

LCC starts to gain its attention in transportation infrastructure as it provided numerous benefits (Hajek et al. 2016, Ramamurthy et al. 2009):

1. Straightforward and quick pouring – the contractor is capable of producing and pouring 400 to 600 m^3 per day to reduce the construction time and cost.
2. Self-compacting
3. Excellent Freeze-Thaw resistance
4. Environmental friendly

Several road constructions have been done with the LCC due to these benefits mentioned above. For instance, LCC has been used in an industrial zone in the UK as subbase material to replace the original layer which consists of peat (Drusa 2016). Illinois also apply LCC in their road construction to provide a solution to the soft organic underlying soil, and it benefits the contractor by lowering unit cost, reduced construction time, and higher quality of material (Drusa 2013). Application of LCC are also found in Canada, Alberta used the LCC as subbase material in bus-lane construction. Ontario also applied LCC in rural road and highway (Maher and Hagan 2016).

Though there are several LCC applications and studies, there are still lacking a complete and thorough guide of LCC. Mechanical properties of LCC at a specific low density (400 to 600 kg/m^3) needs to be adequately examined.

1.1 Definition

The term “cellular concrete.” or “foamed concrete” can be referred to a type of lightweight concrete which contains stable air bubble or gas cell distributed homogeneously in the cement mix (ACI 523). Unlike the traditional Portland cement concrete, LCC does not contain any coarse aggregate in the mix (Maruyama and Camarini 2015). ASTM C796 gives a more detailed definition of the LCC as:

“A lightweight product consisting of Portland cement, cement-silica, cement-pozzolan, lime-pozzolan, or lime-silica pastes, or pastes containing blends of these ingredients and having a homogeneous void or cell structure, attained with gas-forming chemicals or foaming agents (for cellular concretes containing binder ingredients other than, or in addition to Portland cement, autoclave curing is usually employed)”

Another definition that has been widely cited noted that the foamed concrete is:

“A cementitious material having a minimum of 20 percent by volume of mechanically entrained foam in the plastic mortar or grout.”

This definition further narrows down the type of foamed concrete since air entrained concrete has lower entrained air (3-8%) and aerated concrete is formed chemically. (Barnes 2008).

The two definition both mentioned that the cellular concrete or foamed concrete is a cementitious material that contains air bubble or foam in the mix. Since this research is mainly using the material from Canada, it would be better to follow the definition from ASTM and ACI.

Currently, there are no test standards for cellular concrete in Canada. American Concrete Institute (ACI) published two guides for the cellular concrete with unit weight above 800 kg/m3 (ACI 523.3R) and less than 800 kg/m3 (ACI 523.1R). The two guides provide a general concept of the cellular concrete such as concrete properties, mixing procedures, and applications. It is noted that the application mentioned in ACI

523.1R is for roof deck application. American Society for Testing and Materials (ASTM) issued two standards for the foaming agent (ASTM C796 and ASTM C869).

1.2 Constituent Materials

The typical composition of lightweight cellular concrete contains Portland cement, Pozzolan materials, fine aggregate, water, and foam.

1.2.1 Portland cement

Typically Type I Portland cement, blast furnace slag cement and Portland pozzolan cement can be used as the base mix of cellular concrete. Type III and IIIA cement are also used if the mixture requires high early strength (ACI 523). The total cement contents are usually around 300 to 400 kg/m^3. The density of the cement can be adjusted depending on the strength requirements or the design density of the mix (Brady et al. 2001).

1.2.2 Pozzolan materials

The Pozzolan material is a finely divided material which is rich in silica or alumina. Pozzolanic by-product such as fly ash and blast furnace slag could benefit the contractor by reducing the cost, maintain consistency and increase strength in long-term performance (Kearsley and Wainwright 2001). Jones et al. (2017) stated that replacing Portland cement with fly ash up to 40% could significantly reduce the embodied carbon dioxide by 65% compared to the 100% Portland cement mix while has a similar 28 d strengths (0.25 MPa compared to 0.31MPa). However, the drawbacks of using fly ash are the slow rate of strength gain, and it might cause foam instability as the water demand might increase (Ozlutas 2015).

1.2.3 Fine aggregate

Fine sand is the typical fine aggregate to be used in producing cellular concrete. It is found that sand with a maximum size of 2 mm yields higher strength than 5 mm sand. BCA (1994) suggested replacing fine sand with course fly ash in mixes with a plastic density below 600 kg/m3.

1.2.4 Water

Water-cement ratio plays a vital role in cellular concrete. The w/c ratio needs to be determined based on the constituents materials to provide and maintain suitable workability of the mix. The typical range of w/c ratio is from 0.40 to 1.25 (Ramamurthy 2009). Insufficient water in the mix may lead to the collapse of the mix as the mix is too stiff which breaks the bubble while excess water could increase drying shrinkage (Nambiar and Ramurthy et al., 2006).

1.2.5 Foam

Pre-formed foam is essential in cellular concrete as it can control the plastic density of the mix (Wee et al. 2006). It is consists of a foaming agent and compressed air to help produce foam (BCA 1994). The foam should have homogeneous bubble structure to provide concrete with reasonable strength (Brady et al. 2001). The structure should be capable of resisting the pressure of the base mix until the initial setting time is reached as the air bubbles will be surrounded by the strong skeleton of concrete (Ramamurthy et al. 2009).

1.3 Cellular concrete concerning sustainable construction

Cellular concrete is a lightweight construction material that can fill up to 80 to 90 percent of air void at low densities. The high air void content significantly reduced the number of ingredients used and wasted produced (Ozlutas 2015). Jones and McCarthy (2005) noted that the self-flowing ability of cellular concrete excludes the need for compaction, which saves the energy from placement. Furthermore, it is easy to excavate and remove from the site due to its low strength.

Cellular concrete reduces the use of non-renewable primary sources such as coarse aggregate and fine aggregate at densities below 600 kg/m^3 (BCA 1994). Alternatively, Jones et al. (2012) stated that recycled (demolition fines) and industry by-product (fly ash) can be used in cellular concrete as filler. Moreover, fine fly ash could provide several benefits such as reducing embodied carbon dioxide (eCO_2) and drying shrinkage strains, Ozlutas (2015) demonstrated the influence of replacing Portland cement with fly ash up to 40% in Figure 1.

In pavement construction, cellular concrete provides a sustainable solution to soft soil base. This is due to its good constructability such as short install time, reducing excavation time, self-compacting, and can be placed in winter (Maher 2016).

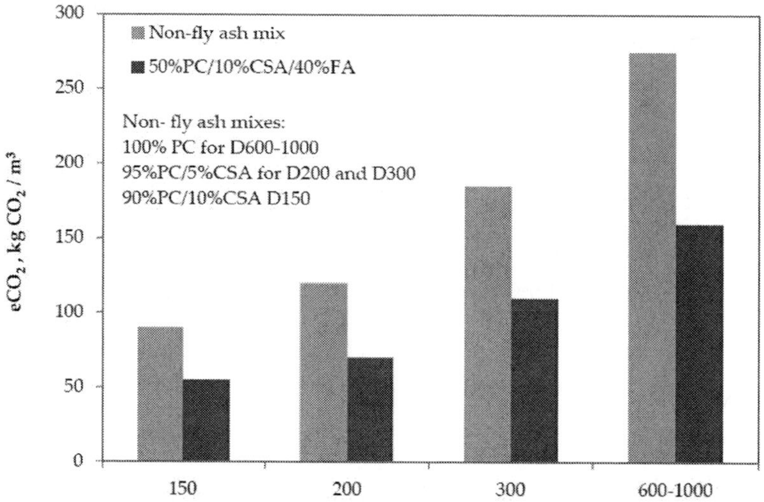

Figure 1 Influence of plastic density and fly ash on the eCO2 of foamed concrete (Ozlutas 2015)

1.4 Application of Lightweight Cellular Concrete

Lightweight cellular concrete has been widely used in civil and structural engineering areas due to its distinctive properties such as reduced density, low thermal conductivity, excellent flowability, self-compaction ability, and its relative cost-effectiveness (Amran et al. 2015). Sari and Sani (2017) summarized the applications of cellular concrete with different density, as shown in Table 1. The typical density of cellular concrete in the application is between 1,000 kg/m^3 to 1,500 kg/m^3, which mainly used for cast-in-place wall, prefabrication, and housing applications. Densities between 300 kg/m^3 to 600 kg/m^3 is related to pavement construction as it provides soil stabilization and road construction functions.

Table 1 Summary of foamed concrete applications based on density (Sari and Sani 2017)

Density (kg/m^3)	Application
300-600	Replacement of existing soil, soil stabilization, raft foundation.
500-600	Currently being used to stabilize a redundant, geotechnical rehabilitation and soil settlement. Road construction.
600-800	Widely used in void filling, as an alternative to granular fill. Some such applications include filling of old sewerage pipes, wells, basement, and subways.
800-900	Primarily used in production of blocks and other non-load bearing building element such as balcony railing, partitions, parapets, etc.
1,100-1,400	Used in prefabrication and cast-in-place wall, either load bearing or non-load bearing and floor screeds.
1,100-1,500	Housing applications.

1,600-1,800	Recommended for slabs and other load-bearing building element where higher strength required.

The application of lightweight cellular concrete become popular in the world. For instance, it has been a solution for the southern US regions that are suffering from the housing shortage or the adverse weather such as hurricanes and earthquakes. In Canada, lightweight cellular concrete is used as a filler for tunnel annulus grouting, flowable fills and in geotechnical applications. The annual market size of lightweight cellular concrete in the UK is estimated to be 250,000 to 300,000 m^3 annually which includes an extensive mine stabilization project. This is the same market size in Korea while it is used as an essential component in a floor heating system. Lightweight cellular concrete was used as a subbase material in Holland due to the low traffic loading and considered cost-effectiveness at the times of repair and rehabilitation. The lightweight cellular concrete also provides resistance to freeze-thaw and frost heave in concrete paving (Amran et al. 2015).

2 CHARACTERIZATION TESTS AND STANDARDS

Over the past ten years, a number of materials characterization tests had been performed in different Canadian laboratories on lightweight cellular concretes with various unit weights. The tests were done to evaluate the stiffness under monotonic loading to determine the shear strength parameters, assess the freeze-thaw resistance, determine the insulation capacity, and measure the hydraulic conductivity of the materials.

Most of the laboratory tests were performed in accordance with standard procedures stated in ASTM or AASHTO standards. In the absence of standard testing protocols for LCC, test methods that had been devised for other materials, which are similar to LCC in various contexts, were adopted based on best practice or expert opinion. Resilient modulus test conducted according to ASTM D 7369 and direct shear test performed as per ASTM D 3080 are among these tests.

The complete list of laboratory tests conducted to characterize the material is shown in Table 2. The results of these tests are presented and discussed in the following section.

Table 2 List of laboratory tests

No.	Test Description	Test Method
1	Resilient Modulus	ASTM D 7369
2	Indirect tensile strength	ASTM D 6931
3	Modulus of elasticity and Poisson's ratio	ASTM C 469
4	Freeze-thaw resistance	ASTM C 666, procedure B
5	Hydraulic conductivity	ASTM D 5084
6	Thermal conductivity	ASTM C 518 and ASTM D 5334
7	Monotonic direct shear test	ASTM D 3080

3 TEST RESULTS AND DISCUSSION

Test results of indirect tensile strength testing and resilient modulus test were shown in Table 3. The indirect tensile test was applied on 15 specimens with three different densities, which are 400, 500 and 600 kg/m^3. ASTM D6931 was used to perform the test. It is clear that the indirect tensile strength of the lightweight cellular concrete increase as density increase, with an average of 0.12, 0.22 and 0.35 MPa for the three densities. Resilient modulus test was also done on 15 specimens with 400, 500 and 600 kg/m3 density. Results showed that the resilient modulus follows the same trend as the indirect tensile strength, with an average of 389, 550 and 725 MPa for three densities. It also exceeds the typical resilient modulus values ranges (34 to 290) for the Unified Soils Classification System (USCS) materials (Papagiannakis, 2008).

Table 3 Test results of Indirect Tensile Strength test and Resilient Modulus test

Parameter	400 (kg/m^3)					500 (kg/m^3)					600 (kg/m^3)				
IDT (Mpa)	0.09	0.13	0.13	0.12	0.14	0.23	0.27	0.24	0.19	0.19	0.4	0.34	0.37	0.33	0.28
Mr (Mpa)	343	384	458	374	388	477	515	557	580	623	696	801	739	752	641

Figure 2 and 3 demonstrates the box plot of the test results which give a clear view of the data trend.

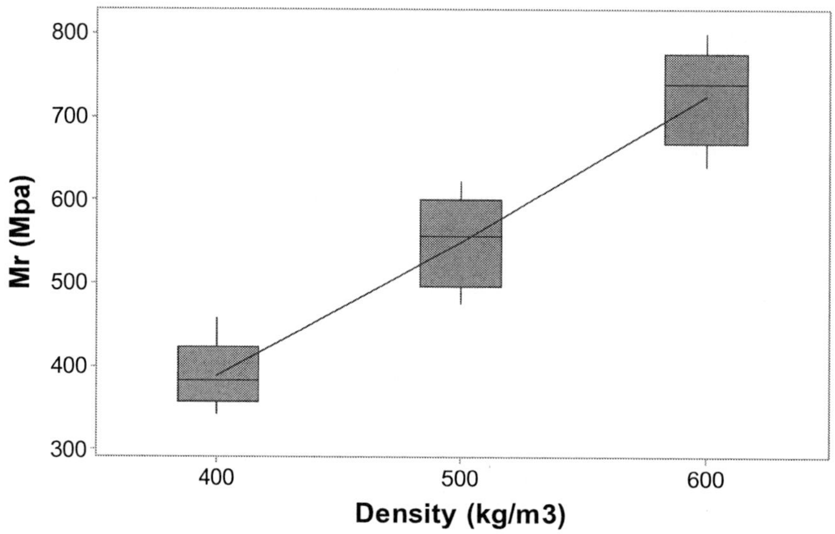

Figure 2 Box plot of the Resilient Modulus test results

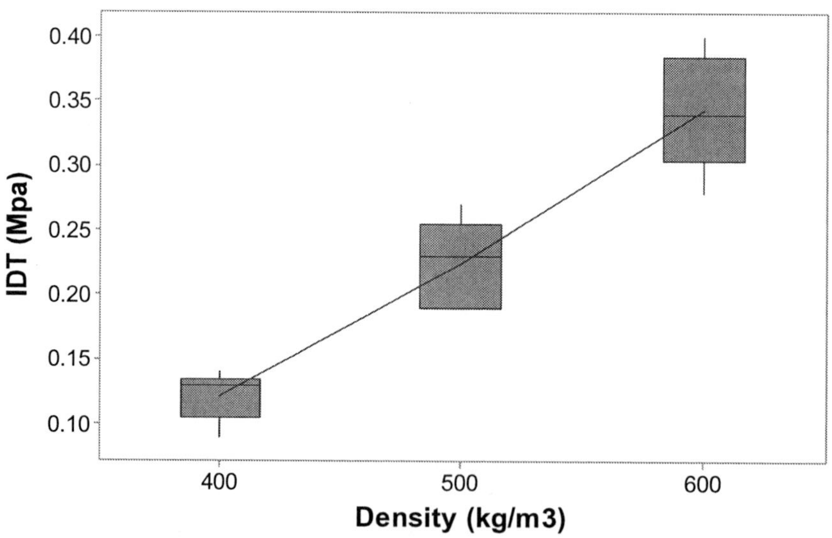

Figure 3 Box plot of the Indirect Tensile Strength test results

Table 4 shows the results of compressive strength test base on ASTM C495. Six 75 mm diameter by 150 mm high cylindrical samples with different density were tested at 65 and 71 days. Results demonstrated that the compressive strength increased as the density increase. Modulus of Elasticity (E) and Poisson's Ratio are tested for 400, 475, and 600 kg/m^3 following the ASTM C469. The E value for 400 and 475 density is similar or even higher. However, the Poisson's ratio range from 0.19 to 0.28 except for the 400 kg/m^3 sample which has a value of 0.47.

Table 4 Test results of Unconfined Compressive Strength, Modulus of Elasticity and Poisson's Ratio

Density (kg/m^3)	UCS (MPa)	E (GPa)	Poisson's Ratio	Age
400	0.72	0.85	0.28	71
400	0.73	1.09	0.47	65
475	1.43	0.89	0.2	71
475	1.28	0.85	0.19	65
600	2.76	1.33	0.24	71
600	2.44	1.27	0.2	65

The freeze-thaw test has been done on 475 and 575 kg/m^3 density specimens at 28-days based on ASTM C666 Procedure B. Figure 3 and 4 demonstrate the results of the freeze-thaw test. According to the specification, the test should continue until 300 cycles or the relative modulus of elasticity drops to 60% of its initial modulus. It is clear that for both density samples, the relative dynamic modulus of elasticity still exceed 80% after 350 cycles, which notes that the samples have good freeze-thaw resistance.

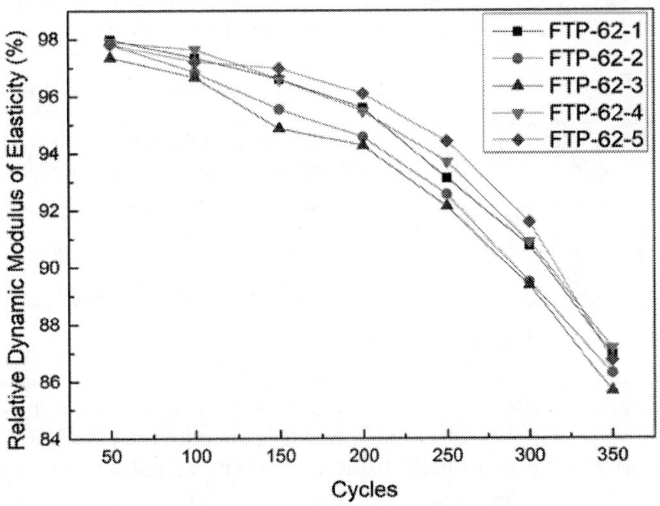

Figure 3 Results of Freeze-thaw test (475 kg/m^3)

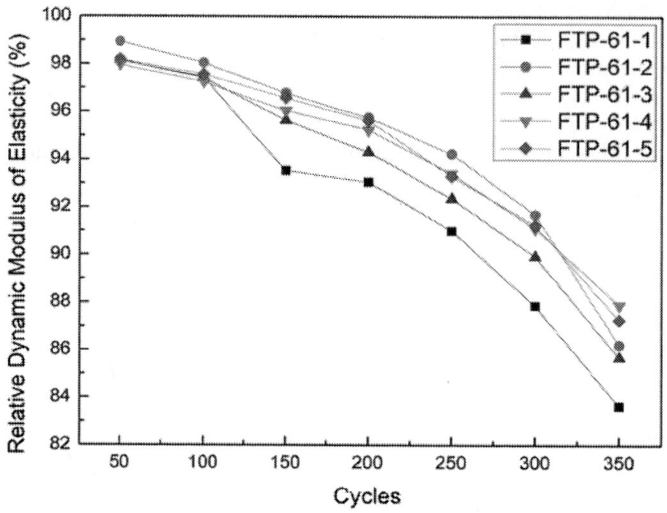

Figure 4 Results of Freeze-thaw test (575 kg/m^3)

The hydraulic conductivity test has been conducted on five specimens with different densities following ASTM D 5084. The results are noted in Table 5. The size of the specimens is 70 mm diameter by 150 mm high cylinder. The test results show there is no direct relationship to density.

Table 5 Test results of Hydraulic Conductivity testings

Density (kg/m^3)	Final Hydraulic Conductivity
400	6.02E-06 cm/s @ 72 mins
405	3.16E-06 cm/s @ 94 mins
475	4.37E-06 cm/s @ 62 mins
475	1.4E-06 cm/s @ 76 mins
600	5.7E-07 cm/s @ 122 mins

Thermal Conductivity tests were done on 400, 475, and 600 kg/m3 specimens. The dimension is 300X300X75 mm for all plate specimens. Two ASTM test standard was applied in this tests, which are ASTM C5334 and C518. Table 6 noted the test results. It is clear that the thermal conductivity of the three density range from 0.05 to 0.09 for ASTM C5334, while test results of ASTM C518 range from 0.076 to 0.101. Both of the results shown the increasing trend as density increase.

Table 6 Test results of Thermal Conductivity test (ASTM C5334 and ASTM C518)

Density (kg /m^3)	Age	C5334	C518
400	38	0.05	0.076
400	38	0.05	0.078
475	38	0.07	0.084
475	38	0.07	0.087
600	38	0.08	0.098
600	38	0.09	0.101

Monotonic direct shear testing had been done on 475 kg/m^3 samples following ASTM D3080. Three 75 mm diameter by 150 mm high cylinder were tested. The monotonic direct shear test had been done on intact, saw cut and cold joint samples with confining stress of 10, 30 and 60 kPa. Table 7 shown the comparison of cohesion and friction angle between LCC and other typical geotechnical materials. It is found that LCC has better cohesion and friction angle than sand, silt, and clay, which could be an alternative material when used in geotechnical applications.

Table 7 Comparison of cohesion and friction angle between LCC and other materials (Ortiz et al. 1986)

Material	Cohesion (kPa)	Friction angle
Sand	0	30-34
Silt	2-3	25-32
Clay	6-10	6-24
LCC (saw cut)	3	43
LCC (cold joint)	33	38
LCC (intact)	7	47
Rock	30,000	49

ANOVA (Analysis of Variances) test has been performed on the above data to examine the influence of density versus other parameters. A significance level of 5% was used in the test, and the results are shown in Table 7. It is clear that the p-value of unconfined compressive strength, Modulus of Elasticity, Resilient Modulus, Indirect Tensile Strength, and Both of the Thermal Conductivity tests have not exceeded 0.05, which indicated that the results are significant and prove that the density of lightweight cellular concrete has an impact on these parameters. This is considered reasonable as these parameter increase as densities increase. However, the p-value of Poisson's ratio is 0.191, this shown that the densities may not influence Poisson's ratio.

Table 7 Results of ANOVA test

UCS vs Density

Source	DF	Adj SS	Adj MS	F-value	P-value
Density	2	3.6417	1.8209	87.4	0.002
Error	3	0.0625	0.0208		
Total	5	3.7042			

E vs Density

Source	DF	Adj SS	Adj MS	F-value	P-value
Density	2	0.2025	0.1013	9.68	0.049
Error	3	0.0314	0.0105		
Total	5	0.2339			

Poisson's ratio vs Density

Source	DF	Adj SS	Adj MS	F-value	P-value
Density	2	0.038	0.0190	3.02	0.191
Error	3	0.019	0.0063		
Total	5	0.057			

Mr vs Density

Source	DF	Adj SS	Adj MS	F-value	P-value
Density	2	283085	141543	49.16	0.000
Error	12	34553	2879		
Total	14	317638			

IDT vs Density

Source	DF	Adj SS	Adj MS	F-value	P-value
Density	2	0.1235	0.0617	51.74	0.000
Error	12	0.0143	0.0012		
Total	14	0.1378			

Thermal Conductivity (C5334) vs Density

Source	DF	Adj SS	Adj MS	F-value	P-value
Density	2	0.0012	0.00062	37	0.008
Error	3	0.0001	0.00002		
Total	5	0.0013			

Thermal Conductivity (C518) vs Density

Source	DF	Adj SS	Adj MS	F-value	P-value
Density	2	0.00052	0.00026	70.41	0.003
Error	3	0.00001	0.000004		
Total	5	0.00053			

4 CONCLUSION

Lightweight Cellular Concrete is an alternative construction material that raised its attention in recent decades. Due to its densities, it could have various application in construction. It has several benefits in construction such as good constructability, self-compacting, environmentally friendly. In this study, tests were performed on low density lightweight cellular concrete (400 to 600 kg/m^3) to evaluate its properties. Findings were listed as below:

- The resilient modulus and indirect tensile strength show good consistency. The test results of resilient modulus shown that the lightweight cellular concrete has a higher value than the typical range for subbase and subgrade materials.
- The result of the freeze-thaw test shown that lightweight cellular concrete still has a value of at least 80% of the relative dynamic of elasticity after 350 cycles. This demonstrates that the lightweight cellular concrete has good freeze-thaw resistance.
- ANOVA test has been performed on the test data, and it is shown that the density of cellular concrete has an influence on Unconfined Compressive Strength, Modulus of Elasticity, Resilient Modulus, Indirect Tensile Strength and Thermal Conductivity, while Poisson's ratio seems does not have a connection with density.

Acknowledgements

The authors gratefully acknowledge the support from CEMATRIX CANADA for providing the testing data.

References

Ozlutas, K. 2015. *Behaviour of ultra-low density foamed concrete* (Doctoral dissertation, University of Dundee).

Valore, R. C. 1954. Cellular Concretes: Composition and Method of Preparation. *Journal of American Concrete Institute*, **2**, 773-795.

Tiwari, B., Ajmera, B., Maw, R., Cole, R., Villegas, D., & Palmerson, P. 2017. Mechanical Properties of Lightweight Cellular Concrete for Geotechnical Applications. *Journal of Materials in Civil Engineering*, **29**(7), 06017007.

Maruyama, R. C., & Camarini, G. 2015. Properties of cellular concrete for filters. *International Journal of Engineering and Technology*, **7**(3), 223.

Mydin, M. A. O. 2014. Models for Prediction the Strength and Stiffness of Foamed Concrete at Ambient Temperature. *European researcher. Series A*, **(1-2)**, 124-129.

Jones, M. R., & McCarthy, A. 2005. Preliminary views on the potential of foamed concrete as a structural material. *Magazine of concrete research*, **57**(1), 21-31.

Drusa, M., Fedorowicz, L., Kadela, M., & Scherfel, W. 2011. Application of geotechnical models in the description of composite foamed concrete used in contact layer with the subsoil. *Zbornik*, **1**, 644-652.

Lee, Y. L., Goh, K. S., Koh, H. B., & Bakar, I. 2009. Foamed aggregate pervious concrete–an option for road on peat. *Proceedings of MUCEET2009 Malaysian Technical Universities Conference on Engineering and Technology*. p. 1.

Hulimka, J., Knoppik-Wróbel, A., Krzywoń, R., & Rudišin, R. 2013. Possibilities of the structural use of foamed concrete on the example of slab foundation. *Proceedings of the 9th Central European Congress on Concrete Engineering*, **Vol. 67**, p. 74).

Kadela, M., & Kozłowski, M. 2016. Foamed concrete layer as sub-structure of industrial concrete floor. *Procedia Engineering*, **161**, 468-476.

Hajek, M., Decky, M., Drusa, M., Orininová, L., & Scherfel, W. 2016. Elasticity Modulus and Flexural Strength Assessment of Foam Concrete Layer of Poroflow. *In IOP Conference Series: Earth and Environmental Science*, **44**, No. 2, p. 022021).

Ramamurthy, K., Nambiar, E. K., & Ranjani, G. I. S. 2009. A classification of studies on properties of foam concrete. *Cement and Concrete Composites*, **31**(6), 388-396.

Maher, M. L., & Hagan, J. B. (2016). MAT-758: Constructability Benefits of the Use of Lightweight Foamed Concrete Fill (LFCF) in Pavement Applications.

American Concrete Institute ACI 523.3R-14. 2014. *Guide for Cellular Concretes above 50lb/ft^3*.

American Concrete Institute ACI 523.1R-06. 2006. *Guide for Cast-in-Place Low Density Cellular Concrete.*

Barnes, R. A. 2008. Foamed Concrete: Application and Specification. *In Excellence in Concrete Construction through Innovation: Proceedings of the conference held at the Kingston University, United Kingdom*, **9-10**, p. 3.

ASTM C796/C796M. 2012. Standard Test Method for Foaming Agents for Use in Producing Cellular Concrete Using Preformed Foam. ASTM International. USA.

Brady, K. C., Jones, M. R., & Watts, G. R. 2001. *Specification for foamed concrete*. TRL Limited.

Kearsley, E. P., & Wainwright, P. J. (2001). The effect of high fly ash content on the compressive strength of foamed concrete. *Cement and concrete research*, **31**(1), 105-112.

Jones, M. R., Ozlutas, K., & Zheng, L. 2017. High-volume, ultra-low-density fly ash foamed concrete. *Magazine of Concrete Research*, 1-11.

BCA 1994. *Foamed concrete - Composition and properties*. First published in 1991, British Cement Association.

Nambiar, E. K., & Ramamurthy, K. 2006. Influence of filler type on the properties of foam concrete. *Cement and Concrete Composites*, **28**(5), 475-480.

Wee, T. H., Babu, D. S., Tamilselvan, T., & Lim, H. S. 2006. Air-void system of foamed concrete and its effect on mechanical properties. *Materials Journal*, **103**(1), 45-52.

Ramamurthy, K., Nambiar, E. K., & Ranjani, G. I. S. 2009. A classification of studies on properties of foam concrete. *Cement and Concrete Composites*, **31**(6), 388-396.

Jones, R., Zheng, L., Yerramala, A., & Rao, K. S. 2012. Use of recycled and secondary aggregates in foamed concretes. *Magazine of concrete research*, **64**(6), 513-525.

Amran, Y. M., Farzadnia, N., & Ali, A. A. 2015. Properties and applications of foamed concrete; a review. *Construction and Building Materials*, **101**, 990-1005.

Sari, K. A. M., & Sani, A. R. M. 2017. Applications of Foamed Lightweight Concrete. *In MATEC Web of Conferences*, **97**, p. 01097). EDP Sciences.

Papagiannakis, A. T., & Masad, E. A. 2008. *Pavement design and materials*. John Wiley & Sons Inc., Hoboken, New Jersey, 46.

Ortiz, J.M.R, J Serra, & C. Oteo. 1986. Curso Aplicado de Cimentaciones. Third ed. Colegio de Arquitectos de Madrid. Madrid

Building Tomorrow's Society

Bâtir la Société de Demain

Fredericton, Canada

June 13 to 16, 2018 / *13 au 16 juin, 2018*

COMPARISON OF DIFFERENT ACCELERATED CORROSION TECHNIQUES

Said, Mahmoud. E[1,3], Hussein, Amgad[1], Hassan, Assem A[1] and Gillis, Nick.[2]
[1] Civil Engineering Department, Memorial University, Canada
[2] SNC Lavalin Inc, Canada
[3] meaeaas2@mun.ca

Abstract: This paper presents different techniques that could be used in conducting an accelerated corrosion experiments in a lab environment. The objective of the techniques is to induce corrosion in full-scale structural reinforced concrete members. The study aims to demonstrate the benefits and drawbacks of each technique. Two full scale reinforced concrete two-way slabs were prepared for this experiment. The slabs had an identical dimension of 1900 mm × 1900 mm × 150 mm. The slabs reached the same level of corrosion, 25% mass loss, using two different accelerated methods: constant voltage and constant current. The induced corrosion in each slab was evaluated. The corrosion state was examined based on current measurement, half-cell potential tests, and mass loss. Both techniques showed a close agreement between the actual mass loss and the theoretical mass loss that was calculated using Faraday's equation. During corrosion, the concrete loses its resistivity which is directly reflected by an increase in the current in the case of constant voltage. Hence, using constant voltage could cause damage more than that by normal environmental conditions. Whereas constant current technique keeps the current intensity uniform over the corrosion process regardless of the resistivity of concrete, leading to uniform corrosion rate which avoid extreme damage.

Keywords: Accelerated corrosion, Constant voltage, Constant current, Two-way Slabs

1 INTRODUCTION

The corrosion of reinforced concrete (RC) elements occurs due to the difference in the electrical potential along the reinforcement. This process is considered as an electrochemical process, where an anode is created at a certain part of the bar. At the anode, electrons are released due to an oxidation reaction. Whereas a cathode part is created by attracting the released electrons due to the reduction reaction that occurs at certain another part of the bar. The released electrons flow through an electrolyte solution from the anodic side to the cathodic side. In the cathodic part, the negative electrons convert to hydroxyl ions (OH^-) (oxidizing agent) in the presence of moisture and oxygen as:

[1] $O_2 + 2H_2O + 4e^- \longrightarrow 4\,OH^-$

These hydroxyl ions travel through the electrolyte solution (concrete pore water) and combine with the ferrous ions to form iron hydroxides($Fe(OH)_2$). In the presence of oxygen and moisture, the ferrous hydroxide ($2Fe(OH)_2$) converts to ferric oxide ($Fe_2O_3H_2O$), i.e. rust (Neville, 2011).

[2] $Fe^{2+} + 2OH^{-} \longrightarrow Fe(OH)_2$ (Ferrous hydroxide)

[3] $4Fe^{2+} + 2H_2O + O_2 \longrightarrow 4Fe(OH)_3$ (Ferric hydroxide)

It is concluded that corrosion needs three elements to exist: a chemical potential difference between the two sides of the steel bar; an electrolyte solution to provide conductivity; and an electron path through the metal to allow the flow of electrons (Neville, 2011). This corrosion mechanism is similar to flashlight battery mechanism. Therefore, accelerated corrosion techniques innovated from the idea of flashlight battery in order to simulate corrosion inside a lab under normal environmental conditions (Amleh, 2000).

The corrosion process naturally needs a long period to occur. Therefore, researchers use accelerated corrosion techniques to speed up the corrosion process to save time and money. Accelerated corrosion tests are used to investigate the behavior of corroded structural elements. The objective of such research is to correlate the accelerated corroded laboratory results and the real corrosion that occurs to existing structures (Davis, 2000). The mass loss due to corrosion is controlled using a direct electrical current, in which Faraday's equation is used to predict the mass loss (Toongoenthong and Maekawa, 2004). Current is applied through the application of constant voltage or constant current. There are several studies in the literature that used constant voltage (Amleh, 2000; Lee *et al.*, 2000; Ahwazi, 2001; Kumar *et al.*, 2012; Deb and Pradhan, 2013; Pellegrini-Cervantes *et al.*, 2013; Al-Swaidani and Aliyan, 2015), and others used constant (impressed) current (El Maaddawy and Soudki, 2003; El Maaddawy, Soudki and Topper, 2005; Kashani, Crewe and Alexander, 2013; Pritzl, Tabatabai and Ghorbanpoor, 2014; Talakokula, Bhalla and Gupta, 2014; Altoubat, Maalej and Shaikh, 2016). Generally, in both techniques, an electrochemical potential is created between the embedded reinforcement inside the concrete that acts as an anode and copper or steel mesh as a cathode. The cathode could be external or internal (embedded inside concrete). An electrolyte solution such as chloride salts solution is used to activate the corrosion process (El Maaddawy and Soudki, 2003).

In constant current technique, an electrical power supply provides constant current densities through the specimen to reach a certain level of corrosion (El Maaddawy and Soudki, 2003). In contrast, the constant voltage technique adjusts the power supply through the application of constant volt. The constant volt in several studies varied from 5V to 30V (Ahwazi, 2001; Yoon *et al.*, 2001; Xia, Jin and Li, 2012). The corrosion is monitored by measuring the current using a computer-controlled data acquisition system.

Altoubat et al. (2016) conducted an experimental study that illustrated the effect of both accelerated corrosion techniques (constant voltage and constant current) on the corrosion behavior of small-scale columns. They noticed that the damage created in the specimen by constant current is more than that by constant voltage. Nonetheless, both techniques caused the same mass loss in the corroded reinforcement. The authors recommended the use of constant current technique to simulate corrosion damage, especially it ensures that the targeted corrosion level is reached within a reasonable amount of time.

Several studies used the impressed current techniques with different current densities that ranged from $45\mu A/cm^2$ as a minimum value to $10400\mu A/cm^2$ as a maximum value (Almusallam *et al.*, 1996; Bonacci *et al.*, 1998). El Maaddawy & Soudki (2003) concluded that the variation in current densities has no effect on the percentage of mass loss that was calculated based on Faraday's law. Their specimens reached mass loss up to 7.27%. However, they noticed that using current density above $200\mu A/cm^2$ would cause a significant increase in crack width and strain response due to the corrosion. The authors also recommended not to use different current densities for to reach different levels of corrosion. It was concluded that changing the current densities to obtain different levels of corrosion may misguide the analysis of test results.

To the best of the authors' knowledge, the literature only covers accelerated corrosion techniques that have been used to induce corrosion in RC beams, but there are no previous studies that have been carried out on two-way slabs using any such techniques. This could be attributed to the challenges that could face researchers to induce corrosion in two-way slabs. The number of corroded bars in two-way slabs are more than those in beam or one-way slabs. Therefore, a corrosion experiment of two-way slabs needs data acquisition systems with large number of free channels to record the current readings and requires long period of time. Lab-space availability for a longer period of time could be a challenge. Furthermore, the reinforcement of any two-way slab consists of two meshes (flexural and compression) and each mesh consists of two layers. It requires special attention to isolate each rebar to ensure uniform corrosion. This paper aims to demonstrate a lab setup that could be used to induce corrosion in two-way slab using constant current or constant voltage. The advantage and disadvantage of using each technique will be demonstrated. The authors believe that this study will provide some knowledge to guide the future researchers to choose the suitable accelerated approach to investigate the effect of corrosion on full-scale RC two-way slabs.

2 EXPERIMENTAL PROGRAM

2.1 Details of Test Specimens

Two RC slabs with the same identical dimension and reinforcement were cast. Each slab had dimensions of 1900 × 1900 × 150 mm with clear cover of 30 mm and 1% flexural reinforcement distributed uniformly using 15M bars that are spaced at 190 mm as shown in Figure 1. The details of the test specimens are shown in Table 1. The reinforcement was weighted before casting to enable the calculation of the mass loss after the corrosion test. All intersection points between bars were carefully isolated to control and ensure uniform corrosion in each corroded bar. An electrical wire was connected at the end of the tested bars from both ends.

Table 1 Details of Test Slabs

Side dimensions (mm × mm)	h (mm)	c (mm)	d (mm)	f_c' (MPa)	Main reinforcement
1900 × 1900	150	250	104	41.1	Two layers of 15M bars (ρ = 1%)

*h is the height; c is column width; d is the effective slab depth; ρ is the reinforcement ratio; and f_c' is the concrete compressive strength.

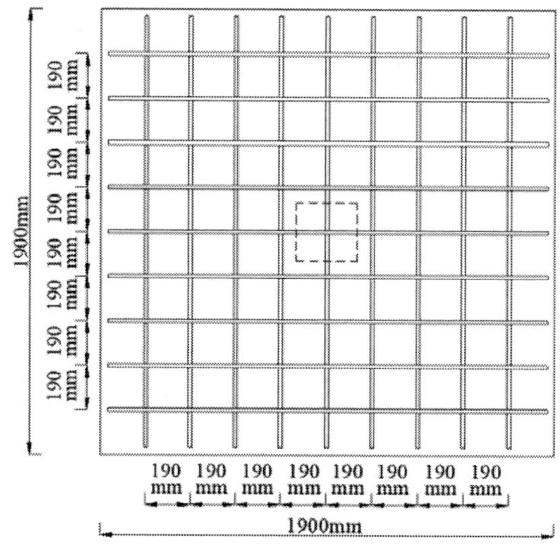

a. Flexural mat reinforcement 15M @ 190 mm

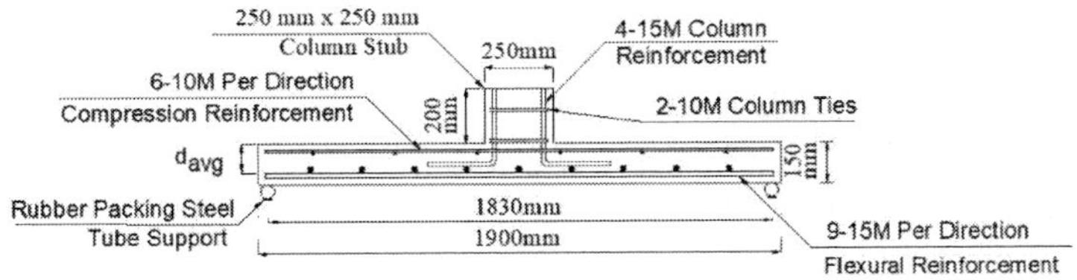

b. Cross-section of the slab

Figure 1: Details of a typical slab

Figure 2 shows a slab specimen and the form work used. Two slabs were under investigation on this study. The first slab (SV) was corroded till reach 25% mass loss using constant voltage accelerated corrosion technique. The second slab (SC) reached the same level of corrosion but by using constant current accelerated corrosion technique.

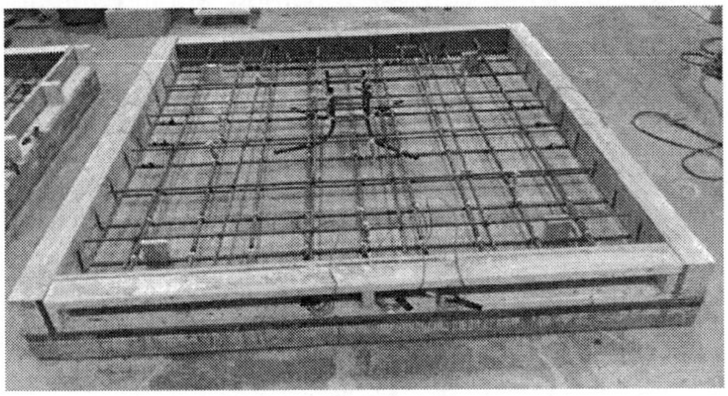

(a)

(b)

Figure 2: Test specimens: (a) reinforcement mesh; (b) wire-bar connection

2.2 Material Properties

Table 2 shows the concrete mixture composition used in casting the slabs. Based on the recommendation of ACI 318-14 concrete horizontal elements exposed to de-icing salts are classified as class F3. Under such exposure, the targeted concrete strength has to be at least 40 MPa and the w/c ratio should not be more than 0.4 (ACI Committe-318, 2014). Table 3 shows the actual properties of the concrete and the reinforcement.

Table 2 Mixture Composition of Concrete Mix

Cement (kg)	Water/cement ratio	Water (L)	Coarse/fine ratio	Coarse aggregate (kg)	Fine aggregate (kg)	Aggregate size (mm)	Fresh density (kg/m3)
350	0.4	140	1.3	1083	833	10	2407

Table 3 Steel and Concrete Properties

	Steel properties			Concrete Properties	
Designation	Area mm^2	f_y MPa	ε_y ($\mu\varepsilon$)	f_c' MPa	f_r MPa
15M	200	475	2200	41.1	3.82

Ready mix concrete was used based on specific mix design. Before casting, the slump was measured, and the prisms and cylinders were prepared. The poured concrete was consolidated using an electrical vibrator. The column stub was cast on the next day. The form work was removed after 24 hrs from casting the column stub. Water curing process continued for four days followed by air curing for 28 days.

2.3 Accelerated Corrosion Setup

After casting the slabs, each slab was placed upside down in leveled position. A dike wall tank was built over the targeted corrosion area using foam sheets. The dike was connected to a hose with a valve to facilitate the drainage of the dike from the electrolyte solution when needed. The hose was located at the lower level point of the dike. A steel mesh was placed inside the dike to act as a cathode. The tank was filled with a chloride salt solution with concentration of 5% by weight of the water. A power supply was used to provide a direct current by connecting the positive charge to the tension reinforcement and the negative charge to the steel mesh. The direct current connected to slab SV through the application of constant voltage and connected to slab SC through the application of constant current. The theoretical mass loss is calculated based on Faraday's equation as shown in Eq. (4):

[4] $$\text{Mass loss} = \frac{t\ i\ M}{z\ F}$$

Where t is the time passed in seconds, i is the current passed in amperes, M is atomic weight (for iron $M = 55.847$ g/mol), z is the ion charge (two moles of electrons) and F is Faraday's constant which is the amount of electrical charge in one mole of electron ($F = 96,487$) (Lachemi *et al.*, 2014).

2.3.1 Constant Voltage Technique Setup

The Constant voltage accelerated corrosion setup consisted of: power supply, data acquisition systems, current transducers, distribution board, electrolyte solution, steel mesh, and computer to process the data. Figure 3 shows the elements used in the setup.

Computer-controlled data acquisition system

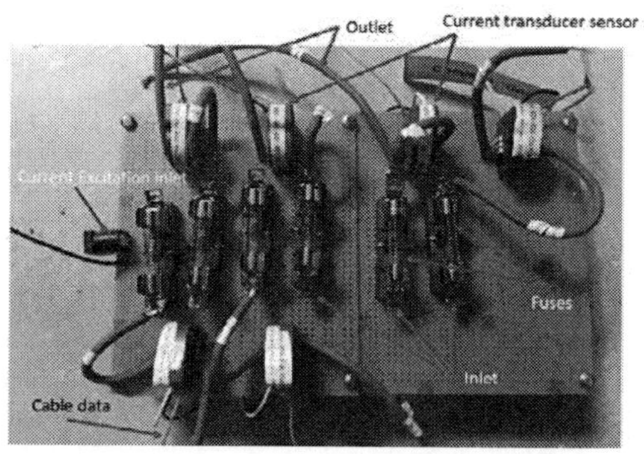

Current transducer board

Figure 3: Constant voltage accelerated test setup

A power supply was adjusted to provide a constant voltage 15 V. The positive terminal was connected to distribution board that distributed the input current into several output terminals with the same constant voltage 15 V, as a parallel circuit. The data acquisition system read the voltage value. On the other hand, the current value was needed to calculate the mass loss according to Faraday's equation. Hence, special current transducers were used to convert the current reading to voltage. Accordingly, each output terminal was connected to a current transducer. The output of the current transducers was recorded using the data acquisition system and stored in a computer. The outlet from each transducer in the current transducer board was connected to the bars under corrosion process. The negative terminal from the power supply was connected directly to the steel mesh that served as a cathode. The steel mesh was immersed inside the electrolyte solution inside the dike above the corroded targeted area.

2.3.2 Constant Current Technique Setup

Constant current accelerated corrosion setup consisted of: power supply, electrolyte solution, and steel mesh. Figure 4 shows the elements used in the setup for constant current technique. Each corroded bar needed a separate power supply in order to control the mass loss and to ensure uniform corrosion in all corroded bars. Each power supply was adjusted to apply current density of $200\mu A/cm^2$. This value was chosen according to the recommendations of El Maaddawy & Soudki (2003) that the current density should not exceed $200\mu A/cm^2$ to avoid any significant increase in crack width and strain response due to the corrosion. The counting of corrosion time started after the half-cell potential read more than -350 mV to ensure that the current overcome the concrete resistant and the occurrence possibility of corrosion was more than 90% (ASTM C876, 2015).

Figure 4: Constant current accelerated corrosion technique setup

3 EXPERIMENTAL RESULTS AND ANALYSIS

3.1 Time-dependent corrosion tests results

3.1.1 Current Results

Slab SC was corroded under the application of constant current. The current intensity applied was $200\mu A/cm^2$ which was equivalent to 0.1 A. This current was applied for 146 days until the targeted corrosion level of 25% mass loss was reached. This technique enable the determination of the exact time. On the other hand, slab SV was corroded under the application of constant voltage 15 V. The current was monitored using a computer-controlled data acquisition system. Figure 5 shows the relation between the current and the elapsed time for a corroded bar. It could be noticed that the current initially slightly dropped down then remained constant for almost 84 days. This was followed by a linear increase till the targeted level of 25% of mass loss was reached. The slight drop down in the current could be ascribed to the presence of the passive film layer around the bar that prevented it against corrosion. Then, the current weakened this layer with the help of chloride ingress, the corrosion started and its rate increased progressively. The sudden jump in the current could be attributed to the bond loss between the corroded bar and the concrete cover in which spalling of concrete covers started to occur.

Figure 6 shows the mass loss rate versus time for a corroded bar. It can be seen that the mass loss rate increased with increasing the current rate, which is associated with increasing the rate of corrosion. At the end of corrosion process, the current intensity exceeded $200\mu A/cm^2$, which surpassed the recommended from El Maaddawy & Soudki (2003). This would cause a significant increase in crack width due to the corrosion.

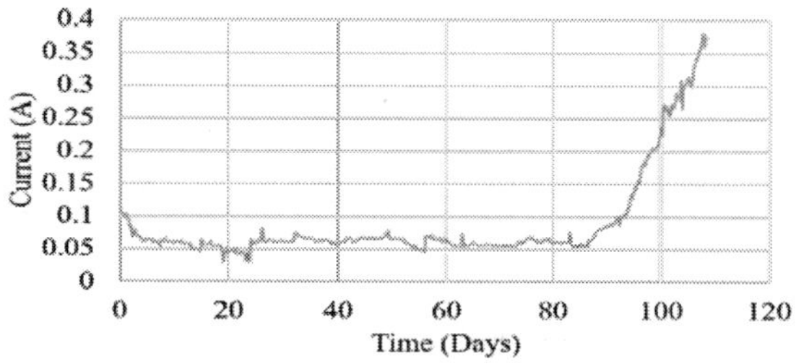

Figure 5: Current time history for SV specimen

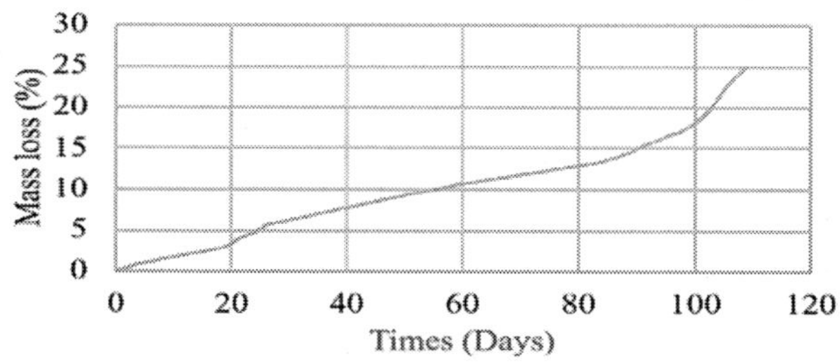

Figure 6: Mass loss rate with time

3.1.2 Half-Cell Potential Measurements

The half-cell potential readings are an indication of the existence of corrosion. Higher probability of existing corrosion is expected when the reading of the half-cell potential gives more negative values. The half-cell potential was used to determine the corrosion of the reinforcement in accordance with (ASTM C876, 2015).

The half-cell potential was used on slab SC at the beginning of the test. It was used to determine the instance of the occurrence of corrosion to count the time lapsed in order to calculate the mass loss based on Faraday's equation. The corrosion time started when the half-cell read -350 mV. The half-cell potential was used to evaluate the corrosion process for slab SV during the entire experiment. The half-cell readings were recoded once every three days at twenty-four different location that cover the entire corroded area. Figure 7 shows the half-cell reading versus time over the corroded area. The potential reading was measured directly above the flexural reinforcement mesh. The mesh consisted of two layers; top and lower ones. Figure 8 shows the reading variation of half-cell versus the level of corrosion. It can be seen from both figures that the corroded bars in the upper layer had more negative values than the lower one. This could be attributed to the concrete cover for the upper layer that was smaller than upper one by 16 mm. Hence, the upper layer was exposed to more chlorides ingress than the lower one. This result is in good agreement with previous studies that the concrete cover has a significant effect on the half-cell potential values (Klinghoffer, 1995). It can be also noticed from both figures that there was no variation in the reading of the half-cell when the corrosion started. Therefore, this test was used only as an indication of the occurrence probability of corrosion for un-corroded members.

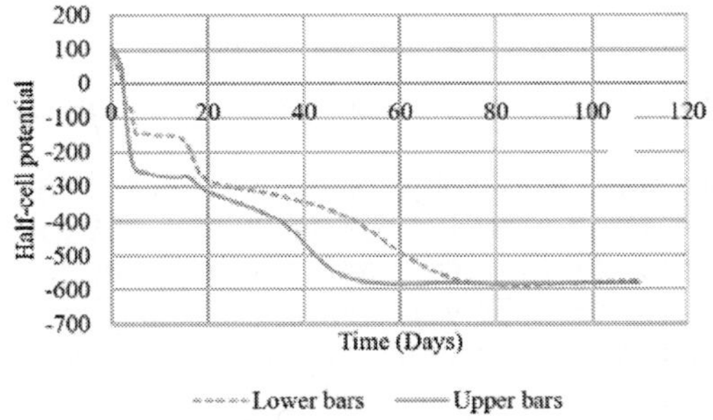

Figure 7: Half-cell potential reading vs. time

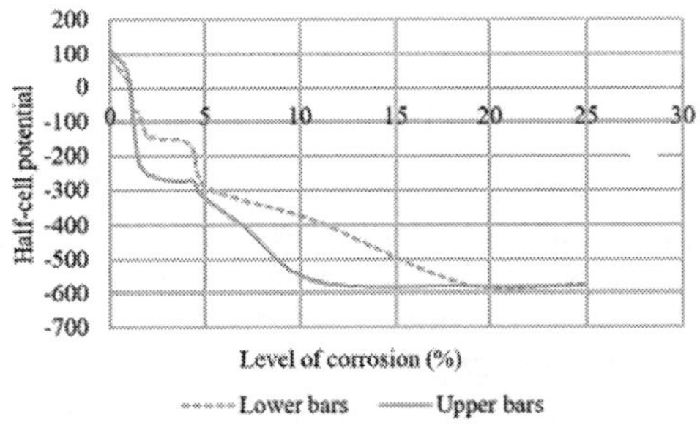

Figure 8: Half-cell potential readings vs. corrosion level

3.2 Test Results after Corrosion

3.2.1 Mass Loss

After the loading test was completed, the concrete cover was removed and the corroded bars were extracted from the slabs. The corrosion products were removed from the bars using a wire brush, then the bars were soaked in a HCL solution according to ASTM Standards G1-03 method (ASTM G1, 2003). Next, the bars were brushed again lightly to remove the remaining loose corrosion products. The cleaned bars were weighted and each bar was compared with its original weight. The reinforcement were weighed and labeled before casting and pouring the concrete. The percentage mass loss for each bar was calculated based on Eq. (5):

$$[5]\ \%\,\text{mass loss} = \frac{(\text{initial weight - final weight})}{\text{initial weight}}$$

The actual mass loss for both slabs showed a close agreement with the theoretical mass loss. It was noticed that the actual mass loss was less than the theoretical ones for both slabs. This could be ascribed to the loss in current through concrete cover to overcome the concrete resistivity. Thus, this amount of current did not contribute to reducing the reinforcement mass due to corrosion. This conclusion was reached by several researchers (Auyeung, Balaguru and Chung, 2000; Spainhour and Wootton, 2008). However, the Faraday's equation gives a close agreement between actual and predicted mass loss regardless to the accelerated corrosion technique used. Figure 9 shows the corrode reinforcement for both slabs after removing the concrete cover.

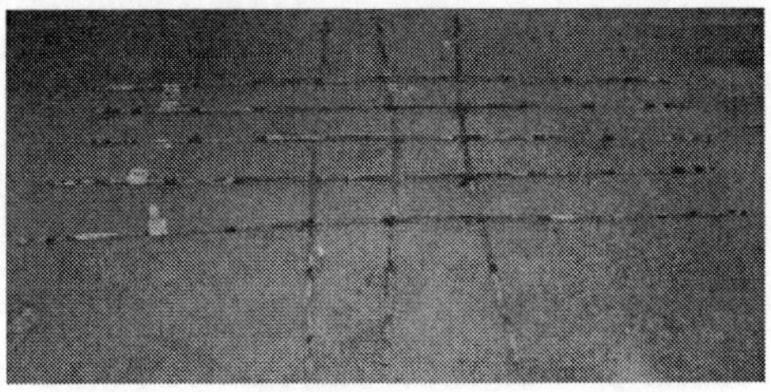

a. Slab SV

b. Slab SC

Figure 9: Extracted corroded bars

4 CONCLUSIONS

In this paper, a lab setup that could be used to induce corrosion in two-way slabs was presented. Two techniques, constant current and constant voltage were used. The following conclusions were reached:

- Constant current technique could be more feasible than constant voltage technique. Most labs are equipped with data acquisition systems that read voltage. Each corroded bar using constant voltage technique will need an individual current transducer to convert it to voltage and a free channel in the data acquisition system. When using constant current, a power supply has to be provided for each. The combined cost of current transducers and data acquisition system could be more expensive than simple power supplies that are used in constant current technique.
- The constant voltage produces large volume of raw data that are needed to be processed every day to calculate the mass loss. Moreover, the data needs to be continuously stored and monitored to avoid any corruption of the file due to the accumulated recorded data.
- In constant current, the time period required for corrosion could be easily estimated before the test. Constant voltage does not have this option as the current changes with time. The constant voltage technique for high levels of corrosion could take less time than constant current because the current could reach very high values, although surge in current values should not be recommended in this type of testing.
- The current can reach very high values in constant voltage technique, which could exceed the recommended value of $200\mu A/cm^2$ according to El Maaddawy & Soudki (2003). This increase could cause an increase in the concrete crack width leading to more damage than that caused by real corrosion.
- The concrete resistance increases with time; therefore, high voltage would be needed to overcome this resistance. Subsequently, the corrosion test for constant voltage should not be used with aged concrete. Rather, it would be recommended to start the test directly after 28 days from casting the concrete to avoid any possible increase that could happen in concrete resistance. Using constant current could be a better option to use with aged concrete since the required voltage also increases while keeping the current constant.

Accordingly, the authors believe that using constant current could be a better option when it comes to available lab facilities, space and time. Moreover, for severe levels of corrosion, constant current could better simulate the performance of real corroded elements compared to constant voltage.

References

ACI Committe-318 (2014) Building code requirement for structural concrete (ACI 318M-14) and commentary (ACI 318RM-14), Building Code Requirements for Structural Concrete.

Ahwazi, B. B. N. (2001) A comparative study of deterioration of bond due to corrosion in different concretes. McGill University.

Al-Swaidani, A. M. and Aliyan, S. D. (2015) 'Effect of adding scoria as cement replacement on durability-related properties', International Journal of Concrete Structures and Materials. Korea Concrete Institute, 9(2), pp. 241–254. doi: 10.1007/s40069-015-0101-z.

Almusallam, A., Al-Gahtani, A. S., Aziz, A. R. and Rasheeduzzafar, A. (1996) 'Effects of reinforcement corrosion on bond strength', Construction and Building Materials, 10(2), pp. 123–129.

Altoubat, S., Maalej, M. and Shaikh, F. U. A. (2016) 'Laboratory simulation of corrosion damage in reinforced concrete', International Journal of Concrete Structures and Materials. Korea Concrete Institute, 10(3), pp. 383–391. doi: 10.1007/s40069-016-0138-7.

Amleh, L. (2000) Bond deterioration of reinforcing steel in concrete due to corrosion. McGill University.

ASTM C876 (2015) Standard test method for corrosion potentials of uncoated reinforcing steel in concrete. doi: 10.1520/C0876-09.2.

ASTM G1 (2003) Standard practice for preparing, cleaning, and evaluation corrosion test specimens.

Auyeung, Y., Balaguru, P. and Chung, L. (2000) 'Bond behavior of corroded reinforcement bars', Materials Journal, 97(2), pp. 214–220.

Bonacci, J., Thomas, M., Hearn, N., Lee, C. and Maalej, M. (1998) 'Laboratory simulation of corrosion in reinforced concrete and repair with CFRP wraps', in Annual Conference of the Canadian Society of Civil Engineering, pp. 653–662.

Davis, J. (2000) Corrosion: understanding the basics. doi: 10.1361/cutb2000p001.

Deb, S. and Pradhan, B. (2013) 'A study on corrosion performance of steel in concrete under accelerated condition', in Proceedings of the International Conference on Structural Engineering Construction and Management, Kandy, Sri Lanka. Available at: http://dl.lib.mrt.ac.lk/handle/123/9365.

Kashani, M. M., Crewe, A. J. and Alexander, N. A. (2013) 'Nonlinear stress-strain behaviour of corrosion-damaged reinforcing bars including inelastic buckling', Engineering Structures, 48, pp. 417–429. doi: 10.1016/j.engstruct.2012.09.034.

Klinghoffer, O. (1995) 'In situ monitoring of reinforcement corrosion by means of electrochemical methods', Nordic Concrete Research, 95(1), pp. 1–13. Available at: http://www.germann.org/Publications/GalvaPulse/2. Klinghofer, O., In Situ Monitoring of Reinforcement Corrosion by means of Electrochemical Methods, Nordic Concrete Research 95 1.pdf.

Kumar, M. K., Rao, P. S., Swamy, B. L. P. and Chandra Mouli, C. (2012) 'Corrosion resistance performance of fly ash blended cement concrete', Internatioanl Journal of Research in Engineering and Technologi, 1(3), pp. 448–454. doi: 10.1016/0140-6701(95)93047-7.

Lachemi, M., Al-Bayati, N., Sahmaran, M. and Anil, O. (2014) 'The effect of corrosion on shear behavior of reinforced self-consolidating concrete beams', Engineering Structures. Elsevier Ltd, 79, pp. 1–12. doi: 10.1016/j.engstruct.2014.07.044.

Lee, C., Bonacci, J. F., Thomas, M. D. a., Maalej, M., Khajehpour, S., Hearn, N., Pantazopoulou, S. and Sheikh, S. (2000) 'Accelerated corrosion and repair of reinforced concrete columns using carbon fibre reinforced polymer sheets', Canadian Journal of Civil Engineering, 27(5), pp. 941–948. doi: 10.1139/cjce-27-5-941.

El Maaddawy, T. A. and Soudki, K. A. (2003) 'Effectiveness of impressed current technique to simulate corrosion of steel reinforcement in concrete', Journal of Materials in Civil Engineering. American Society of Civil Engineers, 15(1), pp. 41–47.

El Maaddawy, T., Soudki, K. and Topper, T. (2005) 'Analytical model to predict nonlinear flexural behavior of corroded reinforced concrete beams', ACI Structural Journal, 102(No. 4), pp. 550–559.

Neville, A. M. (2011) Properties of concrete (Fifth Edition), Cement and Concrete Research. doi: 10.1016/0008-8846(96)82366-0.

Pellegrini-Cervantes, M. J., Almeraya-Calderon, F., Borunda-Terrazas, A., Bautista-Margulis, R. G., Chacón-Nava, J. G., Fajardo-San-Miguel, G., Almaral-Sanchez, J. L., Barrios-Durstewitz, C. P. and Martinez-Villafañe, A. (2013) 'Corrosion resistance, porosity and strength of blended portland cement mortar containing rice husk ash and nano-SiO2', International Journal of Electrochemical Science, 8(8), pp. 10697–10710.

Pritzl, M. D., Tabatabai, H. and Ghorbanpoor, A. (2014) 'Laboratory evaluation of select methods of corrosion prevention in reinforced concrete bridges', International Journal of Concrete Structures and Materials, 8(3), pp. 201–212. doi: 10.1007/s40069-014-0074-3.

Spainhour, L. K. and Wootton, I. A. (2008) 'Corrosion process and abatement in reinforced concrete wrapped by fiber reinforced polymer', Cement and Concrete Composites. Elsevier, 30(6), pp. 535–543.

Talakokula, V., Bhalla, S. and Gupta, A. (2014) 'Corrosion assessment of reinforced concrete structures based on equivalent structural parameters using electro-mechanical impedance technique', Journal of Intelligent Material Systems and Structures, 0(0), pp. 1–16. doi: 10.1177/1045389X13498317.

Toongoenthong, K. and Maekawa, K. (2004) 'Interaction of pre-induced damages along main reinforcement and diagonal shear in RC members', Journal of Advanced Concrete Technology, 2(3), pp. 431–443. doi: 10.3151/jact.2.431.

Xia, J., Jin, W.-L. and Li, L.-Y. (2012) 'Effect of chloride-induced reinforcing steel corrosion on the flexural strength of reinforced concrete beams', Magazine of Concrete Research. Thomas Telford, 64(6), pp. 471–485.

Yoon, S., Wang, K., Weiss, W. and Shah, S. (2001) 'Interaction of reinforced between loading, corrosion, and serviceability of reinforced concrete', ACI Materials Journal, 97(6), pp. 637–644.

`Building Tomorrow's Society

Bâtir la Société de Demain

Fredericton, Canada

June 13 to 16, 2018 / *13 au 16 juin, 2018*

CORRELATION BETWEEN DIFFERENT FACTORS CONTROLLING SHRINKAGE BEHAVIOUR OF GEOPOLYMER MORTAR

Alhamdan, Ahmad[1,3] and Soliman, Ahmed[2]
[1]MASc. Student, Department of Building, Civil and Environmental Engineering Concordia University, Montreal, Quebec Canada.
[3] ahmad.alhamdan@mail.concordia.ca
[2]Assistant Professor, Department of Building, Civil and Environmental Engineering Concordia University, Montreal, Quebec Canada.

Abstract: Cement industry produces large amounts of carbon dioxide causing many environmental problems including global warming and air quality deterioration. As an alternative to cement, many studies have been conducted on alkali-activated materials (so-called geopolymers). Geopolymers showed similar or better mechanical and durability performance compared to that of cement-based materials. However, shrinkage behaviour of geopolymer materials is still one of the main problems that impede its in-situ applications. Hence, this study focuses on investigating the interrelationship between different factors that control shrinkage behaviour of geopolymer materials. These factors include supplementary cementitious material type, activator type and concentration, and curing temperatures. The setting time, workability, compressive strength, autogenous and drying shrinkage of alkali-activated mortars (AAMs) were evaluated. Results showed that the effect of activator type and concentration on the shrinkage behaviour differs according to the curing temperature. Moreover, the relationship between autogenous and drying shrinkage was significantly affected by the type of supplementary cementitious material used. Generally, the curing temperature is the most controlling factor for shrinkage behaviour of AAMs tested and also showed a strong influence on the effects of other factors.

1 INTRODUCTION

Global concrete production has grown widely over the recent decades in line with the population growth and the need for more infrastructures. The high demand for concrete in the construction industry increased ordinary Portland cement (OPC) production. The depletion of natural resources and the significant environmental pollution associated with carbon dioxide (CO_2) emissions generated during cement manufacturing process are of major concerns. Alkali-activated binder (AAB) represents a promising alternative to OPC in many constructional applications (Yang et al. 2013).

AABs rely on alumina-silicate precursors, found naturally occurring (eg. metakaolin) or formed as industrial by-products (eg. fly ash, slag), leading to a significant reduction in carbon emission. AAB production requires an alkaline medium (eg. sodium silicate Na_2SiO_3) and water to chemically activate the precursor forming the cementitious material (Khale and Chaudhary 2007). This process (so-called "geopolymerization") initiates with the dissolution of silicon and aluminum ions from the precursors into the alkaline medium. The dissolved ions react under sufficient alkalinity to produce silicate and aluminate compounds which are converted to a poly (alumina-silicate) structure with interconnected silicate oxide

(SiO_4) and aluminate oxide (AlO_4) tetrahedral. Thereafter, the hardening stage is achieved through the poly-condensation of those hydrolyzed species (Provis and Bernal 2014). This process requires cations like Na+, or K+ to reduce the charge variation which may be due to the coordination of aluminum in whole structure (Duxson et al. 2007).

Despite the various advantages of utilizing AABs in the construction industry, their widespread use has been hindered due to several challenges. One of the main challenges is shrinkage. Shrinkage mechanism of alkali-activated binders is still not clear. Shrinkage can significantly affect AAB performance by inducing cracks under restrained conditions.

2 EXPERIMENTAL PROGRAM

2.1 Materials

In the present study, FA class-F and GGBS were used as the precursor materials. The chemical compositions and physical properties of the FA and GGBS are listed in Table 1. Sodium hydroxide with a purity level of 98% and sodium silicate (SiO2 (30%), Na2O (20%), H2O (50%), specific gravity 1.38 g/mL) were used to prepare the alkali activators. The alkali activator was prepared as the following: Distilled water was used to dissolve the NaOH solid at 3M, 6M, and 12M, and left overnight to cool down to the room temperature while covered to prevent carbonation. Sodium silicate solution was then mixed with the NaOH solution to form the alkali activator solution (mass of sodium silicate solution/ NaOH solution = 2). ASTM standard river sand was used as the fine aggregate material. The oven-dry, saturated surface dry specific gravity, and absorption of the standard sand used were 2.54, 2.55, and 0.08 %, respectively.

2.2 Mix proportion and specimen preparation

The mix proportions of the alkali activated fly ash and slag mortars are shown in **Table 2**. Two precursor blends were utilized in this study Fly ash (FA) and Slag (GGBS). Six AAM mixtures were designed using a solution made of three different NaOH concentrations to cover a wide range of concentrations (3, 6, and 12M) each mixed with a constant sodium silicate concentration. The mass ratio of sand to precursor was fixed at 2:1 for all mixtures. The mass ratio of activator solution to precursor was 0.40 for all mixtures. The mass ratio of the water glass to NaOH solution was 2.0 in all activator mixtures. Two curing regimes were applied: 23°C with relative humidity of 60 ± 5% (ambient) and at 60°C (oven).

Table 1: Properties of precursors used.

Chemical composition (%)	Fly ash (FA)	Slag (GGBS)
CaO	2.60	56.10
SiO_2	46.00	21.00
Al_2O_3	33.00	17.00
Fe_2O_3	10.50	0.62
SO_3	-	0.77
Physical properties		
S.S.D Specific gravity (g/cm^3)	2.38	2.90
Surface area (m^2/kg)	290 (Blaine)	laine)

2.3 Specimen preparation

Initially, fly ash was dry mixed with sand in an electric mixer at a slow speed of (140 ± 5 revs/min) and medium speed (285 ± 5 revs/min) for 2 minutes to ensure homogeneity of the mixture. The prepared alkali activator solution is then added to the mixture and mixed for an additional 2 minutes. Immediately after mixing, 50 mm cubic molds for compressive strength evaluation according to ASTM Standard C109 / C109M-16a, and 25 mm × 25 mm × 285 mm prismatic molds for shrinkage measurements according to ASTM Standard C490 / C490M – 17, were cast. Specimens were compacted using a vibrating table and a tamping rod. All specimens were cured at room temperature of 23 °C and at a relative humidity of 60% ±

5% inside molds for 24 hours. After 24 hours, specimens were demolded and stored in ambient room conditions or in the oven at 60°C until testing age. All reported results represent the average of three replicates.

Table 2: Mix proportion of 1 m^3 alkali-activated mortar (AAM) specimens.

Mixture	FA (kg)	GGBS (kg)	NaOH solution molarity (M)	NaOH solution (kg)	Na_2SiO_3 solution (kg)	Sand (kg)
F3	655	0	3	87.33	174.67	1,310
F6	655	0	6	87.33	174.67	1,310
F12	655	0	12	87.33	174.67	1,310
S3	0	627	3	83.6	167.2	1,254
S6	0	627	6	83.6	167.2	1,254
S12	0	627	12	83.6	167.2	1,254

2.4 Testing program

The workability and flow of mortar mixtures were evaluated using the procedure listed in ASTM Standard C1437-15. Initial and final setting times were measured using a Vicat needle as described in ASTM Standard C191-13. Compressive strength development was evaluated at ages of 1, 7, 14, and 28 days using a 300 kN universal testing machine (UTM) in accordance with ASTM Standard C109 / C109M-16a. Cubes were loaded until failure at a load rate of 345 kPa/s. Cubic specimens (50 × 50 × 50 mm) were prepared for compressive strength measurements. The mortar cubes were compacted using a vibrating table and a tamping rod. Cubes were loaded until failure at a load rate of 345 kPa/s. Autogenous and drying shrinkage strains were measured using prism samples (25.4 × 25.4 × 254 mm) in accordance with ASTM Standard C490 / C490M - 17 using a length comparator with dial gauge accuracy of 0.0001-inch. After demolding, autogenous shrinkage specimens were wrapped with a first layer of polyethylene film and a second layer of aluminum foil to prevent any moisture loss.

3 EXPERIMENTAL RESULTS

3.1 Flowability

Variations in flowability for different alkali-activated mortar (AAM) mixtures with respect to NaOH concentrations are shown in **Table 3**. In general, the flow of AAMs decreased with the increase in the molar concentration of NaOH in the activating solution. For instance, reductions in the average flow of AAF mortars when NaOH concentrations increased from 3M to 6M and 12M were 5.56%, and 14.81%, respectively. Na_2SiO_3 liquid (waterglass) is known to have high viscosity itself hence, the mixture of waterglass with highly concentrated NaOH solutions tend to further increase the viscosity of the activating mixture and thus reduce the flow of AAM (Chindaprasirt et al. 2007). Moreover, the flowability of AAS mortars was lower than that of the AAF mortars at similar NaOH concentration.

3.2 Setting time

Setting time tests were performed in order to ascertain the effects of precursor type and NaOH solution molarity on the stiffing of AAM. Initial and final setting time results of different AAM are shown in **Table 3**. Generally, higher NaOH concentrations accelerated both the initial and final setting behavior of AAM mixtures. For instance, reductions in the initial setting time of AAF mortars when NaOH concentrations increased from 3M to 6M and 12M were 12.5%, and 38.2%, respectively. The corresponding reductions in final setting times were 11.2% and 35%, respectively. Higher activator concentrations tend to accelerate the rate of the hydration reaction hence, directly influencing the stiffness behavior of AAMs. AAS mortar mixtures exhibited faster settings compared to AAF mortar mixtures. For example, the initial setting in 3M NaOH activated slag mortars was achieved 36 minutes earlier than in 3M NaOH activated fly ash.

Table 3: Setting time and flow values of AAM mixtures.

Mixture	Setting time (min)		Flowability (cm)
	Initial	Final	
F3	144	286	108
F6	126	254	102
F12	89	186	92
S3	108	171	94
S6	72	114	87
S12	47	76	76

3.3 Compressive strength

3.3.1 Effect of activator concentrations

The compressive strength development of AAM mixtures activated with different concentrations of NaOH was studied. **Figure 1** shows the compressive strength of AAF mortars. Generally, the compressive strength of AAF mortars increased with higher activator concentrations under both curing conditions. For instance, when NaOH concentration increased from 3M to 6M and 12M, the 28 days compressive strength of ambiently cured AAF mortars increased by 46.2%, and 117.26%, respectively. The increase in 28 days compressive strength of oven cured AAF mortars due to higher NaOH concentrations was almost equal to the one found in ambient cured ones. **Figure 2** shows the compressive strength of AAS mortars. Similar compressive strength increase trend with higher NaOH concentrations was found in AAS mortars.

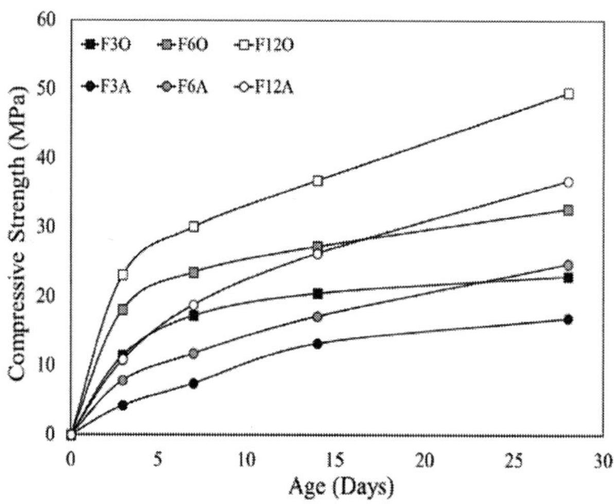

Figure 1: Compressive Strength of AAF mortar specimens (ambient (A) and oven (O) cured).

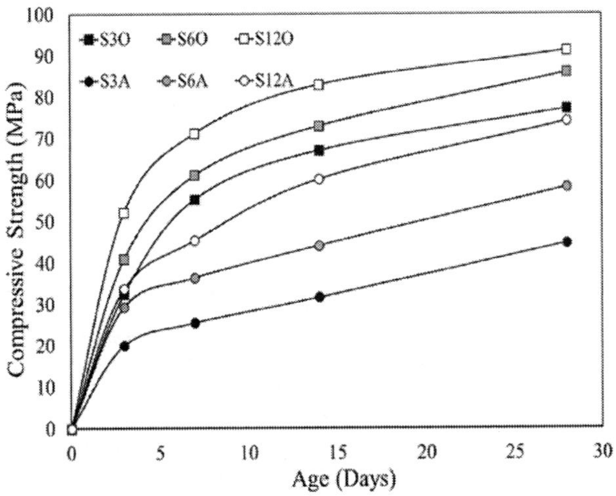

Figure 2: Compressive Strength of AAS mortar specimens (ambient (A) and oven (O) cured).

3.3.2 Effect of precursor type

Compressive strength results for all mortar specimens activated with 3M, 6M, and 12M (NaOH) concentrations and cured at both conditions are plotted in **Figs. 1 and 2**. Generally, AAM specimens exhibited steady strength development between 3 and 28 days. AAS mortar specimens achieved higher 28 days compressive strength than AAF mortars under both curing conditions. For instance, at low activator molarities (3M NaOH), the 28 days compressive strength of ambient cured AAS mortars was 160% higher than that of AAF mortars. The corresponding compressive strength of oven cured AAS mortars was 263% higher than that of AAF mortars.

3.3.3 Effect of curing temperature

Compressive strength results of AAF and AAS mortars activated with different NaOH concentration and cured at ambient and oven conditions are shown **Figs. 1 and 2**. Results indicated that all specimens cured in the oven had acquired extra compressive strength during the 28 days study period. For instance, the 28 days compressive strength values of ambient cured AAF mortars activated with 3M, 6M, and 12M NaOH concentrations were increased by 36, 32, and 35%, respectively when cured in the oven. Normally, the activation reaction of fly ash at room temperature is slow (Shi and Day 1999). On the other hand, previous research showed that GGBS can be activated effectively at room temperature by an alkaline medium (Shi and Day 1999). As expected, ambient cured AAS mortars activated with 3M, 6M, and 12M NaOH concentrations achieved high 28 days compressive strength results of 44.67, 58.22, and 74.11 Mpa, respectively.

3.4 Drying Shrinkage

3.4.1 Effect of activator concentrations

Shrinkage is a critical parameter for long-term serviceability of concrete structures. The drying shrinkage strains for AAM activated with different (NaOH) concentrations and cured at ambient and oven temperatures are plotted in **Figs. 3 and 4**. Typically, the magnitude and rate of drying shrinkage in all AAM specimens increased with higher NaOH concentrations. Around 80-86 % of the total shrinkage was observed within the first 7 days depending on activator concentrations. For instance, the drying shrinkage strain experienced by 12M NaOH activated AAF mortars was 25%, and 34% more than the ones experienced by 6M, and 3M NaOH activated mortars, respectively under ambient curing. The corresponding drying shrinkage strains under oven curing were also 18%, and 27% higher, respectively.

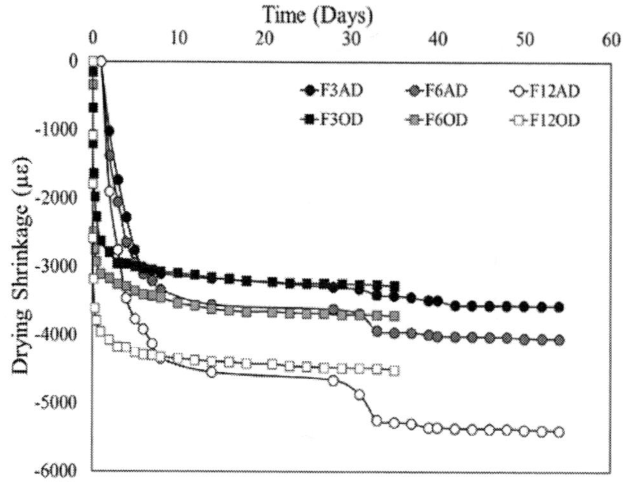

Figure 3: Drying shrinkage of AAF mortar specimens (ambient and oven cured).

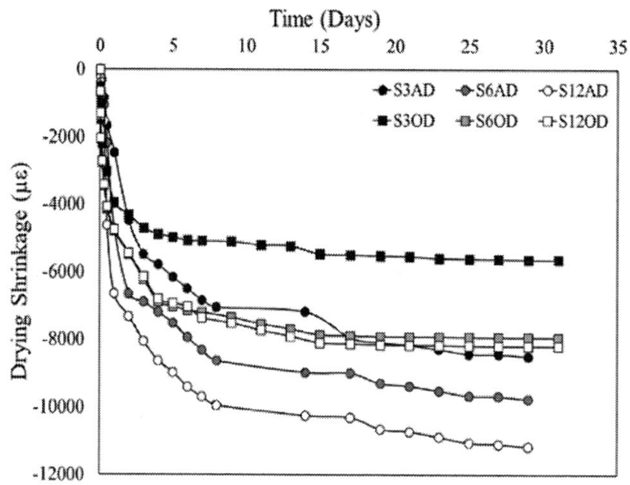

Figure 4: Drying shrinkage of AAS mortar specimens (ambient and oven cured).

3.4.2 Effect of precursor type

The effect of precursor type on drying shrinkage of AAMs was evaluated. Drying shrinkage strains of different mortar specimens activated with 3M, 6M, and 12M (NaOH) concentrations and cured at ambient temperature are plotted in **Figs. 3 and 4**, respectively. Results indicated that AAS mortars experienced higher drying shrinkage strains compared to AAF mortars. For instance, at low activator concentration (3M NaOH), the drying shrinkage strains of AAS mortars ambient and oven cured were 71.8%, and 142% higher than those measured in AAF mortars, respectively. Around 81-85 % of the total drying shrinkage was observed in the first 7 days. These results are in agreement with the literature (Fernández-Jimenez et al. 2006, Chi and Huang 2013).

AAS binders are known to have a high percentage of mesopores compared to AAF and OPC binders (Chen and Brouwers 2007). Consequently, the more refined pores found in AAS resulted in higher capillary stress which in turns significantly increased the drying shrinkage. Moreover, the spherical shape of fly ash particles tends to act as micro-aggregate and fills minor voids hence, increasing the volume stability of AAF mortars (Rashad 2013).

3.4.3 Effect of curing temperature

One of the major obstacles challenging the implementation of AABs in the industry is the requirement for a high curing temperature to obtain the expected superior properties over ordinary PC products. The effect of curing condition on different AAMs at several activator concentrations was studied. Drying shrinkage strains of AAF and AAS mortars cured at both ambient and oven conditions are shown in **Figs. 3 and 4**, respectively. Generally, the drying shrinkage strains were reduced in all mortar specimens cured in the oven at 60 °C compared to the ones cured at ambiance. The value of shrinkage reduction had a correlation with both precursor type and activator concentration. In AAF mortars, the shrinkage of oven cured mortars activated with 3M, 6M, and 12M NaOH concentrations were decreased by around 8.16, 8.31, and 16.41%, respectively compared to ambient cured specimens.

3.5 Autogenous Shrinkage

3.5.1 Effect of activator concentration

Autogenous shrinkage is caused by self-desiccation which is a result of the formation of finer pores in the microstructure of hardened binder and the water consumption by the continued hydration reaction (Lee et al. 2014). The autogenous shrinkage strains of AAF and AAS mortars activated with different (NaOH) concentrations and cured at ambient and oven temperatures are plotted in **Figs. 5 and 6**, respectively.

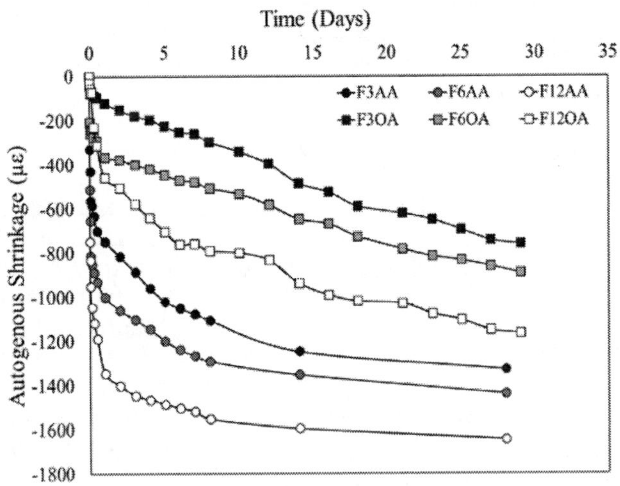

Figure 5: Autogenous shrinkage of AAF mortars (ambient and oven cured).

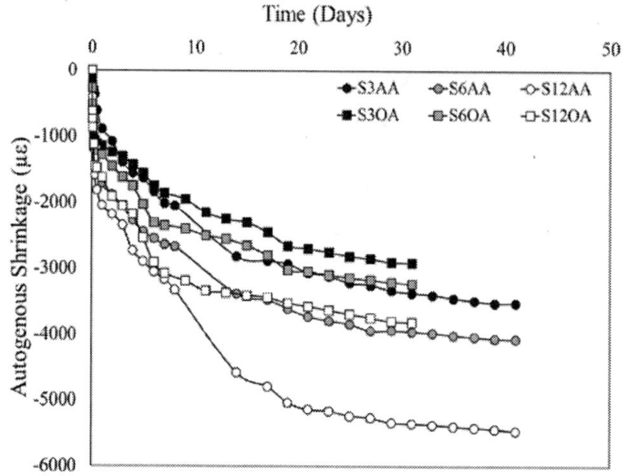

Figure 6: Autogenous shrinkage of AAS mortars (ambient and oven cured).

Generally, autogenous shrinkage rates and strains of AAMs increased with higher activator concentrations. For instance, the autogenous shrinkage strain experienced by 12M NaOH activated fly ash mortars was 9.3%, and 19.3% more than the ones experienced by 6M, and 3M NaOH activated mortars, respectively under ambient curing. The corresponding autogenous shrinkage strains under oven curing were also 28.5%, and 55.7% higher, respectively.

3.5.2 Effect of precursor type

Autogenous shrinkage strains for different mortar specimens activated with 3M, 6M, and 12M (NaOH) concentrations and cured at ambient temperature are plotted in **Figs. 5 and 6**. Generally, AAS mortars experienced the higher autogenous shrinkage strains compared to AAF mortars. For example, the autogenous shrinkage strain of ambient cured AAS mortar was 144% higher than that of AAF mortars when activated with low activator concentrations (3M NaOH). The corresponding autogenous shrinkage strain under oven curing was also 71.8% higher. For AAMs activated with high activator concentrations (12M), the drying shrinkage strain of AAS mortar was 217% higher than the one experienced by AAF mortars under ambient curing. The corresponding drying shrinkage under oven curing was 63%, and 275% higher, respectively. These results give an idea of the much better dimensional stability of AAF mortar compared to AAS mortars.

3.5.3 Effect of curing condition

Autogenous shrinkage results of AAF and AAS mortars cured at both ambient and oven conditions are shown in **Figs. 5 and 6**, respectively. Generally, the autogenous shrinkage strains were reduced in all AAM specimens cured in the oven at 60 °C compared to the ones cured at ambiance. Reduction in autogenous shrinkage strains due to oven curing was slightly more pronounced in AAF mortars and almost equal to AAS mortars. For instance, the autogenous shrinkage strains experienced by 3M, 6M, and 12M NaOH activated AAF mortars were reduced by around 46.4, 40.5, and 30%, respectively when oven curing is adopted. The corresponding reductions in AAS mortars were 17.24, 20.58, and 30%, respectively.

4 CONCLUSION

- Setting time and flowability of AAMs seem to be affected directly by the activator concentrations, as concentrations increase, both formers decrease.
- The strength of the AAMs depends highly on the activator dosage, as it was clear that higher NaOH dosages increased the compressive strength at all ages regardless of precursor type used.
- The autogenous and drying shrinkage strains experienced by AAMs were significantly higher than the ones found in OPC mortar.

- Regardless of the activator concentrations, AAS mortars exhibited higher drying and autogenous shrinkage strains compared to AAF mortars.

References

ASTM C109 / C109M-16a, Standard Test Method for Compressive Strength of Hydraulic Cement Mortars (Using 2-in. or [50-mm] Cube Specimens), ASTM International, West Conshohocken, PA, 2016, www.astm.org

ASTM C1437-15, Standard Test Method for Flow of Hydraulic Cement Mortar, ASTM International, West Conshohocken, PA, 2015, www.astm.org

ASTM C191-13, Standard Test Methods for Time of Setting of Hydraulic Cement by Vicat Needle, ASTM International, West Conshohocken, PA, 2013, www.astm.org

ASTM C490 / C490M-17, Standard Practice for Use of Apparatus for the Determination of Length Change of Hardened Cement Paste, Mortar, and Concrete, ASTM International, West Conshohocken, PA, 2017, www.astm.org

Chen, W., and Brouwers, H. J. H. 2007. The Hydration of Slag, Part 1: Reaction Models for Alkali-Activated Slag. *Journal of Materials Science*, **42**(2): 428-443.

Chi, M., and Huang, R. 2013. Binding Mechanism and Properties of Alkali-Activated Fly Ash/Slag Mortars. *Construction and Building Materials*, **40**: 291-298.

Chindaprasirt, P., Chareerat, T., and Sirivivatnanon, V. 2007. Workability and Strength of Coarse High Calcium Fly Ash Geopolymer. *Cement and Concrete Composites*, **29**(3): 224-229.

Duxson, P., Provis, J. L., Lukey, G. C., and Van Deventer, J. S. 2007. The Role of Inorganic Polymer Technology in the Development of 'Green Concrete'. *Cement and Concrete Research*, **37**(12): 1590-1597.

Fernández-Jimenez, A. M., Palomo, A., and Lopez-Hombrados, C. 2006. Engineering Properties of Alkali-Activated Fly Ash Concrete. *ACI Materials Journal*, **103**(2), 106.

Khale, D., and Chaudhary, R. 2007. Mechanism of Geopolymerization and Factors Influencing its Development: A Review. *Journal of Materials Science*, **42**(3): 729-746.

Lee, N. K., Jang, J. G., and Lee, H. K. 2014. Shrinkage Characteristics of Alkali-Activated Fly Ash/Slag Paste and Mortar at Early Ages. *Cement and Concrete Composites*, **53**: 239-248.

Ma, Y., Hu, J., and Ye, G. 2013. The Pore Structure And Permeability Of Alkali Activated Fly Ash. *Fuel*, **104**: 771-780.

Provis, J. L., and Bernal, S. A. 2014. Geopolymers and Related Alkali-Activated Materials. *Annual Review of Materials Research*, **44**: 299-327.

Rashad, A. M. 2013. Properties Of Alkali-Activated Fly Ash Concrete Blended With Slag. *Iran J Mater Sci Eng*, **10**(1): 57-64.

Shi, C., and Day, R. L. 1999. Early Strength Development and Hydration of Alkali-Activated Blast Furnace Slag/Fly Ash Blends. *Advances in Cement Research*, **11**(4): 189-196.

Yang, K. H., Song, J. K., and Song, K. I. 2013. Assessment of CO2 Reduction of Alkali-Activated Concrete. *Journal of Cleaner Production*, **39**: 265-272.

Building Tomorrow's Society

Bâtir la Société de Demain

Fredericton, Canada

June 13 to 16, 2018 / *13 au 16 juin, 2018*

DAMAGE DUE TO ABRASION IN SCC CONTAINING DIFFERENT SUPPLEMENTARY CEMENTING MATERIALS

Ridgley, Katherine E. [1], Abouhussien, Ahmed A. [1], Hassan, Assem A.A. [1,2] and Colburne, Bruce [1]

[1] Memorial University of Newfoundland, Canada[2] ahassan@mun.ca

Abstract: This investigation evaluates and compares the abrasion resistance of various self-consolidating concrete (SCC) mixtures. A normal concrete (NC) mixture with higher coarse-to-fine aggregate ratio was also tested for comparison. Various supplementary cementing materials (SCMs) were tested under the rotating cutter method for abrasion resistance. The effect of using different SCMs in SCC mixtures including fly ash, metakaolin (MK), silica fume, and slag on the abrasion resistance of SCC was examined. Results from the abrasion tests indicated that the SCC mixture containing MK had the highest abrasion resistance among all tested mixtures. The results also indicated that SCC mixtures showed improved abrasion resistance compared to NC mixture in terms of average wear depth and percent of weight loss.

1 INTRODUCTION

Self-consolidating concrete (SCC) is a highly workable concrete intended for use in heavily reinforced or hard-to-reach areas (Rao et al. 2012, Safiuddin et al. 2012, Aslani 2013, Abouhussien et al. 2015). SCC can be produced through the use of chemical admixtures to reduce the water content, decreasing the proportions of coarse aggregate in the mixture and/or incorporating supplementary cementing materials (SCMs) such as fly ash, silica fume, metakaolin, and/or natural pozzolans (Lachemi et al. 2007, Das and Chatterjee 2012, Karahan et al. 2012, Abouhussien and Hassan 2014). Similar to any concrete, SCC can be subject to various types of deterioration, depending on the type of the structure and its exposure condition. One of the most important types of this deterioration is the abrasion of concrete surfaces.

Abrasion is a form of natural attack on concrete, which can be defined as the process of scraping or wearing away of a material. Concrete abrasion can occur in several ways, including wear on floors due to human traffic, wear due to vehicles, abrasive substances in flowing water, and/or high-velocity waters that create cavitation (Scott and Safiuddin 2015). Abrasion can also occur in areas with heavy ice flow when concrete is used in Arctic environments (Scott and Safiuddin 2015). Resistance to abrasion depends on the hardness of the concrete (aggregate and paste hardness combined) and the aggregate/paste bond (Papenfus 2003). Aggregate hardness is a very important aspect as it makes up the majority of the mix proportions (Papenfus 2003). In the case of SCC, less coarse aggregate is used, putting more dependence for the abrasion resistance on the paste.

Abrasion resistance of concrete can be increased by increasing the binder content (at a given water content), using SCMs (which may depend on the age of testing), using superplasticizers to optimize the water content, applying sufficient consolidation, and/or using more durable aggregates (Papenfus 2003,

Turk and Karatas 2011). Abrasion resistance of concrete can be tested in a variety of ways according to the available ASTM standards. For example, ASTM C779 tests the abrasion resistance of horizontal concrete surfaces in three different test setups depending on the type and degree of abrasive force (ASTM 2012a). Also, the ASTM C418 method evaluates the abrasion resistance of concrete via sandblasting (ASTM 2012b). This procedure simulates the action of waterborne abrasives and abrasives under traffic on concrete surfaces. On the other hand, ASTM C1138 is an underwater concrete abrasion test intended to simulate the behaviour of swirling water containing transported solid objects that produce abrasion of concrete and cause potholes and related effects (ASTM 2012c). The simplest abrasion test setup used in the literature on various concrete types is the ASTM C994 rotating cutter method (ASTM 2012d). This test method has proven successful in past studies of highway and bridge concrete surfaces subject to traffic (Dong et al. 2013).

The utilization of SCMs in the SCC mixtures is anticipated to improve their abrasion resistance. Previous studies on normal-vibrated concretes confirmed that fly ash class C and F showed better abrasion resistance than Portland cement concrete (Liu 1981). Similarly, silica fume concrete has shown to be more resistant to wear with increasing amounts of the material, up to 10% (Ghafoori and Diawara 1999). Turk and Karatas (2011) evaluated the abrasion of SCC with varying dosages of fly ash and silica fume through the use of the ASTM C779 test. Their results showed that the SCC with silica fume had the least amount of wear (Turk and Karatas 2011). Yet, the effect of including other types of SCMs (such as metakaolin and slag) on the abrasion resistance of SCC mixtures is lacking. The main objective of this study is to evaluate the abrasion resistance of variable SCC mixtures (with different SCMs) in comparison with normal-vibrated concrete.

2 EXPERIMENTAL PROCEDURE

2.1 Materials Properties and Mixture Design

Six concrete mixtures were developed and tested in this study. Four of the six mixtures were SCC containing fly ash (FA), metakaolin (MK), silica fume (SF), and ground granulated blast furnace slag (SG). The other two tested mixtures were plain SCC and normal concrete (NC), chosen to act as control mixtures. The four mixtures with SCMs had replacement proportions of 30% FA, 20% MK, 8% SF, and 30% SG. These replacement levels were obtained from a previous study (Abouhussien and Hassan 2015) conducted to evaluate the optimum percentage of each of these SCMs in concrete. All tested mixtures had a total cementitious content of 500 kg/m^3. The water-to-binder (W/B) ratio was set at a constant value of 0.4 in this investigation. The coarse-to-fine aggregate (C/F) ratio was 0.7 for the SCC mixtures and 2.0 for the normal-concrete mixture. The normal-concrete mixture was designed with the same binder content and W/B ratio as the other tested SCC mixtures to highlight the influence of increasing the C/F ratio on the abrasion resistance. A high-range water-reducing admixture (HRWRA) was used in the SCC admixtures to maintain proper workability and to create adequate slump flow in the range of ~700 ± 50 mm, as per ASTM C1611 (ASTM 2009). Each of the six tested mixtures used type GU Canadian Portland cement conformed to ASTM Type I with a specific gravity of 3.15 (ASTM 2017). The two types of aggregate used were natural sand and 10-mm coarse aggregate. Both aggregates had a specific gravity of 2.6 and water absorption of 1%. The specific gravities of FA, MK, SF, and SG were 2.38, 2.5, 2.27, and 2.9, respectively. The HRWRA used was similar to that described in ASTM C 494 Type F (ASTM 2013), with a specific gravity of 1.2 and pH of 9.5. Table 1 shows the details of the six mixtures developed in this study.

Table 1: Mixture proportions

Mixture Type	Cement (kg/m^3)	SCM (%)	SCM (kg/m^3)	C/F	W/B	C.A. (kg/m^3)	F.A. (kg/m^3)	Water (kg/m^3)	HRWRA (l/m^3)
NC	500	0	0	2.0	0.4	1111.5	555.8	200	0
SCC	500	0	0	0.7	0.4	686.5	980.8	200	2.37
FA	350	30% FA	150	0.7	0.4	670.0	957.2	200	2.08
MK	400	20% MK	100	0.7	0.4	677.7	968.1	200	5.42
SF	460	8% SF	40	0.7	0.4	681.3	973.2	200	4.17
SG	350	30% SG	150	0.7	0.4	682.1	974.5	200	2.25

2.2 Test Sample Details and Abrasion Test

For each mixture, three cylinders and three prisms were cast. All specimens were cured at 25°C in a moist-curing room for a period of 28 days before testing. The cylinder samples were used to test the compressive strength according to ASTM C39 (ASTM 2011), while the prism samples were cut into three 100-mm cubes for the preparation of the ASTM C944 abrasion test (ASTM 2012d). The abrasion test consisted of a drill press with a chuck capable of holding and rotating the cutter and constantly applying a force of 98 N on the specimen being tested (Figure 1). The abrasion test was set to run for six rounds, one-minute intervals each, with a total abrasion time of six minutes. The weight loss of the specimens was taken after each of the one-minute intervals to calculate the mass loss due to abrasion. In addition to measuring the weight loss, the depth of wear resulting from abrasion was also estimated using electronic calipers. Each mixture has three samples undergo the abrasion and compression testing to estimate the average result.

Figure 1: Abrasion test setup

3 RESULTS AND DISCUSSION

As previously stated, the weight loss for each sample was measured during the abrasion test at one-minute intervals for a total of six minutes. The wear depth was also measured to further show the physical effects of the abrasion on the concrete. The results from this data collection during testing can be seen in Tables 2 and 3. Figure 2 presents the average weight loss plotted against time after each minute for all tested samples. The average weight loss was calculated by taking the average results of the three replicated samples of each mixture. It appears from Figure 2 that the weight loss followed a linear relationship over time. The tested samples lost more weight as the time increased up to six minutes. This linear relationship could be related to that the increase in the depth of wear was more associated with a wear in the fine materials rather than the coarse aggregate particles.

Table 2: Percent loss due to abrasion during testing

Mixture Name	Sample Number	Percent Loss (%)						Average Percent Loss (%)	Standard Deviation of Average Percent Loss (%)
		1 min	2 min	3 min	4 min	5 min	6 min		
NC	S1	0.21	0.35	0.46	0.53	0.64	0.7		
	S2	0.18	0.35	0.43	0.5	0.58	0.63	0.64	0.04
	S3	0.18	0.31	0.41	0.47	0.53	0.6		
SCC	S1	0.17	0.33	0.46	0.59	0.66	0.76		
	S2	0.09	0.14	0.22	0.27	0.35	0.4	0.59	0.15
	S3	0.16	0.32	0.41	0.5	0.54	0.6		
FA	S1	0.11	0.19	0.27	0.36	0.45	0.53		
	S2	0.17	0.32	0.48	0.61	0.72	0.8	0.68	0.11
	S3	0.13	0.26	0.38	0.5	0.61	0.7		
MK	S1	0.08	0.15	0.21	0.27	0.31	0.38		
	S2	0.09	0.18	0.26	0.34	0.42	0.46	0.43	0.04
	S3	0.11	0.18	0.33	0.36	0.39	0.46		
SF	S1	0.11	0.22	0.31	0.38	0.44	0.5		
	S2	0.11	0.19	0.28	0.35	0.41	0.47	0.50	0.02
	S3	0.13	0.24	0.34	0.4	0.47	0.52		
SG	S1	0.12	0.24	0.32	0.4	0.44	0.48		
	S2	0.13	0.31	0.36	0.46	0.51	0.56	0.55	0.05
	S3	0.17	0.31	0.4	0.48	0.52	0.6		

Table 3: Wear depth and compressive strength

Mixture Name	Sample Number	Wear Depth (mm)	Average Wear Depth (mm)	Standard Deviation of Average Wear Depth (mm)	Compressive Strength (MPa)	Average Compressive Strength (MPa)	Standard Deviation of Average Compressive Strength (MPa)
	S1	1.08			52.1		
NC	S2	0.97	0.99	0.07	51.5	52.5	0.97
	S3	0.92			53.8		
	S1	0.80			54.7		
SCC	S2	0.84	0.93	0.16	53.8	54.4	0.40
	S3	1.15			54.6		
	S1	0.77			50.7		
FA	S2	1.44	1.18	0.29	52.3	51.5	0.65
	S3	1.34			51.4		
	S1	0.61			73.2		
MK	S2	0.70	0.60	0.08	72.5	72.6	0.31
	S3	0.50			72.6		
	S1	0.31			69.8		
SF	S2	0.78	0.66	0.25	67.9	68.7	0.80
	S3	0.88			68.4		
	S1	0.94			60.9		
SG	S2	0.83	0.86	0.06	57.3	58.6	1.59
	S3	0.81			57.8		

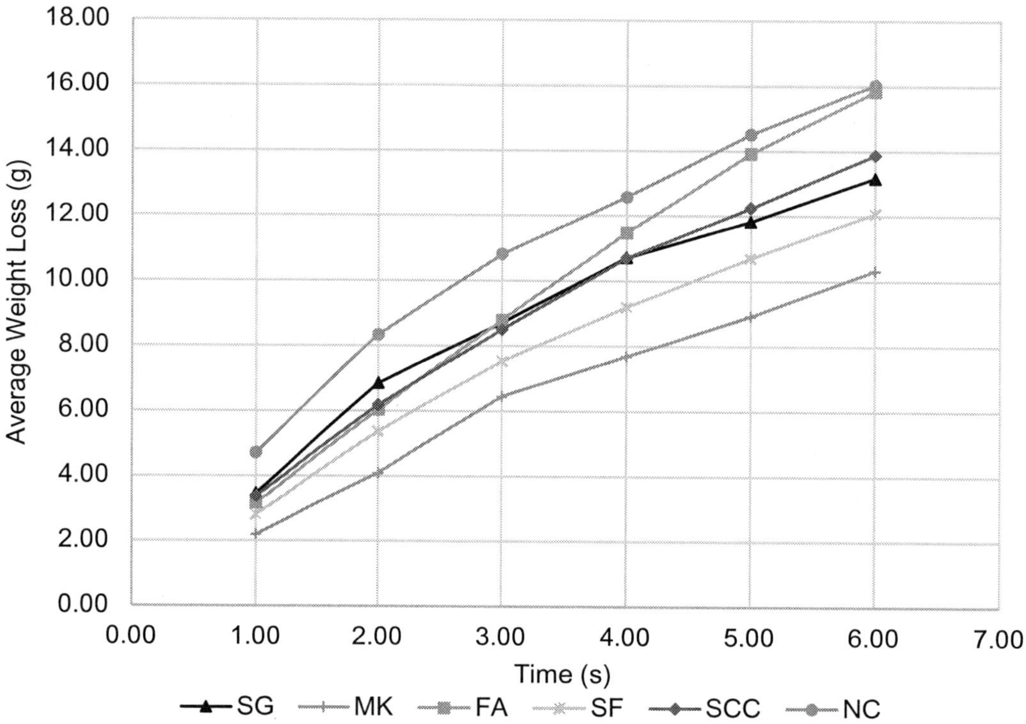

Figure 2: Average weight loss (g) versus time (s)

3.1 Effect of Concrete Type on Abrasion Resistance

Two types of concrete were evaluated in this study: SCC and NC (with a higher C/F ratio). The NC mixture showed an average wear depth of 0.99 mm and average percent loss of 0.64%. Also, the average compressive strength of the NC mixture at 28 days was 52.4 MPa. In comparison, the SCC mixture (without SCMs) had improved results with an average wear depth and percent loss of 0.93 mm and 0.59%, respectively, and an average compressive strength of 54.3 MPa. As seen in Figure 2, above the trend in the SCC mixture line is much lower than the NC mixture line. It should be noted that the percentage of C/F ratio in the mixture plays an important role in concrete abrasion. In this investigation, both SCC and NC mixtures had relatively high cement content (500 kg/m^3) and relatively low W/B ratio (0.4), creating a stronger paste to resist abrasion. And because SCC had higher mortar content (less coarse aggregate), it showed higher resistance to abrasion compared to NC mixtures. This reasoning would only be valid in the case that the strength of the mortar is higher than that of the coarse aggregates. In addition, studies show that optimum compaction is valuable for good abrasion resistance (Papenfus 2003). SCC is inherently more compactable than NC, giving it that added advantage (improved abrasion resistance). It is worth noting that, the increase in the compressive strength of SCC compared to NC mixture could also be another reason of the higher abrasion resistance of SCC over the NC counterpart.

3.2 Effect of SCMs on Abrasion Resistance

Figure 2 shows the average weight loss over time due to abrasion in all tested mixtures. Significant variations in the abrasion damage were observed from this figure among the tested SCC mixtures with different SCMs. The MK mixture proved to have the highest abrasion resistance of all tested mixtures, with a total average percent loss of 0.43% (Table 2). This result was anticipated as the compressive strength is correlated to the abrasion resistance (Scott and Safiuddin 2015); this mixture also had the highest average compressive strength at 28 days (seen in Table 3). Moreover, MK has a very high pozzolanic reactivity, which makes the paste of this mixture much stronger and leads to a higher abrasion resistance. NC and FA mixtures had the least abrasion resistance as they also had the lowest values for average compressive strength. The results of surface wear depth also showed a similar trend of variation as the variation in wear loss. MK samples had the least amount of wear depth (0.60 mm) and FA samples had the most (1.18 mm). The SCC mixtures with SCMs are ranked from the lowest to highest abrasion resistance as follows: FA, SG, SF, and MK. The average percent loss and wear depth due to abrasion of these SCMs mixtures, in that order, were 0.68%, 0.55%, 0.50% and 0.43%, and 1.18, 0.86, 0.66, and 0.6 mm, respectively. In general, the results of SCC with SCMs were significantly better than the NC and SCC control mixtures. These results matched the outcomes of a previous study, which indicated that the addition of silica fume as an SCM to NC and SCC significantly improved its abrasion resistance (Ghafoori and Diawara 1999, Turk and Karatas 2011). The only exception to this rule is the FA mixture, which had higher percentage of weight loss and larger wear depth compared to the NC mixture. However, the compressive strength of FA mixture was slightly lower than NC mixture, which can directly affect the abrasion resistance of a mixture.

4 CONCLUSIONS

The rotating cutter method of abrasion was used to test the abrasion resistance of various concrete mixtures. The types of these tested mixtures were: NC, SCC, and SCC with various SCMs. An extensive analysis of the collected abrasion data was completed post-testing, which determined the effect of concrete type and SCMs on the abrasion resistance. This analysis gave the following conclusions:

- Regarding the effect of concrete type on the materials resistance to abrasion, the control SCC mixture (without SCMs) had a lower average percentage of abrasion mass loss and wear depth as well as an increased compressive strength compared to its NC counterpart.
- Varying the SCM type in the SCC mixtures yielded evident changes in the abrasion resistance as obtained from the tests performed herein. The abrasion resistance of the SCC mixtures containing SCMs can be ranked in order from highest to lowest as MK, SF, SG, and FA.
- The mixture with the highest abrasion resistance overall was the SCC mixture containing 20% MK. This mixture showed the lowest percent loss, wear depth, and highest compressive strength. On the other hand, the SCC mixture with 30% FA exhibited the least performance with respect to the average percent loss and compressive strength.
- This study proved that the usage of SCMs in self-consolidating concrete was beneficial in being capable of resisting abrasion when compared to NC and plain SCC. SCC is an excellent high performing concrete which would be sufficient for use in areas where abrasion damage is a concern.

References

Abouhussien, A.A. and Hassan, A.A.A. 2014. Application of Statistical Analysis for Mixture Design of High-Strength Self-Consolidating Concrete Containing Metakaolin. *Journal of Materials in Civil Engineering*, ASCE, **26**(6): 04014016.

Abouhussien, A.A. and Hassan, A.A.A. 2015. Optimizing the Durability and Service Life of Self-Consolidating Concrete Containing Metakaolin Using Statistical Analysis. *Construction and Building Materials*, **76**: 297-306.

Abouhussien, A.A., Hassan, A.A.A. and Hussein A.A. 2015. Effect of Expanded Slate Aggregate on Fresh Properties and Shear Behaviour of Lightweight SCC Beams. *Magazine of Concrete Research*, **67**(9): 433-442.

Aslani, F. 2013. Effects of Specimen Size and Shape on Compressive and Tensile Strengths of Self-Compacting Concrete with or without Fibres. *Magazine of Concrete Research*, **65**(15): 914-929.

ASTM C1611. 2009. Standard Test Method for Slump Flow of Self-Consolidating Concrete. *ASTM International*, West Conshohocken, PA, USA.

ASTM C39. 2011. Standard Test Method for Compressive Strength of Cylindrical Concrete Specimens. *ASTM International*, West Conshohocken, PA, USA.

ASTM C779. 2012a. Standard Test Method for Abrasion Resistance of Horizontal Concrete Surfaces. *ASTM International*, West Conshohocken, PA, USA.

ASTM C418. 2012b. Standard Test Method for Abrasion Resistance of Concrete by Sandblasting. *ASTM International*, West Conshohocken, PA, USA.

ASTM C1138. 2012c. Standard Test Method for Abrasion Resistance of Concrete (Underwater Method). *ASTM International,* West Conshohocken, PA, USA.

ASTM C944. 2012d. Standard Test Method for Abrasion Resistance of Concrete or Mortar Surfaces by the Rotating-Cutter Method. *ASTM International*, West Conshohocken, PA, USA.

ASTM C494. 2013. Standard Specification for Chemical Admixtures for Concrete. *ASTM International*, West Conshohocken, PA.

ASTM C150. 2017. Standard Specification for Portland Cement. *ASTM International*, West Conshohocken, PA, USA.

Das, D. and Chatterjee, A. 2012. A Comparison of Hardened Properties of Fly-Ash-Based Self-Compacting Concrete and Normally Compacted Concrete under Different Curing Conditions. *Magazine of Concrete Research*, **64**(2): 129-141.

Dong, Q., Wu, H., Huang, B., Shu, X. and Wang, K. 2013. Investigation into Laboratory Abrasion Test Methods for Pervious Concrete. *Journal of Materials in Civil Engineering,* ASCE, **25**(7): 886-892.

Ghafoori, N. and Diawara, H. 1999. Abrasion Resistance of Fine Aggregate Replaced Silica Fume Concrete. *ACI Materials Journal,* **96**(5): 559-569.

Karahan, O., Hossain, K.M.A., Ozbay, E., Lachemi, M. and Sancak, E. 2012. Effect of Metakaolin Content on the Properties Self-Consolidating Lightweight Concrete. *Construction and Building Materials,* **31**: 320-325.

Lachemi, M., Hossain, K.M.A., Patel, R., Shehata, M. and Bouzoubaâ, N. 2007. Influence of Paste/Mortar Rheology on the Flow Characteristics of High-Volume Fly Ash Self-Consolidating Concrete. *Magazine of Concrete Research*, **59**(7): 517-528.

Liu, TC. 1981. Abrasion Resistance of Concrete. *ACI Journal,* **78**(5): 341-350.

Papenfus, N. 2003. Applying Concrete Technology to Abrasion Resistance. *Proceedings of the 7th International Conference on Concrete Block Paving (PAVE AFRICA 2003)*, Sun City, South Africa, ISBN Number: 0-958-46091-4.

Rao, S.V., Rao, S.M.V., Ramaseshu, D. and Kumar, R.P. 2012. Durability Performance of Self-Compacting Concrete. *Magazine of Concrete Research*, **64**(11): 1005-1013.

Safiuddin, M., Salam, M.A. and Jumaat, M.Z. 2012. Flowing Ability of Self-Consolidating Concrete and its Binder Paste Phase Including Palm Oil Fuel Ash. *Magazine of Concrete Research,* **64**(10): 931-944.

Scott, B.D. and Safiuddin, M. 2015. Abrasion Resistance of Concrete – Design, Construction and Case Study. *Concrete Research Letters*, **6**(3): 136-148.

Turk, K. and Karatas, M. 2011. Abrasion Resistance and Mechanical Properties of Self-Compacting Concrete with Different Dosages of Fly Ash/Silica Fume. *Indian Journal of Engineering and Materials Sciences,* **18**: 49-60.

Building Tomorrow's Society

Bâtir la Société de Demain

Fredericton, Canada

June 13 to 16, 2018 / *13 au 16 juin, 2018*

DESIGN AND FORMULATION OF CONCRETE WITH POLYETHYLENE TEREPHTHALATE (PET) USING STATISTICAL MIXTURE DESIGN

Natalia Vivas[1,2], Leonard M. Lye[1], Baiyu Zhang[1]
[1] Faculty of Engineering and Applied Science, Memorial University of Newfoundland, A1B 3X5, St. John's, NL, Canada
[2] Corresponding author: nv6185@mun.ca

Abstract: This paper investigated the incorporation of waste Polyethylene Terephthalate (PET) from plastic bottles into concrete as a replacement for natural fine aggregate and found an optimal combination of components that produces a useful concrete product. Six components were considered: cement, water, coarse aggregate, fine aggregate, superplasticizer, and waste PET. A total of 31 mixes including waste PET were prepared based on a statistical mixture design approach. The responses of these mixtures were workability, compressive strength, and splitting tensile strength. The waste PET was first reduced in volume by shredding and then combined with the rest of the components. The responses from the experiments were statistically analysed and a model fitted to each response. Linear models were found to fit the responses best. Using the desirability function approach, four optimal options were selected and then verified in the lab by comparing the experimental with the predicted values. Except for one, all the values fell within the 95% prediction interval. The average response values obtained with the optimal combination were: (1) compressive strength of 23.8 MPa; (2) slump 123 mm, and (3) splitting tensile strength of 3.33 MPa. This mix can be used in basements foundation walls or slabs, inside buildings not exposed to freezing temperatures. It is recommended that future work should consider method of mixing, time of mixing, volume of mix, curing conditions, and other responses to further understand the characteristics of incorporating waste PET into concrete.

Keywords: Polyethylene Terephthalate (PET), concrete aggregate, statistical mixture design.

1 INTRODUCTION

Polyethylene Terephthalate (PET) is one of the main fractions of the plastic waste stream (Silva et al., 2013). PET is mostly employed as a multi-purpose plastic for bottled water, soft drinks, and as single-use packaging material (Andrady 2015). Products made of PET are generally large in volume and can take approximately one thousand years to decompose under natural environmental conditions (Silva, et al. 2013; de Brito and Saikia 2013).

Incorporating waste PET into the concrete industry is an innovative approach that has been seen as promising in recent research (Gu and Ozbakkaloglu 2016; Ge et al. 2014; de Brito and Saikia 2013). It only requires shredding the waste PET into small particles and adding the shredded particles into the natural aggregate mixture (sand, gravel or crushed stone). Using waste PET as a natural aggregate replacement has two important benefits: the waste PET products that occupy an enormous volume in the waste stream can be dramatically reduced by shredding and disposed of; and natural aggregate can be partially replaced, reducing the impact to natural resource availability. Thus, an important reduction in environmental impact of waste disposal while saving natural resources and energy consumption can be achieved (Gu and Ozbakkaloglu 2016; Frigione 2010). Table 2 summarizes the past research on waste PET incorporation into concrete. The main aspects considered in each study were reviewed. The values achieved in the properties and the tests performed in each study were reviewed with special focus on workability, compressive strength, and splitting tensile strength.

One of the goals of this research is to produce a practical and useful concrete mixture containing waste PET. Thus, the performance criterion was oriented to achieve workability and compressive strength requirements for different applications. Only mixes containing waste PET were prepared in order to focus this research on the comparison among different waste PET percentages. According to Beall (2001), the minimal requirements for compressive strength range between 17.5 and 24.5 MPa. Thus, the optimization will be oriented to maximize the compressive strength as much as possible. Additionally, workability was set to range between 25 and 127 mm according to typical values of workability based on (Beall 2001) and (Mehta and Monteiro 2014).

2 EXPERIMENTAL APPROACH

In the present study, the shredder used was built by MUN Technical Services using plans from an international plastic recycling organization called Precious Plastic (https://preciousplastic.com). This organization is devoted to increasing knowledge about plastic recycling worldwide. A complete guide to building a pilot plastic shredder machine is available on the website.

The waste PET was collected from the waste stream at the St. John's recycling centre. The collected waste PET was mostly water and soft drink bottles. Three main types of waste PET were identified based on the volume and thicknesses of the bottles as shown in Table 3. The bottles were unwashed and not separated by color. Only the labels and the lids were manually removed before shredding. Based on past research, the incorporation of waste PET as a fine aggregate produced more advantages than the incorporation of waste PET as coarse aggregate. Thus, the screen of the shredder was selected to generate as fine a particle as possible. The shredded particles were then reprocessed. It is important to note that, shredding waste PET to produce fine particles consumes higher energy and time than shredding waste PET into coarser particles.

2.1 Reduction in volume of waste PET

The volume obtained after shredding the waste PET was measured and compared with the original volume. The bulk density was measured for all types of waste PET according to ASTM standards. Table 2 shows the results for the reduction of volume after shredding the different types of bottles. The reduction of volume of the waste PET bottles was 29 fold or 96.6% for Type 1, 23 fold or 95.6% for Type 2, and 10 fold or 90% for Type 3. This shows that just shredding the bottles alone provide tremendous savings in landfill or storage space. See Figure 1.

2.2 Grading of waste PET

After shredding, sieve analysis was performed to determine the size and grading of the waste PET particles. The waste PET was mixed for all the experiments using the following proportions: 70% Type 1, 20% Type 2, and 10% Type 3. The waste PET gradation was deficient in particle sizes lower than 1 mm and had a high percentage of particles ranging between 4.75 and 9.5 mm. Figure 2 shows the final grading of the aggregates used in all the experiments.

Table 1: Past studies on waste PET incorporation into concrete.

Reference	Recycling method	Gradation / particle shape	Replacement	Admixtures	Workability	Compr. strength 28 days (MPa)	Splitting tensile (MPa)	Observations/ Other properties
Choi et al. (2005)	Mechanical/ Thermal	5 - 15 mm / rounded	25%, 50%, 75% V	Granulated blast furnace	Slump 100 - 205 mm	21.8 - 37.2	1.94 - 3.32 MPa	Modulus of elasticity (15.6 - 25.5 GPa)
Juki et al (2013)	Mechanical	5 mm / flaky	25%, 50%, 75% V	-	-	15.6 - 31.27	Reductions from 15 to 60% compared to a normal blend	Modulus of elasticity (10.4 - 25.9 GPa)
Choi et al. (2009)	Mechanical/ Thermal	5 - 15 mm / rounded	25%, 50%, 75% V	Water reducer	Slump 100 - 222 mm	21 - 35	1.9 - 3.2	Modulus of elasticity (18 - 30 GPa)
Albano et al. (2009)	Mechanical	Fine 0.26 cm / flaky Coarse 1.14 cm / flaky Mix 50% each one / flaky	10%, 20% V	-	Slump 20 - 90 mm	12 - 27	1.4 - 2.8	Modulus of elasticity (12 - 29 GPa)
Frigione (2010)	Mechanical	<2 mm / flaky	5% W	-	Vebe 37 - 62	40 - 69.7	4.1 - 6.3	Shrinkage 1 year (650 - 987 10^{-6})
Ackaozoglu et al. (2010)	Mechanical-washing	0-4mm / flaky	50% V	Granulated blast furnace	-	22.4 - 27	-	Water absorption of concrete with WPET (11.9 - 22%)
Silva et al. (2013)	Mechanical/ Thermal	Fine 4mm / flaky Coarse 2 - 11.2mm / flaky Pellet 1-4 mm	7.5%, 15% V	-	Slump 133 - 141 mm	19.7 - 36.7	-	Carbonation depth (14 -28.8 mm)
Ferreira et al. (2012)	Mechanical/ Thermal	Fine 4mm, flaky Coarse 2 - 11.2mm, flaky Pellet 1- 4 mm	7.5%, 15% V	-	Slump 120 - 140 mm	22 - 38	1.5 - 3.4	Modulus of elasticity (17- 38 GPa)
Ismail and Al-Hashmi (2008)	Mechanical	0.15 - 4.75 mm / flaky	10%, 15%, 20% V	-	Slump 20 - 80 mm	22 - 43	-	Flexural strength (3 - 6 MPa)
Batayneh et al. (2007)	Thermal	0.15 -4.75 mm / flaky	5%, 10%, 20% V	-	Slump 57 - 78 mm	10 - 34	0.6 - 4	Flexural strength (0.6 - 5 MPa)
This study (2018)	Mechanical	0.15 -4.75 mm / flaky	8.4-17%V	Super plasticizer	Slump 51- 164 mm	7.4 - 27.8	1.4 - 3.2	Used statistical mixture design.

Table 2: Reduction in volume of waste PET

Type of bottle	Thickness (mm)	Weight of 1 bottle (g)	Initial volume (ml)	Bulk density (g/ml)	Final volume of 1 bottle (ml)	Reduction in volume
1	0.10	6	500	0.34	17	29x (96.6%)
2	0.25	30	2000	0.34	88	23x (95.6%)
3	0.40	20	591	0.33	60	10x (90.0%)

(a) (b) (c)

Figure 1: Initial and final volume of the bottles used in the experiments. (a) Type 1, bottle of 500 ml and 0.1 mm of thickness; (b) Type 2, bottle of 2000 ml and 0.25 mm of thickness and (c) Type 3, bottle 591 ml and 0.40 mm of thickness.

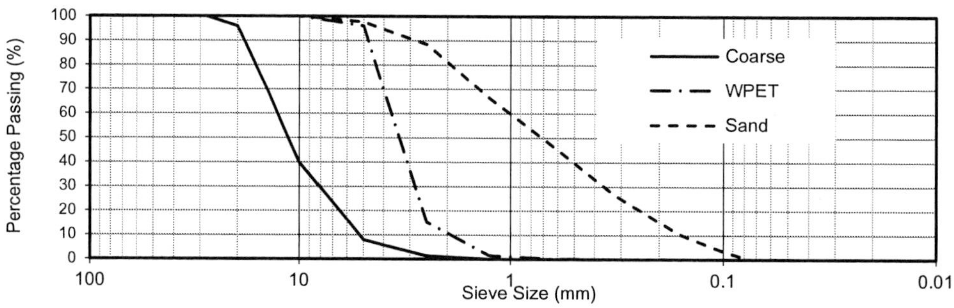

Figure 2: Comparative sieving analysis of sand, coarse aggregates, and waste PET

2.3 Ranges of the components of the mixture

Based on past research the proportions for all the components of the concrete mixture are defined. Table 3 shows the summary of components and their proportions used in the present study. The table shows the proportions in mass for practical calculations and the proportions in volume according to the design requirements.

Table 3: Summary of the proportions of the components in mass and volume.

Component	Mass fraction (kg/m^3) Low	Mass fraction (kg/m^3) High	Volumetric fraction (m^3) Low	Volumetric fraction (m^3) High
Cement (A)	306	365	0.097	0.116
Water (B)	144	177	0.144	0.177
Coarse (C)	1153	1257	0.44	0.48
Fine (D)	506	653	0.193	0.249
Waste PET (E)	24	53	0.018	0.04
Superplasticizer (F)	3	14	0.003	0.014

3 EXPERIMENTAL DESIGN

In mixture design, the variables to be modified are considered ingredients or components and the measured response depends on the proportion of each component (Myers and Montgomery, 2009; Cornell, 2002). When a proportion of a component increases, the proportion of some of the other components should decrease (Myers and Montgomery, 2009). Cornell (2002) also states that a mixture design is considered constrained when the components of a mixture have additional restrictions, such as maximum or minimum limits. The constrained mixture design is appropriate for concrete mix preparations and the responses can be modelled and predicted using a mixture polynomial model which takes the inherent constraint into account (Simon et al. 1997; Kharazi et al. 2013).

3.1 Mixture (Scheffé) polynomials for mixture designs

Mixture polynomial models are also known as Scheffe polynomials after the author who developed them. These models do not have an intercept and quadratic terms from an ordinary polynomial are rolled into the interaction term to create a mixture quadratic term which is a measure of non-linear blending (Cornell 2002). In this study it is assumed that the responses might follow quadratic models, and the mixture polynomial for q components is given by:

[1] $$\hat{y} = \sum_{i=1}^{q} \beta_i x_i + \sum_{i<j}^{q-1} \sum_{j}^{q} \beta_{ij} x_i x_j$$

Where: q is the number of components, and $\hat{y}$ is the predicted response. An optimal IV design was used to obtain the experimental run combinations for the six-component quadratic model. The IV-optimal design seeks to minimize the integral of the prediction variance across the design space which will model the true response surface with greater precision (Myers and Montgomery 2009). Twenty-one (21) experiments would be necessary to determine all the coefficients for a mixture with six components. Additionally, three extra points are necessary to provide information about the error and lack of fit, and three replicated points was included to better understand the behaviour of the data due to repeatability. Finally, four additional center points were added to monitor the process. Hence a total of 31 run combinations were used. Design-Expert V10 from Statease Inc. (2017) is used for both experimental design and subsequent analysis of the results.

4 ANALYSIS OF RESULTS

Table 5 shows the combinations of components on a mass basis and the obtained results for slump, compressive and splitting tensile strengths. The statistical analysis was performed on the data for compressive strength as shown in the Table 4.

Table 4: Analysis of variance for compressive strength

Source	Sum of squares	Degree of freedom	Mean square	F-value	p-value prob> F
Linear vs. Mean	473.11	5	94.62	17.52	< 0.0001
Quadratic Vs. linear	112.91	15	7.53	3.40	0.0280
Sp. Cubic vs. Quadratic	5.89	4	1.47	0.55	0.7100
Residual	16.22	6	2.70		

The level of significance selected for this design was 0.05. Table 6 shows that the linear model presented a p-value < 0.0001. Thus, the null hypothesis (H_0) was rejected and the data of the experiments on compressive strength fitted a linear model. The components A (cement), B (water), C (Coarse aggregate), D (Fine aggregate), E (waste PET), and F (superplasticizer) significantly influenced the compressive strength. On the other hand, the quadratic model has p-value of 0.028 that is also lower 0.05. However it will be shown that the linear model is the correct choice. Once the linear model was fitted to the data, lack of fit test showed that there is no statistically significant lack of fit which meant that the residual error did not exceed the pure error. Therefore, the model was adequate to fit the data from the experiments. The adequacy of the model was next evaluated by the adjusted R-squared and predicted R-squared shown in Table 7. The adjusted R-squared represents the variation of the rest of experiments compared with the mean, while the predicted R-squared explains the accuracy on the predictions of the model (Myers and Montgomery, 2009).

Table 5. Mass fraction of the components and results of the experiments.

Run	Cement (Kg/m³)	Water (Kg/m³)	Coarse (Kg/m³)	Fine (Kg/m³)	WPET (Kg/m³)	Superp (Kg/m³)	w/c	a/c	% V WPET	Slump (mm)	Comp. Str (MPa)	Split. Tens (MPa)
1	365	157	1251	544	37	14	0.43	5.0	11.9	128	17.5	3.0
2	320	176	1151	615	51	10	0.55	5.7	13.9	146	15.1	2.2
3	306	153	1231	649	37	4	0.50	6.3	10.2	102	12	1.5
4	364	157	1211	630	29	3	0.43	5.1	8.4	121	27.8	3.2
5	338	165	1217	581	44	8	0.49	5.4	12.9	87	12.9	2.3
6	365	177	1195	573	26	12	0.48	4.9	8.4	154	23.6	3.1
7	365	177	1256	528	29	3	0.48	5.0	9.8	136	21.89	3.2
8	306	166	1203	635	42	3	0.54	6.1	11.7	128	8.8	1.4
9	338	165	1217	581	44	8	0.49	5.4	12.9	116	11.6	1.7
10	334	144	1196	646	48	10	0.43	5.7	12.7	67	19.3	2.7
11	335	175	1256	513	39	14	0.52	5.4	12.9	142	13.5	2.4
12	332	160	1198	624	34	14	0.48	5.6	9.5	140	12.39	2.7
13	306	168	1214	601	38	14	0.55	6.1	10.9	129	7.46	2.1
14	365	177	1155	592	48	3	0.48	4.9	13.7	140	19.05	2.7
15	338	165	1217	581	44	8	0.49	5.4	12.9	115	16.05	2.4
16	306	165	1169	633	48	14	0.54	6.0	12.9	121	10.1	1.5
17	324	177	1193	625	30	3	0.55	5.7	8.4	155	18.8	2.7
18	328	150	1241	622	28	14	0.46	5.8	8.1	164	15.45	2.8
19	335	175	1256	513	39	14	0.52	5.4	12.9	149	11.5	2.2
20	326	158	1198	594	53	14	0.48	5.7	15.0	99	14.4	2.1
21	359	164	1151	619	43	14	0.46	5.1	11.9	121	19.06	2.9
22	365	164	1244	519	53	7	0.45	5.0	16.8	51	16.6	2.9
23	334	144	1256	600	50	3	0.43	5.7	14.2	84	13.5	1.9
24	365	164	1244	519	53	7	0.45	5.0	16.8	94	17.3	2.7
25	365	177	1183	528	53	14	0.48	4.8	16.5	116	16.3	2.6
26	314	173	1256	575	26	8	0.55	5.9	8.3	154	13.5	1.6
27	338	165	1217	581	44	8	0.49	5.	12.9	117	15.6	2.5
28	365	177	1183	528	53	14	0.48	4.8	16.5	97	15.6	2.6
29	306	157	1256	558	53	13	0.51	6.1	15.8	127	10.65	1.1
30	322	177	1234	540	53	3	0.55	5.7	16.3	103	8.03	1.5
31	328	155	1182	645	53	3	0.47	5.7	14.0	84	16.69	2.1

Table 6: Adjusted R-squared, Predicted R-squared, and PRESS of compressive strength.

Source	Std. Dev	R-squared	Adjusted R-squared	Predicted R-squared	PRESS
Linear	2.32	0.778	0.7336	0.6530	211.01
Quadratic	1.49	0.9636	0.8909	-0.6567	995.33

Table 6 shows that the predicted R-squared value was 0.65 for the linear model whereas it was negative for the quadratic model. A negative predicted R-squared means that the model is no better than using the mean value. Hence the choice of the linear model is the correct choice. All the assumptions of ANOVA were also fulfilled. The linear model was established and the coefficients of the components for the linear predictive equation were defined. These coefficients represent the influence of each component on the compressive strength. The predictive equation of the linear model in real values is shown in Equation [2]:

$$[2]\ Compressive\ strength\ (MPa) = +535.96(A) - 72.61\ (B) - 65.16(C) + 39.76\ (D) - 225.38(E) - 176.51\ (F)$$

Analyzing the predictive equation with real values of compressive strength, the components that influence the most the compressive strength can be identified. The cement (A), waste PET (E), and superplasticizer (F) highly influence the compressive strength. A moderate variation in one of these components will significantly impact the compressive strength outcome. The positive sign in cement content (A) indicates that increasing the cement content will increase compressive strength, whereas the negative sign in the remaining components indicates that increasing these components decrease compressive strength. The same analysis was applied for each one of other properties. All the properties fitted linear models. Table 7 shows the equation for each property.

Table 7: Summary of the predictive models.

Property	Predictive model (real values)	R-squared adjusted	R-squared predicted
Compressive strength	$Comp.strength\ (MPa) = +535.96(A) - 72.61\ (B) - 65.16(C) + 39.76\ (D) - 225.38(E) - 176.51\ (F)$	0.73	0.65
Slump	$Slump\ (mm) = -755.04(A) + 1310.60\ (B) - 4.61(C) + 208.15(D) - 2360.66(E) + 1636.75\ (F)$	0.64	0.52
Splitting tensile strength	$Splitting\ tensile\ (MPa) = +72.59\ (A) - 7.52\ (B) - 7.78\ (C) + 1.37\ (D) - 31.06(E) + 12.2\ (F)$	0.79	0.73

5 NUMERICAL OPTIMIZATION

In the numerical optimization, using the desirability function approach, the components or properties can be optimized. The desirability function allows the experimenter to maximize, minimize or keep within target of the goals. The importance of each goal can also be established. By setting a higher importance, the goal is prioritized over the other goals (Anderson and Whitcomb, 2005). Once the goals and the importance are selected, the desirability function analyses all the possible sets of components that achieve the goals, and ranks them from 0 to 1. This is used to rank the options based on their desirability. Finally, the experimenter tests the predicted combination and verifies the results against the prediction intervals (Anderson and Whitcomb, 2005).

On the basis of the previously mentioned goals, the desirability function proposed a set of optimized options. Four options were selected from the list and further tested. Some components such as cement, waste PET, and superplasticizer content produced a strong influence over the compressive strength. Thus, their proportions were very similar in all the optimization options. Components such as water, coarse aggregate, and fine aggregate had slight variations in the optimized options. Table 8 shows the combination of components proposed by the desirability function to be the most convenient to reach the desired properties.

Table 8: Optimization options ranked based on the desirability function.

Number	Cement	Water	Coarse	Fine	PET	Plasticizer	Slump	Comp 28 days	Splitting 28 days	Desirability
1	0.116	0.157	0.461	0.241	0.022	0.003	120.615	24.860	3.352	0.915
2	0.116	0.160	0.459	0.240	0.022	0.003	124.107	24.759	3.346	0.912
3	0.116	0.163	0.456	0.239	0.022	0.003	126.999	24.527	3.324	0.905
4	0.116	0.157	0.458	0.241	0.024	0.003	115.165	24.489	3.298	0.903

5.1 Verification

According to the combinations proposed by the numerical optimization, the mixes were tested in the lab and the results were compared with the prediction intervals for all the selected options. The tests were performed in the same fashion as the 31 previous mixes.

5.1.1 Verification option 1

Table 9. Components for optimization option 1.

Component	Value m^3	Value mass kg/m^3	Lower limit	Upper limit
(A) Cement (m^3)	0.116	365.4	0.097	0.120
(B) Water (m^3)	0.157	157.0	0.14	0.18
(C) Coarse (m^3)	0.461	1209.0	0.44	0.48
(D) Fine (m^3)	0.241	631.0	0.19	0.25
(E) PET (m^3)	0.022	29.3	0.018	0.040
(F) Superplasticizer (m^3)	0.003	3.0	0.003	0.014

The prediction in option 1 had good agreement with the laboratory tests. The experimental values of the properties fell into the 95% prediction interval. The values of slump (123 mm) and compressive strength (23.8 MPa) are suitable values for the concrete utilization. All four optimization options were tested in the lab, compared with the predicted values and the optimization goals. The option that fulfils both, the workability and compressive strength requirements was option 1. It is important to note that option 4 obtained a higher compressive strength but did not satisfied the workability requirements. Thus, option 1 was selected as the most suitable combination as shown in Table 10.

Table 10. Optimization option 1. Predicted results vs. Experimental results.

Response	Predicted value	Experimental value	Standard deviation	95% Prediction interval	
				Lower limit	Upper limit
Slump (mm)	120.6	123.0	16.337	82.33	158.90
Compressive strength (MPa)	24.9	23.8	2.324	19.41	30.31
Splitting tensile (MPa)	3.4	3.3	0.270	2.72	3.99

6 CONCLUSIONS

The statistical mixture design approach is shown to be a useful and practical tool for the examination of the influences among components in mixtures. The components with the largest influence on the responses were cement, waste PET, and superplasticizer. Based on the predictive equations, high cement contents had a positive influence on compressive strength and splitting tensile strength, as expected and high waste PET contents decreased the measured properties. High superplasticizer contents had a large positive influence on slump of the mixtures. Finally, through multi-objective optimization, the optimal combinations of components were found. The optimized combination with highest compressive strength was: (1) cement= 365 kg/m^3 (2) water=157 kg/ m^3 (3) coarse aggregate= 1209 kg/m^3 (4) fine aggregate= 631 kg/m^3 and (5) waste PET= 29 kg/m^3 The response values achieved by this combination were: (1) compressive strength 23.8 MPa; (2) slump 123 mm; and (3) splitting tensile of 3.33 MPa. This mix can be used in basements foundation walls or slabs, inside buildings not exposed to freezing temperatures. The amount of fine natural aggregate replaced by waste PET was 8.4%, 58 kg/m^3. Additionally, when substituting this natural aggregate 3755 bottles of waste PET can be recycled. Although the incorporation of waste PET into concrete reached the standards for the properties proposed in this study, additional tests, such as chloride penetration, column leach test, and thermal properties, should be tested to determine the feasibility of the use of waste PET as a fine aggregate substitute.

The present study found that all the properties followed a linear model. If this is known ahead of time, the number of experiments required would be about 15 saving resources and time to investigate other factors and responses.

Acknowledgements

Special thanks to Memorial University of Newfoundland for partially funded this project. We also thank the City of St. John's for providing the waste PET for this project and to MUN Technical Services, especially Billy Bidgood for fabricating the shredding machine. The help of Dr. Amgad Hussein who provided advice on concrete mix design and experimental materials is also gratefully acknowledged.

References

Akçaözoğlu, S., Atiş, C., Akçaözoğlu, K. 2010. An investigation on the use of shredded waste PET bottles as aggregate in lightweight concrete. *Waste Management*, 30, 285-290.

Albano, C., Camacho, N., Hernández, M., Matheus, A., Gutiérrez, A. 2009. Influence of content and particle size of waste PET bottles on concrete behaviour at different w/c ratios. *Waste Management*, 29, 2707-2716.

Anderson, M.J., & Whitcomb, P.J. 2005. RSM simplified: optimizing process using response surface methods for design of experiments. Productivity Press.

Andrady, A. 2015. Plastics and environmental sustainability. Wiley and sons. New Jersey, NY.

Batayneh, M., Iqbal, M., Asi, I. 2007. Use of selected waste materials in concrete mixes. *Waste Management.* 27. 1870–1876.

Beall, C. 2001. Masonry and concrete. McGraw Hill Professional. New York, NY.

Choi, Y., Moon, D., Kim, Y., Lachemi, M. 2009. Characteristics of mortar and concrete containing fine aggregate manufactured from recycled waste polyethylene terephthalate bottles. *Construction and Building Materials*, 23, 2829-2835.

Choi, Y. W., Moon D.J., Chung J. S., Cho, S. 2005. Effects of waste PET bottles aggregate on the properties of concrete. *Cement and Concrete Research.* 35, 776-781.

Cornell, J. A. 2002. Experiments with mixtures (4th ed.), Wiley & Sons, New York, NY.

de Brito, J., Saikia, N. 2013. Recycled aggregate in concrete, use of industrial, construction and demolition waste. Springer-Verlag London.

Ferreira, L., de Brito, J., Saikia, N. 2012. Influence of curing conditions on the mechanical performance of concrete containing recycled plastic aggregate. *Construction and Building Materials*, 36, 196-204.

Frigione, Mariaenrica. 2010. Recycling of PET bottles as fine aggregate in concrete, Waste Management, Vol. 30, pp. 1101-1106.

Ge, Z., Huang, D., Sun, R., Gao, Z. 2014. Properties of plastic mortar made with recycled polyethylene terephthalate. *Construction and Building Materials,* 73, 682-687.

Ge, Z., Sun, R., Zhang, K., Gao, Z., Li, P. 2013. Physical and mechanical properties of mortar using waste polyethylene terephthalate bottles. *Construction and Building Materials*, 44, 81-86.

Gu, L., & Ozbakkaloglu, T. 2016. Use of recycled plastics in concrete: a critical review. *Waste Management.* 51, 19-42.

Precious plastic – build the shredder. Retrieved from: https://preciousplastic.com/en/videos/build/shredder.html

Ismail, Z. & Al-Hashmi, E. 2007. Use of waste plastic in concrete mixture as aggregate replacement. *Waste management*, 28. 2041-2047.

Juki, M. I., Khairunnisa, M., Mohd, K., Annas, M., Koh, H.B., Norzila, O. , Aeslina, A. K., Muhammad Roslan, A., Faisal, S. K. 2013. Development of concrete mix design nomograph containing polyethylene terephthalate (PET) as fine aggregate. *Advance Materials Research.* 701. 12-16.

Kharazi, M., Lye, L. M., and Hussein, A. 2013. Designing and optimizing of concrete mix proportions using statistical mixture design methodology. Proceedings Canadian Society for Civil Engineering Annual Conference, Montrel. Quebec, GEN-71, 1-10.

Myers, R., Montgomery, D., Anderson-Cook, C. 2009. Response surface methodology, process and product optimization using design of experiments. (3rd ed.), Wiley & Sons, New York, NY.

Mehta, P. K., Monteiro, J. M. 2014. Concrete: microstructure, properties, and materials, (4th ed.). Proportioning Concrete Mixtures, McGraw-Hill Professional. New York, NY.

Silva R.V., de Brito J., Saikia N. 2013. Influence of curing conditions on the durability-related performance of concrete made with selected plastic waste aggregates. *Cement & Concrete Composites*, 35, 23-31.

Simon, M. J., Lagergren E.S., and Snyder, K.A. 1997. Concrete mixture optimization using statistical mixture design methods. *International Symposium on High Performance Concrete*, New Orleans.

Statease Inc. 2017. Design Expert (version 10) (software). Minneapolis. MN.

Building Tomorrow's Society

Bâtir la Société de Demain

Fredericton, Canada

June 13 to 16, 2018 / *13 au 16 juin, 2018*

DURABILITY OF ADVANCED CEMENT-BASED MATERIALS USED AS REPAIRS FOR DETERIORATED CONCRETE

Tahmureszadeh, Keikhosrow[1,3], Shehata, Medhat[2]
[1] Ryerson University, Canada
[2] Ryerson University, Canada
[3] ktahmure@ryerson.ca (corresponding author email)

Abstract: With advanced technology and knowledge in the concrete industry, production of strong and durable construction material becomes more feasible. However, use of such materials as repairs on top of existing concrete (substrate) is challenging. There are limited research on the durability and sustainability of bond strength under cycles of freezing and thawing in the presence of de-icing salt. Hence, this research is designed to evaluate and compare the bond strength of two repair materials – commercial specialized repair material with self-consolidating properties (SRM) and normal conventional concrete (NC) – casted on top of NC substrate. In order to achieve this objective, the study adopts three different evaluating test methods (pull-off, slant shear, and splitting tensile) with different surface texture (rough and smooth) through four cycles of freezing and thawing with de-icing salt (0, 5, 15, 25). The results showed that although the bond strength for SRM initiated at the higher value for all the evaluating tests, its value started to decline significantly compared to the NC under the effect of freezing and thawing in the presence of de-icing salt. Running statistical T-test analysis demonstrated that the difference between rough and smooth surface is statistically significant with the rough surface showing higher bond strength in all.

1 INTRODUCTION

Many aged concrete buildings and infrastructures in North America are suffering severe deterioration due to external causes like aggressive environmental exposure. These types of structures are in high demand for sustainable and long-lasting repair materials. Nowadays, with advanced technology, production of strong and durable construction material becomes more feasible. Therefore, there is more motivation in the industry to replace the normal conventional concrete (NC) with new advanced cement-based materials although NC for several decades was the primary selection for restoration of deteriorated structures. The main reason behind this motivation is that NC did not perform satisfactorily in some conditions mainly due to its pores structure such as harsh environmental exposure – e.g. freezing and thawing cycles with or without de-icing salt (Cai and Liu 1998; Hanjari, Utgenannt, and Lundgren 2011). However, the use of advanced cement-based materials as repair on top of existing deteriorated concrete (substrate) is challenging, since repair relies not only on the quality of the material but also on the interaction and compatibility of such material with the substrate. Various factors directly influence the behaviour of bond interface strength between repair product and substrate such as material characteristics of repair products, substrate surface texture, the moisture content of substrate at the time of applying repair, type of stress acting on the bond interface, and the conditions of specimens.

Mechanical, chemical, electromechanical, and permeability characteristics of the new advanced cement-based material (repair overlay) compare to the old concrete (substrate) are significantly essential in the selection of repair. Properties such as level of alkalinity, electrical resistivity, compressive and tensile strength, compressive strength, modulus of elasticity and rupture, drying shrinkage, thermal movement, creep, etc., directly affect the behaviour of repair products and the bond interface strength between repair and substrate (Emmons, Vaysburd, and McDonald 1993).

Substrate preparation in terms of surface texture, moisture content, and deterioration conditions before and at the time of applying repair is another major factor that directly influences the bond interface strength and durability of repair product. Increasing the substrate surface texture up to certain degree was proven by most of the researchers that have a direct positive impact on improving the bond interface performance. This bond strength enhancement is mainly due to increasing the mechanical interlock, enlarging the contact area of the repair product with a substrate, and providing the repair material with a higher amount of voids to penetrate and bond (Wang, Xu, and Liu 2016; Duarte, Nuno, and Santos 2011; Courad 2000). On the other hand, some researchers believe that the bond strength is not significantly affected by the surface roughness. They claim that preparation of substrate surface texture in most cases could damage the substrate and produce micro-cracks which reduce the bond performance overall (Lukovic et al. 2013).

The second significant parameter in the surface preparation is the moisture content of the substrate at the time of applying repair. Shin and Wan (2010) examined the bond performance of new and old concretes under two different substrate conditions: (i) dry and (ii) saturated surface dry (SSD). They concluded that substrate with SSD conditions produces almost twice bond strength compared to the substrate in dry condition. In the case of dry substrate, part of mixing water which was optimized for the repair mix will be taken by the substrate, reducing the level of hydration. On the other hand, wet substrate is not recommended for two main reasons: (a) the capillary voids are filled with water; therefore, the hydration products of repair materials will be significantly reduced at bond interface, and (b) the extra water available at the surface of substrate increases the water to cement ratio (W/C) of the adjacent repair products at bond interface, as a result reduce the bond strength (Austin and Robins 1995). In an effort to mimic field conditions, substrate conditioning in the laboratory is another important aspect of substrate preparation.

There are different types of tests available to measure the bond strength between repair and substrate. The tests are divided into three main categories based on the type of stress acting on the specimen by Saucier et al. (1991): (1) direct tensile stress (e.g., pull-off test), (2) indirect tensile stress such as splitting tensile method, and (3) combine shear and compression stress like slant shear method. It is important to mention that the results measured by each of these tests might be completely different from one to another due to the nature of acting force and geometry of samples (Momayez et al. 2005). It was reported by Momayez et al. (2005) that the slant shear and pull-off test provide the highest and lowest bond interface strength, respectively.

In the rehabilitation and repair of deteriorated concrete structures, the bond between the substrate (old concrete) and new repair is most of the time the weakest interface in the structure. Therefore, it requires a considerable attention and testing (Momayez et al. 2005). This paper aims to provide investigation and comparison on durability and sustainability of the bond interface behaviour of two different repair materials – commercial specialized repair material with self-consolidating properties (SRM) and normal conventional concrete (NC) – under freezing and thawing cycles in the presence of de-icing salt. The three different assessment testing methods explained above were adopted to better evaluate the bond interface strength between new and old concrete.

2 MATERIALS AND EXPERIMENTAL DETAILS

The experimental program was divided into two major phases: (I) mechanical characteristics, and (II) bond strength. Phase-I examines and compares the mechanical characteristics of two repair materials (SRM and NC) including compressive strength, and static modulus of elasticity to have a basic understanding of the repair products compatibility with their substrate. The second major phase (Phase-II) was designed to study the bond strength behaviour of repair materials with the substrate under cycles of freezing and thawing in the present of de-icing salt. This phase represents one of the main sources of deterioration in Canada which includes exposure to severe weather in winter seasons.

2.1 Materials

2.1.1 Concrete Substrate

Normal conventional concrete mix was selected for substrate with the class exposure of C2 (non-structurally reinforced concrete exposed to chloride and freeze-thaw cycles), nominal maximum size of coarse aggregate 20 mm crushed limestone, design slum of 75 mm, cement type GU (Table 3), W/C of 0.42, air category 1 (5-8% air entrainment), and minimum specified compressive strength of 32 MPa (CSA A23.1/CSA A23.2 2014).

2.1.2 Repair Products

Normal conventional concrete (NC): The similar mix design used for substrate was implemented for this repair. The only change was through the nominal maximum size of coarse aggregate (NMSA) which was reduced from 20 mm to 16 mm. CSA 23.1 (2014) defines the allowable NMSA with certain limitation. As it was stated by the CSA 23.1 the NMSA should be less than one-third of the repair thickness. The thickness of repairs was selected to be 50 mm; therefore, 20 mm is higher than the limitation (one-third of 50 equals to 16.67). To avoid this problem, the aggregates passing sieve 16 mm were used. The aggregate properties and the mix proportions of concrete substrate and NC are summarized in Tables 1 and 2, respectively. Table 3 lists the chemical composition of General Use Portland cement used in this study (GU PC).

Commercial specialized repair material with self-consolidating properties (SRM): the SRM was selected to have a comparison between the repair materials under research and available repair product in the market. The NMSA was reported by the commercial company as 10 mm.

Table 1: Aggregate properties

Mixture ID	Coarse Aggregate					Fine Aggregate				
	NMSA[1] (mm)	Abs[2] (%)	MC[3] (%)	SG[4] Dry	DRD[5] (kg/m^3)	NMSA (mm)	Abs (%)	MC (%)	SG Dry	FM[6]
Substrate	20	0.81	0.07	2.73	1623.54	5	1.30	0.25	2.54	2.8
NC	16	0.81	0.08	2.73	1581.82	5	1.30	0.28	2.54	2.8

[1] Nominal Maximum Size of Aggregate – [2] Absorption – [3] Moisture Content – [4] Specific Gravity – [5] Dry Rodded Density – [6] Finesse Modulus

Table 2: Mix design properties (kg/m3)

Mixture ID	W/C	Cement	Water	Coarse Aggregate	Fine Aggregate	Air[1] (%)
Substrate	0.42	440	184	1023	584	7
NC	0.42	440	184	949	652	7

[1] Air content, both mixtures were designed for 7%; however, the range of air content between 5 to 8% was accepted.

Table 3: Chemical properties of General Use Portland cement (GU PC)

Chemical Composition	(%)	Chemical Composition	(%)
Calcium Oxide (CaO)	62.39	Sulfur Trioxide (SO_3)	4.03
Silicon Dioxide (SiO_2)	19.54	Free Lime	1.33
Aluminium Oxide (Al_2O_3)	5.21	Total Alkali (Na_2O eq.)	0.95
Ferric Oxide (Fe_2O_3)	2.16	Loss on Ignition (LOI)	2.36
Magnesium Oxide (MgO)	2.39		

2.2 Substrate Preparation

2.2.1 Substrate Casting and Surface Texture

To evaluate the effect of surface roughness on the bond strength for slant shear and splitting tensile tests, two different surface textures – rough and smooth – were tested. To minimize the damage caused by surface texture preparation such as grinding, hammering, milling, sand-water blasting, wire-brushing, etc., samples were textured while they were fresh. In order to achieve this goal, special moulds were prepared as illustrated in Figure 1. Two different technics of surface preparation for fresh cylindrical specimens were implemented: (1) the surface of specimens were trowelled by a wooden float and broomed to slightly texture the substrate according to MTO LS-412 (1997). Figure 2a demonstrates the final surface fabrication of substrates which was classified as smooth texture. (2) For the rough texture, the specimens were not trowelled; So, some portion of cement coated coarse aggregates were left on the surface to provide extra roughness (Figure 2b). For the direct pull-off test, slab specimens (300 x 300 x 75 mm) with only smooth surface texture were used considering the insignificant impact of roughness under direct tensile force.

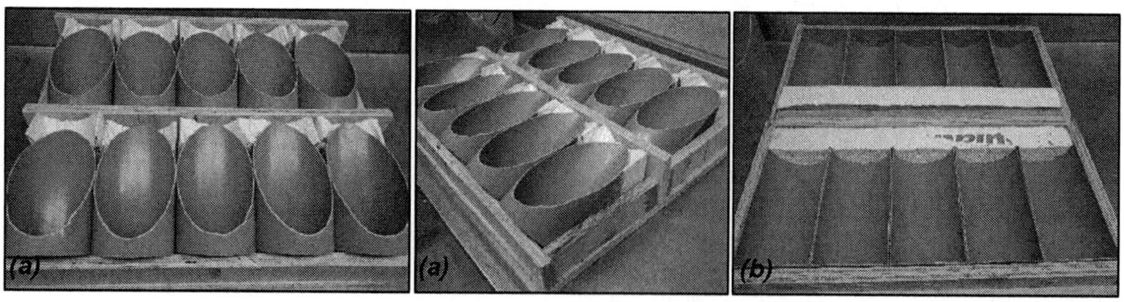

Figure 1: Substrate moulds preparation for surface texture (a) slant shear and (b) splitting tensile test

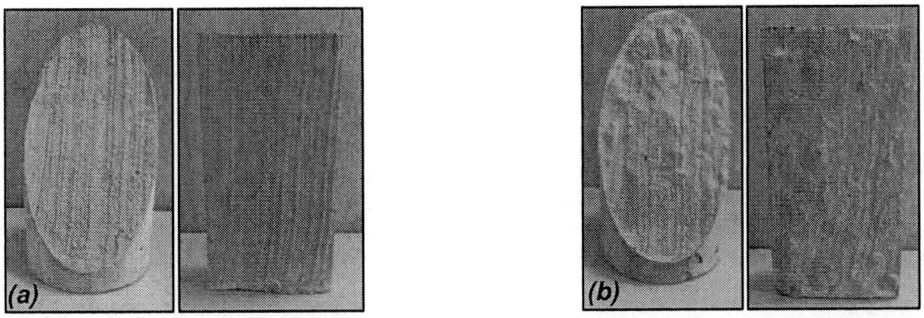

Figure 2: (a) Smooth and (b) rough surface texture for slant shear and splitting tensile tests

2.2.2 Stabilizing the Substrate Shrinkage

Allowing the substrate to shrink in a laboratory is essential since deteriorated concrete (substrate) under service condition is usually experienced most of its shrinkage. Therefore, to better simulate the life condition of substrate and stabilize its shrinkage, substrate was casted with conventional concrete, stored in the moist curing room (RH>98%) for 14 days to gain its specified strength, then relocated to the shrinkage room (46%<RH<54% and 21^{0}C<Temperature<25^{0}C) for 2 months before applying repair materials. The condition of shrinkage room and rate of evaporation was checked as specified by CSA A23.2-21C (2014).

2.2.3 Substrate Moisture Condition at the Time of Applying Repairs

During the casting of repair materials, it was decided to keep the substrate in SSD condition. Hence, the substrate was sprayed with water one day prior to casting and transferred to curing room. Finally, an hour before casting to make sure the substrate is in SSD condition, it was moved outside of curing room.

2.3 Applying Repairs and Conditioning

After finishing the substrate preparation, repair products were casted on top of the substrate. Afterward, samples were kept in the moist curing room for 14 days to cure the repair materials. Immediately following, they stored in the shrinkage room for 2 months to allow repairs to shrink. Then the initial bond strength was evaluated on the specimens and named as a zero cycle of freezing and thawing. Next, it was followed by freezing and thawing cycles in the presence of 3% sodium chloride (NaCl) solution. The cylindrical samples were completely submerged in the salt solution while slab specimens were covered by 6 mm salt solution (Figure 4a) according to MTO LS-412 (1997). In order to ensure that the specimens experience complete cycles of freezing and thawing, the samples were stored in a freezer at -18 ± 2 ^{0}C for 24 ± 2 hours and thawed at room temperature (23 ± 2 ^{0}C) for another 24 ± 2 hours. At the end of thawing cycles, if any loss in solution due to evaporation was observed, the de-icing salt solution was added to ensure that samples are completely submerged. Finally, to evaluate the durability of repair products, the bond strength testing (slant shear, splitting tensile, and pull-off) was performed on the samples after 5, 15, and 25 cycles of freezing and thawing in the presence of de-icing salt.

2.4 Mechanical Characteristics (Phase-I)

Compressive strength was measured following the method specified by ASTM C39 (2017). For each test, three standard cylindrical specimens were used and the average value was reported. The static modulus of elasticity was measured according to ASTM C469 (2014) using three cylindrical specimens (100 x 200 mm) with two longitudinal strain gauges at two sides of the cylinder with another transverse strain gauge at the middle of the specimen.

2.5 Bond Strength under Cycles of Freezing and Thawing with De-icing Salt (Phase-II)

The bond strength was evaluated by three methods, namely: slant shear, splitting tensile, and pull-off test.

The slant shear test was performed per ASTM C882 (2013). Three cylindrical specimens (100 x 200 mm) at the loading rate of 0.25 ± 0.05 MPa/s were tested for each repair products (Figure 3a). The substrate bond interface in the slant shear test is prepared with a 60-degree angle from the horizontal line. According to the ASTM C882, slant shear strength is equal to applied force divided by slant bond interface area.

Splitting tensile method is the test used for measuring the indirect tensile strength of concrete as outlined in ASTM C496 (2011). The value calculated by this method is approximately 10 to 15 percent higher than the direct tensile strength (Mehta and Monteiro 2006). This test was carried out using standard cylindrical specimens (100 x 200 mm) at the loading rate of 0.7 to 1.4 MPa/min (Figure 3b).

Figure 3: (a) Slant shear and (b) splitting tensile test

Pull-off test is a more practical test on site as a quality control tool. This test is appropriate for field and laboratory to measures the direct tensile strength of thin layer of overlay materials if the failure happens exactly at the bond interface (Bonaldo, Barros, and Lourenc-O 2005). Coring of repair material and part of the substrate prior to testing is one of the major disadvantages of the pull-off test. Excessive vibration of the coring machine, not placing it perpendicular to the surface of sample, size, and depth of coring can significantly affect the results. To eliminate the coring disadvantages and investigate the impact of the freezing and thawing cycles with de-icing salt at the bond interface, four small cylindrical repair specimens with a diameter of 75 mm and height of 50 mm were casted on top of slab substrate (Figure 4a). Slabs with repair samples were sealed with dense Styrofoam on each side (Figure 4a) to be able to condition them with the cycles of freezing and thawing in the presence of de-icing salt (Figure 4b). At the end of each cycle (0, 5, 15, and 25), the pull-off test was performed as specified by ASTM C1583 (2013).

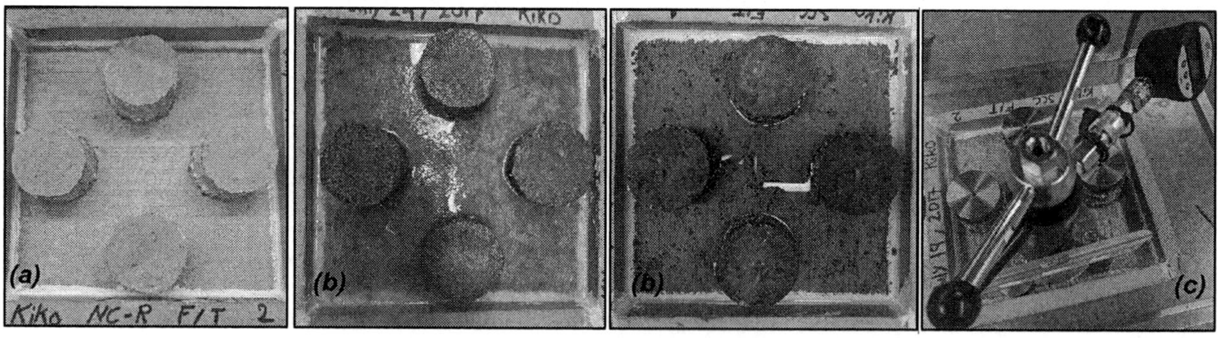

Figure 4: (a) Preparation, (b) conditioning of slab – freezing/thawing, and (c) testing the bond strength

3 RESULTS

3.1 Mechanical Characteristics (Phase-I)

Figure 5a illustrates the difference between the SRM and NC in terms of compressive strength development. SRM was able to almost reach twice the strength of NC at each curing period. For both repair products, the compressive strength development almost stops increasing at 56 days of moist curing. Figure 5b shows that the SRM and NC have a very close stress-strain behaviour, particularly before failure. The static modulus of elasticity for SRM and NC were calculated and found to be 31.94 and 33.66 GPa, respectively. This close behaviour of these two materials in terms of static modulus of elasticity is mainly due to having coarse aggregates of similar properties.

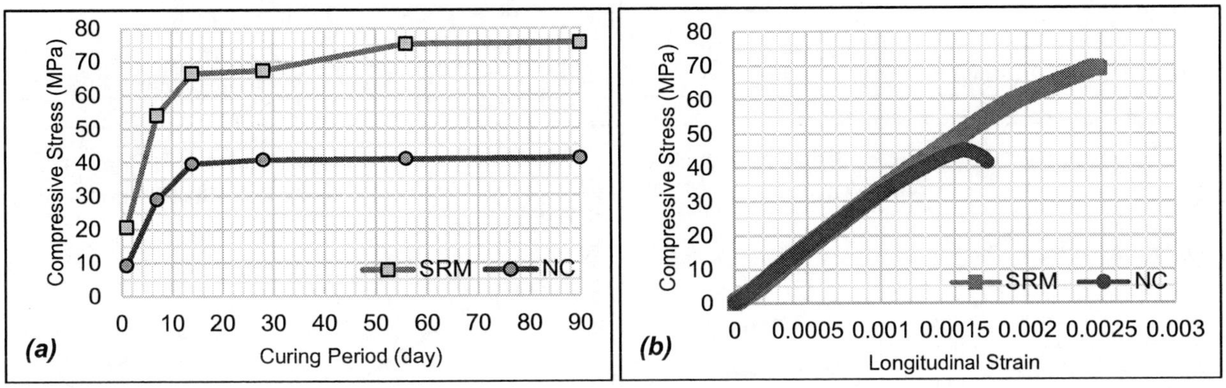

Figure 5: (a) Compressive strength development and (b) modulus of elasticity (Phase-I)

3.2 Bond Strength under Freezing and Thawing Cycles with De-icing Salt (Phase-II)

The results of slabs and cylindrical samples are summarized in Figure 6 for pull-off, slant shear, and splitting tensile test. Zero cycle of freezing and thawing in Figure 6 correspond to initial reading which was measured immediately after 14 days of moist curing and 56 days of being maintained in the shrinkage room.

According to Figure 6a, for the pull-off test, the NC had almost half the bond strength of that for SRM before exposure to cycles of freezing and thawing. With the cycles, however, the bond strength of SRM declined at the faster rate reaching half that of the NC at 15 cycles. For slant shear test (Figure 6b), the rough surface had the higher bond strength for both materials after any number of cycles. The reduction in the strength after freezing and thawing cycles occurred at the higher rate for SRM applied on the smooth surface. For splitting tensile test, rough and smooth, shown in Figures 6c and 6d, the rough surface demonstrated higher bond strength for both repair products. In addition, the degradation in strength of the SRM was higher than that of NC. Hence, NC in comparison to SRM showed better durability under the freezing and thawing cycles in the presence of de-icing salt although SRM started with higher initial bond strength measurement. In addition, since the change in the bond strength between SRM and NC appeared at faster rate for pull-off test (Figure 6a), and later for slant shear (Figure 6b), the testing methods can be arranged from high to low based on the effect of freezing and thawing cycles with de-icing salt on the bond strength as follow: (1) pull-off, (2) slant shear, and (3) splitting tensile test.

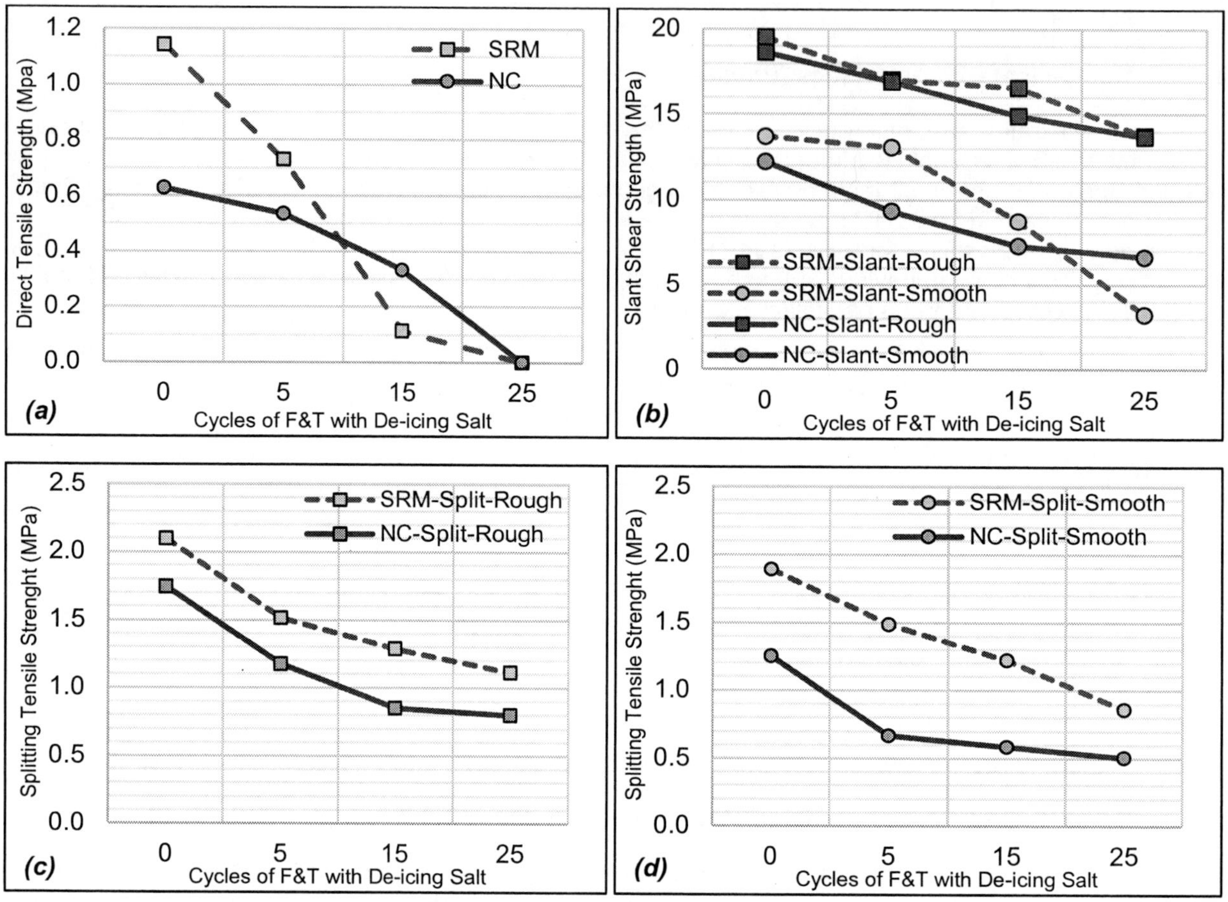

Figure 6: Test results of (a) pull-off, (b) slant shear, (c) splitting tensile rough, & (d) splitting tensile smooth

The stress loss (σ_L) was calculated as a percentage of bond strength measurement reduction at three different cycles of freezing and thawing (5, 15, and 25) in comparison to the initial bond strength (measured at zero cycles of freezing and thawing). The stress loss (σ_L) results listed in Tables 4 and 5 mathematically confirm the observation from the graphs for pull-off (Figure 6a) and slant shear test (Figure 6b). For instance, the stress loss for NC under pull-off test at 15 cycles was calculated to be much lower than SRM (47.1% compared to 90.1%) and similar behaviour was seen for slant shear test (rough and smooth) at 25 cycles of freezing and thawing which is another reason that NC performs better under higher cycles of freezing and thawing.

To demonstrate the effect of surface roughness on the bond strength, the stress improvement ($\sigma_{Im.}$) between rough and smooth surface was calculated for both slant shear and splitting tensile tests (Table 5). As it can be seen, the rough texture in all cases enhances the bond strength. The surface roughness effect is more significant at the higher cycles of freezing and thawing. For example, at 25 cycles the stress improvement for slant shear test was reported to be 323.5% and 210.3% for SRM and NC, respectively. By comparing the stress improvement between these two different tests, slant shear test is influenced more by the surface roughness in comparison to splitting tensile test. This significant difference between these two tests is mainly due to the nature of the acting load on the specimens.

For all the tests in this paper, the coefficient of variation (CV) between specimens within the samples was calculated and reported in Tables 4 and 5. The reported CVs for slant shear and splitting tensile tests were less than 11.5% and 16%, respectively which suggests a very close degree of variation between results. Moreover, all the calculated CVs for the pull-off test are reported to be less than 20%. The increasing value of the CVs between the slant shear, splitting tensile, and pull-off tests could be due to the different geometry, type of load, and human errors (Momayez et al. 2005). For the pull-off test, in particular, not being able to keep the applying load at the constant rate or properly aligning the instrument influence the range of results.

In addition to CV, the P-value (calculated probability) based on T-test between rough and smooth surface texture for the same evaluating test (slant shear or splitting tensile) and each cycle of freezing and thawing was calculated. Since the bond strength results after each cycle of freezing and thawing measured to be less than the initial reading, one tail and unpaired data were selected. According to the calculated P-value reported in Table 5, the slant shear and splitting tensile tests for both SRM and NC demonstrate the statistically significant difference between rough and smooth texture with 95% level of confident ($\alpha < 5\%$). The only exception was for SRM under splitting tensile during 5 and 15 cycles where non-statistically significant results were observed. This is mainly because of very close bond strength between rough and smooth texture for splitting tensile test particularly at 5 and 15 cycles of freezing and thawing (Figure 6a). The small stress improvement ($\sigma_{Im.}$) between the smooth and rough surface at these two cycles (1.9% and 5.4%) confirms this behavior.

Moreover, the P-value was calculated between two different materials (SRM and NC) for each method of testing and results are listed in Table 4. The results indicate that there is no statistically significant difference between repair products under the slant shear test for both rough and smooth texture. For pull-off test the results were significantly different except for cycle #15 and beyond. This indicates that the use of SRM over NC does not necessarily provide any advantages under higher cycles of freezing and thawing. However, splitting tensile tests showed the statistically significant difference between the two repair materials throughout the tested cycles (up to 25 cycles).

Table 4: Pull-off test results and statistical analysis between different repair products

Cycles	Direct Tensile (Pull-off) Test							P-Value[4]			
	SRM			NC			P-Value[4]	Slant Shear Test		Splitting Test	
	σ_L[1] (%)	CV[2] (%)	N[3]	σ_L[1] (%)	CV[2] (%)	N[3]		Rough Texture	Smooth Texture	Rough Texture	Smooth Texture
0	0.0	12.99	4	0.0	11.91	4	0.001	0.313	0.214	0.031	0.004
5	36.1	13.49	4	14.7	11.64	4	0.027	0.915	0.009	0.009	0.000
15	90.1	20.00	3*	47.1	10.41	3*	0.116	0.103	0.128	0.005	0.000
25	100.0	n/a	0**	100.0	n/a	0**	n/a	0.965	0.630	0.002	0.006

[1] Stress Loss – [2] Coefficient of variation – [3] Number of tested samples – [4] P-value calculated by running T-test between SRM & NC

* One samples failed before test & did not considered – ** All four samples failed before finishing the 25 cycles of freezing & thawing

Table 5: Results and statistical analysis of slant shear and splitting tensile tests

Repairs	Cycles	Slant Shear Test[a]						Splitting Tensile Test[a]					
		Rough Texture		Smooth Texture		$\sigma_{Im.}$[1] (%)	P-Value[2]	Rough Texture		Smooth Texture		$\sigma_{Im.}$ (%)	P-Value
		σ_L (%)	CV (%)	σ_L (%)	CV (%)			σ_L (%)	CV (%)	σ_L (%)	CV (%)		
SRM	0	0.0	6.64	0.0	5.20	42.5	0.003	0.0	2.44	0.0	7.26	10.9	0.049
	5	12.8	5.13	4.7	8.35	30.4	0.005	27.8	7.24	21.5	5.95	1.9	0.370
	15	15.2	6.63	36.3	9.86	89.7	0.001	38.7	8.05	35.4	3.09	5.4	0.196
	25	29.7	8.93	76.4	7.28	323.5	0.002	47.0	3.91	54.6	13.36	29.7	0.024

Repairs	Cycles	Slant Shear Test[a]						Splitting Tensile Test[a]					
		Rough Texture		Smooth Texture		$\sigma_{Im.}$[1] (%)	P-Value[2]	Rough Texture		Smooth Texture		$\sigma_{Im.}$ (%)	P-Value
		σ_L (%)	CV (%)	σ_L (%)	CV (%)			σ_L (%)	CV (%)	σ_L (%)	CV (%)		
NC	0	0.0	1.97	0.0	11.29	52.6	0.006	0.0	10.61	0.0	10.34	39.5	0.012
	5	9.0	8.81	23.7	9.96	81.4	0.002	32.7	4.99	46.7	8.69	76.3	0.000
	15	19.8	5.34	40.5	4.07	105.2	0.001	51.3	9.97	53.3	6.65	45.6	0.009
	25	26.4	8.38	45.9	13.65[b]	210.3	0.022	54.2	8.07	59.7	2.96	58.6	0.006

[1] Improvement between rough & smooth texture for the same test – [2] P-value calculated by running T-test between rough & smooth

[a] 3 samples were tested for each rough & smooth texture – [b] One of the smooth samples was failed before test & did not considered

4 CONCLUSIONS

Based on the experimental program and the materials tested in this research, following conclusion are drawn:

1. Increasing the substrate surface roughness was proven to enhance the bond strength no matter the method of testing, type of repair materials, and cycles of freezing and thawing with de-icing salt. The improvement of substrate surface roughness on bond strength at higher cycles of freezing and thawing was found to be more significant.
2. The surface roughness effect ($\sigma_{Im.}$) on the slant shear test observed to be much higher than the splitting tensile test.
3. For both repair products, the minimum stress loss (σ_L) due to freezing and thawing cycles in the presence of de-icing salt was found for the slant shear test with the rough texture.
4. The impact of freezing and thawing cycles with de-icing salt determined to be more severe and critical on the pull-off test carried out here.
5. The NC was proven to perform better than SRM under higher cycles of freezing and thawing with de-icing salt under both pull-off and slant shear tests. However, the same behaviour was not concluded for splitting tensile test method. For this particular test, similar behaviour can be expected at higher cycles of freezing and thawing.
6. The P-value results comparing the rough and smooth texture under slant shear test for both repair products provide statistically significant difference with a level of confidence of 0.05.
7. The P-value results comparing two repairs for slant shear indicate no statistically significant difference. The same can be said for pull-off but only under higher cycles of freezing and thawing.

Acknowledgment

This research project is supported by the Natural Sciences and Engineering Research Council of Canada (NSERC) grant. The authors highly acknowledge the financial support of NSERC.

References

ASTM C1583. 2013. "Standard Test Method for Tensile Strength of Concrete Surfaces and the Bond Strength or Tensile Strength of Concrete Repair and Overlay Materials by Direct Tension (Pull-off Method)." *American Society for Testing and Materials*, 1–4. doi:10.1520/C1583_C1583M-13.

ASTM C39. 2017. "Standard Test Method for Compressive Strength of Cylindrical Concrete Specimens." *American Society for Testing and Materials*, 1–7. doi:10.1520/C0039.

ASTM C469. 2014. "Standard Test Method for Static Modulus of Elasticity and Poisson's Ratio of Concrete in Compression." *ASTM International*, 1–5. doi:10.1520/C0469.

ASTM C496. 2011. "C496/C496M-11. Standard Test Method for Splitting Tensile Strength of Cylindrical Concrete Specimens." *American Society for Testing and Materials*, 1–5. doi:10.1520/C0496.

ASTM C882. 2013. "Standard Test Method for Bond Strength of Epoxy-Resin Systems Used With Concrete by Slant Shear." *American Society for Testing and Materials*, 1–4. doi:10.1520/C0882.

Austin, Simon, and Peter Robins. 1995. "Tensile Bond Testing of Concrete Repairs." *Materials and Structures* 28: 249–59. https://journals-scholarsportal-info.ezproxy.lib.ryerson.ca/pdf/13595997/v28i0005/249_tbtocr.xml.

Bonaldo, Everaldo, Joaquim A O Barros, and Paulo B Lourenc-O. 2005. "Bond Characterization between Concrete Substrate and Repairing SFRC Using Pull-off Testing." *International Journal of Adhesion & Adhesives* 25: 463–74. doi:10.1016/j.ijadhadh.2005.01.002.

Cai, H, and X Liu. 1998. "FREEZE-THAW DURABILITY OF CONCRETE: ICE FORMATION PROCESS IN PORES." *Cement and Concrete Research* 28 (9): 1281–87. https://journals-scholarsportal-info.ezproxy.lib.ryerson.ca/pdf/00088846/v28i0009/1281_fdocifpip.xml.

Courad, L. 2000. "Parametric Study for Creation of the Interface between Concrete and Repair Products." *Materials and Structure* 33: 65–72. https://orbi.uliege.be//bitstream/2268/17611/1/Parametric study.pdf.

CSA A23.1/CSA A23.2. 2014. *A23.1-14 Concrete Materials and Methods of Concrete Construction / Test Methods and Standard Practices for Concrete*. Mississauga: CSA Group.

Duarte, Miguel, Eduardo Nuno, and Brito Santos. 2011. "Factors Affecting Bond between New and Old Concrete." *ACI Materials Journal* 108 (August): 449–57.

Emmons, Peter H., Alexander M. Vaysburd, and James E. McDonald. 1993. "A Rational Approach to Durable Concrete Repairs." *Concrete International* 15 (9): 40–45. http://imcyc.com/biblioteca/ArchivosPDF/Reparacion de Estructuras/4 A rational approach to durable concrete repair.pdf.

Hanjari, Kamyab Zandi, Peter Utgenannt, and Karin Lundgren. 2011. "Experimental Study of the Material and Bond Properties of Frost-Damaged Concrete." *Cement and Concrete Research* 41: 244–54. doi:10.1016/j.cemconres.2010.11.007.

Lukovic, M., E. Schlangen, G. Ye, and B. Savija. 2013. "Impact of Surface Roughness Onthe Debonding Mechanism in Concrete Repairs." *Proceedings of the 8th International Conference on Fracture Mechanics of Concrete and Concrete Structures, FraMCoS 2013.*

Mehta, P. Kumar, and Paulo J M Monteiro. 2006. *Concrete: Microstructure, Properties, and Materials. The McGraw-Hill Companies*. 3rd Editio. The McGraw-Hill Companies. doi:10.1036/0071462899.

Momayez, A., M. R. Ehsani, A. A. Ramezanianpour, and H. Rajaie. 2005. "Comparison of Methods for Evaluating Bond Strength between Concrete Substrate and Repair Materials." *Cement and Concrete Research* 35 (4): 748–57. doi:10.1016/j.cemconres.2004.05.027.

MTO LS-412. 1997. "LS-412 Method of Test for Scaling Resistance of Concrete Surfaces Exposed to Deicing Chemicals." *Ministry of Transportation, Ontario (MTO)*, no. 17: 5.

Saucier, F, J Bastien, M Pigeon, and M Fafard. 1991. "A COMBINED SHEAR-COMPRESSION DEVICE TO MEASURE CONCRETE-TO- CONCRETE BONDING." *Experimental Techniques*, 50–55.

Shin, Hak-Chul, and Zhifu Wan. 2010. "Interfacial Properties between New and Old Concretes." *Second International Conference on Sustainable Construction Materials and Technologies.*

Wang, Bing, Shilang Xu, and Fei Liu. 2016. "Evaluation of Tensile Bonding Strength between UHTCC Repair Materials and Concrete Substrate." *Construction and Building Materials* 112. Elsevier Ltd: 595–606. doi:10.1016/j.conbuildmat.2016.02.149.

Building Tomorrow's Society

Bâtir la Société de Demain

Fredericton, Canada

June 13 to 16, 2018 / *13 au 16 juin, 2018*

EFFECT OF AGE HARDENING ON MODULUS OF WARM MIX ASPHALT MIXTURES

Tavassoti-Kheiry, Pezhouhan[1,4], Boz, Ilker[2], Solaimanian, Mansour[1], and Barzegari, Saman[1]

[1] The Pennsylvania State University, U.S.A.

[2] Michigan State University, U.S.A.

[4] pejoohan@alumni.psu.edu

Abstract: Within the past two decades, various warm mix asphalt technologies (WMA) have been introduced with the goal of mix production at lower temperature as compared to the traditional hot mix asphalt (HMA). Effect of the WMA additives on the rheology of binders and mechanical properties of the mixes have been well investigated by several researchers. However, studying the effect of age hardening on the mechanical properties of the compacted WMA has not been adequately researched. This paper presents the results of an experimental study on changes in shear and Young's dynamic modulus of three popular WMA technologies in the United States as well as a control HMA over a period of four years. To this end, small scale cylindrical specimens were prepared and studied in the laboratory. A chemical additive, a wax-based additive, and water foaming process were utilized in preparation of the specimens. The specimens were tested for properties at various time intervals. Modulus was measured right after preparation and laboratory short term conditioning, then at subsequent intervals after production: two years, and four years. The results indicate different trend of hardening with respect to the modulus across the technologies within the range of this study. Impact resonance (IR) tests at three temperatures of 10, 25, and 40°C were conducted on the replicates to non-destructively measure the shear and Young's modulus values of the specimens. IR results indicate that the trend of changes in modulus with time is different among the mixes.

Keywords: Dynamic Modulus, Impact Resonance Test, Warm Mix Asphalt, Age Hardening

1 INTRODUCTION

Warm-mix asphalt (WMA) technologies offer several advantages over the conventional hot-mix asphalt (HMA) due to their lower energy consumption, extended construction season, longer allowable hauling, and lower levels of construction related emissions (Hasan et al. 2017, Tavassoti Kheiry 2015). Resistance of WMA to thermal cracking, fatigue, rutting, and moisture susceptibility has been the focus of several studies in the past, which includes performance evaluation both in the laboratory and in the field (Hasan et al. 2017, Goh and You 2008, Hurley et al. 2009, Bower et al. 2016). It is expected that the use of WMA must not compromise the structural performance of the pavement. WMA technologies affect the mix properties directly through their effects on binder and reduced aging. Therefore, in addition to the aforementioned performance measures, aging related properties of the WMA should be further researched since these technologies are relatively new (Bower et al. 2016). Such studies can provide a better insight of the trend of changes in WMA properties with respect to time. A survey of the literature shows several indications of studies on the effect of WMA additives on binder rheology and aging properties (Bonaquist 2011, D'Angelo et al. 2008, Hurley and Prowell 2005a, 2005b, 2006). Aging of the compacted warm mix specimens has also been assessed in some studies, however, to a lesser extent as compared to the binder aging. In order to simulate the short and long-term aging in the laboratory, standard practices have been developed. A promising method seems to be conditioning the loose mix in forced-draft oven prior to mixture compaction for a given period of time and temperature rather than conditioning the compacted specimen in the oven (Elwardany et al. 2017).

When studying the changes in mechanical properties of asphaltic mixtures over time, it should be noted that part of the hardening may be contributed to mechanisms other than chemical aging. For example, due to the presence of binder phase in asphaltic mixtures, some mechanisms such as steric hardening may be reflected in the properties of specimens stored for a relatively long period of time. In such cases, studying the changes in material properties with respect to temperature can reveal valuable information (Brown et. al 1957).

The use of non-destructive tests (NDTs) to evaluate the asphaltic mixtures in the laboratory has gained more popularity recently. Techniques such as ultrasonic pulse velocity (UPV), impact resonance (IR), resonant column (RC), and acoustic emission (AE) testing are some examples that have been used with asphalt concrete characterization (Boz and Solaimanian 2015, Tavassoti Kheiry et al. 2017, 2018). The growing popularity of these techniques stems from their applicability to a wide range of materials in a relatively simple, economic, fast, and repeatable manner. The non-destructive nature of these tests makes it possible to monitor the trend of changes in the properties of a given specimen as a function of time; therefore, eliminating the concerns regarding the inherent variability among the replicate specimens in destructive tests.

In this paper, a different approach is employed with respect to studying the age hardening of compacted asphalt concrete specimens. The results of the study are presented for a laboratory scale study on three WMA and one control HMA mixes. The specimens were stored in the laboratory environment for a period of four years undergoing identical conditioning. The trend of changes in Young's and shear modulus was studied across the WMA and HMA technologies as a function of time to deliver an estimate of age hardening related properties of these mixes.

2 OBJECTIVES AND METHODOLOGY

While there is a considerable amount of information about performance testing of warm mix asphalt mixes in the laboratory, research data on monitoring the trend of changes in modulus of such mixes with time is limited, and the main reason for pursuing the research presented in this paper. The objective of this research is to assess the age hardening changes of the WMA that have been maintained in the laboratory environment for extended period of time, through monitoring damping ratio, shear, and Young's modulus of elasticity. For this purpose, the same replicate specimens are retested at different intervals using the Impact Resonance (IR) test. The non-destructive nature of IR test makes it possible to monitor the trend of changes in dynamic modulus of the same specimen at a range of temperatures.

3 MATERIALS AND SAMPLE PREPARATION

For the purpose of mix design and sample preparation, one source of aggregate and one source of binder were used for both the WMA and HMA mixtures in this study. The aggregate was 100% crushed dolomite limestone. The target particle size distribution of the aggregate is presented in Table 1. Superpave mix design was followed for a standard 9.5 mm dense graded hot mixed asphalt at 75 gyrations. The binder content was established at 5.4% and was used for all of the mixes investigated in this paper. The asphalt binder was graded as PG 58-22.

Superpave Gyratory Compactor (SGC) was used to prepare cylindrical specimens. The specimens were then trimmed and cored to retrieve small-scale specimens of 60 mm in diameter and 120 mm in height for impact resonance testing. Three replicates were prepared for each of the mixes of interest, resulting in a total of 12 specimens for Impact Resonance (IR) testing. An air void level of 7.5±0.5% was targeted and successfully achieved during the sample preparation.

The warm mix additives, i.e. an organic wax-based additive and a chemical additive, were first blended in the binder using a high shear blender at 135°C prior to preparing the mixtures. The organic and the chemical additives were blended at 1,000 and 500 revolutions per minute (rpm), respectively, as recommended by the manufacturers. Table 2 provides details of sample preparation. It can be seen that the mixing and compaction temperatures, i.e., 147 and 138°C for the conventional HMA mix in this study, are lowered by about 15 to 17°C depending on the WMA technology.

Table 1: Particle size distribution of aggregate used

Sieve and % Passing									
½" (12.5 mm)	3/8" (9.5 mm)	#4	#8	#16	#30	#50	#100	#200	Pan
100	95	74.8	44.5	28.35	19.5	12.25	8.2	5.65	0

Table 2: WMA sample preparation parameters

WMA Technology	Mixing Temperature (°C)	Compaction Temperature (°C)	Application Rate (%)
Organic (Wax) Additive	132	121	1.5
Chemical Additive	132	121	0.4
Water Foaming	138	128	2.0

The asphalt binder (either the neat for HMA, or the modified for WMA) was subsequently mixed with preheated aggregate. The loose mixture was then conditioned in an oven at 121°C for 2 hours. The mixture was then transferred to the SGC molds for compaction to achieve the desired air voids level. Table 3 provides a summary of the average volumetric properties for each of the mixes investigate in this study.

Table 3: Compacted asphalt concrete sample properties

Technology	Average G_{mb}	Average Air Void (%)
Organic (Wax) Additive	2.361	7.5
Chemical Additive	2.357	7.7
Water Foaming	2.353	7.9
HMA	2.360	7.8

4 IMPACT RESONANCE TEST

An approximately free-free boundary condition was maintained for the compacted specimens during the impact resonance testing. To this end, the experimental set-up outlined by ASTM C215 (2014) were adapted and the specimen was placed on a soft sponge rubber mat. Figure 1 shows schematics of the IR test in this study. The specimens were kept in a precise temperature control chamber with the ability to control the temperature by 0.1°C. IR tests were conducted at three temperatures of 10, 25, and 40°C. Each IR test was repeated five times on any of the specimens to mitigate the effect of test variability to the possible extent. Fast Fourier Transform (FFT) algorithm was used to process the time domain data into frequency domain data. Therefore, the resonant frequency and damping ratio of each vibration mode (longitudinal and transverse) were determined for each test at each temperature using the processed frequency domain data.

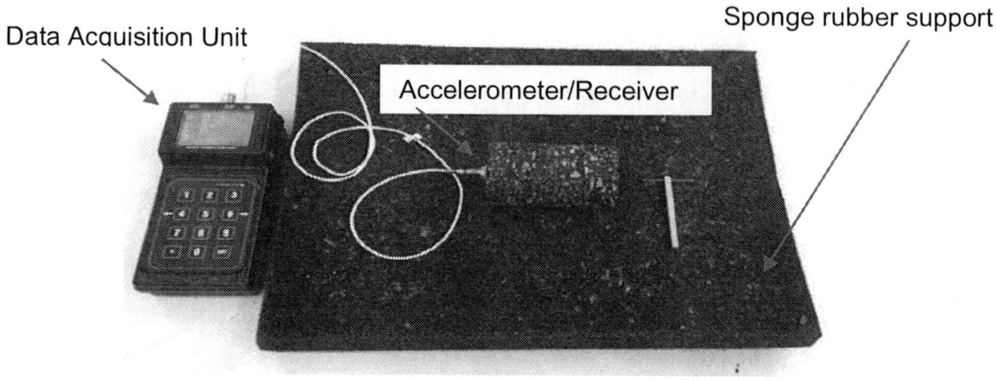

Figure 1: Schematics of impact resonance test of asphalt concrete

Given the fact that the IR test is a truly non-destructive test, same specimens were stored in the laboratory environment for the purpose of performing a second and third round of testing after two and four years. Thus, the specimens were retested using IR at the same temperatures at the time intervals of interest. The measured resonant frequencies at each temperature along with the specimen size, mass, and bulk specific gravity were used to calculate the shear modulus, dynamic modulus and damping ratio.

5 TEST RESULTS AND DISCUSSIONS

5.1 IR Resonance and Repeatability Validation

IR tests were performed in both transverse and longitudinal modes, which makes it possible to calculate shear and Young's modulus and Poisson's ratio for each of the specimens. Upon obtaining the IR responses, Hankel matrix was utilized to verify that the responses follow the second order of motion as expected for IR testing. The Hankel matrix was also employed as a filtering algorithm to eliminate unwanted noise with respect to the second order equation of motion. It was confirmed that the responses did not violate the equation of motion assumptions within the range of temperatures investigated in this study. Therefore, the filtered data was utilized to proceed with further analysis toward modulus calculation. Details of the Hankel matrix analysis is not within the scope of this paper, and can be found elsewhere (Boz et al. 2017a).

On the other hand, one of the crucial aspects in order to study the changes in properties of asphalt concrete specimens over a relatively long period of time would be repeatability of the test results. IR test is believed to be very repeatable, which makes it a reliable tool for the purpose of monitoring the changes in mechanical properties of the specimens over a given period of time. Table 4 provides a summary of the test results obtained for testing one of the HMA replicate specimens, which exhibited the highest level of variability among the HMA replicates at 10°C. Detailed results for every replicate specimen are not presented due to the space limitations, and instead the average results are provided in this paper. Inspecting the raw data obtained at the range of the test temperatures in this study revealed that even for the highest test temperature, i.e. 40°C, the coefficient of variation remained smaller than 10%. The variability was significantly lower at temperatures below 40°C.

Table 4: An example data set showing excellent repeatability of IR at 10°C

Mode of Test	1st test (Hz)	2nd test (Hz)	3rd test (Hz)	4th test (Hz)	5th test (Hz)	Average (Hz)	Std. (Hz)	COV (%)
Longitudinal	12,718	12,150	12,150	12,150	12,264	12,286	246.3	2.0
Transverse	7,608.1	7,721.7	7,608.1	7,721.7	7,267.4	7,585.4	186.6	2.5

5.2 High Frequency Modulus and Master Curve

Generally, the resonant frequency of specimens in IR testing ranges from 2 to 20 kHz, depending on the geometry and density of the asphalt concrete specimens, which is considerably beyond the conventional testing frequencies (Tavassoti-Kheiry et al. 2018). However, the results from IR tests conducted at different temperatures can be used to construct the modulus master curve through the use of established theories and assumptions. An important assumption that would be needed to construct the full spectrum master curve would be the binder shift factor assumption (Boz et al. 2017b). Several studies have shown that the mixture shift factors are not, generally, significantly different than the binder shift factor. Therefore, binder shift factor can be used in combination with IR test results to provide a meaningful modulus master curve over the full spectrum of the frequencies. The binder, therefore, was tested at a range of temperatures and at 16 frequencies (ranging from 0.1 to 100 rad/s). The generalized logistic model (Equation 1), which provides a better fit as compared to the conventional symmetric sigmoid function, was used in this study. Using the polynomial shift factor function (Equation 2) along with Equation 1 and replacing the E* with G* of the binder would result in the binder master curve.

[1] $$Log|E^*| = \delta + \frac{\alpha}{\left[1+\lambda e^{\beta+\gamma[\log f_r]}\right]^{1/\lambda}}$$

where |E*| is the dynamic modulus, f_r is the reduced frequency in Hz, and δ, λ, α, β, and γ are treated as fitting parameters that specify the shape of the sigmoid function. The limiting minimum modulus (δ), so-called the low asymptote of the sigmoid function, can be estimated using the improved Hirsch model, which uses the VMA, VFA, G_{sb}, strain levels and G* values. Further details can be found in Boz et al. (2017b).

[2] $$Log[a(T)] = a_1(T)^2 + a_2(T) + a_3$$

where T is the reference temperature for the purpose of master curve construction, a(T) is the shift factor, and a_1, a_2, and a_3 are the fitting parameters.

Upon determination of the shift factor parameters and minimum limiting modulus, the fitting parameters of the sigmoid function in Equation 1 were determined through the numerical optimization at a reference temperature of 20°C. This process was repeated for each of the mixes investigated in this study to construct the associated modulus master curves based on the binder data and IR test results. Figure 2 shows the modulus master curve for the chemical additive-based WMA in this study at a reference temperature of 20°C.

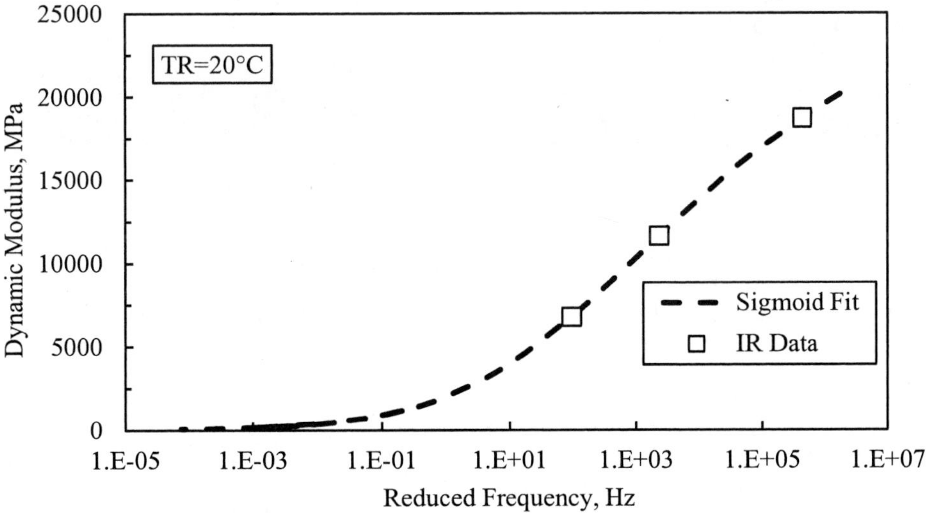

Figure 2: Modulus master curve using IR data for the chemical additive-based WMA

Similar analysis was conducted on the data from the organic additive-based, water foaming and conventional HMA mixes. As mentioned before, although the constructed master curves provide a reasonable estimation of the modulus of the mixes over the full spectrum of frequencies, there are several assumptions that introduce a level of uncertainty when comparing the modulus of different mixes. Regardless of the promising capabilities of the full spectrum master curve constructed from IR data, only the high frequency modulus values, which are the direct measurement through the IR tests, are used for studying the age hardening of different mixing technologies in this paper. Details are presented in the following sections.

5.3 Age Hardening and Modulus Changes

Measuring the modulus at different temperatures and different time intervals (i.e., after sample preparation, after 2-years, and after 4-years of storage in laboratory environment) can provide an estimate of the age-related hardening of the WMA in this study. Figure 3 provides a summary of the dynamic modulus results measured at 10, 25, and 45°C over the period of four years from sample preparation for the specimens prepared through water foaming and organic wax-based additive. It can be seen that the hardening has a steeper rate for the first two years from the time of sample preparation compared to the rate of hardening for the last two years. The trend of changes, however, is different among the four mixes. The fact that the storage conditions were identical for all the mixes indicates that such difference can be contributed to the inherent differences in age-related hardening of the mixes. For example, the WMA specimens prepared using water foaming exhibit the highest age hardening as compared to the rest of the mixes. It should be noted that by comparing the initial modulus values among the four mixes in this study it can be recognized that water foaming provided a mix that was initially softer as compared to the other investigated technologies. This might be associated with a lower level of aging during the mix production and is found to be consistent with the findings of the previous NCHRP studies (NCHRP 843 2017). However, the focus of this paper is studying the trend of changes in the stiffness of these four mixes as a function of time, which is shown to be different than the initial stiffness of mixes. In spite of the magnitudes, similar trends were observed for both the shear modulus and Young's modulus of elasticity.

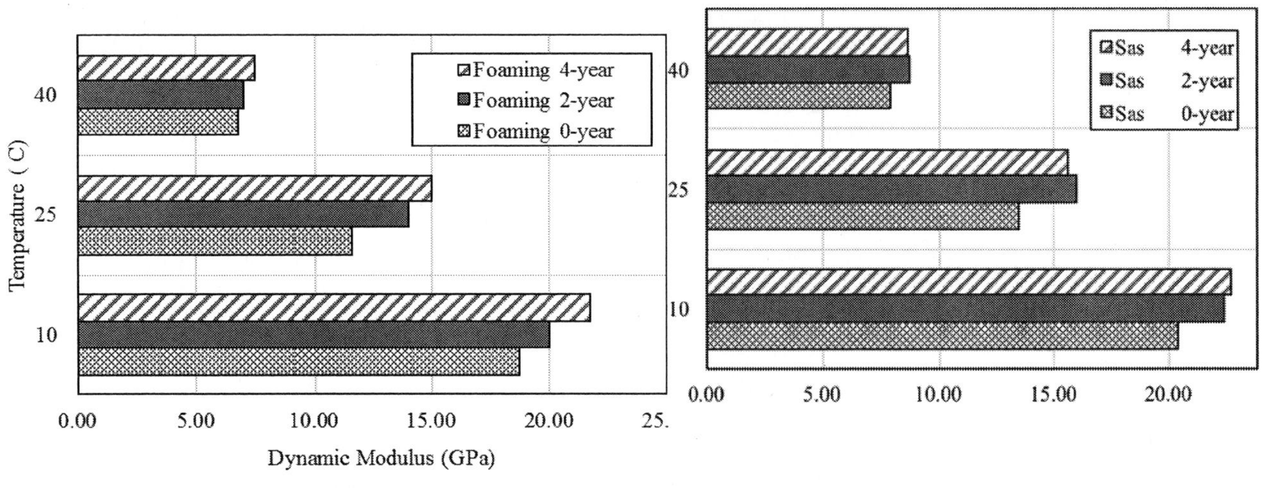

Figure 3: Changes in Young's modulus of water foaming (left) and organic wax-based (right) WMA specimens as a function of time at 10, 25, and 40°C

Figure 4 illustrates the changes in shear modulus for the chemical additive WMA specimens and the control HMA mix. With respect to the percent changes in modulus as a function of time, it can be seen that water foaming WMA specimens exhibited the largest age hardening over four years (i.e., rate of 31%), followed by the organic wax-based specimens (i.e., rate of 26.9%). On the other hand, HMA and the chemical additive based WMA specimens showed maximum rates of 15.4 and 16.3%, respectively.

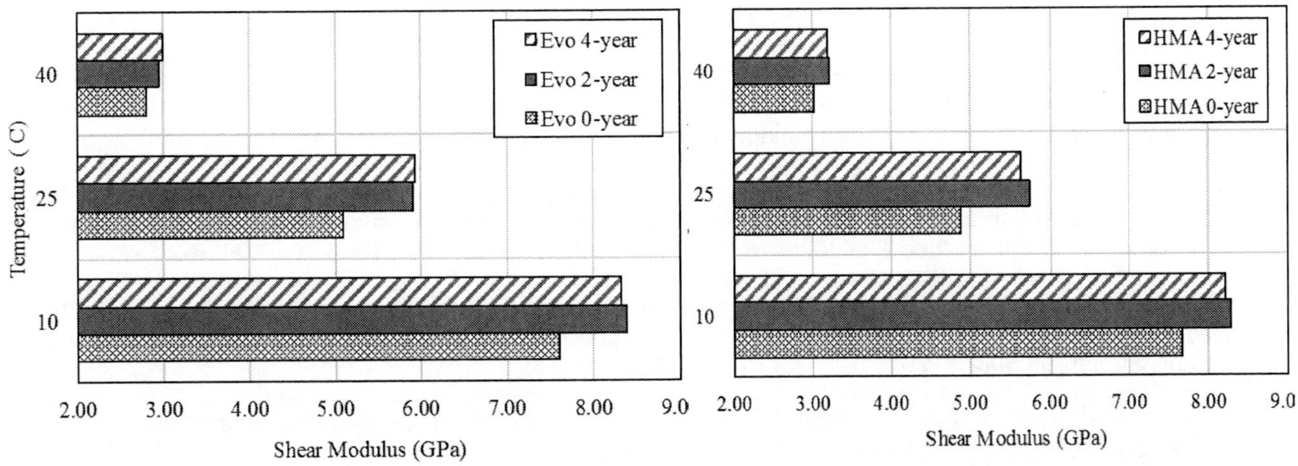

Figure 4: Changes in shear modulus of the chemical-additive based WMA (left) and HMA (right) specimens versus time at 10, 25, and 40°C

6 SUMMARY AND CONCLUSIONS

Four series of replicate specimens, consisting of three sets of warm-mix asphalt (WMA) and one set of control hot mix asphalt (HMA) mixtures, were prepared in the laboratory. The specimens were tested using the impact resonance (IR) tester at different time intervals over four years and the shear and Young's modulus of elasticity values were measured at 10, 25, and 40°C each time. The specimens were first tested within a week after sample preparation and were stored afterward for two years prior to retesting. The final set of testing was conducted when the specimens were four years old. The same specimens were used and tested at the same temperatures. The following conclusions can be drawn based on the results of this study:

It was shown that for a given gradation, binder, and mix design, the initial modulus of elasticity was consistently larger for the WMA prepared by the wax-based additive followed by the mix prepared with the chemical additive. The WMA prepared through water foaming technology as well as the HMA mix exhibited statistically similar stiffness, while foaming showed slightly softer response.

The trend of changes associated with age hardening of the mixes was different over the four-year study period. Foaming technology exhibited the highest rate of changes as compared to the other mixes in this research. HMA specimens, overall, exhibited the least changes due to the age-related hardening as compared to the rest of the mixes in this study.

Except for the WMA specimens prepared using water foaming, the rate of changes in modulus values can be considered very slow after 2 years. This conclusion is drawn based on the fact that no significant difference was found between the modulus of 2-year old specimens versus that of 4-year old specimens.

Acknowledgements

The authors would like to thank the financial support provided by the Northeast Center of Excellence for Pavement Technology (NECEPT) of the Pennsylvania State University.

References

ASTM C215-14, Standard Test Method for Fundamental Transverse, Longitudinal, and Torsional Resonant Frequencies of Concrete Specimens, ASTM International, West Conshohocken, PA, 2014,

Bonaquist, R. 2011. Mix Design Practices for Warm Mix Asphalt. NCHRP Report 691. Transportation Research Board, National Research Council, Washington, D.C.

Bower, N., Wen, H., Wu, S., Willoughby, K., Weston, J. and DeVol, J. 2016. Evaluation of the performance of warm mix asphalt in Washington state. Int. Journal of Pavement Engineering, **17**(5): 423–434.

Boz, I. and Solaimanian, M. 2015. The Effect of Aspect Ratio on the Frequency Response of Asphalt Concrete in Impact Resonance Testing. Journal of Testing and Evaluation, **44** (1): 543-554.

Boz, I., Bekiroglu, K., Solaimanian, M., Tavassoti-Kheiry, P. and Lagoa, C. 2017a. Validation of Model Order Assumption and Noise Reduction Method for the Impact Resonance Testing of Asphalt Concrete. Journal of Nondestructive Evaluation, **36**(58). https://doi.org/10.1007/s10921-017-0436-2.

Boz, I., Tavassoti-Kheiry, P., and Solaimanian, M. 2017b. The Advantages of Using Impact Resonance Test in Dynamic Modulus Master Curve Construction through the Abbreviated Test Protocol. Materials and Structures **50**(3): 176.

Brown, A.B., Sparks, J.W. and Smith, F.M. 1957. Steric Hardening of Asphalts. Association of Asphalt Paving Technologists, **26**: 486.

D'Angelo J. et al. 2008. Warm-Mix Asphalt: European Practice. Publication FHWA-PL-08-007. FHWA Office of International Programs, Washington D.C.

Elwardany, M.D., Yousefi, F., Castorena, C. and Kim, R. 2017. Evaluation of asphalt mixture laboratory long-term ageing methods for performance testing and prediction. Journal of Road Materials and Pavement Design, **18**: 28-61.

Goh, S.W. and You, Z. 2008. Mechanical Properties of Warm Mix Asphalt Using Asphamin.

Hasan, M.R.M., You, Z., and Yang, X. 2017. A Comprehensive Review of Theory, Development, and Implementation of warm mix asphalt using foaming techniques. Journal of Construction and Building Materials,**152**: 115-133.

Hurley, G. C. and B. D. Prowell. 2005b. Evaluation of Sasobit® for Use in Warm Mix Asphalt. NCAT Report 05-06, National Center for Asphalt Technology, Auburn, Alabama.

Hurley, G.C. and Prowell, B.D. 2005a. Evaluation of Aspha-Min® Zeolite for Use in Warm Mix Asphalt. NCAT Report 05-04. National Center for Asphalt Technology, Auburn, Alabama.

Hurley, G.C. and Prowell, B.D. 2006. Evaluation of Evotherm® for Use in Warm Asphalt Mixes. NCAT Report No. 06-02. National Center for Asphalt Technology, Auburn, Alabama.

Hurley, G.C., Prowell, B.D. and Kvasnak, A.N. 2009. Michigan Field Trial of Warm Mix Asphalt Technologies: Construction Summary Report No. 09-10X. National Center for Asphalt Technology.

in: Proceedings of Annual Transportation Research Board Meeting, Washington, D.C, National Research Council, National Academy of Sciences.

Tavassoti Kheiry, P., Qiu, T. and Solaimanian, M. 2015. Characterization of Warm Mix Asphalt through Resonant Column Testing. J. Mater. Civ. Eng., **27**(10): 05015001-6.

Tavassoti-Kheiry, P., Boz, I., Chen, X. and Solaimanian, M. 2017. Application of ultrasonic pulse velocity testing of asphalt concrete mixtures to improve the prediction accuracy of dynamic modulus master curve. Proc., ASCE International Conference on Highway and Airfield Technology, DOI: 10.1061/9780784480939.014

Tavassoti-Kheiry, P., Boz, I., Solaimanian, M., and Qiu, T. 2018. Evaluation of Nondestructiveness of Resonant Column Testing for Characterization of Asphalt Concrete Properties. ASTM Journal of Testing and Evaluation, DOI: JTE20160247. ISSN 0090-3973.

Washington State University, Louisiana Transportation Research Center, and Pennsylvania State University—Altoona. 2017. Long-Term Field Performance of Warm Mix Asphalt Technologies. Pre-publication draft of NCHRP Research Report 843. Transportation Research Board, Washington, D.C

Building Tomorrow's Society

Bâtir la Société de Demain

Fredericton, Canada

June 13 to 16, 2018 / *13 au 16 juin, 2018*

EFFECT OF AGGREGATE GRADATION, AGING, AND COMPACTION ENERGY ON SWELLING CHARACTERISTICS OF RUBBER MODIFIED ASPHALTS MIXTURES

Solaimanian, Mansour[1,2], Chen, Xuan[1]
[1] The Pennsylvania State University, United States
[2] Corresponding Author: msolaimanian@engr.psu.edu

Abstract: Use of ground tire rubber (GTR) in asphalt has been experienced for decades. There has been extensive research on rubber modified asphalts and many pavements have been built using this material. In general, use of rubber in asphalt has been shown to be beneficial both from environmental point of view (landfill saving) and field performance (reduction in rutting and increase in fatigue resistance). A challenge faced with the use of rubber modified asphalts for certain types of aggregate gradation is the swelling of rubber particles, and the rebound of the material after compaction. This phenomenon is more severe in the dry process (when rubber is first blended with the aggregate); however, it does also pose a challenge in the wet process (when rubber is first blended with asphalt). Such behavior affects mechanical properties of the mix, and results in reduction of density in the field. A laboratory experiment was undertaken to quantify the level of this swelling and to determine the magnitude of specimen swelling as affected by the aging level and compaction energy when the wet process is utilized. Specimens were prepared under short-term and long-term aging processes and compaction was conducted using a gyratory compactor at different energy levels. The experiment was also conducted in a way to determine the gradation effect, and to separate the effect of rubber swelling on specimen rebound from the effect of compaction energy on this rebound. It was found that the rebound and swelling are highly affected by the aggregate gradation, with denser gradations resulting in higher swell. The study also indicated that extended aging reduces the specimen swelling significantly. Similarly, higher rebound was observed under higher compaction energy. Finally, the mix swelling and rebound were correlated with the rubber modified binder characteristics through multiple stress creep and recovery (MSCR). The findings of this study are significant in assisting the mix designer with a tool for better control of the mix behavior and reduction of swelling both in the laboratory and in the field.

1 INTRODUCTION

Millions of used vehicle tires are disposed each year and this number keeps increasing due to growing demand of vehicles and traffic. Discarded tires pose a severe environmental problem in the United States. Normally, discarded tires are reused, resold, retreated, or landfilled. Among possible uses of scrap tires, only two methods have shown potential for the greatest benefit: use as combustion fuel and crumb rubber modifier (CRM) for paving industry (Heitzman 1992). CRM use in asphalt paving has a relatively long history and its first use goes back 1840 (Heitzman 1992). However, a rational approach for use did not come into play until McDonald introduced the wet process in 1960s.

Mixing crumb rubber with virgin asphalt binder results in the improved performance of blended binder. One challenge faced with CRM binder, however, is the swelling of rubber particles when blended with asphalt binder, and the rebound of the compacted specimens. Swelling is a diffusion – not a chemical – process that results from the liquid moving into the internal matrix of the polymer (Stroup-Gardiner et al. 1993). This phenomenon affects mechanical properties of asphalt mixes and potentially results in reduction of density in the field. When asphalt binder and crumb rubber particles are blended together, the aromatic components of the asphalt binder get absorbed into the polymer chains in crumb rubber particles to form a gel-like material. Such absorption results in swelling of the rubber particles to two to three times their original volume (Heitzman 1992). The interaction is affected by a number of variables such as the temperature at which the blending-reaction occurs, the length of time the temperature remains elevated, the type and amount of mechanical mixing energy, the surface size and texture of the crumb rubber, and the aromatic component contents of the asphalt binder (Heitzman 1992). The swelling causes an enormous increase in the mass viscosity of the system due to reduced free space between the swollen rubber particles (Heitzman 1992), as compared with neat asphalt or asphalt with un-swollen rubber early in the interaction process.

Airey et al. studied the absorption of light fraction of asphalt binder into crumb rubber particles using a basket drainage method (Airey et al. 2003). They examined the chemical composition and rheological properties of residual binder as well as swelling of the rubber particle. Overall, rubber mass increases with the increase of interaction time, but increasing rate decreases with the increase of interaction time, regardless of binder source. Their test results indicated the rate of adsorption is directly related to the viscosity as well as the chemical composition of the bitumen, and the total amount of absorption is controlled by the nature of the crumb rubber, such as grinding process and contents. Peralta et al. studied the influence of binder type on the rubber morphology (Peralta et al. 2010). The authors used the basket drainage method to recover the swelled rubber and residual binder, then calculated their density against interaction time. Based on their calculations, the density of rubber first decreases then increases with the increase of interaction time. They authors believed that a large quantity of air and moisture among the rubber particles expand their volume thus reducing the rubber density at the beginning of the interaction, then air and moisture are replaced by liquid bitumen, causing density of rubber to increase.

Although a number of studies are available on the swell of CRM binder, few can be found on the subject of rebound of compacted CRM asphalt specimens. Therefore, a laboratory experiment was undertaken to quantify the level of rebounding in asphalt mixes after compaction, and to determine if the magnitude of specimen rebounding was affected by the aging level, compaction energy, and gradation when using wet process. Dimension (height and diameter) changes were logged after certain time intervals post compaction to determine the rebound ratio of compacted specimens. Finally, the mix rebound was correlated with the CRM binder characteristics through the results of multiple stress creep and recovery (MSCR) binder test.

2 RESEARCH OBJECTIVES AND SCOPE

The main objective of this study was to investigate the effect of aging, compaction energy, and gradation on the rebound of compacted CRM asphalt mixes using the wet process. This was accomplished by measuring the volume change on compacted CRM specimens. CRM specimens were manufactured using different gradations, and they were exposed to a range of compaction energies and different aging conditions. Additionally, changes in specimen dimension after compaction of CRM asphalt specimens were compared with changes in specimens manufactured with only virgin binder. The MSCR test results on CRM binder were also evaluated to correlate properties of aged binders with the rebounding of compacted specimens.

3 MATERIAL AND EXPERIMENTAL PROGRAM

3.1 3.1 Materials

The dolomite/limestone aggregate used in this study came from a local source, with 55 percent coarse aggregate (retained on #4 sieve), and 45 percent fine aggregate (passing #4 sieve). Apart from one 9.5-millimeter dense gradation, two additional gap gradations were also included. All specimens and binder samples were prepared using a PG58-28 binder and one source of crumb rubber passing #30 mesh. All gradations are shown in Fig. 1.

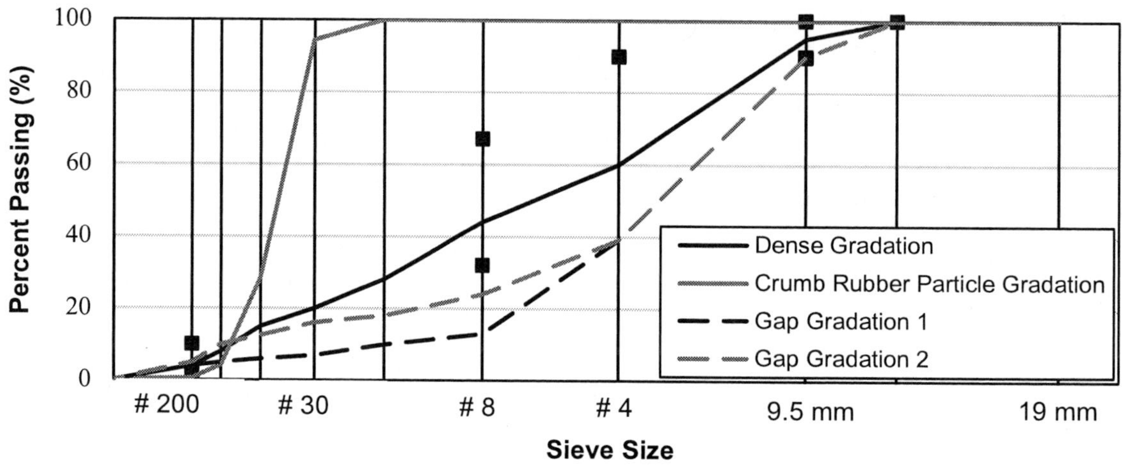

Figure 1: Aggregate gradation and crumb rubber particle gradation

The following procedure was followed to prepare CRM binder:

1. Heat PG58-28 virgin binder at 150°C for 60 minutes.
2. Continue heating virgin binder in a temperature-controlled container until it stabilizes at 165°C.
3. Add crumb rubber into the heated virgin binder gradually and within a period of five minutes. Blending occurs at 3000 RPM while CRM is added. The amount of added CRM will be 15 percent of the total mass (mass of binder plus mass of CRM).
4. Agitate the CRM binder for an hour at a reduced shear rate of 700 RPM after all rubber particles are added.
5. The blended CRM binder is cooled to room temperature.
6. When it is time to use the modified binder, it must be heated at 150°C for one and a half hours.

The temperature of the blending container was maintained at 170 °C for the entire shear blending process. The CRM binder was used no later than one day after manufacturing. Finally, all specimens were mixed and compacted at 150°C.

3.2 Mix Design

A dense graded Superpave 9.5 millimeter mix design (case 1) was used throughout the study with design binder contents of 5.2 percent and 5.7 percent for virgin binder and CRM binder, respectively. When performing mix design with CRM binder, virgin aggregate gradation was adjusted to make room for the added crumb rubber particles. The adjusted gradation with crumb rubber has the same gradation as the virgin aggregate gradation. As a part of the study to evaluate the impact of gradation, six different cases were considered as presented in Table 1.

Table 1 Different Cases for Mix Considered for This Study

Case No.	Binder	Gradation	Binder Content, %
1	Virgin	Dense	5.2
2	CRM	Dense with adjustment for rubber	5.7
3	CRM	Dense as Case 2	5.2
4	CRM	Dense, as Case 1	5.2
5	CRM	Gap Gradation 1	7.0

6	CRM	Gap Gradation 2	6.2

3.3 Test Program

3.3.1 General Scope

The laboratory program was established to capture the effect of aging, gradation, and compaction level on the specimens. After mixing, specimens were placed into two groups. Specimens of Group I were short-term aged (STOA) in loose condition (not compacted) for two hours at 150°C. Specimens of Group II were not only short-term aged but also long-term aged (LTOA) in loose condition at 85°C for 120 hours. Afterwards, they were re-heated at 150°C for two hours before compaction. Specimens were compacted at three energy levels: 70, 90, and 110 gyrations.

3.3.2 Swell Measurements

The change of diameter and height were logged 0, 10, 60, 120, 240, 1200 minutes after compaction. The diameter and height data at 0 minutes were obtained from the data collected by the software of the Superpave Gyratory Compactor (SGC). This data was used to calculate the initial volume of specimens. Data from the remaining the time intervals was measured using two digital calipers (Fig. 2), averaging three measurements. The expansion ratio was calculated as the ratio of volume at a certain time after compaction over the initial volume.

(a) (b)

Figure 2: Measuring diameter (a) and height (b) after compaction

Since swell and rebound occurs right after release of pressure, immediate measurement of air void of specimens before swelling begins is difficult. Thus, a unique method was used to calculate the initial and final air void (AV) of compacted CRM specimens. First, a volume correction factor between ideal cylinder (calculated using height readings from the SGC) and actual specimen volume (measured through AASHTO T 166) was established. The correction factor offsets the irregularities observed on the surface of the asphalt specimens. It is air void dependent and was established using data from specimens made with no rubber as such specimens have negligible volume change with time after compaction. Next, the initial volume (before pressure relief) of CRM specimens was back-calculated using the correction factor and height readings of the corresponding CRM specimens. Here, initial volume refers to the actual volume of the specimen at the completion of compaction and before release of pressure. This initial volume was then used to calculated the initial bulk specific gravity (G_{mb}) and additionally, initial air void. The initial air void was further compared with the actual air void (with swell), measured 20 hours after compaction.

3.3.3 Evaluating Effect of Modified Binder Characteristics

It is evident that swelling of compacted specimens containing rubber modified asphalt is influenced by properties of the asphalt binder after modification. Specifically, it is reasonable to believe that stiffness and elastic recovery characteristics of the rubber modified asphalt play a role in this respect. Therefore, our testing program included determination of the magnitude of stiffness and recovery of modified binders after short-term and long-term aging. For this purpose, the Multiple Stress Creep Recovery (MSCR) test (AASHTO 2014) was utilized. The test determines the percent recovery and non-recoverable creep compliance of asphalt binders. The percent recovery value is intended to provide a means for determining the elastic response and stress dependency of the modified asphalt binders, while the non-recoverable creep compliance is an indicator of the resistance of an asphalt binder to permanent deformation. For this study, short-term aging was achieved through rolling thin-film oven (RTFO) and long-term aging through pressure aging vessel (PAV). The aged CRM binders were tested using the MSCR protocol at 72°C under two stress levels, 0.1 and 3.2 kPa. One exception to the protocol was setting the gap between plates in the dynamic shear rheometer. The standard requires this setting at one millimeter. However, considering the upper size of rubber particles (0.6 millimeter), the gap was set at two millimeters to accommodate the size of crumb rubber particles and maintain a ratio of at least three between the particle size and the plate gap.

4 RESULTS AND ANALYSIS

4.1 Volume Change Comparison

4.1.1 Effect of Aging and Compaction Energy

The volume change of specimens with different aging conditions and compaction energies are shown in Fig. 3 for dense graded mixes. Plots include results from specimens with different aging conditions and compaction energies. The first figure (3a) is for STOA virgin mix and is added as a reference for comparison with CRM mixes. This is a dense graded mix made with PG58-28 binder without any crumb rubber.

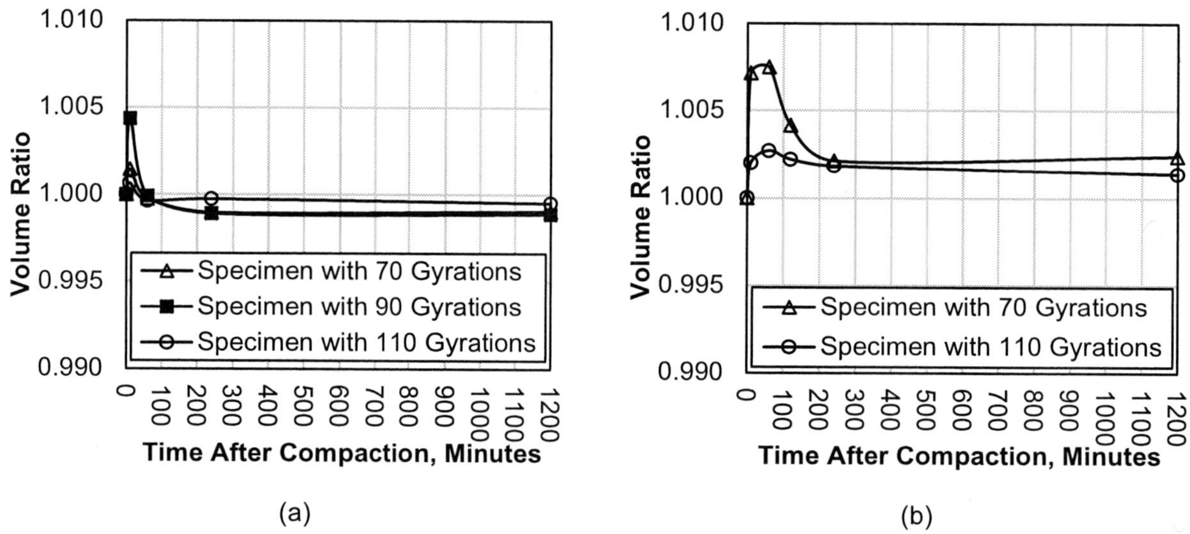

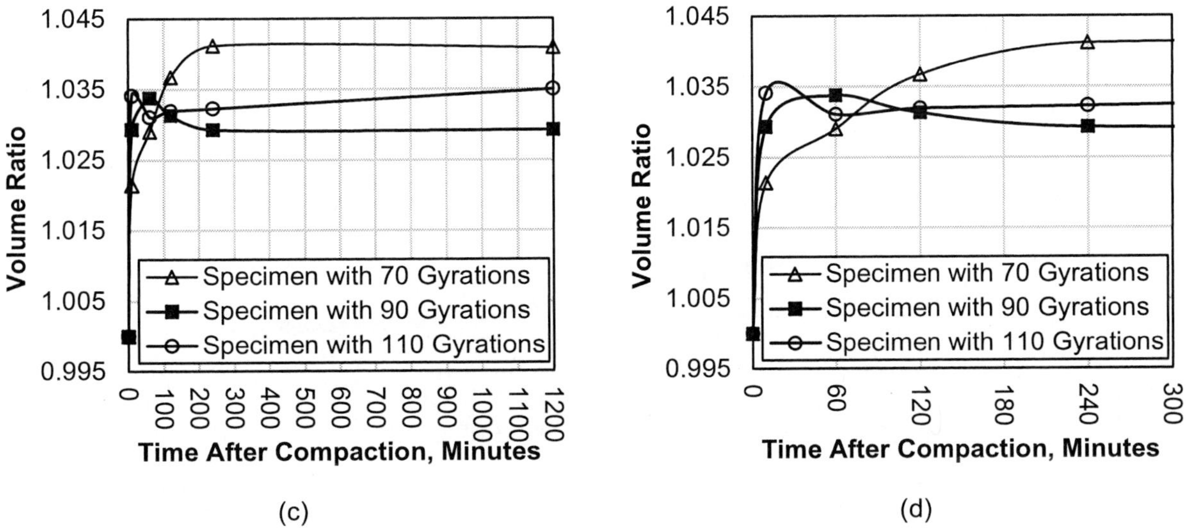

Figure 3: Expansion ratio comparison, (a) STOA virgin mix, (b) LTOA CRM mix, (c) STOA CRM mix, and (d) STOA CRM mix in detail

Regardless of compaction energy, there was no significant volume change for STOA virgin mix (Fig. 3a), except for slight volume increase (less than 0.5 percent) right after compaction. However, after two hours cooling at room temperature, this slight increase in volume vanished and the volume did not change until the end of the measuring cycle. The final volume ratio was almost one, indicating no change in volume with time, after compaction, Similar to the STOA virgin mix, volume change of LTOA CRM mix (Fig. 3b) was not significant. There was also a slight increase in volume right after compaction, but the change of volume vanished after specimen temperature dropped. The final volume increase of LTOA CRM mix was less than 0.3 percent regardless of compaction energy.

Contrary to the virgin mix and the long-term aged mix, the increase in volume for STOA CRM mix was significant. There was a three to four percent increase in volume within a short time after compaction (Fig. 3c). This is an important observation indicating that CRM mixes in dense gradation expand significantly after compaction, and that long-term conditioning of the mixes significantly reduces the potential for expansion.

Finally, there will be the effect of compaction level. Fig. 3c indicates that the specimen with the lowest level of compaction (70 gyrations) has a higher level of rebound compared with the specimen with the highest compaction energy (110 gyrations). The specimen with the median compaction level (90 gyrations) exhibited the lowest expansion level. To understand this behavior, one has to consider the compactability of the mix, and how the compaction path of the mix follows as the number of gyrations increases. Fig. 4 presents the compaction patterns for these three compaction levels using the exact same mix. One could see that, for this specific mix, after 60 gyrations, the change in height is minimal (less than one millimeter per 10 gyrations) for all three specimens. Any further compaction effort beyond this point is not contributing to compaction; rather it is probably simply degrading the mix. This phenomenon is probably the reason for the observed behavior in the magnitude of the rebound for different compaction levels.

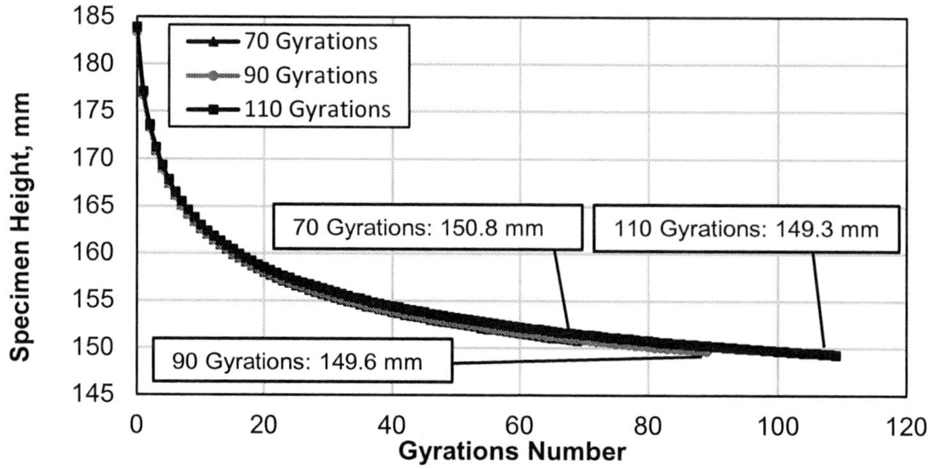

Figure 4: Compaction pattern of STOA CRM mix with three compaction energy levels

Based on the observations discussed above, it is clear that LTOA CRM binder mix has similar rebound characteristics as STOA virgin binder mix, while STOA CRM mix has very different rebound properties after compaction.

4.1.2 Effect of Gradation Change

It was mentioned previously that two gap graded mixes were included in the study to determine the impact of the gradation change on volume expansion after compaction. The results are shown in Fig. 5. All specimens were prepared with the same binder and rubber contents, short-term aged, and compacted at 70 gyrations. Although gap-graded aggregate 2 has a slightly denser gradation compared to gap-graded aggregate 1, both share a similar final volume increase of roughly two percent. This expansion is half the volume increase of the dense gradation. It is obvious that the significant expansion observed in dense graded mixes could be highly reduced by generating a gap in the gradation. This observation is expected as a gap-graded mix provides more space for rubber particles. This increased void space allows the rubber particles to be exposed to lower level of "squeezing pressure" during compaction and hence lower rebound after compaction. It can also be seen that once sufficient gap is generated (see gap-graded 2); further increase in void space (see gap-graded 1) would not result in reducing the volume expansion any further.

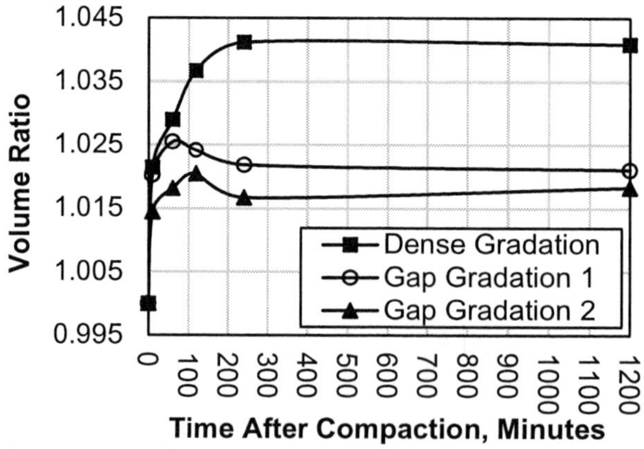

Figure 5: Effect of gradation on volume ratio changes

4.2 Air Void Change Comparison

The significance of volume expansion in CRM mixes is finally reflected in its impact on the pavement density and air void. Hence, it is important to quantify the impact of this expansion on air void. The corresponding air void change of mixes discussed in section 4.1 is shown in Fig. 6. Only partial results are shown here due to space limitation, but similar trends can be observed for the data not shown. Regardless of compaction energy, only slight decrease in air void level can be observed for STOA virgin mix after specimens had been compacted for 20 hours (Fig. 6a). Additionally, the effect of compaction energy is as expected, i.e. higher compaction level leads to lower air voids. LTOA CRM mix showed similar characteristics to STOA virgin mix, with a slight increase in air void for the specimen compacted with 70 gyrations (Fig. 6b). However, significant air void change is observed for short-term aged CRM mixes (Fig. 6c).

For all specimens, regardless of gyration level, the air void increased significantly 20 hours after compaction. Calculations put this void increase as 50 percent over the initial air void, indicating a considerable level of rebound 20 hours after compaction. It is also worth noticing that for STOA mixes, the air void of specimens under 90 gyrations is almost identical to that under 110 gyrations. This observation is further indication of a threshold value for compaction energy beyond which no further increase in density is achieved. This phenomenon also ties into the concept of "locking point" in compaction. Locking point refers to the point in compaction process where aggregate particles "lock" together, and additional gyrations (i.e. more compaction energy) does not contribute to further densification, rather it begins to degrade the aggregate (Prowell and Brown 2007, Honbeck 2008).

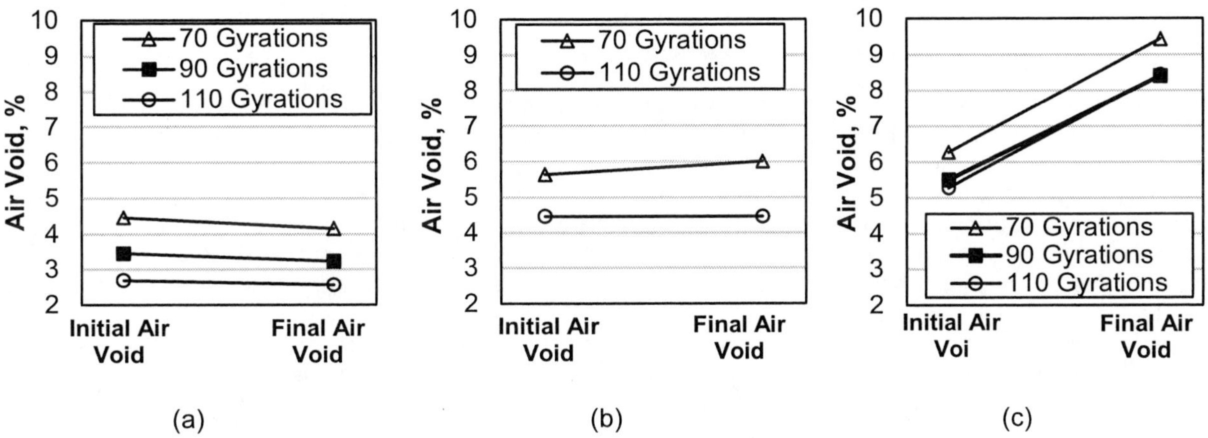

Figure 6: Air void change, (a) STOA virgin mix, (b) LTOA CRM mix, and (c) STOA CRM mix

To further investigate the effect of gradation and compaction energy on the rebound behavior of CRM mixtures, a second phase of study was conducted in which two groups of specimens were prepared using different batch weights but compacted to the same height. For each corresponding specimen of Group 1, an additional mass of 150 grams was added to the batch weight to make the specimens of Group 2. Since all specimens were compacted to a fixed height of 150 millimeter, the additional mass of Group 2 specimens required higher number of gyrations to result in the same height as specimens of Group 1. Obviously, higher mass of Group 2 resulted in lower initial air void as shown in Fig. 7. Within each group, two cases are considered: Case 2 with 5.7% binder based as designed, and Case 3 with 5.2% binder content. Case 1, previously discussed, was with virgin binder (No CRM) and not included in this part of the study. Even though the batch weights within each group are the same, initial air level varies due to different gradations and different theoretical maximum specific gravity (G_{mm}). It must be noted that all specimens were CRM modified and all were short-term oven aged. As discussed previously, the expansion was found to be significant only for CRM mixes exposed to short-term aging.

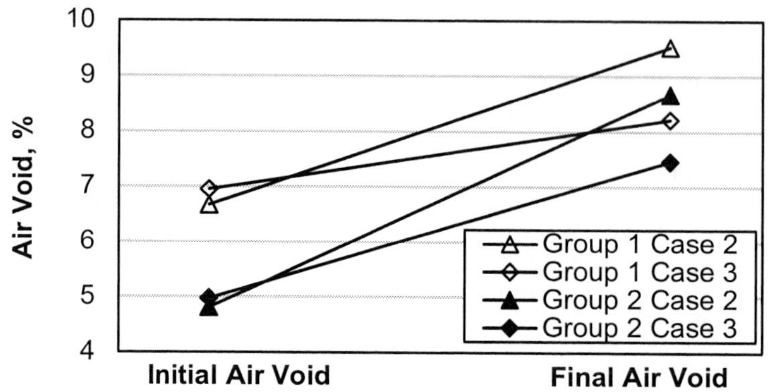

Figure 7: Air void change of different cases,

Note: Group 2 Specimens were made with 150 grams more material compared with Group 1,

Case 2: CRM mix with 5.7 % binder

Case 3: CRM mix with 5.2% binder

There are several observations worth noticing as reflected in Fig. 7. As expected for STOA CRM mixes, all specimens show significant rebound 20 hours after compaction regardless of the initial air void level. More interestingly, the specimens of Group 2 (i.e. with higher mass at the same height) show significantly higher rebound compared with specimens of Group 1. This is true for both types of gradations presented in Fig. 7. Table 2 presents a quantitative comparison of the expansion level for both cases. Group 1 specimens starting at higher air void compared with Group 2 had an air void increase of 43 percent and 18 percent for cases 2 and 3. Group 2 specimens showed approximately 80, 50 percent increase in air void for the corresponding cases. It is reasonable to believe that higher rebound of Group 2 specimens is the result of elastic rubber feeling higher pressure under compaction in the case of Group 2. It is also important to note that within each group, the specimens with higher binder content (i.e. Case 2) had a higher rebound compared with specimens with lower binder content (i.e. Case 3).

Table 2: Expansion level comparison between cases

Group	Case	Initial Air Void, %	Final Air Void, %	Percent Increase in Air Void, %
1 (lower mass, same height)	Case 2	6.67	9.53	43
2 (higher mass, same height)	Case 2	4.81	8.67	80
1 (lower mass, same height)	Case 3	6.95	8.21	18
2 (higher mass, same height)	Case 3	4.98	7.46	50

4.3 Correlation between Specimen Rebound and Properties of Modified Binder

The MSCR binder test results are shown in Fig. 8. The percent recovery value (Fig. 8a) determines the elastic response of modified asphalt binders, while the non-recoverable creep compliance (Fig. 8b) is an indicator of the resistance to permanent deformation under repeated load.

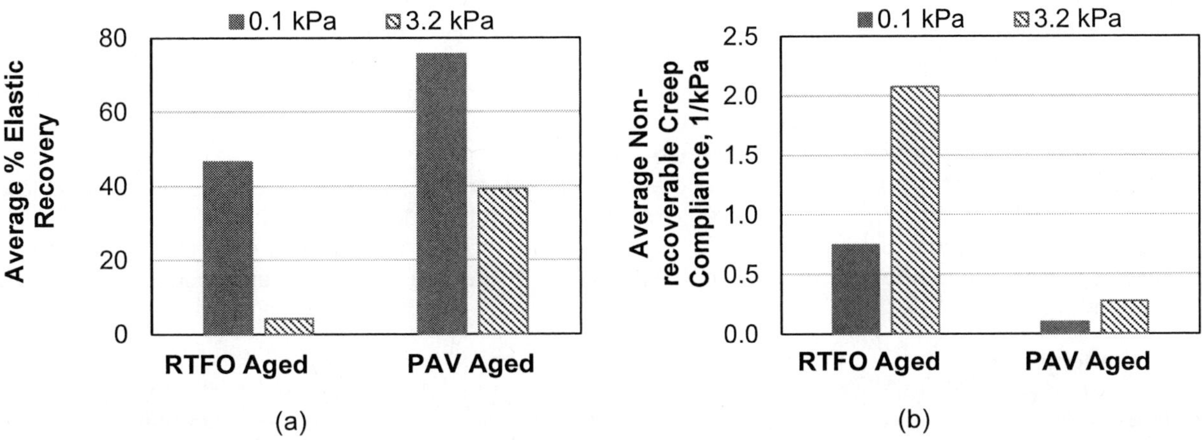

Figure 8: MSCR results of CRM binder, (a) average percent recovery and (b) average non-recoverable creep compliance

The elastic response of CRM binder was improved significantly from 47 percent to 76 percent at 0.1 kPa stress level and from 4 percent to 39 percent at 3.2 kPa stress level, after the binder had been long-term aged. Correspondingly, non-recoverable creep compliance of CRM binder also decreased tremendously after long-term aging.

There are two results worth noting as reflected in Fig. 8: 1) the STOA binder has higher compliance compared to the LTOA binder, 2) the STOA binder exhibits significantly lower recovery compared to the LTOA binder. Both observations are expected as long-term aged binder is obviously stiffer than the short-term aged binder. Furthermore, as the binder gets long-term aged, it pulls away from viscous behavior and moves more towards elastic behavior compared to short-term aged binder due to significant loss of volatiles and oxidation.

At the first look, the higher recovery of the LTOA binder seems contrary to the CRM mix behavior, as LTOA mix exhibited significantly lower swell compared with the STOA mix. However, one should not ignore the high impact of the binder stiffness on the rebound magnitude. While LTOA binder has considerably higher recovery, it is also much stiffer than STOA binder, hence under the same compaction energy (gyrations), it deforms very little compared to STOA binder. As a result, once compaction is complete and pressure is released, the LTOA mix rebounds very little. In other words, the original squeeze for LTOA CRM binder is very small to begin with. Hence, the rebound is small in spite of the fact that it demonstrates such higher recovery in MSCR.

5 SUMMARY AND RECOMMENDATIONS

A laboratory experiment was conducted to quantify the level of rebounding and volume expansion in crumb rubber modified (CRM) asphalt mixes after compaction. In this study, the effects of aging, compaction energy, and aggregate gradation were considered. The mesh 30 asphalt rubber was added to a PG 58-22 binder through laboratory shear blending at elevated temperatures. Rubber was added at 15 percent by the mass of the modified binder. The rebound of compacted specimens was quantified using two parameters: the volume change and the air void change. Additionally, MSCR binder test results were analyzed to determine the impact of binder aging on the rebound behavior of compacted specimens. The following conclusions can be drawn from the observations and analyses under this study:

- There was no significant volume change for mixes produced with virgin binder without rubber particles. Long-term conditioned (LTOA) CRM mixes exhibited a similar behavior with no volume expansion. Short-term conditioned (STOA) CRM mixes, on the other hand, demonstrated considerable rebound after compaction. For these mixes, three to four percent volume increase was observed for the dense graded CRM mixes.

- Corresponding to volume change, significant air void increase was observed for STOA CRM mixes after compaction, while no dramatic change in air void was found in STOA virgin mixes and LTOA CRM mixes.
- The rate of volume expansion was not affected by the level of compaction energy. For all, higher compaction energy resulted in lower initial volume. Expansion after this initial volume was only observed for short-term aged specimens of CRM mix.
- Specimens at lower air void exhibit higher expansion ratio.
- Using MSCR results, it was demonstrated that the swell of long-term aged mixtures is significantly lower than the swell of short-term aged mixtures due to significant change in the binder stiffness.
- Careful attention must be paid to the design of these mixes in terms of aggregate gradation and compaction as the asphalt mixes produced with dense aggregate gradations are susceptible to severe rebounding after compaction.

Acknowledgements

Financial support of Penn State Northeast Center of Excellence for Pavement Technology is greatly appreciated. The authors express their sincere thanks to Mr. Scott Milander, NECEPT technical coordinator, for his assistance with the laboratory work of this research.

References

AASHTO T 350-14, 2014. Standard method of test for multiple stress creep and recovery (MSCR) test of asphalt binder using a dynamic shear rheometer (DSR). American Association of State Highway and Transportation Officials, Washington D.C.

Airey, G.D., Rahman, M.M., and Collop, A.C., 2003. Absorption of bitumen into crumb rubber using the basket drainage method. *International Journal of Pavement Engineering*, **4**(2): 105-119.

B.D. Prowell, E.R. Brown, "Superpave Mix Design: Verifying Gyration Levels in the Ndesign Table", NCHRP Report 573, National Cooperative Highway Research Program, Washington DC, 2007.

M.A. Heitzman, 1992. "*State of the Practice- Design and Construction of Asphalt Paving Materials with Crumb Rubber Modifier*," U.S. Department of Transportation, Federal Highway Administration, FHWA-SA-92-022, Washington, D.C.

N.C. Hornbeck, "Effect of Compaction Effort on Superpave Surface Course Materials", Master's Thesis, West Virginia University, 2008.

Peralta, Joana, Hugo MRD Silva, Ana V. Machado, Jorge Pais, Paulo AA Pereira, and Jorge B. Sousa. 2010. "Changes in rubber due to its interaction with bitumen when producing asphalt rubber." *Road Materials and Pavement Design,* **11**(4), 1009-1031.

Stroup-Gardiner, M., Newcomb, D.E., and Tanquist, B., 1993. Asphalt-rubber interaction. *Transportation research record*, **1417**: 99-108.

Building Tomorrow's Society

Bâtir la Société de Demain

Fredericton, Canada

June 13 to 16, 2018 / *13 au 16 juin, 2018*

THE EFFECT OF MIXING WATER TEMPERATURE ON CONCRETE PROPERTIES IN HOT WEATHER CONDITIONS

Assal, Mennatallah[1] and Abou-Zeid, Mohamed Nagib[1]
[1]Construction Engineering Department, American University in Cairo, Egypt

Abstract: Environmental factors and hot weather conditions can drastically impact the properties, strength and durability of concrete structures. Whereby, higher slump loss rate, faster hydration with accelerated setting, reduced long term compressive strength and plastic shrinkage are more common. Literature review on the impact of hot weather on concrete is not sufficiently elaborated while focusing on curing temperature on overall concrete quality. This study is carried out to investigate the direct impact of mixing water temperature and concrete temperature on the concrete properties serving in hot weather conditions, with special emphasis to the Egyptian hot climate throughout the year. Thus, the scope of this work is to study mixing water temperature's impact on concrete properties being a key ingredient to the concrete mix, with direct impact on workability, strength and durability. To achieve the work objective, four categories of concrete mixtures with various cement contents and admixtures were studied while changing the mixing water temperature at 5, 25 and 45°C to yield a total 12 different test sets. All other concrete ingredients were heated to 45°C ahead of mixing to simulate hot weather conditions. Two sets of results were analyzed starting with fresh concrete tests of slump, unit weight and concrete temperature; in addition to hardened concrete tests of compressive and flexural strengths. Results showed an average 15% increase in the 28 days concrete compressive strength and enhanced workability in 5 °C temperature water Vs. 45 °C suggesting the strong impact of water temperature on concrete properties and need to add cool mixing water to concrete cast in hot weather conditions. Recommendations are provided towards the incorporation of cold water in concrete.

1 INTRODUCTION

Weather conditions either of high or low temperatures directly influence the behavior, performance and properties of the concrete structures during mixing, transport, casting and curing. Such impacts make it of high concern to the concrete manufacturers and final users given its impact on both the technical and economical manufacturing aspects for the structures under design and study (Ortiz et. al. 2010).

Of the most common issues with high weather concrete is plastic shrinkage and cracks. Those induced cracks could appear in all structural members' types, however more likely in the components where the surface area to thickness ratio is large, which is the case with slabs structures. Furthermore, plastic shrinkage cracking is seen as the main cause of initiating concrete structures deterioration in hot weather conditions, that is because those cracks allow and promote the diffusion of the harsh chemical species inside the concrete matrix and mix with direct impact on the reinforced concrete causing its deterioration (Nasir et. al. 2016).

High temperatures and hot weather conditions specifically cause high water demand for concrete increasing the fresh concrete temperatures. This causes increased rate of loss of slump, faster hydration which in return leads to accelerated setting and reduces the long-term strength of concrete accompanied with more evaporation rate (Neville 1995).

Hover highlighted the deteriorating factors in hot weather that can negatively impact concrete structure to include: high temperature, high concrete temperature, low relative humidity, solar radiation and wind velocity (Hover 2005).Whereby, those conditions cause difficulties for fresh concrete as:

- Increased water demand
- Accelerated slump loss which leads to more water addition in the jobsite
- Faster setting, thus difficulties in placing and finishing
- Higher tendency for plastic cracking
- Prompt early curing is critically needed
- Entrained air controlling difficulties
- Higher concrete temperature leads to long term strength losses
- Higher likelihood of thermal cracking (ACI Committee 305 2010).

2 EXPERIMENTAL PROGRAM

2.1 Materials Selection

The materials used for the concrete mixes under study are mainly sourced from local producers and queries as per below:

Cement: Ordinary Portland Cement (ASTM C150), manufactured by Suez Cement company in compliance with international standards (EN 197/1-2011) and Egyptian standards (ES 5756/1-2013).

Coarse Aggregates: Crushed dolomite surface – dry stones for from OCI Crusher, Attakah. The dolomite had a maximum nominal size of 25

Fine Aggregates: Siliceous sand was used in all concrete mixtures. Fine aggregates were obtained from natural a local query near Suez. The sand had a fineness modulus of 2.57, saturated surface dry specific gravity of 2.54 and a percent absorption of 0.58%

Water: Clean drinking water is used as mixing water for the concrete and for cleaning purposes associated with the experimental work hereunder.

Water Reducing Admixtures: Sika Viscocrete – 10 used is a fourth-generation super-plasticizer for concrete and mortar. This admixture meets the requirements of super plasticizers according to SIA 162 (1989) prEN-934-2 and ASTM-C-494 Types G and F.

Silica Fumes: Sourced from Sika to enhance the durability and strength of concrete mixes.

2.2 Concrete Mix Design

Four different mix designs were used in scope of the study, with key variable of the w/c ratio, cement quantities and higher quality concrete mix with superplasticizer added. Mix one with 300 kg cement, w/c ratio at 0.5; second with 350 Kg, w/c ratio of 0.45, third with 450 kg, w/c 0.4 and finally fourth mix with 450 kg with 50 kg Silica fumes. Concrete mixes illustrated in Table 1 below.

Table 1: Mix Designs Used

Item	Mix 1	Mix 2	Mix 3	Mix 4
Cement (kg)	300	350	450	450

Item	Mix 1	Mix 2	Mix 3	Mix 4
Water (kg)	150	157.5	180	180
w/c	0.50	0.45	0.40	0.36
Fine Aggregates(kg)	681	658	605	594
Coarse Aggregates(kg)	1226	1184	1089	1069
Admixtures	-	2 Liters Plasticizer type "A"	5 Liters Plasticizer type "F"	10 Liters Plasticizer type "F" + 50 kg Silica Fume

For every mix design, three different mixing water temperatures were used (5°C, 25°C and 45°C) to yield a total 12 sets of specimens, where each test specimen is as follows;

- 12 cubes of (15°cm x 15°cm x 15°cm) divided into sets of 3 cubes broken at 4-time intervals (1 day, 3 days, 7 days and 28 days)
- beams (75°cm x 15°cm x 15°cm) where one is broken for flexural strength after 28 days and the other used for heat of hydration measurement connected by thermocouple

2.3 Tests

For the tests; fresh concrete tests of unit weight were carried out for every mix design in accordance with ASTM 138 in addition to slump test as per ASTM C143. Temperature was tested through the thermocouple apparatus connected by the data logger and computer, where wires embedded inside the concrete beam form recorded the temperature of the concrete for the 8 hours' span.

For hardened concrete, the strengths assessment of the beams after curing was carried out through universal testing machine and the beams flexural strength was conducted through the three-point loading machine in accordance with ASTM C 293/C78 using ELE machine. In addition, compressive strength test was carried for cubes sets (three in each test time and average recorded) for 1,3,7 and 28-day strength using universal testing machine. Tests carried out summarized in below Table 2.

Table 2: Tests Carried out Summary

Item	Tests	Cubes	Beams	Age of Testing
Fresh Concrete	Slump			
	Unit Weight			
Hardened Concrete	Compressive Strength	X		1,3,7 & 28 days
	Flexural Strength		X	28 days
Other	Concrete Temperature Monitoring		X	24 hours

3 TEST RESULTS AND DISCUSSION

3.1 Fresh Test Results

3.1.1 Slump and Unit Weight

Unit weight, Slump results were measured and summarized below for all 12 concrete mixes as per previously illustrated experimental method in below Table 3 as follows;

Table 3: Slump and Unit Weight Test Results

Mixes	Mix 1 (300 kg Cement)			Mix 2 (350 kg)			Mix 3 (450 kg)			Mix 4 (450 kg)		
Temperature (°C)	5	25	45	5	25	45	5	25	45	5	25	45
Slump (cm)	4	0	1	5	3	1	20	19.5	18	24	22	22
Unit Weight (kg/m3)	2477	2324	2496	2436	2371	2508	2460	2372	2413	2347	2390	2339

For mixes 1&2 slump and workability was very low ranging between 0-5cm with different mixing water temperatures, however showing an increase in workability and slump with lower temperature water added (5°C). While for mixes 3 &4 with superplasticizers added a significant increase in slump was seen recording 18 – 24 cm results within the tests carried out. Similar behavior was observed in the 5°C mixing water temperature of relatively higher slump and workability. Unit weight results for all mixes were recorded between 2323 & 2508 kg/ m^3 ranges.

3.1.2 Concrete Temperature Test

Following the temperature recording by the thermocouple nodes inserted in the beams, the data flows directly to the connected data logger, and graphs can be exported and plotted for the concrete temperature progress over the 24 hours time span of the experimental work. Below graphs show the temperature recorded for the different mixes of concrete at different mixing water temperatures.

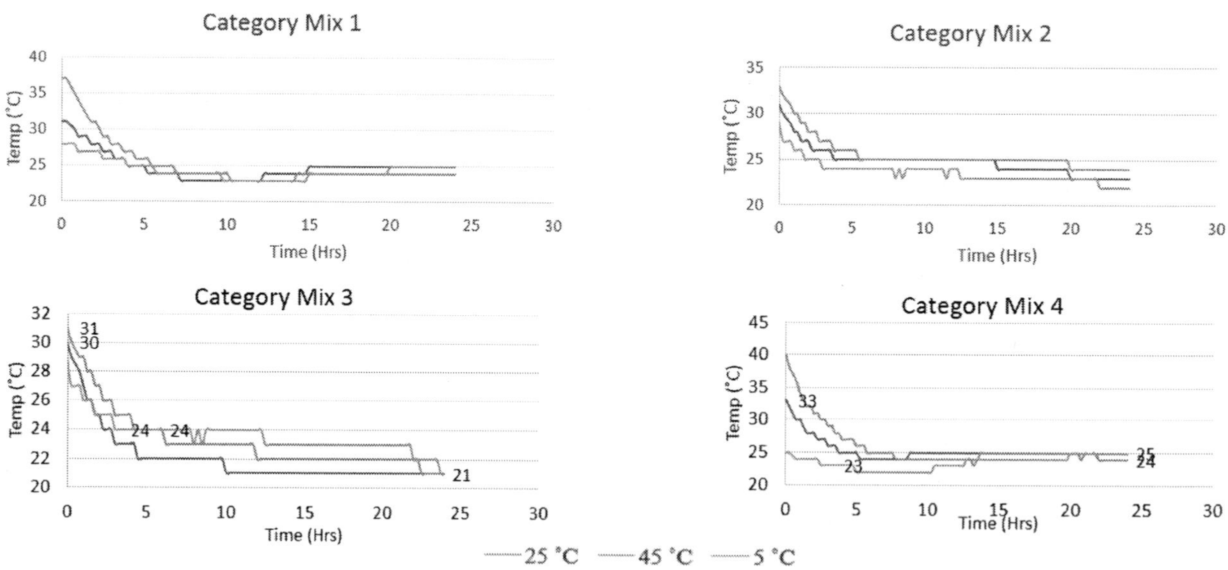

Figure 1: Concrete Temperature Results

For the first concrete mix initial recorded temperatures were at 37, 31 and 28˚C for the 45, 25 and 5 ˚C mixes respectively. Results shows a 9 ˚C drop in the concrete mix temperature with cold water when compared with the highest temperature mix. 5 ˚C concrete mix was first to reach the minimum temperature in both mixes of 22 ˚C in mix 3 in 5.25 hrs. 45 ˚C reached its min ambient temperature in 8 hrs.

Mix 2 showed a similar initial recorded temperature behavior with 45, 25 and 5 ˚C concrete mixes marking 33, 31 and 29 ˚C temperatures at time 0. 5 ˚C concrete mix showed highest rate of heat loss and reaching initial plateau of 24 ˚C after 3.25 hrs. as opposed to 45 ˚C mix reaching a similar lowest temperature after 20.25 hrs. from initial casting. 25 ˚C concrete mix should intermediate behavior in comparison reaching same temperature of 24 ˚C after 15.25 and a total minimum temperature of 23˚C after 20 hrs. from casting.

Results summarized for mix 3 in Figure 1 showedsmaller variation in initial temperatures of 31, 30 & 29 ˚C respectively after casting and at the start of temperature monitoring. This is due to a delay in casting after materials preparations which accounts for experimental error in mix 3. Concrete mix of 25 ˚C showed faster heat loss and reached plateau relatively faster than other mixes reaching a min temperature of 21 ˚C in 10 hrs. followed by 45 ˚C mix at 22.5 hrs. and finally 5 ˚C in 23.75 hrs.

For mix 4; highest initial concrete temperature was recorded at 40 ˚C for the 45 ˚C mix, followed by 33 and 25 ˚C for the 45, 25 and 5 ˚C mixes respectively. 5 ˚C concrete mix was first to reach the minimum temperature in this mix (21˚C in 10 hrs.) 45 ˚C mix reached its min ambient temperature in 10.5 hrs

3.2 Hardened Concrete Tests

3.2.1 Compressive Strength

For the hardened concrete test results, cube specimen sets of 3 each were prepared and tested after 1,3,7 and 28 days for the 12 different concrete mixes under study. The average compressive strength of the 3 cubes in every time interval was recorded to yield the below results illustrated for every concrete mix.

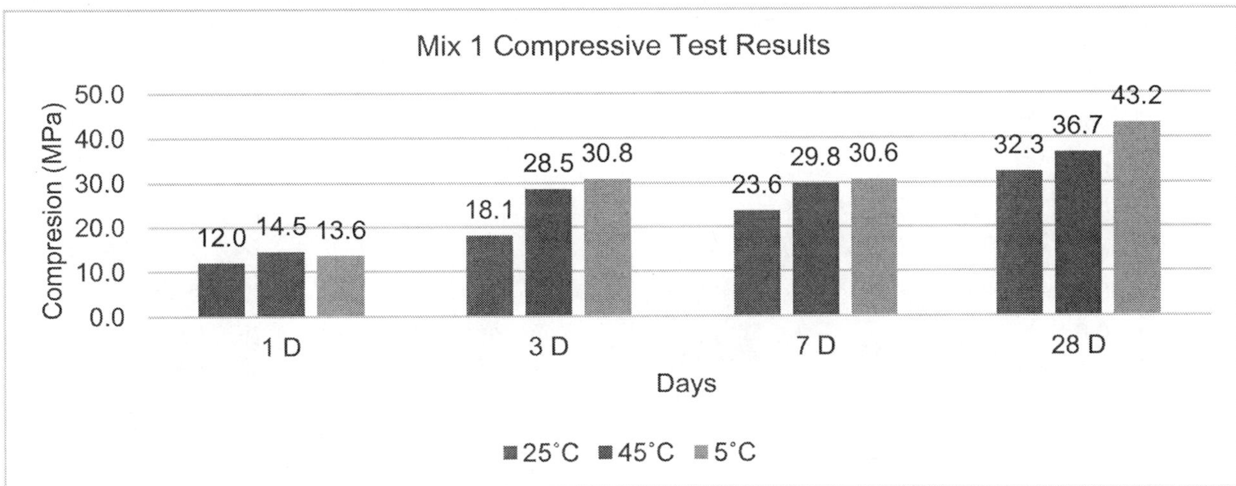

Figure 2: Mix 1 Compressive Test Results

Figure 2 shows results of Mix 1 with 300 kg cement, w/c ratio at 0.5 and no admixtures revealing 21% higher early strength at 45 ˚C mixing water of 14.5 MPa at Day1 when compared with the 25 ˚C mixing water control mix. Higher compressive strength of the 45 ˚C mix is further developed with the 3 and 7 days' cubes tested when compared with 25 ˚C, however slightly higher compressive strength is yielded with the lowest water temperature mix of 5 ˚C. Cold water concrete (5 ˚C) showed the highest % increase in strength at 3 days Vs. 1-day test with 126% increase in strength reaching 30.8 MPa when compared to 1-day strength of 13.6. Looking at the 28-day compressive test results as indicative of long term performance, cold mixing water (5 ˚C) scored highest compressive strength at 43.2 MPa at 17.7% and 25% higher strength than 25 ˚C and 45 ˚C water temperature mixes respectively.

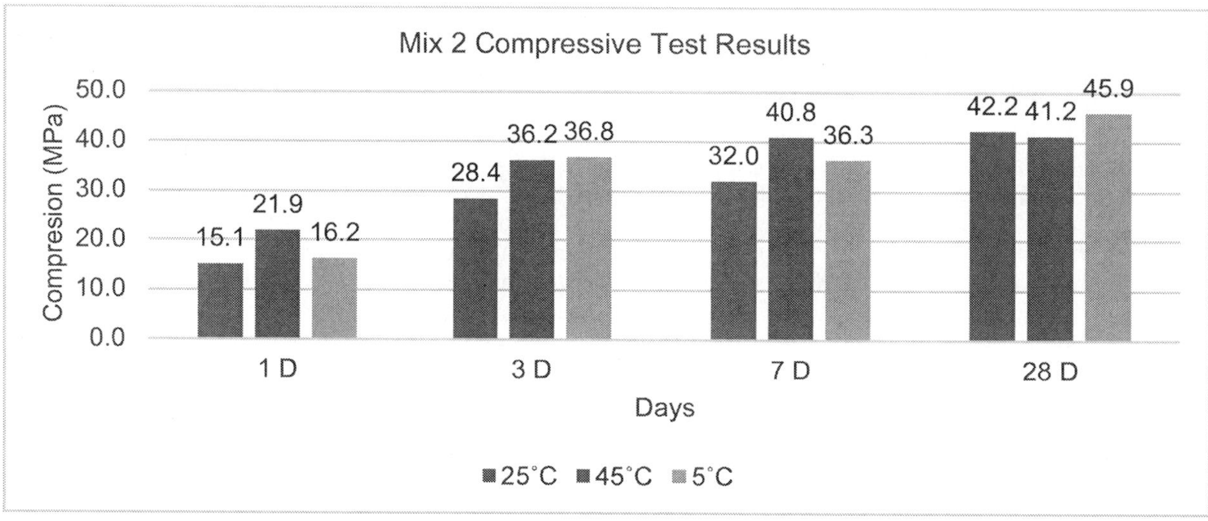

Figure 3: Mix 2 Compressive Test Results

Figure 3 give the results for the second concrete mix studied with 350 kg cement, w/c ratio at 0.4, results of the cubes tested at the different time intervals showed highest early compression at 1-day of 21.9 MPa with the 45 ˚C mixing water temperature which is 45% higher than the 25 ˚C temperature water and 35% higher than 5 ˚C water temperature cubes. Higher compressive strength of the 45 ˚C mix is further developed with the 3 and 7 days' cubes tested when compared with 25 ˚C, however slightly higher compressive strength is yielded with the lowest water temperature mix of 5 ˚C. For the high temperature water of 45 ˚C a smaller rate of strength increase is seen of 96% going from 14.5 to 28.5 MPa in 3-day test. Looking at the day-28 compressive test results, cold mixing water (5 ˚C) scored highest compressive strength at 43.2 MPa at 17.7% and 25% higher strength than 25 ˚C and 45 ˚C water temperature mixes respectively.

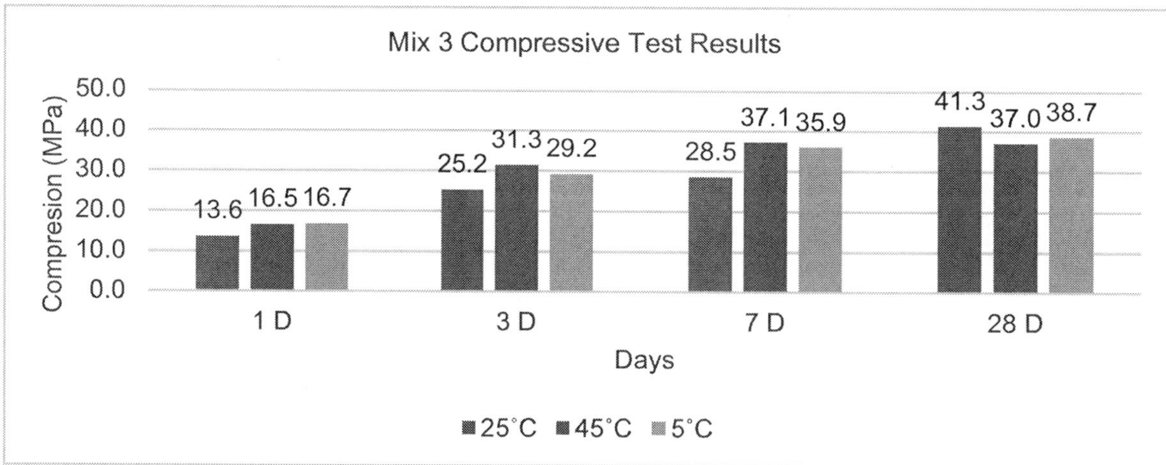

Figure 4: Mix 3 Compressive Test Results

Mix 3 with 450 kg cement showed somehow different trends from the previous two mixes studied. For 1-day compressive strength cubes tested gave comparable strength for the 5 and 45 ˚C cubes with 16.7 and 16.5 MPa respectively. For the day-3 and day-7 cubes, compressive strength increased significantly across all different temperature water mixes reaching a maximum of 31.3 MPa with 45 ˚C temperature mix followed by the 5 ˚C one with 29.2 MPa and overall average 45% Vs. 1-day results across all three mixes. Finally, for the 28 compressive test results, 25 ˚C mix yielded highest compressive results of 41.3 MPa followed by

5 ˚C mix with 38.7 MPa and minimum compressive strength was recorded for the 45 ˚C mix at 37 MPa where test specimen is shown in Figure 4 above.

While results of mix 3 showed some inconsistent and different results from the other 3 mixes in scope of this experimental work, some sources of errors could have contributed to this. This could be attributed to a time lag in concrete casting and mixing after constituent materials heating to 45 ˚C in the oven versus the other mixes. Also, the mix was casted at a lower temperature day that could have also impacted the temperature and material results yielding the results discrepancies vs. the rest of the results data sets.

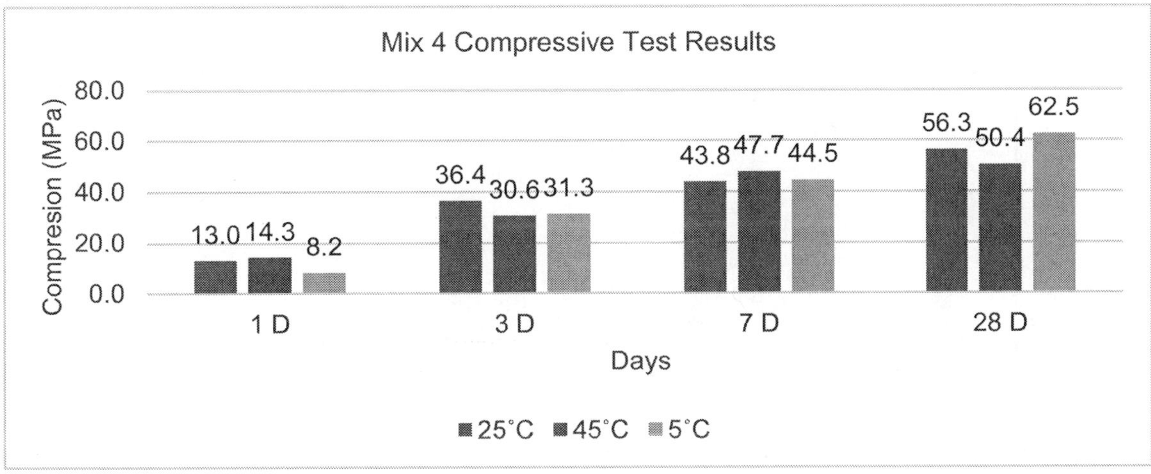

Figure 5: Mix 4 Compressive Test Results

Figure 5 summarizes the test results of the last concrete mix of 450 kg cement, superplasticizer and silica fumes showed the highest concrete results across the different time intervals. Highest early compressive strength was recorded at 14.3 MPa at 1-day with the 45 ˚C concrete mix. Followed by the 25 ˚C mix with 13 ˚C and finally lowest early strength developed with the low water temperature of 5 ˚C giving 8.2 MPa only. Similar strength development high rate was seen with the day-3 strength like in previous 3 mixes, giving an overall average 64% compressive strength increase for the 3 different mixes. The strength results of the individual mixes showed comparable results, highest of 36.4 MPa for the 25 ˚C mix followed by 31.3 MPa for the 5˚C one at 31.3 MPa. It is noted that the mix with 5 ˚C showed the highest temperature increase from 1-day to 3-day of 282% when compared with 180% increase for the 25 ˚C and lowest increase of 114% for the hot water mix of 45 ˚C.

Further strength development is yielded for the day-7 concrete mix with lower rate of strength development that that of 3-day cubes, with average % increase of 28%. For the 28 -day strength cubes, results have shown highest recorded strength of 62.5 MPa which is also the highest across all 12 concrete nixes in scope of this experimental work.

3.2.2 Flexural Test Results

For the flexural test, beams were tested for failure using the three point test machine, though applying loads in the midspan of the beams and testing failure to deliver below results;

Flexural Strength Results (MPa)

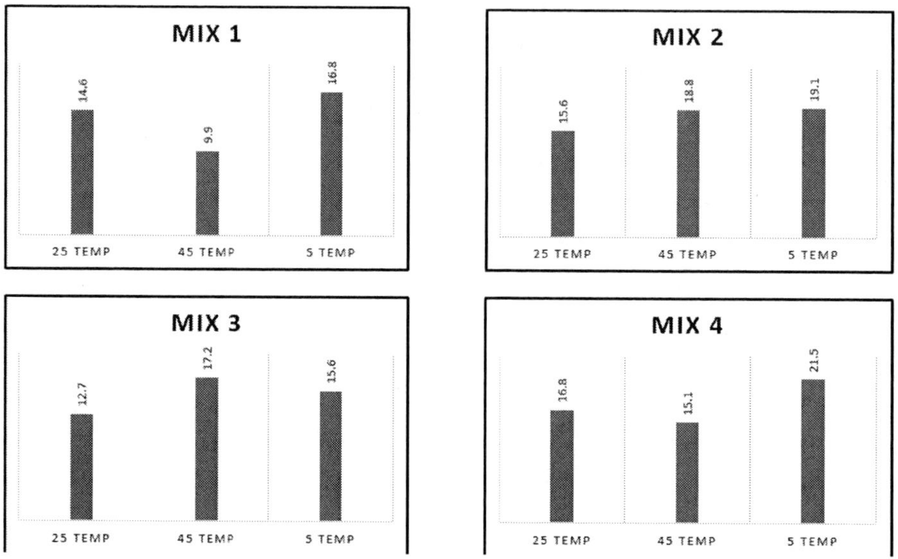

Figure 6: Flexural Test Results

Figure 6 shows the Flexural beams test results for mix 1 of 300 kg cement giving an average of 9.9, 14.6 and 16.8 MPa for the 45, 25 and 5 ˚C water respectively. This reflected that with the increase in water temperature long term structural integrity of the beams increase with 70% higher strength in the 5 ˚C water mix when compared with the 45˚C mix. Those results are consistent with the compression results of the cube showing highest 28-days strength for the 5 ˚C mix as well when compared with both other temperature mixes.

Mix 2 with 350 kg cement, slightly different data results were projected for the 28 days' test of the concrete beams under study. Whereby, highest late concrete strength was recorded for the 5 ˚C temperature mix at 19.1 MPa, followed by the 45 ˚C temp mix with flexural strength of 18.8 MPa and finally 25 ˚C temp mid with 15.6 MPa. The overall data however is showing high developed strength for the colder water mix on the long term 28-days test as opposed to hot water mixes .

Mix 3 with 450 kg cement giving higher average strength among the test results. It recorded average of 17.2, 12.7 and 15.6 MPa for the 45, 25 and 5 ˚C water respectively as shown in Figure 6. Test results however are not consistent in findings with those of the other 3 mixes, giving the highest long term strength (28 days) with the highest temperature mixing water (45 ˚C), followed by the 5 ˚C water and finally 25 ˚C mix. This is however consistent with the compressive test results carried out on the different cubes specimens in the previous section also scoring highest strength at the 45 ˚C water unlike the rest of results. Similar experimental errors would have contributed to those results, with a time duration lag between heating the water and materials, and mixing and pouring the concrete mix.

For mix 4 with 450 kg cement, flexural strength data results for the 28-days test of the concrete beams under showed highest late concrete strength for the 5 ˚C temperature mix at 21.5 MPa, followed by the 25 ˚C temp mix with flexural strength of 16.8 MPa and finally 5 ˚C temp with 15.11. Results come in line with the compressive test results of this mix which gave highest recorded strength for the cubes at the 5 ˚C. results of the mix reinforce the different results delivered throughout the test carried out of having higher long term strength when using cold water recording 28% higher than 25 ˚C mix and 38% higher than hot water mix of 45 ˚C.

4 CONCLUSIONS

In light of materials and procedures used, illustrated results, experimental work and analysis, the following conclusions can be warranted:

1. The fresh concrete mixes with lower mixing water temperature demonstrated a relative increase in the slump test results as recorded in the mixes studied. Such slump increase reflects a general enhanced workability for the concrete mix.

2. Thermocouple data results showed that decreasing mixing water temperature can induce a notable reduction in initial fresh concrete mix temperature during casting and mixing, making it a key parameter in controlling / influencing concrete mix temperatures; especially in hot weather conditions.

3. Thermocouple data further showed higher heat of hydration developing in higher mixing water temperature of 45°C studied, reaching plateau temperatures in relatively higher times than low temperature water mixes of 5°C.

4. Compressive and flexural test results reflected a clear impact of mixing water temperature on the concrete strength developed for the different stages of concrete strength development.

5. For early compressive strength of 1-day, results suggest that higher early strengths developed when increasing the mixing water temperature. This is likely due tp faster exothermic reaction of water with cement material and stronger initial bonding for the different concrete constituents.

6. For compressive strength of 28 days, colder mixing water yielded much higher compressive strength for the concrete studied when compared with higher temperature mixing water. This is attributed to more cracks developed due to higher initially yielded heat with higher mixing water temperature affecting concrete properties.

7. Flexural beams strengths are also reflecting higher developed strengths for the lower mixing water temperature when compared with high mixing water of 45°C used in testing, suggesting better properties, strength and durability of concrete structures.

8. Superplasticizers, admixtures and silica fumes can directly enhance workability and strength of concrete when coupled with lower temperature water in concrete mix.

5 RECOMMENDATIONS

This experimental work conducted in the work scope herein has identified some areas where recommendations can be made and further studies to be conducted including the following:

9. Increasing water temperature variability for a sensitivity analysis to reach optimal mixing temperature that can be later used as reference for concrete manufacturing in hot weather conditions.

10. Using more controlled temperature simulation throughout the experimental work, possibly constructing an environmental chamber to mimic the heat and sun exposure of concrete during casting in elevated temperatures and severe summer conditions.

11. Monitoring the concrete strength for longer period of time (90 days) to further study the structural behavioral and durability of the concrete structure under analysis.

12. Studying the mixed effect of cooling water in addition to other concrete materials (cement or aggregates) during mixing and casting and see any incremental benefits reflected on concrete properties.

Acknowledgements

This publication is based on the work conducted in the Construction Engineering Materials Lab in the American University in Cairo by Mennatallah Assal and under the supervision of Professor Mohamed Nagib Abou-Zeid.

References

Hover, K. (2005). Hot Weather Concrete. STRUCTURE, 27.

Nasir, M., Al-Amoudi, O. S. B., Al-Gahtani, H. J., &Maslehuddin, M. (2016). Effect of casting temperature on strength and density of plain and blended cement concretes prepared and cured under hot weather conditions. Construction and Building Materials, 112, 529-537.

Neville, A. M. (1995). Properties of concrete (Vol. 4). London: Longman.

Ortiz, J., Aguado, A., Agulló, L., & García, T. (2010). Influence of environmental temperatures on theconcrete compressive strength: Simulation of hot and cold weather conditions. *Cement and concrete research*, *35*(10), 1970-1979.

Building Tomorrow's Society

Bâtir la Société de Demain

Fredericton, Canada

June 13 to 16, 2018 / *13 au 16 juin, 2018*

EFFECT OF SHRINKAGE REDUCING ADMIXTURES ON SHRINKAGE BEHAVIOUR OF GEOPOLYMER MORTAR

Alhamdan, Ahmad[1,3] and Soliman, Ahmed[2]
[1]MASc. Student, Department of Building, Civil and Environmental Engineering Concordia University, Montreal, Quebec Canada.
[3] ahmad.alhamdan@mail.concordia.ca
[2]Assistant Professor, Department of Building, Civil and Environmental Engineering Concordia University, Montreal, Quebec Canada.

Abstract: Alkali-activated binder (so-called geopolymer) is an eco-friendly binder alternative for cement. It is characterized by a very low carbon footprint. However, shrinkage of geopolymer materials is still one of the main challenges that impede its in-situ applications. The purpose of this study is to investigate the shrinkage behaviour of geopolymer mortars activated with different activator concentrations and incorporating different dosages of shrinkage reducing admixture (SRA). Moreover, it explores the influence of SRA on the correlation between shrinkage behaviour and different factors including the concentration of activator and curing temperature. Geopolymer mortars were prepared using the conventional two-part alkali activation mechanism. Specimens were exposed to different curing conditions. A series of tests including setting time, flowability, drying shrinkage, and compressive strength of geopolymer mortar were conducted on mixtures with and without SRA. Regardless of the activator dosage, results showed that similar to cement-based materials, SRA was effective in reducing the shrinkage strains experienced by geopolymer mortars. However, the compressive strength was decreased. Furthermore, SRA effectiveness increased with higher curing temperatures and activator concentrations.

1 INTRODUCTION

Global concrete production has grown widely over the recent decades parallel with the population growth and the need for more infrastructures. The high concrete demand in the construction industry increased ordinary Portland cement (OPC) production. The significant environmental pollution associated with carbon dioxide (CO_2) emissions generated during cement manufacturing process are of major concerns. Alkali-activated binders represent a promising alternative to OPC in many constructional applications (Yang et al. 2013).

The acceptance of alkali activated materials in the industry has been hindered by concerns related to their shrinkage. Shrinkage is a critical parameter affecting the durability of the alkali activated material and its resistance to cracking. Shrinkage can be defined as a reduction in the volume of the material due to either drying or chemical reactions taking place in the system. These reductions in volume induce stresses in the material matrix causing deformations. As deformations are restrained in the hardened binder, cracking occurs. The formation of cracks allows the ingression of aggressive chemicals into the concrete, therefore, significantly reducing its service life.

The world's first shrinkage-reducing admixture (SRA) was developed in Japan in 1982 in a partnership between Nihon Cement Co., Ltd. and Sanyo Chemical Industries, Ltd. (Sato et al. 1983). Several different SRAs are commercially available, and although they do not exactly have the same chemical composition, they are similar in nature and all serve to reduce the surface tension of concrete's pore fluid (Nmai et al. 1998). It is commonly understood that by reducing the surface tension of concrete's pore fluid, SRA significantly reduces the magnitude of capillary stresses and shrinkage strains that occur when concrete is losing moisture (i.e., through drying or self-desiccation) (Shh et al. 1992, Tazawa and Miyazawa 1995). It has been shown that by lowering the surface tension of pore fluid, SRA can reduce drying shrinkage of concrete by up to 50% (Shh et al. 1992). In addition, SRA was shown to reduce the capillary stresses that are generated by autogenous shrinkage in high-performance low w/c concretes (Weiss et al. 1999, Bentz et al. 2001). These attributes are extremely beneficial in reducing the risk of shrinkage cracking in concrete elements (Radlinska et al. 2007).

The addition of SRA was also shown to improve the resistance of concrete against plastic shrinkage cracking (Lura et al. 2007). In contrast to the significant benefits of SRA in improving the shrinkage performance of concrete, some negative side effects have also been reported. Namely, it has been observed that SRA reduces the rate of cement hydration and strength development in concrete (He et al. 2006, Bentz 2006). Concrete containing SRA exhibits a delay in setting (Brooks et al. 2000) and a reduction in the compressive and tensile strength for well-cured specimens (Weiss et al. 1999), compared to control specimens with similar mixture proportions. Moreover, it has been reported that the addition of SRA may cause loss of entrained air, which may negatively affect the resistance of concrete to freezing damage (Berke et al. 2003).

Although extensive research has been performed on measuring the shrinkage and mechanical properties of concrete containing SRA, few studies addressed the effect of SRA in alkali-activated concrete. Alkali concentration, binder type, curing regime, and SRA dose, are all important parameters that would give an image of the effect of SRA on fresh, and hardened properties of alkali-activated concrete.

2 EXPERIMENTAL PROGRAM

2.1 Materials

In the present study, FA class-F was used as the precursor. The chemical compositions and physical properties of the FA class-F are listed in **Table 1**. Sodium hydroxide with a purity level of 98% and sodium silicate (SiO_2 (30%), Na_2O (20%), H_2O (50%), specific gravity 1.38 g/mL) were used to prepare the alkali activators. The alkali activator was prepared as the following: Distilled water was used to dissolve the NaOH solid at 3M, 6M, and 12M, and left overnight to cool down to the room temperature while covered to prevent carbonation. Sodium silicate solution was then mixed with the NaOH solution to form the alkali activator solution (mass of sodium silicate solution/ NaOH solution = 2). Commercial shrinkage reducing admixture conforming to the requirements of ASTM Standard C494 / C494M - 17, Type S was used in this study. ASTM Standard C109 / C109M-16a river sand was used as the fine aggregate material. The oven-dry, saturated surface dry specific gravity, and absorption of the standard sand used were 2.54, and 2.55, 0.08 %, respectively.

Table 1: Properties of the precursor used.

Chemical composition (%)	Fly ash
CaO	2.60
SiO_2	46.00
Al_2O_3	33.00
Fe_2O_3	10.50
SO_3	-
Physical properties	
S.S.D Specific gravity (g/cm^3)	2.38
Surface area (m^2/kg)	290 (Blaine)

2.2 Mix proportion and specimen preparation

The mixture proportions of the alkali-activated fly ash (AAF) mortars are shown in **Table 2**. Nine AAF mortar mixtures were designed using a solution made of three different NaOH concentrations to cover a wide range of concentrations (3, 6, and 12M) each mixed with a constant sodium silicate concentration. The mass ratio of sand to precursor was fixed at 2:1 for all mixtures. The mass ratio of activator solution to precursor was 0.40 for all mixtures. The mass ratio of the water glass to NaOH solution was 2.0 in all activator mixtures. Two curing regimes were applied: 23°C with relative humidity of 60 ± 5% (ambient) and at 60°C (oven). Specimens were cured under oven conditions to enhance the alkali activation and increase the degree of the polymerization reaction in AAF mortars.

2.3 Specimen preparation

Initially, fly ash was dry mixed with sand in an electric mixer at a slow speed of (140 ± 5 revs/min) and medium speed (285 ± 5 revs/min) for 2 minutes to ensure homogeneity of the mixture. The prepared alkali activator solution is then added to the mixture and mixed for an additional 2 minutes. Immediately after mixing, 50 mm cubic molds for compressive strength evaluation according to ASTM Standard C109 / C109M-16a, and 25 mm × 25 mm × 285 mm prismatic molds for shrinkage measurements according to ASTM Standard C490 / C490M – 17, were cast. Specimens were compacted using a vibrating table and a tamping rod. All specimens were cured at room temperature of 23 °C and at a relative humidity of 60% ± 5% inside molds for 24 hours. After 24 hours, specimens were demolded and stored in ambient room conditions or in the oven at 60°C until testing age. All reported results represent the average of three replicates.

Table 2: Mix proportion of AAF mortar specimens

Mixture	NaOH solution molarity (M)	Ratio NaOH: water glass	Sand/binder ratio	SRA dosage (% of binder mass)
F3	3	1:2	2:1	0
F6	6	1:2	2:1	0
F12	12	1:2	2:1	0
F3R1	3	1:2	2:1	1
F6R1	6	1:2	2:1	1
F12R1	12	1:2	2:1	1
F3R2	3	1:2	2:1	2
F6R2	6	1:2	2:1	2
F12R2	12	1:2	2:1	2

2.4 Testing program

The workability and flow of mortar mixtures were evaluated using the procedure listed in ASTM Standard C1437-15. Initial and final setting times were measured using a Vicat needle as described in ASTM Standard C191-13. Compressive strength development was evaluated at ages of 1, 7, 14, and 28 days using a 300 kN universal testing machine (UTM) in accordance with ASTM Standard C109 / C109M-16a. Cubes were loaded until failure at a load rate of 345 kPa/s. For each mortar and at each age, 3 cubes were tested, and the results were averaged. Drying shrinkage was measured using prism samples (25.4 × 25.4 × 254 mm) in accordance with ASTM Standard C490 / C490M - 17 using a length comparator with dial gauge accuracy of 0.0001-inch.

3 EXPERIMENTAL RESULTS

3.1 Flowability

Flow table results for the mortar mixtures with and without SRA are presented in **Table 3**. Generally, the flow values of AAF mortar mixtures without admixtures are high, this was attributed to the spherical shape and large specific surface area of FA particles (Fernández-Jimenez et al. 2006). Compared to mortars without admixtures, the flow values were slightly increased by the addition of SRA admixture, hence showing a positive effect on mortar workability. For example, the flowability values for AAF mortars activated with 3M, 6M, and 12M NaOH concentrations, and containing 1% SRA were increased in the range of 1-4%, compared to AAF mortars without SRA. Increasing the dose of SRA to 2% by mass of fly ash, induced even further increase in the flowability. The addition of liquid SRA does increase the specific liquid to solid ratio hence, directly influencing the flowability of mortar specimens (Palacios and Puertas 2005). The flowability values for AAF mortar specimens containing 2% SRA and activated with 3M, 6M, and 12M NaOH concentrations were increased by 1-2%, compared to specimens containing 1% SRA.

3.2 Setting time

The initial and final setting times of AAF mortar specimens are given in **Table 3**. It is well known that a binder should have an acceptable setting time to allow its transportation and casting. If setting time is too long, it is not economical. Contrary to that, irreversible mistakes might happen if the setting time is very short. Thus, the effect of SRA admixtures on the initial and final setting time of AAF mortars was noted in this study. Generally, the setting times of AAF mortars specimens decreased with increasing sodium hydroxide concentrations in the used activator as shown in **Table 3**. The addition of SRA admixture retarded both the initial and final setting times of AAF mortars. The retardation effect is even more pronounced when the percentage of SRA is increased from 1% to 2% (by mass of binder). For instance, the addition of 1% SRA admixture retarded the initial setting time of AAF mortars activated with 3M, 6M, and 12M NaOH by 6, 12, and 13 min, respectively compared to specimens without SRA. The corresponding retardation in final setting time were 8, 10, and 10 min, respectively. Moreover, increasing the dose of SRA admixture to 2% caused an initial setting time retardation of 11, 15, 19 min, respectively and a final setting time retardation of 16, 17, and 18 min, respectively compared to specimens without SRA.

Table 3: Setting time and flowability index of AAF mortar specimens

Mixture	Setting time (min)		Flowability (cm)
	Initial	Final	
F3	144	286	108
F6	126	254	102
F12	89	186	92
F3R1	150	294	112
F6R1	138	262	104
F12R1	102	196	93
F3R2	155	302	114
F6R2	141	271	105
F12R2	108	204	94

3.3 Compressive strength

3.3.1 Effect of SRA on compressive strength of AAF mortars under ambient curing

The development of compressive strength for all AAF mortars, with and without SRA, was studied according to activator concentration, curing condition, SRA content, and age. **Figure 1** shows the compressive strength values over time of 3M, 6M, and 12M NaOH activated fly ash mortar specimens ambient cured. Generally, the addition of SRA reduced the compressive strength of AAF mortars. The addition of 1% SRA admixture caused AAF mortars activated with 3M, 6M, and 12M NaOH to encounter a reduction of 5.88, 11.29, and 9.8% in the 28-day compressive strength values respectively, compared mortars without SRA. The corresponding reduction percentages for the 2% SRA containing AAF mortars were 11.99, 18.02, and 17.2% compared to mortars without SRA.

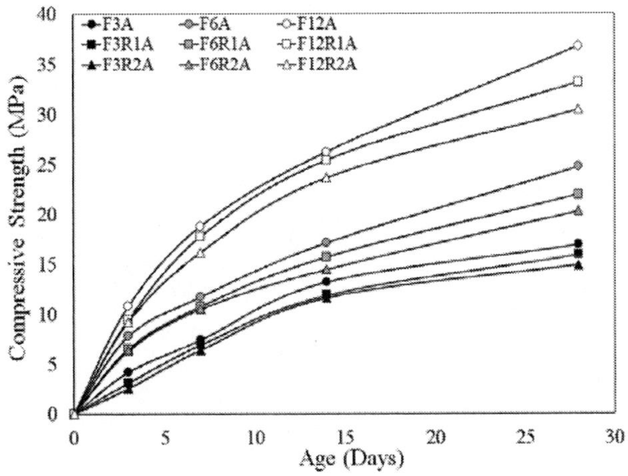

Figure 1: Compressive strength of AAF mortars containing 0, 1, and 2% SRA ambient cured.

Figure 2 shows the compressive strength values over time for 3M, 6M, and 12M NaOH activated fly ash mortars oven cured. The addition of SRA reduced the compressive strength of oven cured mortar specimens. Reductions in compressive strength varied in the range of 6.25-17.23% depending on activator concentration and SRA dosage used. For example, the addition of 1% SRA admixture caused AAF mortars activated with 3M, 6M, and 12M NaOH to encounter a reduction of 9.34, 10.47, and 8.05% in the 28-day compressive strength values respectively, compared to mortars without SRA. The corresponding reduction percentages for the 2% SRA containing AAF mortars were 16.55, 17.23, and 13.97% compared to specimens without SRA.

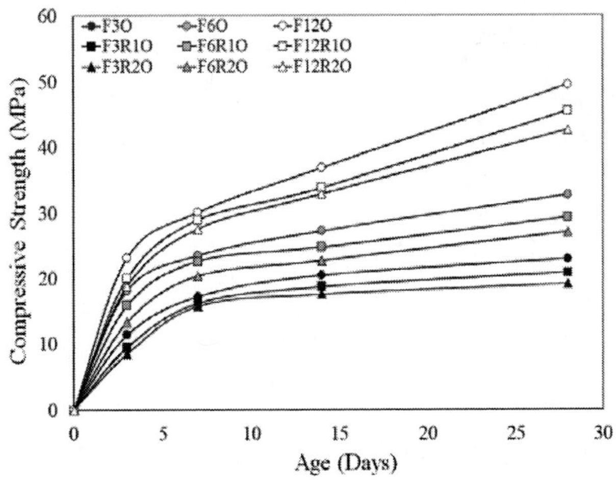

Figure 2: Compressive Strength of AAF mortars containing 0, 1, and 2% SRA oven cured.

3.4 Drying Shrinkage

3.4.1 Effect of SRA on drying shrinkage of AAF mortars under ambient curing

The effect of SRA addition on the drying shrinkage strains of AAF mortar specimens was studied. Drying shrinkage strains of ambient cured fly ash mortars containing 0, 1, and 2% SRA and activated with 3M, 6M, and 12M (NaOH) concentrations are plotted in **Figs. 3-4,** respectively. Results indicated reductions in the drying shrinkage of AAF mortars with the addition of 1% SRA compared to samples without SRA. Reductions in drying shrinkage were more pronounced when SRA dose was increased to 2%. Reduction percentages in the total drying shrinkage of AAF mortars activated with 3M, 6M, and 12M NaOH, caused by the addition of 1% SRA were 26.79, 16.50, and 25.94% respectively, compared to mortars without SRA. The corresponding reduction percentages in total drying shrinkage, when 2% SRA was used were 36.71, 26.58, and 35.96% respectively.

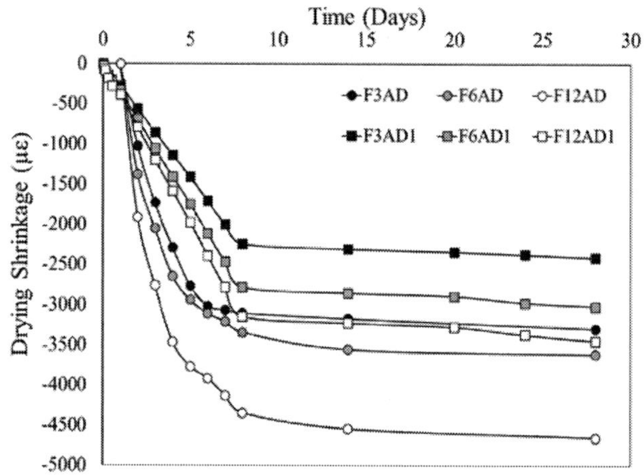

Figure 3: Drying shrinkage of AAF mortars containing 1% SRA (ambient cured).

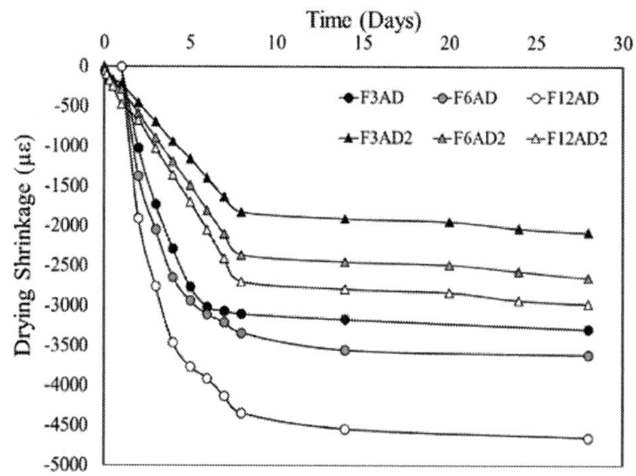

Figure 4: Drying shrinkage of AAF mortars containing 2% SRA (ambient cured).

3.4.2 Effect of SRA on drying shrinkage of AAF mortars under oven curing

Drying shrinkage strains of oven cured Fly ash mortars containing 0, 1, and 2% SRA and activated with 3M, 6M, and 12M (NaOH) concentrations are plotted in **Figs. 5-6**, respectively. Results indicated reductions in the drying shrinkage of AAF mortars with the addition of 1% SRA compared to samples without SRA. Reductions in drying shrinkage were higher when SRA dose was increased to 2%. Reduction percentages in the total drying shrinkage of AAF mortars activated with 3M, 6M, and 12M NaOH, caused by the addition of 1% SRA were 42.42, 45.31, and 46.04% respectively, compared to mortars without SRA. The corresponding reduction percentages in total drying shrinkage, when 2% SRA was used were 52.45, 51.95, and 54.52% respectively.

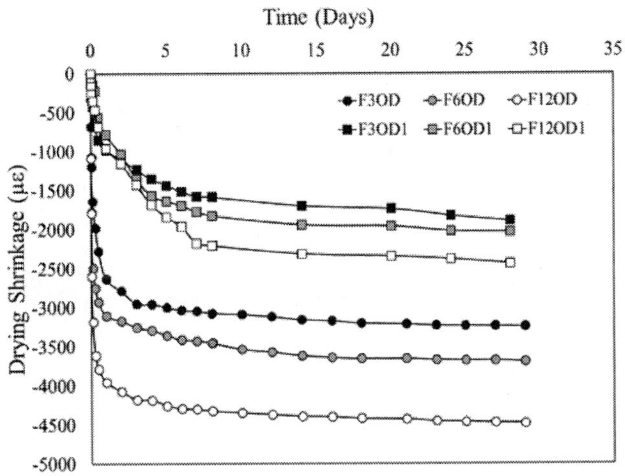

Figure 5: Drying shrinkage of AAF mortars containing 1% SRA (oven cured).

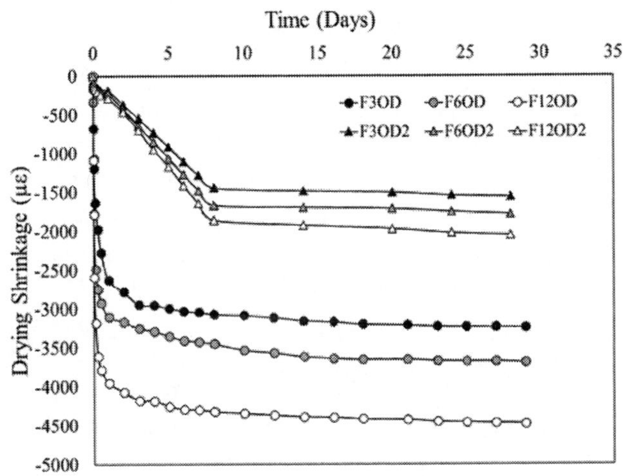

Figure 6: Drying shrinkage of AAF mortars containing 2% SRA (oven cured).

4 CONCLUSION

- Shrinkage reducing admixtures increased the flowability in all alkali activated mortar specimens tested.
- Addition of SRA retarded both the initial and final settings of AAF mortars.
- The addition of SRA reduced the overall compressive strength of alkali activated mortar specimens. Reductions were higher when SRA dose was increased from 1 to 2%.
- The addition of SRA significantly reduced the drying shrinkage of mortar specimens. Reductions were more pronounced when SRA dose was increased from 1 to 2%.

- Reductions in drying shrinkage reached up to 67.08%. The main mechanism of SRA that is responsible for the beneficial shrinkage reductions in AAM, is found to be the reduction in surface tension of pore water prompted by the admixture.

References

ASTM C109 / C109M-16a, Standard Test Method for Compressive Strength of Hydraulic Cement Mortars (Using 2-in. or [50-mm] Cube Specimens), ASTM International, West Conshohocken, PA, 2016, www.astm.org

ASTM C1437-15, Standard Test Method for Flow of Hydraulic Cement Mortar, ASTM International, West Conshohocken, PA, 2015, www.astm.org

ASTM C191-13, Standard Test Methods for Time of Setting of Hydraulic Cement by Vicat Needle, ASTM International, West Conshohocken, PA, 2013, www.astm.org

ASTM C490 / C490M-17, Standard Practice for Use of Apparatus for the Determination of Length Change of Hardened Cement Paste, Mortar, and Concrete, ASTM International, West Conshohocken, PA, 2017, www.astm.org

ASTM C494 / C494M-17, Standard Specification for Chemical Admixtures for Concrete, ASTM International, West Conshohocken, PA, 2017, www.astm.org

ASTM C596-09 (2017), Standard Test Method for Drying Shrinkage of Mortar Containing Hydraulic Cement, ASTM International, West Conshohocken, PA, 2017, www.astm.org

Bentz, D. P. 2006. Influence of Shrinkage-Reducing Admixtures on Early-Age Properties of Cement Pastes. *Journal of Advanced Concrete Technology*, **4**(3): 423-429.

Bentz, D. P., Geiker, M. R., and Hansen, K. K. 2001. Shrinkage-Reducing Admixtures and Early-Age Desiccation in Cement Pastes and Mortars. *Cement and Concrete Research*, **31**(7): 1075-1085.

Berke, N. S., Li, L., Hicks, M. C., and Bae, J. 2003. Improving Concrete Performance with Shrinkage-Reducing Admixtures. *Special Publication*, **217**: 37-50.

Brooks, J. J., Johari, M. M., and Mazloom, M. 2000. Effect of Admixtures on the Setting Times of High-Strength Concrete. *Cement and Concrete Composites*, **22**(4), 293-301.

Fernández-Jimenez, A. M., Palomo, A., and Lopez-Hombrados, C. 2006. Engineering Properties of Alkali-Activated Fly Ash Concrete. *ACI Materials Journal*, **103**(2): 106.

He, Z., Li, Z. J., Chen, M. Z., and Liang, W. Q. 2006. Properties of Shrinkage-Reducing Admixture-Modified Pastes and Mortar. *Materials and Structures*, **39**(4): 445-453.

Lura, P., Pease, B., Mazzotta, G. B., Rajabipour, F., and Weiss, J. 2007. Influence of Shrinkage-Reducing Admixtures on Development of Plastic Shrinkage Cracks. *ACI Materials Journal*, **104**(2): 187-194.

Ma, Y., Hu, J., and Ye, G. 2013. The Pore Structure and Permeability of Alkali Activated Fly Ash. *Fuel*, **104**: 771-780.

Nmai, C., Tomita, R., Hondo, F., and Buffenbarger, J. 1998. Shrinkage-Reducing Admixtures. *Concrete International*, **20**(4): 31-37.

Palacios, M., and Puertas, F. 2005. Effect of Superplasticizer and Shrinkage-Reducing Admixtures on Alkali-Activated Slag Pastes and Mortars. *Cement and Concrete Research*, **35**(7): 1358-1367.

Radlinska, A., Pease, B., and Weiss, J. 2007. A Preliminary Numerical Investigation on the Influence of Material Variability in the Early-Age Cracking Behavior of Restrained Concrete. *Materials and Structures,* **40**(4): 375-386.

Sato, T., Goto, T., and Sakai, K. 1983. Mechanism for Reducing Drying Shrinkage of Hardened Cement by Organic Additives. *CAJ Rev*, **5**: 52-55.

Shh, S. P., Krguller, M. E., and Sarigaphuti, M. 1992. Effects of Shrinkage-Reducing Admixtures on Restrained Shrinkage Cracking of Concrete. *Materials Journal*, **89**(3): 289-295.

Tazawa, E. I., and Miyazawa, S. 1995. Influence of Cement and Admixture on Autogenous Shrinkage of Cement Paste. *Cement and Concrete Research*, **25**(2): 281-287.

Weiss, W. J., Borischevsky, B. B., and Shah, S. P. 1999. The Influence of a Shrinkage Reducing Admixture on the Early-Age Behavior of High Performance Concrete. *In Fifth International Symposium on the Utilization of High Strength/High Performance Concrete*, Sandefjord, Norway, **2**: 1418-1428.

Yang, K. H., Song, J. K., and Song, K. I. 2013. Assessment of CO2 Reduction of Alkali-Activated Concrete. *Journal of Cleaner Production*, **39**: 265-272.

Building Tomorrow's Society

Bâtir la Société de Demain

Fredericton, Canada

June 13 to 16, 2018 / *13 au 16 juin, 2018*

EFFECTIVE USE OF RECYCLED ASPHALT SHINGLES FOR GRANULAR MATERIALS

MacEachern, Matthew[1, 2], Sanchez-Castillo, Xiomara[1] and Oh, Won-Taek[1]
[1.] University of New Brunswick, Canada
[2.] mmaceac1@unb.ca

Abstract: The province of Prince Edward Island (PEI) imports 20,000 to 100,000 tons of aggregates per year for construction purposes. It is estimated that close to 8000 tons of tear-off shingles (end of life roofing shingles) are sent to local landfills annually. To reduce aggregate import costs and to utilize Recycled Asphalt Shingles (RAS), the University of New Brunswick (UNB) has been working closely with the province to investigate the use of RAS as a partial replacement of high quality quarried aggregates (Granular Class A) for its potential use as a granular base in pavements. The main objective of the present study is to suggest an optimum combination of RAS and Granular Class A. The individual gradation curves for each material were obtained and the theoretical design indicated that a replacement of up to 30% of RAS by weight could meet provincial specifications. Three suitable combinations of Granular Class A/RAS were selected (90/10, 80/20, and 70/30) for testing and quantifying the impact of RAS in the mechanical properties of the material. Proctor tests were conducted to determine the Optimum Moisture Content (OMC) for each combination. Based on the standard proctor test results, California Bearing Ratio (CBR) tests were then carried out for the mixtures compacted at OMC. Results showed that the addition of RAS significantly reduced CBR results by making the material up to 85% weaker than that of just Granular Class A material alone.

1 INTRODUCTION

The province of Prince Edward Island (PEI) decided to explore the possibility of using recycled materials as an additive to their granular material that is used for road construction and backfilling. The province has reached out to the University of New Brunswick (UNB) to study the effects of the addition of Recycled Asphalt Shingles (RAS) to their granular material. A full testing plan has been laid out to potentially put to use some of the approx. 8000 tonnes of shingles being disposed of in local landfills annually. It is the goal of this research to find an optimum mixture of shingles that meets the physical requirements that the current granular A performs at; as well as assuring the mixture meets all material characteristics outlined in the provinces contract specifications. As a first step of this research, this paper focuses on the California Bearing Ration (CBR) testing results with the addition of 10, 20 and 30% RAS to the currently used granular material.

2 BACKGROUND AND LITERATURE REVIEW

The recycling and subsequent use of old asphalt shingles have been gaining more popularity in the past few years. RAS is recently being incorporated into hot mix asphalt designs across North America. This is mainly because the incorporation of RAS can reduce not only cost but also conserves significant amounts of materials without sacrificing the strength and quality of the asphalt material. However, increased asphalt hardness and the extra amount of money spent on cleaning and preparing the asphalt shingles for use are considered as concerns with using RAS in asphalt mixes. According to Soleimanbeigi et al. (2012), RAS has great potential to be used as structural fill due to its better dust control and reduced maintenance requirements. However, its current use is restricted to field testing and research due to it being a relatively new material for fill applications. RAS has a potential application to improve roadbeds; however, their effectiveness will depend on the characteristics of the granular material and the percentage of replacement.

Warner et al. (2012) discussed the possible use of recycled asphalt shingles (RAS) in aggregate base (AB) and aggregate subbase (ASB) in roadway construction. In this research, different ratios of RAS and soils native to Wisconsin, US were blended and tested for characteristics such as California Bearing Ratio, Proctor, Resilient Modulus along with others. The addition of fly ash and asphalt emulsion to attempt to improve test results was also discussed. The results of the study found that there are large inconsistencies in the RAS depending on the source as there was no mandatory method to create the RAS in the US at the time of the study. It was also found that just the simple addition of RAS to the aggregate proved to be insufficient for AB; however, a 50:50 ratio of backfill material and RAS proved to be acceptable for backfilling applicatons. Results found that even though the addition of fly ash to help stabilize the RAS showed improvement in the resilient modulus, it was still not enough to deem the mixture acceptable.

Yang et al. (2014) investigated the use of RAS in combination with recycled asphalt pavement (RAP) to form hot mix asphalt (HMA). The investigation looked at different combinations of RAS and RAP and their effects on such asphalt characteristics as stiffness, resilient modulus, fracture temperature, air void percentage, surface characteristics and life expectancy. It was found that the addition of RAS and RAP caused the asphalt to perform more stiffly at higher temperatures. The resilient modulus was lowest when the mix contained 12% RAP and 3% RAS. In the end, it was determined that mixtures containing RAS and RAP preforms well in the short term.

Townsden et al (2007) investigated the effects of using RAS in construction materials on the environment regarding asbestos contents in the shingles. The paper looked at samples taken from several states in the US. It was found during an asbestos case study that out of 27,694 samples taken from across six states, an average of 1.48% contained asbestos. However, in that study, it showed that some states had no samples containing asbestos while others had up to 16.7% of samples containing asbestos. This shows that although there is a low chance or encountering asbestos there is still a chance; therefore, RAS manufactures need to continue testing samples for asbestos. It was discussed that asbestos was found more in roll out shingles than tabbed shingles. Aside from the focus being on asbestos contents, the study also found that RAS has been used in Iowa in combination with gravel to be used on unpaved roads. The mixture was proven to yield less dust and have a longer life than regular gravel roads.

Shrestha et al. (2008) investigated not only combining RAS with granular materials but also selecting the size of the RAS particles. The study focused on two grades of RAS: a fine sample that had all particles passing the 4.75mm sieve and the other which had a nominal aggregate size of 75mm with only 40% passing the 4.75mm sieve. This unique research looked at using the RAS for more than just the replacement of fine particles. The RAS was tested while being mixed with five different aggregate materials. The materials used were three types of recycled concrete aggregate (RCA), quarried crushed limestone as well as crushed gravel containing 72% crushed rock particles. The study focused mostly on how the different mixes preformed in the CBR tests. During CRB testing, the study looked at mixing the five granular materials with 0, 3, 5, 8, 10 and 15% RAS contents. CBR testing, proved that the RAS had no benefit and, in some cases, reduced the CBR of the material. However, the reduction of the CBR was not significant in any case and still passed necessary requirements for the Ontario standard. It was also found that with the addition of 5% and 8% in certain cases made the material more stable. As well, the drainage capability of the different materials was not deemed to be significantly altered with the addition of RAS.

Lee et al. (2010) analyzed a road project in Wisconsin focusing on how the use of recycled materials saved in both the environmental and economic aspects. The study found that the use of recycled material in the base and subbase layers during road construction could result in a 20% reduction in the potential for global warming. Aside from that, reductions in energy use, water consumption and hazardous waste generations were all said to be reduced by 16%, 11% and 6% respectively. The study showed that the use of recycled materials could result in a 21% reduction in the overall life-cycle cost.

3 MATERIALS AND TESTING METHODS

For this research, PEI supplied UNB with the necessary materials to be tested. The granular material was a sample of their current stock being used on the island classified under their specifications as Granular A. The sample of RAS was supplied by a local island contractor that was importing samples from a produced in the Unites States. All testing was performed under the assumption that the RAS acted as a granular material, so the same testing standards were used for both the granular and RAS material. The following sub sections explain the different stages of testing performed in order to complete CBR testing.

3.1 Mixing Combinations

Based on the individual size distribution of the materials it was determined that a maximum theoretical percentage of replacement of up to 30% RAS could be incorporated and still meet the specified requirements for granular A. In order to evaluate the optimum mix percentage of the combination of granular and RAS particles, three different mixing combinations by weight were used in the present study (Table 1).

Table 1: Mixing Combinations (by weight)

Test Condition Number	Granular A Material (%)	Recycled Asphalt Shingles (%)
1	100	0
2	90	10
3	80	20
4	70	30

3.2 Particle Distribution

Sieve analysis was conducted following ASTM C136/C136M (2014). Figure 1 shows the grain size distribution curves of Granular A and RAS. After sieve analyses, necessary amounts of Granular A and RAS were taken and then blends were prepared based on the predetermined percentages shown in Table 1. Blends were hand mixed thoroughly to maximize the uniformity and consistency in the samples used for standard proctor and CBR tests. Blended samples can be seen in Figure 2.

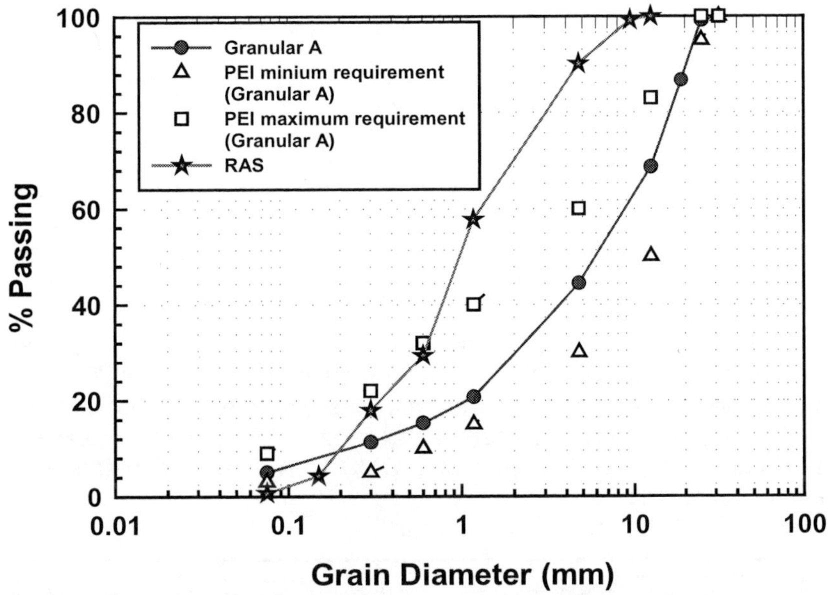

Figure 1: Granular A Particle Distribution

Figure 2: Samples Left to Right (Granular A - RAS) 100-0, 90-10, 80-20, 70-30

3.3 Specific Gravity

Specific gravity tests were performed following ASTM C127 (2015) and ASTM C128 (2015) for coarse and fine particles, respectively. This test was necessary to determine correction factors for oversized materials in standard proctor tests. The granular material supplied by PEI was already estimated to meet their specifications for construction purposes. Summary of test results for Granular A and RAS are shown in Table 2 and Table 3, respectively. Average apparent specific gravity was determined to be 2.64 and 1.84 for Granular A and RAS, respectively.

Table 2: Summary of Specific Gravity Test Results for Granular A

Sample No.	Oven dry	Saturated-surface dry	Apparent specific gravity
1	2.59	2.61	2.64
2	2.60	2.62	2.65
3	2.59	2.61	2.64
Average	2.59	2.61	2.64
Standard Dev.	0.00696	0.00656	0.00605

Table 3: Summary of Specific Gravity Test Results for RAS

Sample No.	Oven dry	Saturated-surface dry	Apparent specific gravity
1	1.55	1.70	1.81
2	1.55	1.72	1.86
3	1.50	1.69	1.84
Average	1.54	1.70	1.84
Standard Dev.	0.023	0.013	0.018

3.4 Standard Proctor Test

Standard proctor tests were carried out following ASTM D698 (2012) to determine optimum moisture content (OMC). Figure 3 shows uncorrected proctor test results for each blend. The OMC and maximum dry density ($\gamma_{d(max)}$) in Figure 3 were then corrected for oversized granular materials as shown Table 4. Corrected maximum dry density linearly decreases with increasing RAS content used in the present study (Figure 4). However, as shown in Figure 4, corrected OMC converges to constant value (approximately 7.33%) as RAS content approaches 20%.

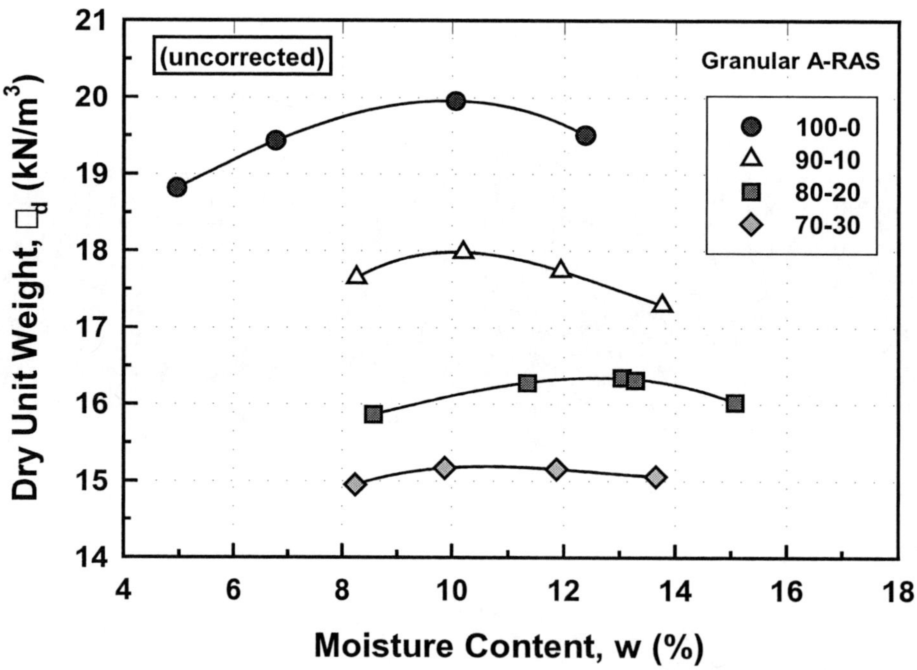

Figure 3: Proctor test results (uncorrected)

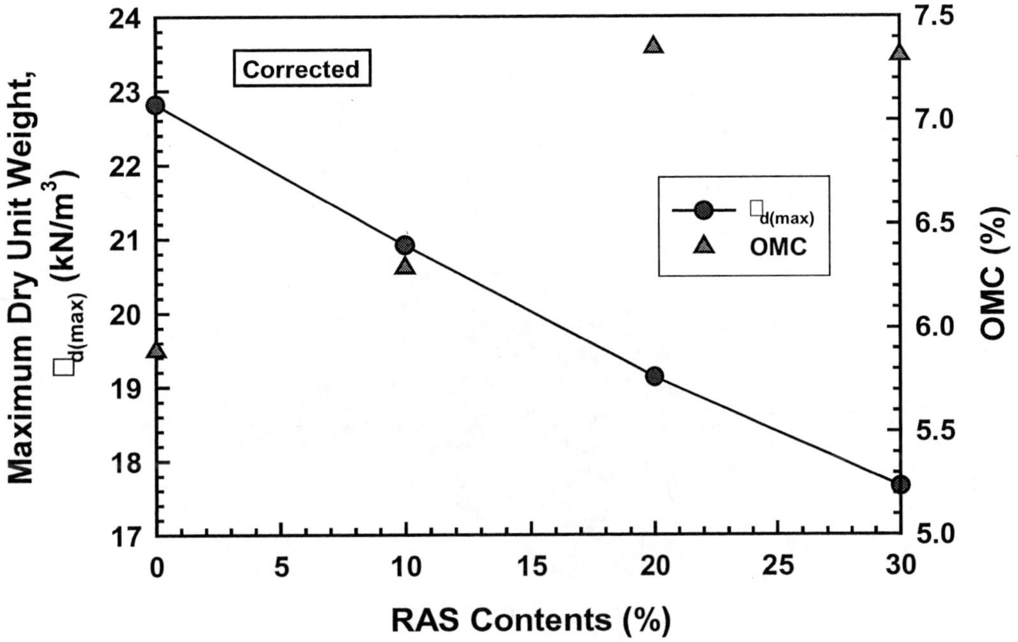

Figure 4: Proctor Results

Table 4: Corrected OMC and $\gamma_{d(max)}$

Mix (Granular A-RAS)	Corrected OMC	Corrected $\gamma_{d(max)}$
100-0	5.89%	22.81
90-10	6.29%	20.91
80-20	7.35%	19.13
70-30	7.31%	17.65

3.5 California Bearing Ratio Tests

For the purposes of these tests, both the granular and RAS materials are considered to behave the same, which leads to the same standards being used for both material. ASTM D1883 (2016) standard was followed in conducting CBR tests for each blend compacted at the corrected OMC. Due to the characteristics of the materials used, the compacted materials fell while flipping the CBR mold. Hence, a different metal base plate was attached to the mold prior to flipping the CBR mold to remove the spacer disk (Figure 5). Load was applied on the penetration piston at the rate of 1.27 mm/min. The stress levels at 2.54mm and 5.08mm were determined for each test scenario. The stress values of 2.54mm and 5.08mm were divided by 6.9 MPa and 10 MPa, respectively. The larger value is chosen for the final CBR result.

Figure 5: Modified CBR Test Assembly

4 CBR TEST RESULTS AND ANALYSIS

Figure 6 shows the uncorrected stress on piston versus penetration curves for each blend from CBR tests. Based on the results in Figure 6, corrected CBR values were determined as shown in Table 5. It was observed that the CBR decreases exponentially with the increase of RAS replacement (Figure 7).

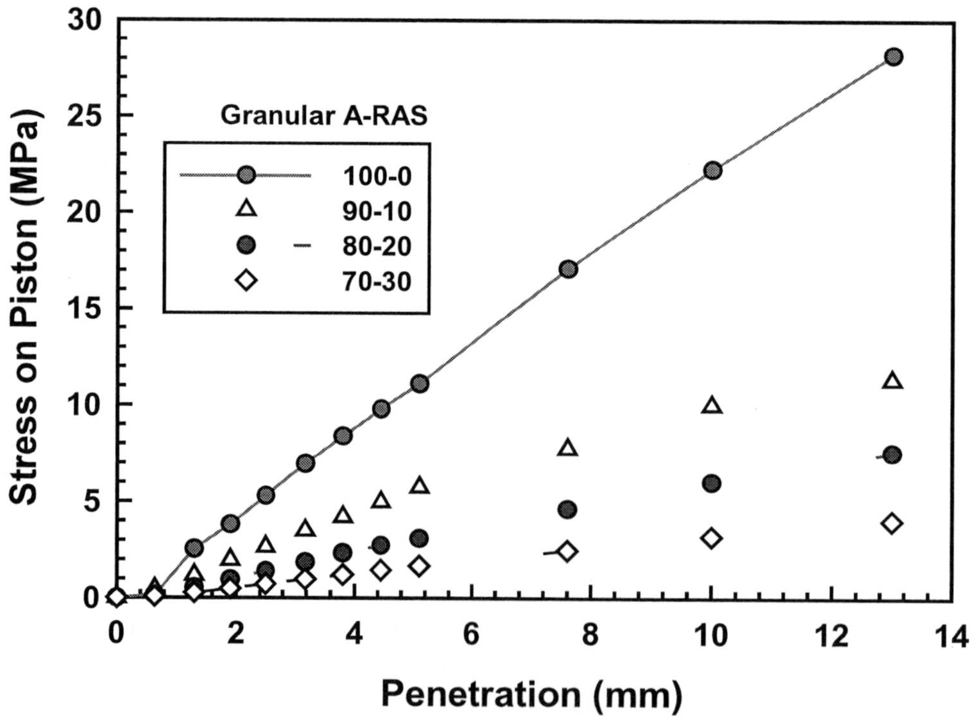

Figure 6: Stress on piston vs penetration Curves from CBR Tests

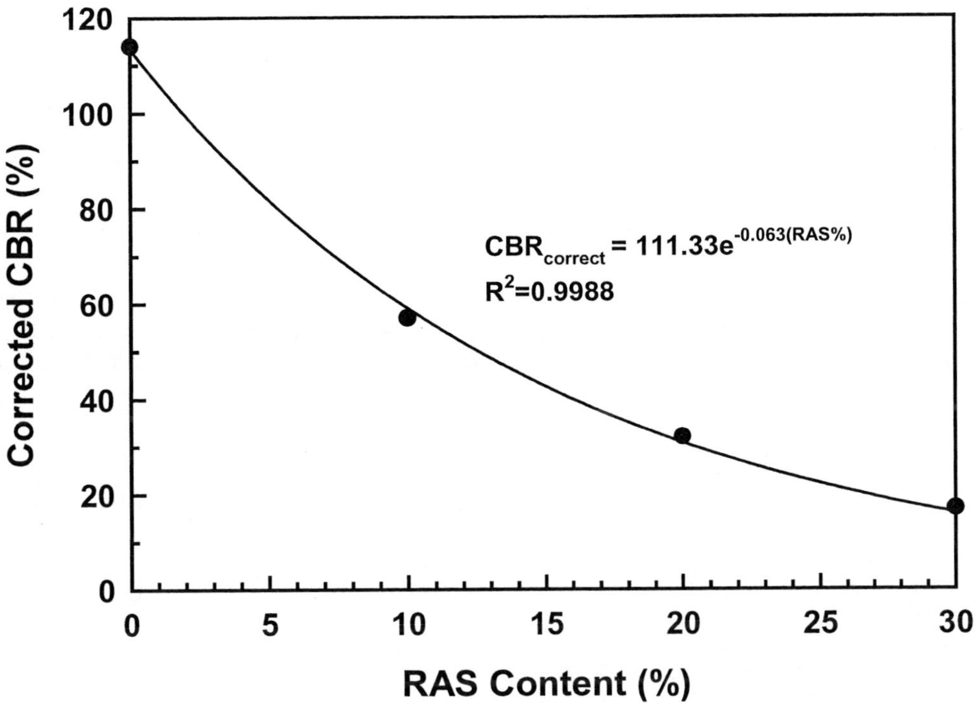

Figure 7: CBR vs Percentage of RAS Replacement

Table 5: California Bearing Ratio Test Results

Mix (Granular A-RAS)	CBR Uncorrected	CBR Corrected
100-0	111%	114%
90-10	57%	57%
80-20	31%	32%
70-30	17%	17%

5 CONCLUSION AND RECOMMENDATIONS

In the present study, a series of CBR tests were carried out for the mixtures of Granular A and RAS, compacted at the OMC. To investigate the variation of CBR with respect to the replacement of Granular A with RAS, three different mix combinations of the two materials were designed. Each mix combination was compared to how the granular material behaved on its own.

The results showed that the addition of RAS significantly decreases the strength of the soil compared to the currently used granular material. With the decrease in strength this would require a thicker layer of granular subbase in roadway construction to achieve similar strength and have the ability to withstand similar loading conditions. As the whole objective of using RAS is to decrease the cost of road construction, this is not ideal. Therefore, it is recommended that if a high percentage of RAS would like to be used, additives should be looked at to stabilize the blend to increase strength. These additives could include:

asphalt emulsion, lime like additives and geosynthetics. Additives come with a price and would need to consider a balanced design that optimizes the use of RAS and additives while still saving money.

This paper is based on the preliminary results of an ongoing study at the University of New Brunswick, aimed at evaluating the effective use of recycled asphalt shingles for granular materials. This paper focuses on the California Bearing Ratio of the granular material partially replaced with different percentages of RAS. One source of crushed stone (Granular A) from the province of Prince Edward Island and a single source of RAS with nominal maximum aggregate size of 4.75mm were used in this study. Hence, the results may vary depending of the quality of the materials, then a case by case analysis may be necessary.

Acknowledgements

The authors would like to thank the province of Prince Edward Island's Transportation Department and its director, Steve Yeo, for the opportunity to complete this research. As well the authors would like to thank the technical staff at UNB including, Chris Forbes and Andrew Sutherland for their help with testing procedures.

References

ASTM. 2015. C127 Standard Test Method of Relative Density (Specific Gravity) and Absorption of Coarse Aggregate. *ASTM Compass.*

ASTM. 2015. C128 Standard Test Method of Relative Density (Specific Gravity) and Absorption of Fine Aggregate. *ASTM Compass.*

ASTM. 2014. C136 Standard test method for Sieve Analysis of Fine and Coarse Aggregates. *ASTM Compass.*

ASTM. 2012. D698 Standard Test Methods for Laboratory Compaction Characteristics of Soil Using Standard Effort. *ASTM Compass.*

ASTM. 2016. D1883 Standard Test Method for California Bearing Ratio (CBR) of Laboratory-Compacted Soils. *ASTM Compass.*

Lee, Edil, et al. 2010. Quantitative assessment of environmental and economic benefits of recycled materials in highway construction. *Transportation Research Record: Journal of the Transportation Research Board,* **2158:** 138-142.

Shrestha, Narayan, et al. 2008. Use of Processed Tear-off Roof Shingles to Improve Performance of Roadbase Materials. *Recycled Materials and Recycling Processes for Sustainable Infrastructure Annual Conference of the Transportation Association of Canada*, Toronto, Ontario

Soleimanbeigi, Ali, Tuncer B. Edil, and Craig H. Benson. 2011. Recycled asphalt shingles mixed with granular byproducts as structural fills. *Journal of ASTM International,* **9**(1): 1-19.

Townsend, Timothy, Jon Powell, and Chad Xu. 2007. Environmental issues associated with asphalt shingle recycling. *Construction Materials Recycling Association, US EPA Innovations Workgroup.*

Warner, J. D., and T. B. Edil. 2012. An Evaluation of Reclaimed Asphalt Shingles for Beneficial Reuse in Roadway Construction. STP, **1540**

Yang, Ddamba, et al. 2014. Investigation on use of Recycled Asphalt Shingles in Ontario Hot Mix Asphalt: A Canadian Case Study. *Canadian Journal of Civil Engineering,* **41**(2): 136-143

Building Tomorrow's Society

Bâtir la Société de Demain

Fredericton, Canada

June 13 to 16, 2018 / *13 au 16 juin, 2018*

THE EFFECTS OF FREEZE-THAW CYCLES AND DEICER SALT BRINE ON THE TENSILE STRENGTH OF RECYCLED ASPHALT MIXTURES

Shahab Moeini F. [1,2], Xiomara Sanchez [1], and Bruce Wilson [1]
[1] University of New Brunswick, Canada
[2] smoeini1@unb.ca

Abstract: In places of the world with severe seasonal variations, such as the province of New Brunswick, asphalt concrete mixtures are subjected to cyclic freezing and thawing during the cold months. Moreover, different types of deicers are frequently used to mitigate the effects of snow, ice, and freezing rain on the pavements to increase the levels of service of roadways. Despite having many safety advantages, the use of deicers has raised concerns over the effects on pavement materials. Salt is the most commonly used deicing product in New Brunswick and understanding its effect on the durability of asphalt concrete would help pavement managers to improve the determination of maintenance, preservation, rehabilitation, and the estimation of the life-cycle costs of pavement assets. This study investigated the effects of Freeze-Thaw Cycles (FTCs) and Deicer Salt Brine (DSB) on the tensile strength of recycled asphalt mixtures and conventional Hot Mix Asphalt (HMA). In partnership with the New Brunswick Department of Transportation and Infrastructure (NBDTI), five different mix designs of plant-produced asphalt mixtures, including recycled hot mixture and conventional HMA were collected from different projects in New Brunswick. These samples were subjected to different conditions simulating extreme weather in New Brunswick, including rain, freeze-thaw cycles, and the use of deicers over the pavement. The Tensile Strength Ratio (TSR) was determined to estimate the stripping susceptibility for each source material. The results from each source were then compared to evaluate the general performance differences between recycled and conventional HMA in New Brunswick, and this characterization is applicable to the development of new cost-effective pavements in the province.

1 INTRODUCTION

Many efforts have been made to address the issues of air pollution, global warming, and emissions of greenhouse gasses. It cannot be ignored that recycling is one of the best ways to a sustainable future. There are some methods of recycling in the asphalt paving industry, including hot mix asphalt recycling, hot in-place recycling, cold in-place recycling, and full depth recycling (Plescan and Plescan 2015). Recycling hot mix asphalt (HMA) material results in a reusable mixture of aggregate and asphalt binder known as reclaimed asphalt pavement (RAP) (Al-Qadi et al. 2007). According to the definition of the Federal Highway Adminstration (FHWA 2001), reclaimed asphalt pavement (RAP) is material containing asphalt and aggregates removed from a pavement and can be applied to any project which requires rehabilitation or major reconstruction. These materials are generated when asphalt pavements are removed for reconstruction, resurfacing, or to obtain access to buried utilities. At the end of the service life of hot mix asphalt pavement, milled materials (RAP) still maintain considerable value, which can be reused in virgin HMA to reduce the amount of new material needed. The design of HMA with RAP is based on a three-tier system (Al-Qadi et al. 2007). Less than 15% of RAP can be used without changing the virgin binder grade. When RAP content is between 15% and 25%, the high and low-temperature grades of the virgin binder are both reduced by one grade. Finally, if HMA has more than 25% RAP, blending charts are used to determine the grade of the binder.

Recycling has numerous advantages compared to conventional HMA, including economic, environmental, and social benefits (Hoppe et al. 2015). This study evaluated the tensile strength of recycled asphalt pavement including RAP after exposure to freeze-thaw cycles and brine in the province of New Brunswick.

2 FREEZE-THAW CYCLES AND ICE CONTROL METHODS IN NEW BRUNSWICK

Several studies have determined the effects of freeze/thaw cycles on flexible and rigid pavements. The annual average daily minimum and maximum temperatures in Fredericton, New Brunswick are -15°C and +25°C respectively (Living in Canada 2018). In addition, the precipitation of snow in New Brunswick generally occurs over a period of 6 months from November to April (Avirneni et al. 2016). Lamothe et al. (2017) revealed that freeze-thaw cycles damage the materials of asphalt mixtures dramatically. They found that this effect is more pronounced after the 30th cycle.

Snow and ice control methods are generally classified into three categories: mechanical, thermal and chemical (Hossain and Fu 2015). The most effective is the chemical method, which is subcategorized into deicing and anti-icing. Also, abrasives such as sand are deployed to improve traction. Different types of deicers are frequently used in cold regions including North America for the control of snow, ice and freezing rain to increase the levels of service and safety on winter roadways. The benefits of anti-icing are well documented in previous studies (TRS 2009; Hossain and Fu 2015). However, the use of these materials has raised concerns over their effects on pavements. Salt is one of the most common deicing products in New Brunswick and understanding the degree of deterioration caused by deicer treatments would help pavement managers to anticipate maintenance, rehabilitation, replacement actions and calculate life-cycle costs for the pavements more accurately.

Currently, research on the effects of deicers on the performance of flexible pavement is rare; however, there are numerous papers in relation to rigid pavements. The deicer salts accumulated on the surface of asphalt pavement can cause damage to asphalt material due to erosion and crystal formation after dehydration (Yu et al. 2013). Generally, deicing salt can be divided into three categories. The first category is chloride salt, such as calcium chloride, magnesium chloride, and sodium chloride. The second type is non-chloride salt such as calcium magnesium acetate (CMA), and potassium acetate. And, the third category is hybrid (Li et al. 2012).

In 2013, researchers from the Utah Department of Transportation Research Division published a literature review on the physical and chemical effects of deicers on concrete pavement (Sumsion and Guthrie 2013). They found that deicers can affect concrete both physically and chemically. Physical effects are typically manifested as cracking and salt scaling, while the chemical effects of deicers are often overshadowed by the physical effects. Sutter et al. (2006) revealed that exposure of various mortar specimens to calcium and magnesium chloride solutions at 40°F led to severe expansion, with deterioration first noticed at 56 days. Shi et al. (2011) soaked samples of rigid pavement in different types of deicer brine for around 340 days at room temperature. They found that continuous exposure to non-inhibited NaCl, inhibited NaCl and the inhibited $CaCl_2$ deicer led to limited levels of strength gain of the concrete, whereas exposure to inhibited $MgCl_2$ led to significant strength loss.

Moreover, some studies have been devoted to enhancing the performance of asphalt mixture through physical or chemical modifications. In 2010, research was done in the United States on the effect of different deicing solutions on the tensile strength of asphalt mixture (Goh et al. 2011), which demonstrated that weaker solutions of NaCl and stronger solutions of MgCl2 and CaCl2 are associated with higher tensile strengths of asphalt mixtures. Lamothe et al. (2017) published a journal paper about the degradation of hot mix asphalt samples subjected to freeze-thaw cycles and partially saturated with water or brine. A saturation process under vacuum was used and adapted from the test method on their research. Table 1 illustrates some details of the previous studies on the effects of deicers on the concrete and asphalt mixtures.

Table 1: Previous Studies on the Effects of Deicers on the Concrete and Asphalt Mixtures

Researcher/s and Year	Material Considering	Method	Period	Type of Salt	Dosage of Salt
Peterson 1995	Concrete	Soaking	22 months	CMA, NaCL	-----
Cody et al. 1996	Concrete	F/T, W/D, Soaking	222 Days	CaCl2, MgCl2, NaCl	-----
Sutter et al. 2006	Concrete	Soak@4°C	84 days	CaCl2, MgCl2, NaCl	3% by vol. of liquid chemicals, 3% by mass of solids
Shi et al. 2011	Concrete	Soak@25°C	330-347 days	CaCl2, MgCl2, NaCl	23-30% (by weight of solution)
Goh et al. 2010	Asphalt Mixture	-----	-----	CaCl2, MgCl2, NaCl	15% or less
Lamothe et al. 2017	Asphalt Mixture	Saturation	90min	-----	7,14,19%

3 RESEARCH GOAL AND OBJECTIVES

The main goal of this study is to evaluate the performance of conventional and recycled asphalt mixtures in New Brunswick in terms of tensile strength after exposure to water, freeze-thaw cycles and deicer salt brine. The main research objective which will be used to achieve this goal is measuring the effects of moisture on the recycled and conventional asphalt concrete mixtures using TSR.

Table 2 illustrates the materials used in this study which have been collected by NBDTI from different paving projects in New Brunswick in the summer and fall 2017.

Table 2: Description of Samples Collected

Source Code	Date, receiving Materials	Mix Type	AC Type	AC %	Antistrip
1 R	Jun. 23/2017	HRD	PG 58-28	6.05	Adhere 6500
1 C	July. 31/2017	D	PG 58-28	5.80	Adhere 6500
2 R	July. 31/2017	HRD	PG 58-28	6.30	No
3 R	July. 31/2017	HRD	PG 58-28	6.40	No
4 R	Sep. 25/2017	HRD	PG 58-28	6.00	Adhere 6500

4 METHODOLOGY

In this study, five different samples of recycled asphalt mixture and HMA from different asphalt plants in New Brunswick were collected. The Marshall compaction method was used for preparing the specimens. These samples were then subjected to different simulated weather conditions. Finally, samples were subjected to a TSR test to determine the effects of those conditions on the asphalt mixtures.

The following are the laboratory stages of this research:

1. Determination of theoretical maximum specific gravity (G_{mm}) of uncompacted mixtures (ASTM-D2041 2011).
2. Preparation and compaction of 4 inches diameter cylindrical asphalt mixtures specimens using Marshall apparatus (ASTM-D6926 2017).
3. Determination of bulk specific gravity of compacted asphalt mixtures (ASTM-D2726 2017).
4. Sorting the specimens into 5 subsets and measuring their heights.
5. Determining the degree of saturation (DoS) and Volume of the specimens after partially saturating them. The following equation has been used to find the DoS (ASTM-D4867 2014):

[1] $DoS = Vol._{aw} / Vol._{Va}$

where, DoS: Degree of Saturation

$Vol._{aw}$: Volume of the absorbed water, cc

$Vol._{Va}$: Volume of the air voids, cc

6. Conditioning the specimens by simulating extreme weather in New Brunswick
7. Measuring the heights of specimens
8. Determining water absorption and DoS of the specimens after conditioning
9. Calculate the tensile strength ratio based on the following equation: (ASTM-D4867 2014)

[2] $S_t = 2000\ P / \pi\ t\ D$ (kPa)

where, S_t: Tensile strength, kPa

P: Maximum load, N

t: Specimen height immediately before tensile test, mm

D: Specimen diameter, mm

5 SIMULATION OF EXTREME WEATHER

Based on the extreme weather in New Brunswick, four different conditions including moisture, single freeze-thaw cycle (FT), multiple freeze-thaw cycles (FTCs), and brine saturation were simulated in this research. (Table 3)

5.1 Moisture

According to ASTM-D4867 (2014), the moisture-conditioned specimens were soaked in distilled water at +60°C for 24 hours. After that, all the specimens, unconditioned and moisture-conditioned were soaked in the water bath at +25°C for 1 hour.

5.2 Freeze-Thaw Cycle (FT)

Based on ASTM-D4867 (2014), each specimen was wrapped tightly with two plastic layers. Each of them was then placed into a leak-proof plastic bag containing 3 mL of water and sealed with a tie. Next, they were placed into a freezer at -18°C for 15 hours, before placing into a water bath at +60°C for 24 hours. The bag and wrapping plastic were then removed, and the temperature of the specimens was adjusted by soaking in a water bath for an hour at 25°C.

5.3 Freeze-Thaw Cycles (FTCs)

The specimens were wrapped and then placed into a plastic bag, in the same way as the FT conditioning. For the minimum and maximum cyclic temperature, due to the lack of a standard test method for exposing asphalt mixtures to rapid freezing and thawing, the standard test method for concrete has been used in this study. According to this standard (ASTM-C666 2015), the nominal FTCs are lowering the temperature from +4°C to -18°C and raising it from -18°C to +4°C in a cabinet (Figure 1). For the Number of FTCs, some studies revealed that in the province of Quebec, the number of cycles is between 40 and 51 annually (Lamothe et al. 2017). So, 50 cycles of freezing and thawing has been chosen in this study as the assumed number of FTCs in the province of New Brunswick per year. The variation of FTCs cabinet temperature in 1 cycle can be found in Figure 1.

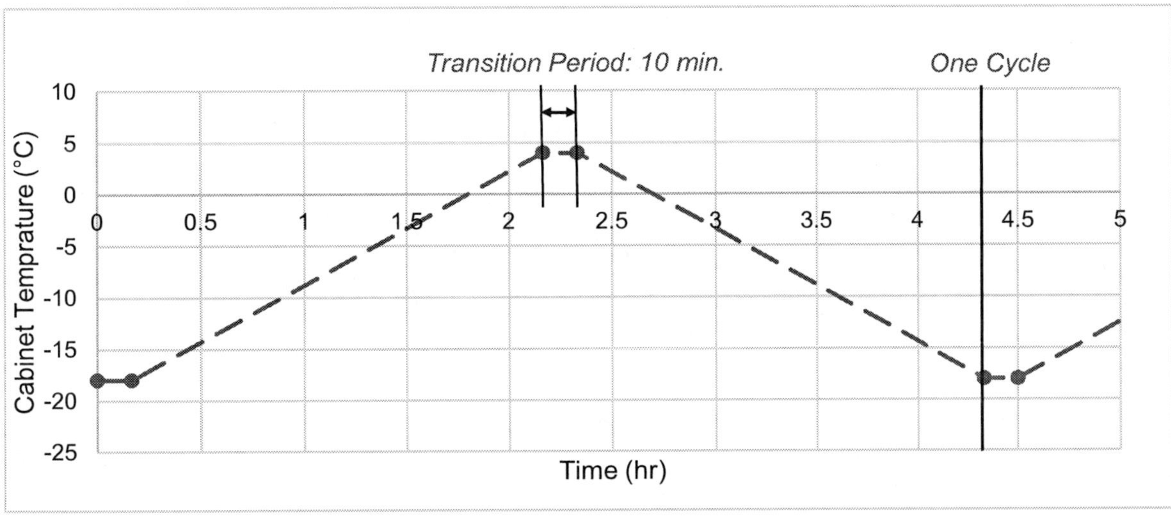

Figure 1: FTCs cabinet temperature in 1 cycle

5.4 Partial Saturation with Brine

Considering Lomothe *et al.* (2017), the specimens were placed into the flask of rice test filled with brine (15% salt by weight of solution) at the vacuum pressure of 30 mmHg for 90 minutes.

Table 3: Summary of different simulations of extreme weather considered in this study

Conditions	Summary of Procedure
Moisture conditioning	24 hours at 60±1°C
FT	15 hours at -18±2°C, then 24 hours at 60±1°C
FTCs	50 cycles between -18°C and +4°C
Saturation with brine	1.5 hrs at 30 mmHg, 15% salt by weight of solution

6 ANALYSIS AND RESULTS

6.1 Effect of Extreme Weather Conditions on the Asphalt Mixtures

Various tensile strengths of asphalt mixtures were derived for the different samples (Figure 2). Sample 4R has the highest and sample 1R has the lowest tensile strength among the other samples. It is obvious that effect of multiple freeze-thaw cycles is more than the other conditions, especially in the sample 4R. In addition, tensile strength of 2R and 3R dropped more than the other samples, which probably is because of the lack of antistrip in the mix. Moreover, the tensile strength of the sample 2R after FT was apparently higher that the moisture conditioned samples; however, subsequent analysis demonstrated that the difference is not statistically significant. To better understand, the effects of different conditions on the tensile strength of asphalt mixtures is plotted on figure 3.

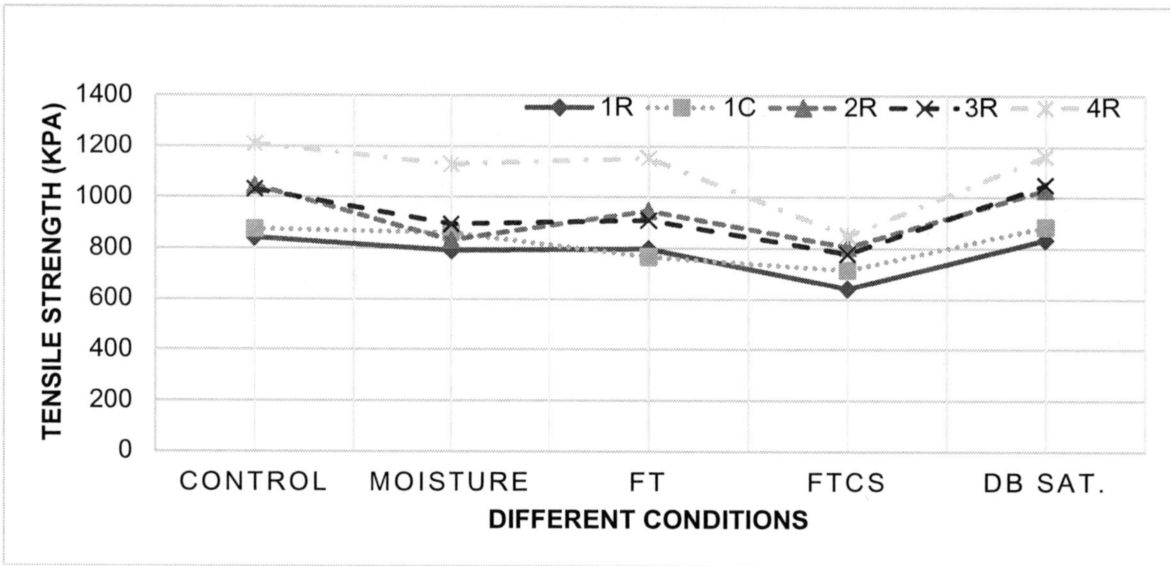

Figure 2: Tensile Strength of Asphalt Mixtures for Different Samples

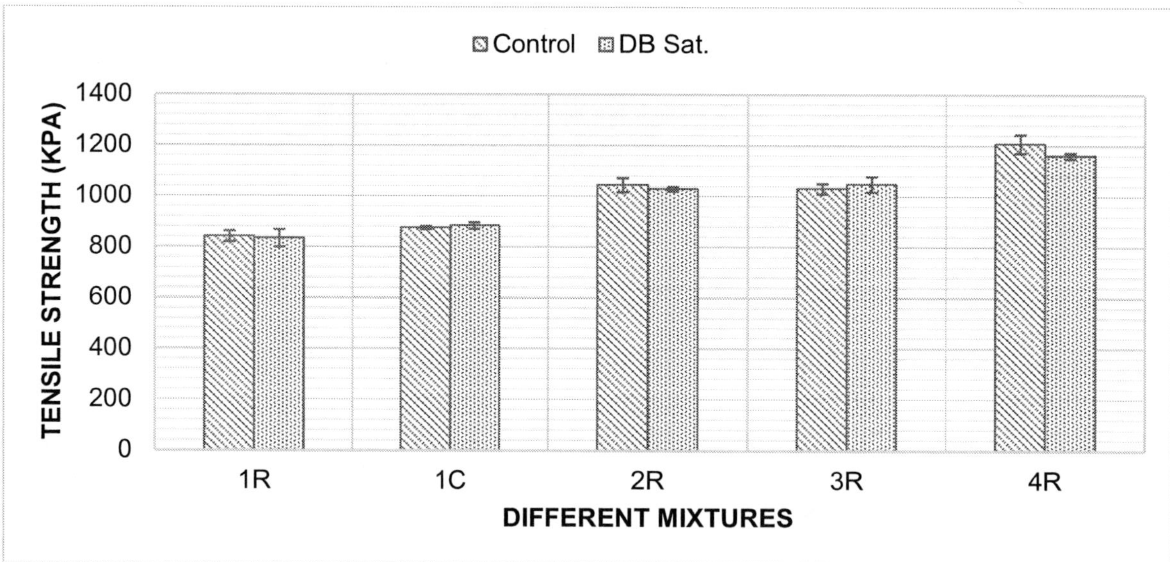

Figure 3: Tensile Strength of Five Samples Considered in This Research

The statistical analysis (t-Test) were done to identify significance of the effects of DSB saturation on different samples. It can be seen that the effects of DSB saturation on the mixtures is not significant (Table 4).

Table 4: t-Test, Paired Two Sample for Means

	1R		1C		2R		3R		4R	
	Con.	DSB	Con.	DSB	Con.	DSB	Con.	DSB	Con.	DSB
Mean	841	834	875	884	1046	1030	1031	1049	1208	1161
Variance	2092	5087	188	659	3424	383	1772	3990	6124	673
Observations	3	3	3	3	3	3	3	3	3	3

t Stat	0.16	-0.5	0.35	-0.3	0.79
P(T<=t) one-tail	0.44	0.33	0.38	0.38	0.26
t Critical one-tail	2.92	2.92	2.92	2.92	2.92
P(T<=t) two-tail	0.89	0.66	0.76	0.76	0.51
t Critical two-tail	4.3	4.3	4.3	4.3	4.3

6.2 Comparison between Recycled and Conventional Mixtures

A comparison of tensile strength between conventional asphalt mixture (1C) and Recycled asphalt mixture (1R) coming from the same source is plotted in figure 4. It can be seen that the control mixture type D has the higher tensile strength compared to the control HRD.

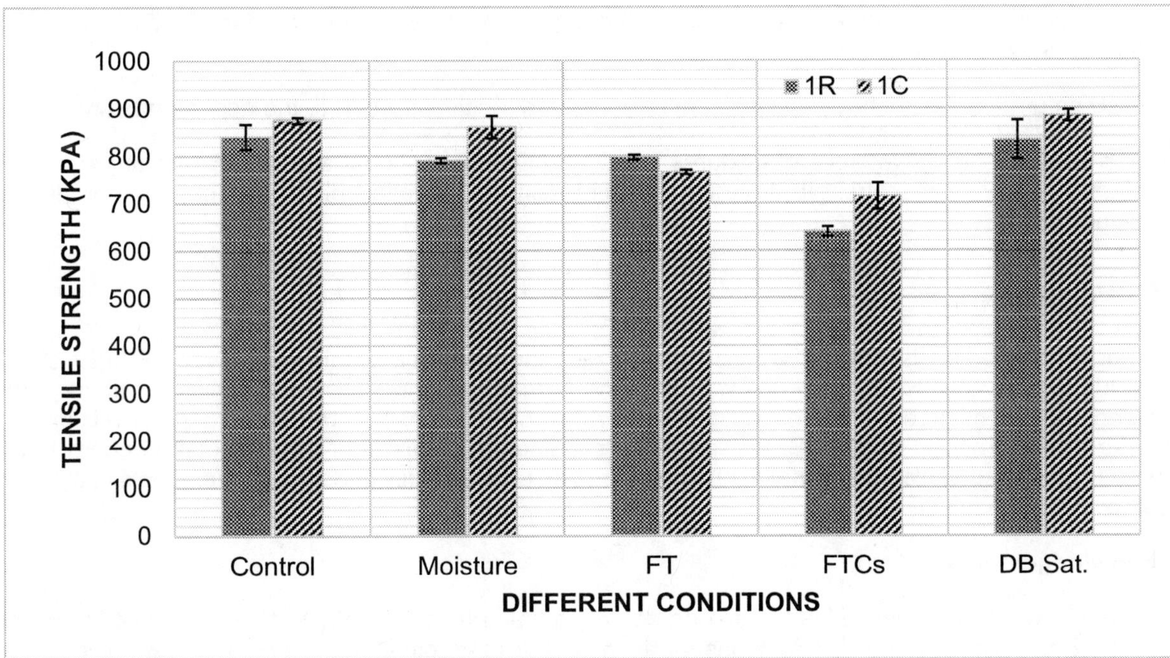

Figure 4: Comparison of Tensile Strength between Conventional Mixture and Recycled Mixture

t-Test analyses show that the difference between tensile strength of conventional and recycled mix is not significant, for all the conditions except for FT where the analysis indicated that the tensile strength of the conventional mix is significantly different (lower) than the strength of the recycled hot mix (Table 5).

Table 5: t-Test for Comparison of Conventional (D) and Recycled (HRD) Mixtures

	Control		Moisture		FT		FTCs		DSB	
	D	HRD	D	HRD	D	HRD	D	HRD	D	HRD
Mean	875	841	861	791	766	797	715	641	884	834
Variance	188	2092	2514	103	98.6	83.4	3470	296	659	5087
Observations	3	3	3	3	3	3	3	3	3	3
t Stat	1.84		2.31		-9.6		1.83		0.9	
P(T<=t) one-tail	0.10		0.07		0.01		0.1		0.23	
t Critical one-tail	2.92		2.92		2.92		2.92		2.92	
P(T<=t) two-tail	0.21		0.15		0.01		0.21		0.46	
t Critical two-tail	4.30		4.30		4.3		4.3		4.3	

7 CONCLUSIONS

Indirect tensile strength tests were subjected to four recycled and one conventional asphalt mixtures. All five materials were then conditioned, exposed to moisture, freeze-thaw, freeze-thaw cycles, and saturation in salt brine. The following conclusions are drawn from this research:

6. After moistening, the reduction of tensile strength of mixture for the samples 2R and 3R without antistrip in the mix were 20% and 13% respectively, while the tensile strength of the other samples containing antistrip was 5% approximately.
7. One Freeze-thaw cycle reduced the tensile strength of asphalt mixtures by approximately 10%.
8. The highest reduction in tensile strength of the asphalt mixtures happened after 50 FTCs conditioning by the range of 18% to 30%.
9. DSB saturation did not have a significant effect on the tensile strength of the asphalt mixtures.
10. In a comparison of two types of mixtures (HRD and D), conventional asphalt mixture apparently had a slightly higher tensile strength than the recycled mixture, however the difference was not statistically significant.

8 RECOMMENDATIONS

This paper is based on the preliminary results of an ongoing study at the University of New Brunswick, aimed at evaluating the effects of freeze/thaw cycles and deicer salt brine on the durability of plant produced recycled asphalt mixtures. This research concentrated on the tensile strength of asphalt mixtures. Four sources in the province of New Brunswick were considered in this study. It is recommended that more sources of recycled and conventional asphalt mixtures be tested to fully understand the effects of the extreme weather on the material.

Acknowledgements

The authors want to acknowledge the in-kind contributions and collaboration of the Construction Branch of the New Brunswick Department of Transportation and Infrastructure. In particular we want to thank the support of Ryan Barclay P.Tech., Senior Technical Advisor / Regional Asphalt Technologist and Matthew Sweezie, P.Eng., Paving Engineer. We also want to extend our gratitude to the Harrison McCain Foundation for funding part of this study.

References

AlQadi, I. L., M. Elseifi, and S. H. Carpenter. 2007. Reclaimed Asphalt Pavement-A Literature Review: 23.

Anti-Icing in Winter Maintenance Operations: Examination of Research and Survey of State Practice. 2009. Minnesota, USA.

ASTM C666. 2015. Standard Test Method for Resistance of Concrete to Rapid Freezing and Thawing. West Conshohocken, PA, USA.

ASTM D2726. 2017. Standard Test Method for Bulk Specific Gravity and Density of Non-Absorptive Compacted Bituminous Mixtures. West Conshohocken, PA, USA.

ASTM D4867. 2014. Standard Test Method for Effect of Moisture on Asphalt Concrete Paving Mixtures. West Conshohocken, PA, USA.

ASTM D6857. 2011. Standard Test Method for Maximum Specific Gravity and Density of Bituminous Paving Mixtures. West Conshohocken, PA, USA.

ASTM D6926. 2017. Standard Practice for Preparation of Asphalt Mixture Specimens Using Marshall. West Conshohocken, PA, USA.

Avirneni, D., Preddinti, P. R. T. and Saride, S. 2016. Durability and Long Term Performance of Geopolymer Stabilized Reclaimed Asphalt Pavement Base Courses. *Construction and Building Materials,* **121**: 198–209.

Cody, R. D., Cody, A. M., Spry, P. G. and Gan, G. L. 1996. Experimental Deterioration of Highway Concrete by Chloride Deicing Salts. *Environmental & Engineering Geoscience*, **2**: 14.

Federal Highway Administration (FHWA). 2001. HMA Pavement Mix Type Selection Guide. Washington, DC, USA.

Goh, S. W., Akin, M., You, Z. and Shi, X. 2011. Effect of Deicing Solutions on the Tensile Strength of Micro- or Nano-Modified Asphalt Mixture.*Construction and Building Materials,* **25**(1): 195–200.

Hoppe, E. J., Lane, D. S., Fitch, G. M. and Shetty, S. 2015. Feasibility of Reclaimed Asphalt Pavement (RAP) Use As Road Base and Subbase Material. *Innovation & Research.*

Hossain, K. and Fu, L. 2015. Optimal Snow and Ice Control of Parking Lots and Sidewalks. Ontario, USA.

Lamothe, S., Perraton, D. and Di Benedetto, H. 2017. Degradation of Hot Mix Asphalt Samples Subjected to Freeze-Thaw Cycles and Partially Saturated with Water or Brine. *Road Materials and Pavement Design,* **18**(4): 849–864.

Lamothe, S., Perraton, D. and Di Benedetto, H. 2017. Deterioration of HMA Partially Saturated with Water or Brine Subjected to Freeze-Thaw Cycles. *Road Materials and Pavement Design,* **18**(4): 705–717.

Li, P., Fu, G., Yang, D., Hu, G. and Zheng, J. 2012. Research on Effects of Deicing Salt on Adhesion Properties between Asphalt Mortar and Aggregate. *Changsha University of Science and Technology,* **88**: 127–31.

Peterson, O. 1995. Chemical Effects on Cement Mortar of Calcium Magnesium Acetate as a Deicing Salt. *Cement and Concrete Research*, **25**(3): 617-626.

Plescan, E-L. and Plescan, C. 2015. Asphalt Pavement Recycling. *Bulletin of the Transilvania University of Braşov,* **8**(1): 265–270.

Shi, X., Fay, L., Peterson, M. M., Berry, M. and Mooney, M. 2011. A FESEM/EDX Investigation into How Continuous Deicer Exposure Affects the Chemistry of Portland Cement Concrete. *Construction and Building Materials*, **25**(2): 957- 966.

Sumsion, E. S. and Guthrie, W. S. 2013. Physical and Chemical Effects of Deicers on Concrete Pavement. *UDOT Research.* Utah, USA.

Sutter, L., Van Dam, T., Peterson, K. and D. Johnston. 2006. Long-Term Effects of Magnesium Chloride and Other Concentrated Salt Solutions on Pavement and Structural Portland Cement Concrete: Phase I Results. *Transportation Research Record*, **1979**: 60-68.

The Climate and Weather of Fredericton, New Brunswick. [accessed February 10, 2018] https://www.livingin-canada.com/climate-fredericton.html

Yu, X., Wang, Y. and Luo, Y. 2013. The Effects of Salt on Rheological Properties of Asphalt after Long-Term Aging. *The Scientific World Journal* :1–10.

Building Tomorrow's Society

Bâtir la Société de Demain

Fredericton, Canada

June 13 to 16, 2018 / *13 au 16 juin, 2018*

EFFECTS OF LOW TEMPERATURE AND HIGH STRAIN RATE ON THE COMPRESSIVE BEHAVIOUR OF CONCRETE

MacLean, Thomas J.[1,3] and Lloyd, Alan[2]
[1] University of New Brunswick, Canada
[2] University of New Brunswick, Canada
[3] tom.maclean@unb.ca

Abstract: Explosions and large impacts are dangerous to structures due to their high force and impulsive nature. When combined with the effects of extreme temperature, the structure can be pushed to its limits, which could result in a catastrophic failure. The research presented here is a preliminary investigation into this largely unexplored design scenario, focusing primarily on the effects of low exposure temperature and high strain rates on concrete materials. The primary purpose of the research was to experimentally determine strength increase factors that incorporate the combined effects of low temperature and high strain rates on the compressive behaviour of concrete. A total of eighteen 101.6 mm x 203.2 mm (4" x 8") concrete cylinders were tested in uniaxial compression under static and dynamic strain rates at the University of New Brunswick. Static loads were applied by a compression machine and dynamic loads were applied by a drop mass impactor. The specimens were exposed to four different temperatures from +20°C to -70°C by means of placing them in a regulated freezer or a dry ice environmental chamber to cool. The vertical strain of the specimens during testing was determined by using Digital Image Correlation (DIC). A high-speed camera was used to monitor the dynamic tests. Stress – strain curves were generated for all static and dynamic samples, allowing for direct comparisons of peak stress and strain between the test conditions. Design equations and relationships between the dynamic increase factor (DIF), temperature increase factor (TIF) and combined increase factor (CIF) are presented.

1 INTRODUCTION

Concrete is a material that has been used in structures for centuries, yet there is still uncertainty concerning its material properties in certain situations. Many researchers have performed high exposure temperature research for the purpose of fire resistance. However, there has not been sufficient experimental work to confidently determine the resulting changes in the behaviour of structural properties due to the exposure of low temperatures, especially in tandem with high strain rates. Examples of some of the applications of the research would be dynamically loaded concrete structures exposed to the cold Canadian climate, where temperatures can be found to be less than -50 °C (National Research Council of Canada, 2015) or in storage facilities where liquid natural gas and other industrial materials are contained (Dahmani et al. 2009).

Literature shows that increasing strain rates results in an increase of the compressive strength of concrete (Bischoff and Perry 1991 & 1995, Hughes and Watson 1978, Ross et al. 1996). Additionally, this strength increase is also shown when exposure temperatures are lowered (Lee et al. 1988, Xie et al. 2014). However, there is little information about the combined effects of both high strain rates and low temperatures, especially at strain rates around 1 s^{-1}. 1 s^{-1} is an accepted typical far field flexural strain rate for flexural concrete design and will be used as a reference magnitude in this study (CSA Group 2012).

This paper describes an experimental research program designed to investigate this scenario. A summary of results of concrete tested under static and dynamic loads while exposed to low temperatures is presented. Design equations and relationships between the dynamic increase factor (DIF), temperature increase factor (TIF) and combined increase factor (CIF) are presented. The factors determine the relative increase in compressive strength observed due to temperature and strain rate influenced tests in comparison to static tests performed at room temperature.

2 EXPERIMENTAL PROGRAM

Eighteen 101.6 mm x 203.6 mm (4"x8") concrete cylinders were tested in uniaxial compression for the experimental program. Twelve concrete cylinders were tested under static conditions while being exposed to various low temperatures at the University of New Brunswick, while the remaining six were loaded dynamically. The testing temperatures used were 20, -10, -40 and -70 °C.

2.1 Specimen Preparation

A concrete mix design was tailored to produce a design target compressive strength (f_c') of 30 MPa. Portland cement and fly ash were combined with water to produce a water/cementitious material ratio of 0.45. Fine aggregate accounted for 43% of the total aggregate for the mix, which had a maximum size of 19.05 mm. The mix design proportions used are shown in Table 1.

Table 1: Mix design proportions

Material	Mix density (kg/m^3)
Portland cement	180
Fly ash	180
Water	162
Coarse aggregate	1035
Fine aggregate	781

Standard 101.6 mm x 203.6 mm concrete cylinder molds were used for mold preparation. However, as being able to monitor the internal temperature of specimens was necessary, modifications to the cylinder molds were required. A 3.4 mm steel wire was cut into 127 mm rods and placed through two small drilled holes located at mid height on opposite sides of the sleeve to stay in position. Two T-type high performance thermocouple wires were then put through one of the holes in the side of the mold and were taped to the rod so that the thermocouple contact points were located at the centre of the cylinder and at half of the radius, as shown in Figure 1. This spacing allowed a thermal gradient to be determined during the testing of each specimen. This was primarily incorporated for static tests, as the room temperature increased the surface temperature of chilled specimens much quicker than the core temperature. To overcome this difference, the measured temperature of both thermocouples throughout the test were averaged to determine the overall test temperature.

2.2 Exposure Temperatures

The desired temperature groups for the experiment were +20, -10, -40 and -70 °C. The 20 °C group was used as control samples, as it represented typical room temperature. Specimens required to reach -10 °C were placed in a freezer until their temperature stabilized. The colder temperatures of -40 and -70 °C were outside the working range of the freezer, therefore a dry ice environmental chamber was created. The chamber submerges cylinders in dry ice as shown in Figure 2. Dry ice sublimates at a temperature of -78.5 °C, therefore it is an ideal material for this test. The exposed ends of the embedded thermocouples protrude the cover of the environmental chamber, therefore temperatures can be monitored and specimens can be removed once desired temperatures are reached.

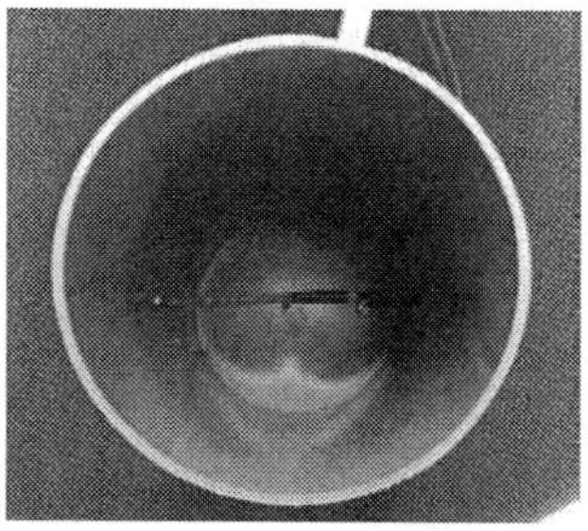

Figure 1: Concrete sleeve layout

Figure 2: Dry ice environmental chamber

2.3 Static Compression Test Setup

Twelve static uniaxial compression tests were performed on cylinders by a compression machine, shown in Figure 3. The specimens were loaded with a continuous stress rate of 0.25 MPa/s, resulting in an average static strain rate of 20 $\times 10^{-6}$ s^{-1}. The exposure temperatures used were 20, -10, -40 and -70 °C. Three compression tests were performed at each temperature interval and average responses were obtained. The static compression tests performed at 20 °C were considered as the control specimens for the entire experimental program. A typical static test setup is shown in Figure 4.

Figure 3: Compression machine

Figure 4: Static test setup

Stress-strain relationships were obtained for each test. Vertical strain data was acquired by using 3D DIC. Load values were obtained from the compression machine at the same rate and time as the images taken, allowing for simple data alignment of stress and strain. The thermocouples of the specimens were connected to a multimeter and the temperature from both thermocouples was taken at the beginning of the test and after failure. The freezer and dry ice environmental chamber specimens underwent an increase of roughly 1.5 °C and 4 °C throughout the tests respectively due to interaction with room temperature.

2.3.1 3D Digital Image Correlation

The vertical displacement of the specimens during compression testing was determined by 3D DIC. DIC is a non-contact optical technique that can be used to monitor strain and displacement that is simple to use and more accurate than manual measurement methods (McCormick and Lord 2010). DIC works by comparing numerous digital photographs taken of a specimen throughout the expected deformation period of a test by tracking the movement of specific blocks of pixels over time (McCormick and Lord 2010). The DIC system that was implemented with static testing was a two camera system that allowed for 3D imaging, as seen in Figure 5. The pair of cameras working in unison provides full-field 3D deformation and strain data. The cameras were set to operate at 5 fps.

DIC systems work most effectively when each of the individual pixel blocks being monitored are random with a large range of contrast and intensity levels (McCormick and Lord 2010). The randomness and contrast levels were obtained by painting the surface of the test samples a matte white and then speckling the surface with flat black spray paint. Matte colours are used to reduce the reflections seen and optimize the data collection process. An example of a speckled sample is shown in Figure 6.

Figure 5: 3D DIC camera setup

Figure 6: Static - speckled cylinder

2.4 Dynamic Compression Test Setup

Six dynamic uniaxial compression tests were performed on concrete cylinders that were loaded by a drop mass impactor, shown in Figure 7. A 740 kg steel sled shown in Figure 8 was raised and then allowed to freefall 1.5 m onto a concrete cylinder with a 101.6 mm x 38.1 mm cylindrical piece of wood on top. The wood was inserted to reduce the applied strain rates of the system by reducing the overall system stiffness. It was intended to have the low velocity impacts produce strain rates of 0.5 – 2 s^{-1}. Figure 9 shows a typical test setup of the cylinder and wooden interface material above one of the load cells for the dynamic tests.

Figure 7: Drop mass impactor

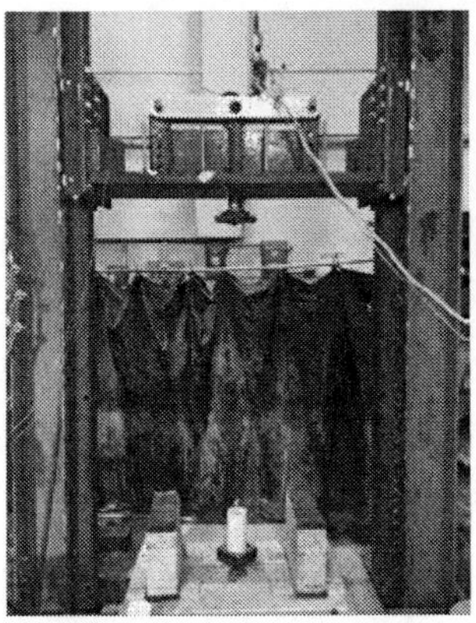

Figure 8: Drop mass sled

Figure 9: Dynamic test setup

2.4.1 Load Data

The applied load produced from the impact of the sled and the specimen was measured using two load cells. The load cells were designed to have a capacity of 1450 kN and were calibrated based on the compression machine used for static testing. Data acquisition software was used to monitor both load cells at a rate of 25 kHz. One load cell was placed below the specimen, while the other was attached to the bottom of the drop mass sled. An average of both load cells was taken to determine the average applied load to the specimen over time.

2.4.2 High Speed Camera

The vertical strain of the specimens during the dynamic compression testing was determined by using 2D DIC with a high speed camera. The single camera was set to take images at a rate of 20 kHz. A distinct difference between the speckling patterns of the cylinder for static and dynamic tests can be observed in Figure 6 and Figure 9. Due to the high frame rate of the high speed camera, the image resolution had to be reduced. This resulted in larger speckles needing to be used, as there were fewer pixels to monitor for DIC purposes. The speckle patterns on the dynamic specimens were applied manually by randomly applying dots using a dry permanent marker dipped in matte black paint.

3 EXPERIMENTAL RESULTS

The experimental results focus on the change in material properties of concrete due to the varying factors of temperature and strain rate. The primary objective is to determine the effects that these factors have on the stress – strain response of ordinary concrete.

3.1 Static Results

A typical stress-strain relationship obtained from the axial compression testing of cylinders at static load rates is shown in Figure 11. The curve has four different points labelled on the plot, which correlate to the adjacent images in Figure 10. The different colour gradients show the applied strain on the concrete at that point of the test. Virtual extensometers were placed vertically along the specimen and an average vertical strain was determined for each photograph. The average strain was paired with the analog stress data obtained from the testing machine to create the stress-strain plots for each test.

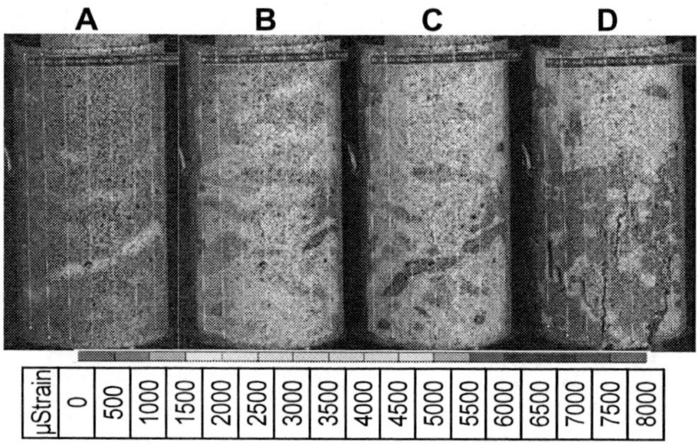

Figure 10: Strain data images during testing

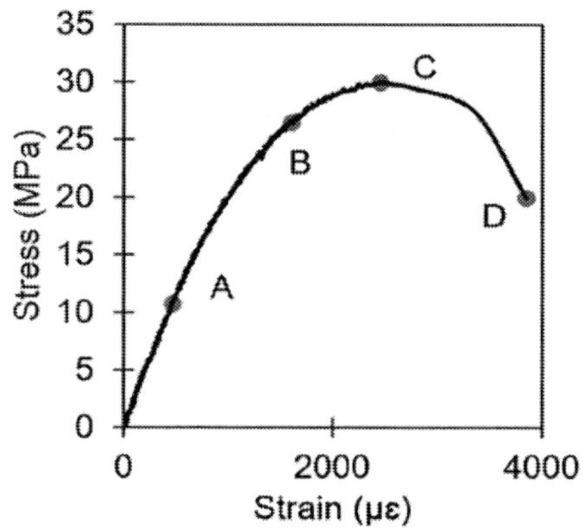

Figure 11: 20C-S3 - Stress vs. strain

The four temperature groups each had three specimens tested statically and the resulting stress-strain curves were plotted together in Figure 12. The concrete material properties of interest are the peak stress and associated peak strain. In addition to the experimental data, Hognestad's curve was included in each plot of Figure 12 as a solid line (Hognestad et al. 1955). Equation 1 was used to develop this theoretical curve, which was calculated using the average peak values for stress and strain of each temperature group.

$$[1]\ \sigma = \begin{cases} f_c'\left[\frac{2\varepsilon}{\varepsilon_0} - \left(\frac{\varepsilon}{\varepsilon_0}\right)^2\right], & \varepsilon < \varepsilon_0 \\ f_c'\left[1 - 0.15\left(\frac{\varepsilon-\varepsilon_0}{\varepsilon_u-\varepsilon_0}\right)\right], & \varepsilon_u \geq \varepsilon \geq \varepsilon_0 \end{cases}$$

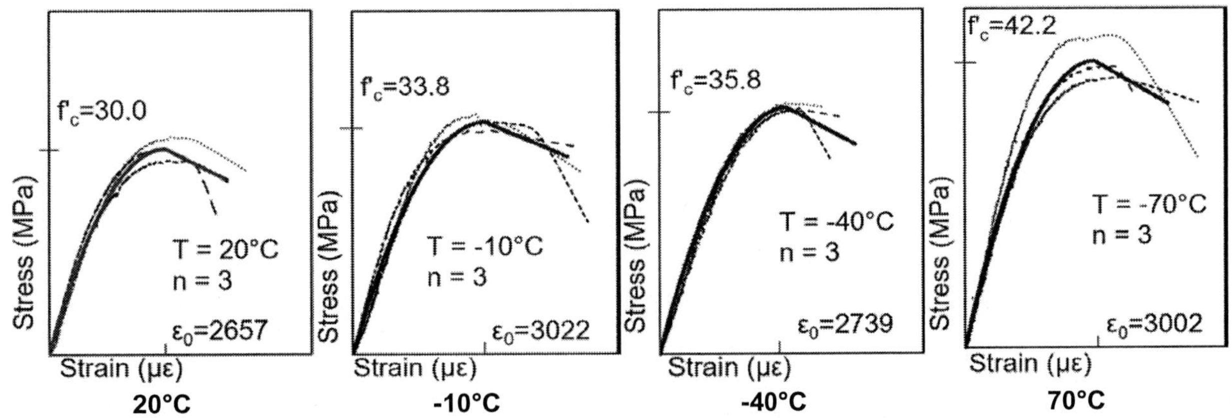

Figure 12: Static compression stress vs. strain plots

The four plots in Figure 12 are based off the same horizontal and vertical scales, which show the significant increase in compressive strength observed as temperature decreased. The largest strength gain occurred when temperatures were decreased to -70 °C, resulting in an average strength gain of 41% compared to the control samples. The calculation of the temperature increase factor and static peak strain ratio are shown in Equation 2 and Equation 3 respectively. It should be noted that there was not a linear increase in brittleness due to the lower exposure temperatures as previous literature suggested (Xie et al. 2014). A summary of the average group properties from the static tests is shown in Table 2.

[2] $\mathrm{TIF} = \frac{f'_{cT}}{f'_{c20}}$

[3] $\text{Temperature Peak Strain Ratio} = \frac{\varepsilon_{0,T,s}}{\varepsilon_{0,20,s}}$

In Equation 2, f'_{cT} is the average compressive strength of the desired temperature group and f'_{c20} is the average compressive strength of the 20 °C group. In Equation 3, $\varepsilon_{0,T,s}$ is the average peak strain of the desired temperature group and $\varepsilon_{0,20,s}$ is the average static peak strain of the 20°C group.

Table 2: Static axial compression test results summary

Average Group Property	20 °C	-10 °C	-40 °C	-70 °C
f'_c (MPa)	29.97	33.83	35.79	42.24
ε_o (με)	2657	3022	2872	3002
AVG temperature (°C)	22.6	-10.1	-41.2	-68.1
TIF	1	1.13	1.20	1.41
Temperature peak strain ratio	1	1.14	1.03	1.13

A linear increase was observed in the temperature increase factor for each group as temperature decreased. The peak values for stress and the associated average testing temperature for each test is shown in Figure 13. The linear relationship shows the increase in strength due to decreasing the exposure temperature of concrete at static rates.

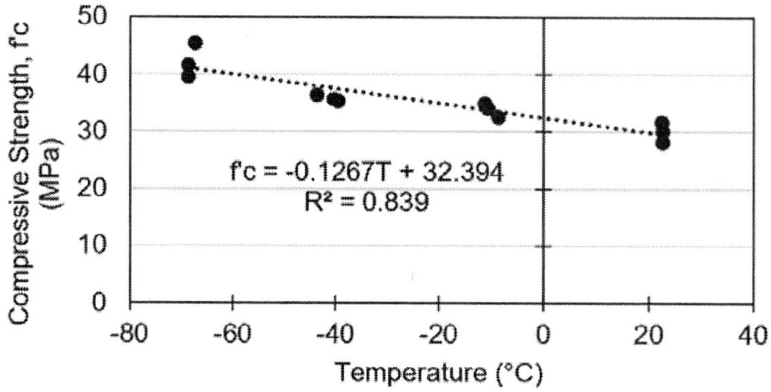

Figure 13: Static compressive stress vs. temperature

3.2 Dynamic Results

Figure 17 shows the typical member response to dynamic loads by the Drop Mass Impactor. Stress was measured over time by the two load cells and the average value was taken. Strain was measured using DIC over time by placing a minimum of three virtual extensometers on the concrete surface and taking the average for each photograph. A sample of images taken by the high-speed camera and the strain data extraction of a typical test is shown in Figure 14. The average strain values from each of the four images are shown in Figure 15. The two measurement systems produced Figure 15 and Figure 16 independently. The two systems were aligned by determining the frame of impact and the first load increase. This time alignment allowed for the generation of the stress-strain curve of a cylinder loaded dynamically.

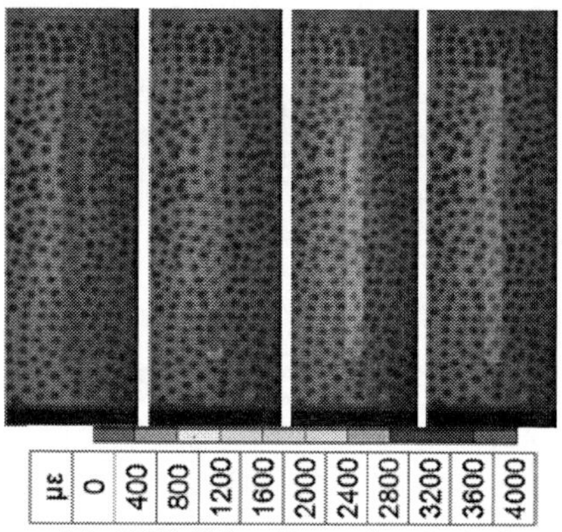

Figure 14: 20C D1 – Sample strain images

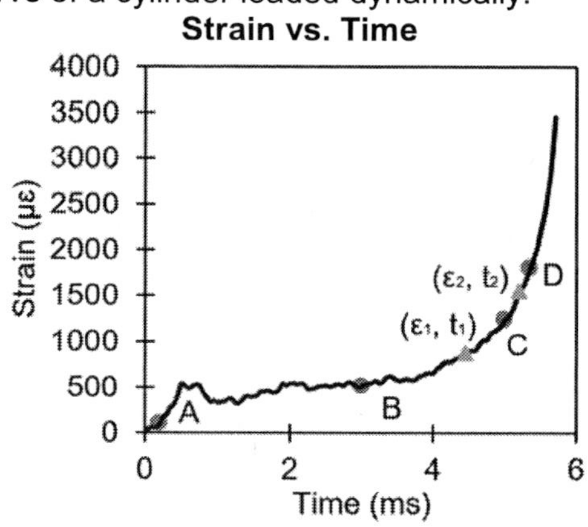

Figure 15: 20C D1 - Strain vs. time

Stress vs. Time

Stress vs. Strain

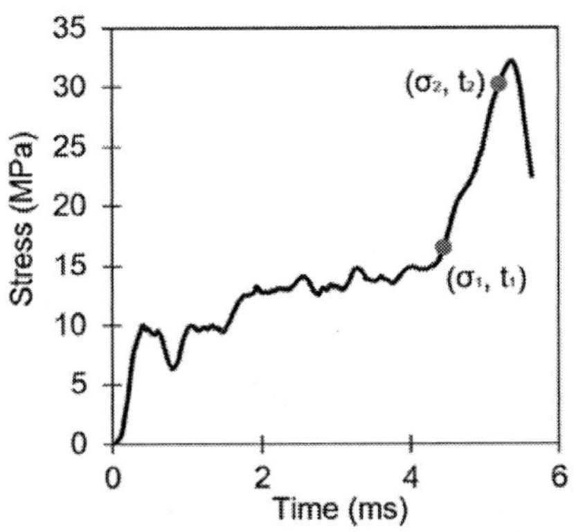

Figure 16: 20C D1 - Stress vs. time

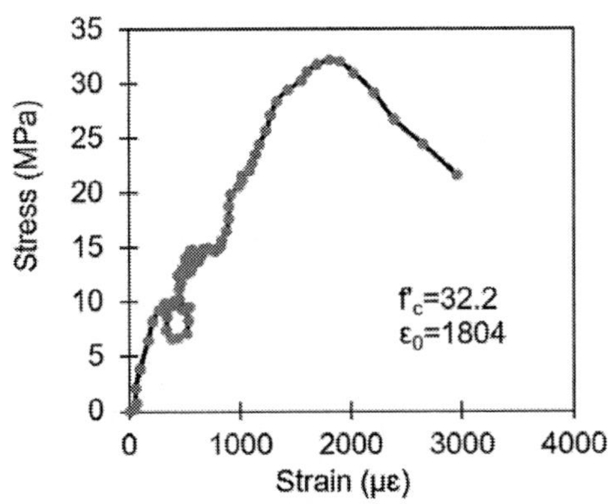

Figure 17: 20C D1 - Stress vs. strain

The horizontal plateaus in Figure 15 and Figure 16 from roughly 1-4 ms are due to the impact with the wooden interface material. This interface material caused noise in the stress-strain data due to the high density of points. However, the material was required to reduce the overall stiffness of the setup to maintain strain rates in the magnitude of 0.5-2 s^{-1}. The average strain rate for dynamic testing was 1.21 s^{-1}. One additional trial was conducted using a steel interface instead; this resulted in the strain rates being above 10 s^{-1} due to steel's high stiffness. The effects of the wooden material can be observed until roughly 15 MPa in the previous plots. After this point, the wooden material had reached its deformation limit and only the concrete deformed. The DIC strain measurements only consider concrete deformation during the test.

Strain rates were determined by taking two points on the stress vs. time plots after the plateau from the wood crushing when the response was essentially linear due to concrete being the sole contributor to deformation. The associated strain values were taken and divided by the time between the points. This is a linear approximation, as the strain rate was not constant. The calculation used to determine strain rate is shown in Equation 4. Sample points for calculating strain rates are shown in Figure 15 and Figure 16.

[4] $\dot{\varepsilon} = \frac{\varepsilon_2 - \varepsilon_1}{t_2 - t_1}$

Temperature, dynamic and combined increase factors can be determined through examining the peak stress values of the dynamic and static tests at various temperatures. Equation 5 and Equation 6 show the calculations of the dynamic and combined increase factors.

[5] DIF = $\frac{f'_{cTd}}{f'_{cTs}}$

[6] CIF = $\frac{f'_{cTd}}{f'_{c20s}}$

In Equation 5, f'_{cTd} is the average compressive strength of a desired temperature group loaded dynamically and f'_{cTs} is the average compressive strength of the same temperature group loaded statically. In Equation 6, f'_{cTd} is the peak compressive strength of a desired temperature group loaded dynamically and f'_{c20s} is the peak compressive strength of the 20 °C group loaded statically.

Figure 18 shows the temperature increase factors for static loads and the combined increase factors for dynamic loads. The highest average combined increase factor was 1.61, which occurred to the dynamically loaded specimens exposed to -70 °C. Dynamic loads for the 20 °C samples had an average dynamic increase factor of 1.15.

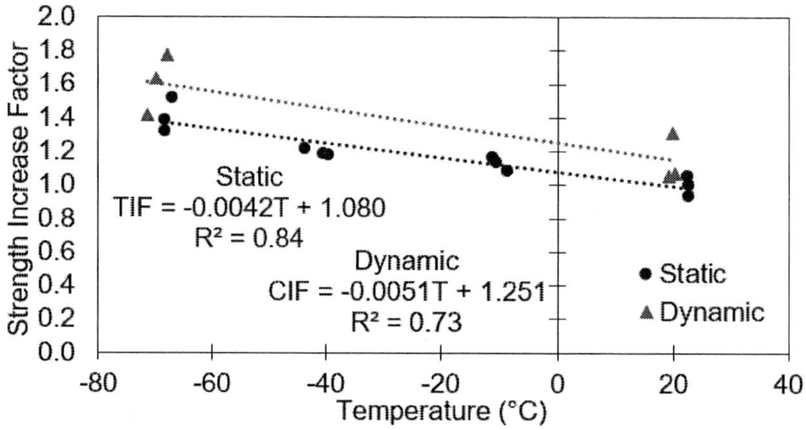

Figure 18: Increase factor vs. temperature

Even though there was not any significant change in peak strain for static tests exposed to different temperature, there was an apparent increase in brittleness due to the application of dynamic loads. Equation 7 shows the calculation for the dynamic peak strain ratio. Relative to the static tests of the same temperature, the dynamic tests for 20 and -70 °C had dynamic peak strain ratios of 0.71 and 0.72 respectively. There were variations in the dynamic strain rates of tests by $\pm$ 0.45 s^{-1}. This small difference was deemed acceptable to allow for direct comparisons between the tests. Table 3 shows a summary of all dynamic test properties monitored and the associated factors.

[7] Dynamic Peak Strain Ratio = = $\frac{\varepsilon_{0,T,d}}{\varepsilon_{0,T,s}}$

In Equation 7, $\varepsilon_{0,T,d}$ is the average peak strain of a temperature group loaded dynamically and $\varepsilon_{0,T,s}$ is the average peak strain for the same temperature group loaded statically.

Table 3: Dynamic axial compression test results summary

Average Group Property	20 °C	-70 °C
f'_c (MPa)	34.38	45.71
ε_0 (με)	1896	2171
DIF	1.15	1.14
Temperature (°C)	19.8	-69.8
TIF	1	1.40
CIF	1.15	1.61
Dynamic peak strain ratio	0.71	0.72
Strain rate (s^{-1})	1.34	1.08

4 ANALYSIS

Analysis was performed by comparing the experimentally determined strength increase factors amongst themselves to determine apparent relationships. The experimental test results correlated to produce design equations that are presented. In addition, the test results are compared to past literature.

4.1 Strength Increase Factors

Important results from the experimental program are the magnitudes of the temperature, dynamic and combined strength factors. Table 4 shows a summary of all of the strength increase factors obtained in the experimental program for 20 and -70 °C.

Table 4: Strength Increase Factors

Load Type	Property	20 °C	-70 °C
Static	TIF_S	1	1.41
	DIF	1.15	1.14
	TIF_D	1	1.40
Dynamic	$(DIF)(TIF_D)$	1.15	1.60
	CIF	1.15	1.61

Table 4 shows that the magnitude of the DIF was not dependent on exposure temperature, as the samples had essentially the same increase factor of 1.15 and 1.14 for the 20 °C and -70 °C groups respectively. Additionally, the TIF was not dependent on the applied strain rate, as the samples had essentially the same increase factor of 1.41 and 1.40 for static and dynamic loads respectively. Lastly, an interesting relationship can be determined for the CIF, as the value obtained experimentally can be effectively obtained by multiplying the DIF and TIF, as shown in Equation 8. This was applicable for both data sets quite accurately, as shown in Table 4. This would allow for design scenarios involving the combined effects of strain rate and low temperature to be tested independently and then combine the strength increase factors based on Equation 8. This finding would greatly simplify future test procedures.

[8] $CIF = (DIF)(TIF)$

4.2 Design Equations

Equation 9 and Equation 10 are the two design equations presented for determining the increase in compressive strength of a specimen due to low temperatures, impact strain rates and the combined effects.

[9] $TIF = -0.0042T + 1.080$

[10] $CIF = -0.0051T + 1.251$

These design equations are applicable for exposure temperatures of 20 °C to -70 °C, strain rates of the magnitude of 1 s^{-1} and a static compressive strength of roughly 30 MPa.

4.3 Comparisons to Literature

The experimental program results reinforce the notions that compressive strength increases when exposed to temperature and strain rates are increased independently (Bischoff and Perry 1991 & 1995, Hughes and Watson 1978, Ross et al. 1996, Lee et al. 1988, Xie et al. 2014). It should be noted that there has been large variations in magnitudes of increases for both scenarios in literature.

Shourky et al. (2011) presented a linear equation for static compressive strength based on exposure temperatures from -30°C to +60 °C, which had a slope of -0.13. The experimental trend line for static compressive stress as a function of temperature in Figure 13 showed a very similar slope of -0.127. It is reasonable to assume that slope of the equation found by Shourky et al (2011) would continue if tested at lower temperatures, thus agreeing with the experimental values found for static temperature effects.

Filiatrault and Holleran (2001) and Qiao et al. (2016) have both performed dynamic compression tests exposed to low temperatures. Filiatrault and Holleran (2001) saw an average CIF of 1.91 when a rate of 0.1 s^{-1} was applied to a specimen while exposed to a temperature of -40 °C. Meanwhile, Qiao et al. (2016) saw an average CIF of 1.5-1.63 when strain rates of 33-63 s^{-1} were applied while exposed to temperatures of -75 °C. The obtained CIF results from the experimental program land between values of both studies.

5 CONCLUSIONS

Based on the preliminary test results obtained, the following conclusions can be made about the combined effects of low exposure temperature and hard impact strain rates.

- The dynamic increase factor does not appear to be effected by temperature changes. The strength increase between the dynamic and static tests of the same temperature was the same regardless of temperature, with factors of 1.15 and 1.14 respectively due to average strain rates of 1.2 s^{-1}.
- The temperature increase factor does not appear to be influenced by varying strain rates. The strength increase observed between the 20 °C and -70 °C samples remained the same regardless if it was loaded statically or dynamically, with increase factors of 1.40 and 1.41 respectively.
- The combined effects of low exposure temperature and increasing strain rates was essentially equal to multiplying the observed DIF and TIF. The largest increase resulted in a CIF = 1.61.
- The dynamic peak strain ratio comparing the dynamic peak strain of a cylinder to the static peak strain of the same temperature was the same regardless of temperature. The 20 °C and -70 °C samples had peak strain ratios of 0.71 and 0.72 respectively, resulting in more brittle failures.

Further testing is required to determine the combined effects of low exposure temperature and high strain rates. The effects of applying higher strain rates (10 – 100 s^{-1}) to the test specimens at low temperatures, the effects of the moisture content of specimens during combined effects tests and the influence of static compressive strength on the magnitude of strength increases observed should be investigated in the future.

References

Bischoff, P.H. and Perry, S.H. 1991. Compressive Behaviour of Concrete at High Strain Rates. *Materials and Structures,* **24**(6): 425-450.

Bischoff, P.H. and Perry, S.H. 1995. Impact Behaviour of Plain Concrete Loaded in Uniaxial Compression. *Journal of Engineering Mechanics,* **121**(6): 685-693.

CSA Group. 2012. Design and Assessment of Buildings Subjected to Blast Loads. Mississauga, Ontario, Canada.

Dahmani, L., Khennane, A. and Kaci, S. 2007. Behavior of the Reinforced Concrete at Cryogenic Temperatures. *Cryogenics,* **47**(9): 517-525.

Filiatrault, A. and Holleran, M. 2001. Stress-Strain Behaviour of Reinforcing Steel and Concrete Under Seismic Strain Rates and Low Temperatures. *Materials and Structures,* **34**(4): 235-239.

Hughes, B.P. and Watson, A.J. 1978. Compressive Strength and Ultimate Strain of Concrete under Impact Loading. *Magazine of Concrete Research,* **30**(105): 189-199.

Hognestad, E., Hanson, N.W., and McHenry, D. 1955. Concrete Stress Distribution in Ultimate Strength Design. Journal of the American Concrete Institute, 52(12), 455-480.

Lee, G.C., Shih, T.S. and Chang, K.C. 1988. Mechanical Properties of Concrete at Low Temperature. *Journal of Cold Regions Engineering,* **2**(1): 13-24.

National Research Council of Canada. (2015) *National Building Code of Canada 2015*, National Research Council of Canada, Ottawa, Ontario, Canada.

McCormick, N. and Lord, J. 2010. Digital Image Correlation. *Materials Today,* **13**(12): 52-54.

Qiao, Y., Wang, H., Cai, L., Zhang, W. and Yang, B. 2016. Influence of Low Temperature on Dynamic Behaviour of Concrete. *Construction and Building Materials,* **115**: 214-220.

Ross, A.C., Jerome, D.M., Tedesco, J.W. and Hughes, M.L. 1996. Moisture and Strain Rate Effects on Concrete Strength. *Materials Journal,* **93**(3): 293-300.

Shourky, S.N., William, G.W., Downie, B. and Raid, M.Y. 2011. Effect of Moisture and Temperature on the Mechanical Properties of Concrete. *Construction and Building Materials,* **25**(2): 688-696.

Xie, J., Li, X. and Wu, H. 2014. Experimental Study on the Axial-Compression Performance of Concrete at Cyrogenic Temperatures. *Construction and Building Materials,* **72**: 380-388.

Building Tomorrow's Society

Bâtir la Société de Demain

Fredericton, Canada

June 13 to 16, 2018 / *13 au 16 juin, 2018*

ELECTRICALLY CONDUCTIVE CONCRETE

Aboutaleb, Dina[1], Gohar, Reham[1], Ibrahim, Kareem[1], Mahran, Mahitab[1], Zaghloul, Yomna[1, 2], Ghaly, Athnasious [1], Hamza, Ahmed [1], Hussam, Ahmed[1], El Nahas, Eman [1] and Abou-Zeid, Mohamed[1]

[1] The American University in Cairo, Egypt

[2] yomnaar@aucegypt.edu

Abstract: Concrete has been known for decades as a non-conductive material. However, electrically conductive concrete is an innovative type of concrete, which incorporates conductive material to form a contiguous network of current flow. Such materials include graphite powder, steel shavings and steel fibers. This new mix offer a spectrum of applications such as de-icing on bridges and runways, cathodic protection, radiant heating and electromagnetic shielding. This study aims at exploring means to prepare conductive concrete together with achieving better understanding of the properties and performance of conductive concrete made with three potentially conductive materials at different dosages. To meet this objective, concrete mixes were prepared with various presumably conductive materials. The testing scheme includes conventional fresh and hardened concrete tests as well as specialized conductive and permeability tests. Results reveal that producing conductive concrete is possible through various means. However, care has to be taken that such mixes require thorough mixing and incorporation of materials. Initial findings suggest that conductivity is a function of both the type of the materials and dosage incorporated. The findings of this study are promising and should be built upon in order to propose such new concrete types in conductive media construction applications.

1 INTRODUCTION

Conventional concrete is known and used as an insulating material, however nowadays by making concrete conductive several advantages were seen that could be applied to several applications. Conductive concrete was first investigated in 1970's and 80's. Over the years, several mixes were experimented using different electrically conductive materials. Some of these materials are graphite powder, carbon particles and steel shavings, which is a waste material from steel industries. Adding any of these materials to the concrete mix will change the concrete's properties. These properties include but not limited to compressive strength, flexural strength, water permeability, electrical conductivity and heating rate. Moreover, new properties are expected to be found in this concrete, which will able us to use it in new applications such as de-icing on bridges and runways, cathodic protection, radiant heating, electromagnetic shielding and electrical grounding. To produce conductive concrete, the concrete is mixed with a suitable dosage of the conductive material so that a contiguous network of current flow is formed. The dosage should be enough to allow the concrete to be electrically conductive while maintaining adequate mechanical properties of the concrete.

2 LITERATURE REVIEW

2.1 Carbon and Graphite products for conductive concrete

The Roca Spur Bridge was the first implementation that used conductive concrete in bridge deck de-icing, Seven commercial carbon and graphite products were used to create 10 different mixtures with different dosages for comparison and evaluation, the mixtures included black diamond which is the trade name of a natural graphite crystalline in the form of pellets, Earth Link (EL) which is the trade name of graphite cement, which is made up of approximately 70% Portland cement and 30% graphite powder. Finally, EC, which is elemental carbon, was used with different particle sizes.

2.2 Steel Shavings and steel fibers for conductive concrete

A research was conducted in the University of Nebraska to test the de-icing capabilities of conductive concrete. In the research Christopher Y. Tuan, 2004, constructed a simulated bridge deck using conventional reinforced concrete of an area of 1.2 x 3.6 m (4 x 12 ft.) and 152 mm-thick (6 in.). After the bridge deck was constructed, conductive concrete was casted over the reinforced concrete slab to form a new slab of thickness 90 mm. The conductive concrete mix, which was developed at the University of Nebraska, contained 1.5% of steel fibers and 20% of steel shavings per volume. The steel fibers and shavings that were added to a regular concrete mixture aimed to achieve the required electric resistivity for electrical resistance heating. Two 64 mm-wide (2.5 in.) and 6 mm- thick (0.25 in.) steel plates were embedded along the length of the overlay for electrodes. The steel plates had perforations greater than or equal to the 13 mm (0.5 in.) maximum aggregate size to allow concrete to flow through to provide good anchorage in the overlay.

2.3 Conductive concrete applications

Xie et al, 2004, used carbon fibers and particles for conductive materials in de-icing applications. Shintani and Nakamura developed a conductive concrete mixture for electromagnetic wave shielding. In 1998 at the University of Nebraska, Yehia et al., 2004, developed a conductive concrete mixture specifically for bridge deck de-icing. Other applications are radiant heating where Conductive concrete is used as an indoor radiant heat option, electrical grounding and the first practical project using HCFA concrete was electrically grounding a telecommunications tower in Rudolph, WI. and finally cathodic protection to prevent corrosion of reinforcing steel where the reinforcement becomes in a stable passive state when using a negative protective current since the formation of hydroxide ions on the reinforcement restores the protective passive layer.

3 OBJECTIVE & SCOPE

In this research, the main objective was to evaluate the potential of producing conductive concrete that can be implemented with various conductivity levels in diverse applications. To meet this objective, a conventional concrete mix was used as a control to test three different materials, which are steel fibers, steel shavings and graphite powder with different dosages for each material and compare all the results to get a better understanding of conductive concrete.

4 EXPERIMENTAL PROGRAM

4.1 Material Properties

- Fine Aggregates: well graded sand, sieved on sieve no. 4, used as fine aggregate with specific gravity
- Coarse Aggregates: well graded crushed limestone used as coarse aggregates with specific gravity
- Water: normal tap water was used for the mixes and curing.
- Cement: Ordinary Portland cement, Type I
- Super plasticizer: Super plasticizer Type F
- Steel Fibers: 50 mm x 1 mm end-hooked steel fibers with minimum tensile strength 1000 N/mm2

- Steel Shavings: Steel Shavings of different sizes and shapes contaminated with oil and other foreign elements.
- Graphite: very fine graphite powder produced by a local factory.

4.2 Concrete Mix design

The concrete mix used in the experimental work of this study consists of nine different mixes including the control mix and three mixes with different dosage for the steel fibers and steel shavings, and two mixes for the graphite powder. The nine mixes are shown in table 1. The cement dosage was 400 kg/m3 and a fixed water-to-cement ratio of 0.5 was used throughout all the mixes. The material added replaced the aggregates by volume with a fixed ratio of 1:1.8 of the fine aggregate to the coarse aggregate. An air content of 3% was assumed before the testing and adjusted according to the super plasticizer needed.

Table 1: Mix Design

Mix	Property	Cement (kg)	w/c	Water (kg)	Coarse (kg)	Fine (kg)	Material (m3)	Super Plasticizer (liters)
1	Conventional concrete (Conv.)	400	0.5	200	1099	611	0	4
2	Steel Fibers 1% (SF1)	400	0.5	200	1050	583	10	4
3	Steel Fibers 1.5% (SF1.5)	400	0.5	200	1042	579	15	8
4	Steel Fibers 2% (SF2)	400	0.5	200	1033	574	20	8
5	Steel Shavings 10% (SS10)	400	0.5	200	899	499	100	12
6	Steel Shavings 15% (SS15)	400	0.5	200	815	453	150	12
7	Steel Shavings 20% (SS20)	400	0.5	200	724	402	200	6
8	Graphite Powder 5% (GP5)	400	0.5	200	983	546	50	8
9	Graphite Powder 15% (GP15)	400	0.5	200	822	456	150	4

4.3 Standard Tests

4.3.1 Fresh Tests

These tests included slump, unit weight, air content and temperature. Slump was done to measure the workability of the mix, Unit weight was used to measure the density of the freshly mixed, Air content is used to determine the air content of freshly mixed concrete without any air voids that may exist within the aggregates. Finally, temperature test measures the temperature of the concrete using a temperature-measuring device

4.3.2 Compressive Strength

This test was conducted to evaluate the compressive strength of each type of concrete using 150x150x150 mm cubes. For each of the nine types of cement, three cubes samples were taken on each age of testing.

4.3.3 Flexural Strength

Beams of 750x150x150 mm were tested to find the flexural strength after 28 days. For each mix, two beams were tested under center-point loading.

4.4 Specialized Tests

4.4.1 Water Permeability

In accordance with the German standard DIN 1048, three cubes of 150x150x150 mm were tested to determine the susceptibility of an unsaturated concrete to the penetration of water. The test is carried out after 28 days by putting the specimen over a water jet for 72 hours.

4.4.2 Conductivity Test

This test was conducted in order to determine the electrical resistivity and hence conductivity of the different mixes. This test has no standardized reference. Therefore, the specimen dimensions and test setup were empirically selected. The specimens for the test were 90x30x10 cm slabs (1 for each mix). Two nails were cast on opposite sides within the slab. The setup works by connecting the main AC power supply to a step-down & up transformer. An ammeter is connected in series, followed by a light bulb which was introduced in order to determine the minimum voltage required to induce sufficient current in order to light a bulb. This wire is then connected to one of the nails in the slab, and another wire is connected to the opposite nail, which goes back to the transformer to end the circuit. Through studied literature review, it was expected that the resistivity of the material would increase throughout time. Accordingly, readings for the voltage and current were taken every week for a period of three months to monitor the change in the material properties.

Figure 1: Slab for testing

Figure 2: Laser Gun

4.4.3 Heating Rate test

This test aimed to determine the rise in temperature per time for one of the best conducting mixes. The initial setup was composed of a thermocouple wire, which, was cast within every slab in order to measure the internal temperature of the various mixes. However, this set-up did not prove effective because the current moved in the thermocouple wire itself and caused the reader to give false results. The second method used the help of a laser gun, which can measure the surface temperature of any material. The purpose of this test was to study the effect of passing current through one of the best conducting slabs.

4.4.4 Graphite water absorption

The test was carried out by mixing a small amount of the graphite powder with water, and then allowing it to dissipate the water through a filter paper. The powder was weighed before and after the test and an absorption percentage was determined.

5 RESULTS & ANALYSIS

5.1 Fresh Tests

5.1.1 Unit Weight

The unit weight as illustrated by the results was found to be slightly lower in the graphite mixes than the control mix as the graphite has lower specific gravity than the aggregates it replaced, and also because the water added for the graphite absorption. The steel shavings seem to increase the unit weight significantly due to their high specific gravity. On the other hand, the steel fibers increased the unit weight of the concrete in the first mix, however by increasing the fibers percentage, the unit weight decreased. This is consistent with the air content values, as the percent of the material increases, more air is entrapped within the mix decreasing its unit weight.

5.1.2 Slump

In order to get good workability levels of the mixes, the super plasticiser was added with different dosages to each mix as shown in the mix design. All the slump results were in an acceptable range but more research have to be done in order to reach the optimum workability level of each mix.

5.1.3 Air Content

Results showed that the air content in the graphite mixes were within the normal range, however for the steel fibers it was higher due to insufficient compaction and in the steel shavings the air entrapped was the highest due to the shape of the shaving as it is spiral and may have caused the air to be entrapped. Also, insufficient compaction might be a factor.

5.1.4 Temperature

All the mixes temperatures are within the normal range of fresh concrete's temperature

Table 2: Fresh tests results

	Conv.	SF1	SF1.5	SF2	SS10	SS15	SS20	GP5	GP15
Unit weight (kg/m3)	2471	2575	2437	2109	2604	2843	2614	2397	2142
Slump (cm)	11.5	2	17.5	13	15	5	10	22	7
Air content (%)	1.5	1.9	2.5	3	1.1	4	3.8	1.8	1.6
Temperature (°C)	23.7	25.7	23.9	23.8	22.9	22.9	25	25	25.4

5.2 Hardened Tests

5.2.1 Compressive Strength

The results of the compressive strength test are shown in figure 1. It shows that the concrete mix with steel fibers have the highest compressive strength reaching more than 45 MPa after 28 days. It also shows that the mixes containing steel shavings or graphite powder decrease the strength of the concrete relative to the control mix. The high percentage of water absorption of the graphite powder is the main reason for the decrease in the strength. None of the mixes containing steel shavings or fibers reached a strength more than 25 MPa. The minor irregularities in the results could be due to insufficient compaction.

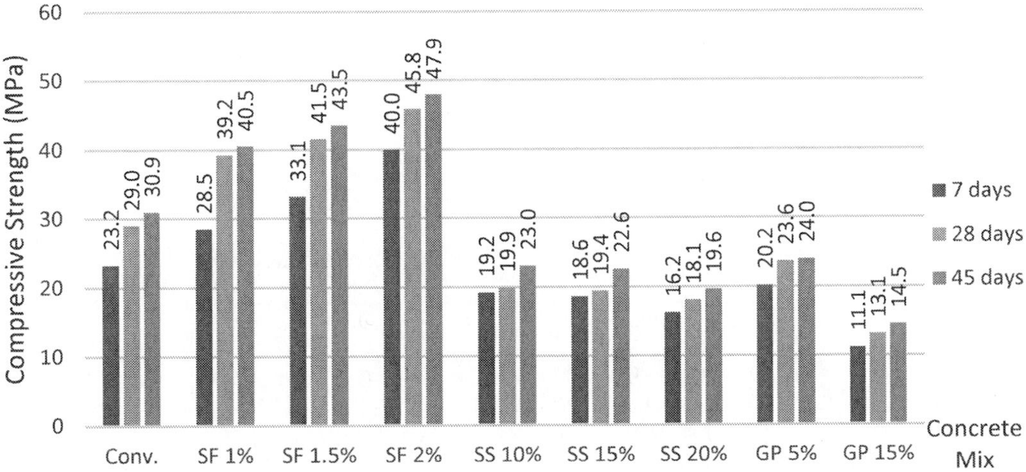

Figure 3: Compressive strength results

5.2.2 Flexural Strength

Figure 4 illustrates the flexural strength in MPa for the mixes, which is similar to the results of the compressive strength showing that the mixes containing steel fibers attain the highest strength of 13.8 MPa compared to 8.8 MPa for the conventional concrete. The fibers mixes exhibited ductile failure while the other mixtures collapsed and split. On the other hand, the steel shavings decrease the flexural strength of the concrete mix reaching similar results ranging between 6.1 & 6.3. The graphite also decreased the flexural strength significantly reaching 3MPa for the graphite 15% mix. The high percentage of water absorption of the graphite powder is the main reason for the decrease in the strength.

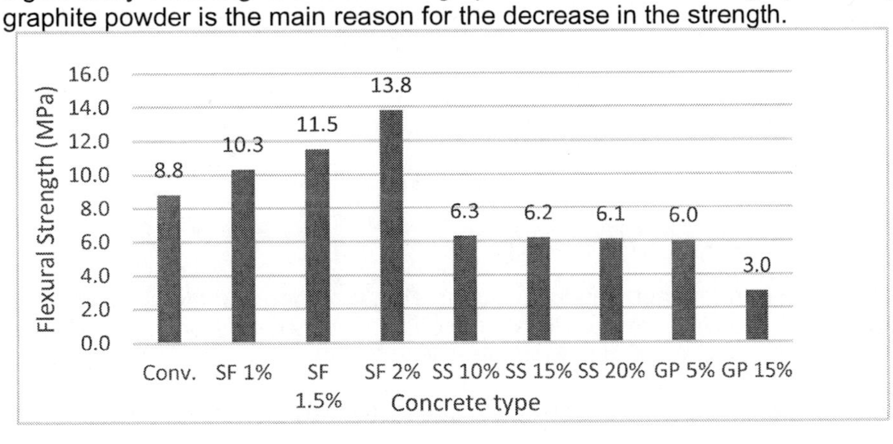

Figure 4: Flexural strength results

5.3 Specialized Tests

5.3.1 Water Permeability Test

The water penetration test results show a clear trend for all the materials. Given the conventional concrete mix as our datum, the graphite seems to lower the water penetration of the concrete due to its fineness as it fills out all the gaps and the voids in the mix. As graphite percentage increases in the mix, the water penetration decreases.

As for the steel shavings, they resulted in a high water penetration reaching 7cm penetration for the mix containing 20% shavings. This result could be due to the air entrapped inside the mixes, which is caused by the different sizes and shapes of the steel shavings.

Finally, the steel fibers show promising results. Compared to the conventional concrete, they also lower the water penetration reaching 0.5 cm for the 1% fibers.

Table 3: Water permeability test results

Mix	Water Penetration (cm)
Conv.	3.5
SF 1%	0.5
SF 1.5%	0.8
SF 2%	1.3
SS 10%	6.5
SS 15%	6.5
SS 20%	7
GP 5%	1.9

5.3.2 Conductivity Test

Displayed below are the equations used in order to determine σ, the electrical conductivity, for each mix.

[1] $R = \frac{V}{I}$

Where V is the voltage in volts, I is the current in amperes and R is the resistance in ohms.

[2] $\rho = \frac{V}{I}(\frac{wt}{L})$

Where ρ is the electrical resistivity in ohm-cm, w, t, & L are the width, thickness, and length of the specimen respectively

[3] $\sigma = \frac{1}{\rho}$

σ is the electrical conductivity in 1/ohm-cm

In an attempt to further understand conductive concrete properties and the application of different materials together, a new slab, which includes two maximum dosages of two different materials (15% graphite powder & 2% steel fibres), was cast and tested for conductivity.

Displayed below are the results for the electrical conductivity for all tested materials with time.

Table 4: Conductivity after 28 Days

Mix	V (V)	I (A)	R (Ω)	***ρ*** (Ωcm)	σ (1/Ωcm) x10-3
CC	310	0.08	3875	15500	0.065
GP 5%	200	0.09	2222	8889	0.113
GP 15%	110	0.09	1222	4889	0.205
SS 10%	188	0.09	2089	8356	0.12
SS 15%	182	0.09	2022	8089	0.124
SS 20%	135	0.09	1500	6000	0.167

Mix	V (V)	I (A)	R (Ω)	ρ (Ωcm)	σ (1/Ωcm) x10-3
SF 1%	130	0.09	1444	5778	0.173
SF 1.5%	38	0.09	422	1689	0.592
SF 2%	27	0.09	300	1200	0.833
GP 15% SF 2%	9.9	0.09	110	440	2.273

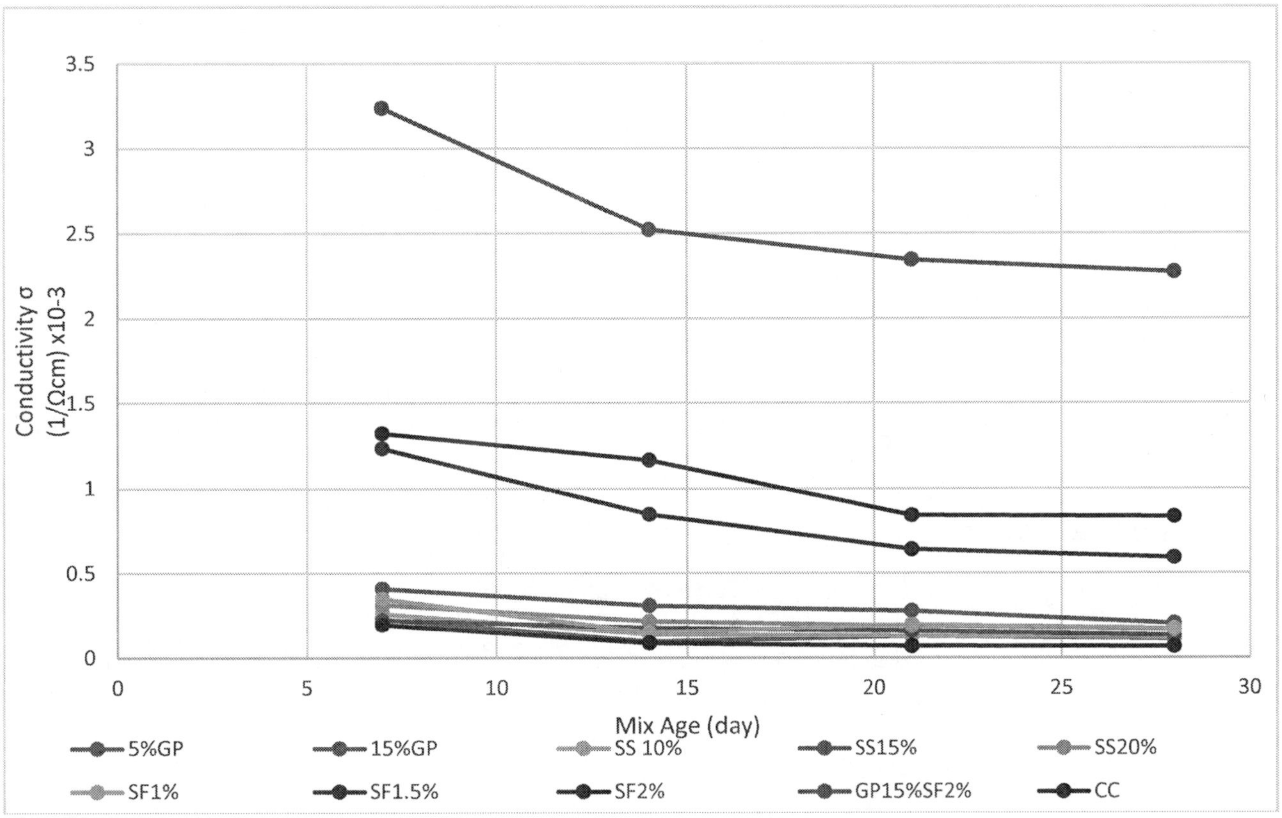

Figure 5: Electrical conductivity results for all mixes

Analysis:

The graph shows the conductivity vs. time for the different mixes. Overall, all tested materials (graphite powder - steel fibres - steel shavings) yielded higher electrical conductivity results than conventional concrete. The most conductive mixes from those containing one material were the steel fibres with dosages of 1.5% and 2%. The results for the mix containing 1% steel fibres are farther from the higher dosages but are still higher than the results for different materials. The results for graphite powder and steel shavings show less conductive behaviour and relatively lie in the same area, which is close to conventional concrete. The graph also shows the results for the special slab, which contains 15% graphite powder and 2% steel fibers, compared to the rest of the mixes, which contain only one type of material. This mix yielded much higher results for electrical conductivity, yet, further testing needs to be carried out in order to determine its mechanical properties and durability.

Although the results have a common trend and continue to decrease with time, all tested materials yielded higher electrical conductivity than conventional concrete. This phenomenon occurs due to the loss of water (which is an electrolyte) from the concrete and its replacement by air.

5.3.3 Heating Rate Test

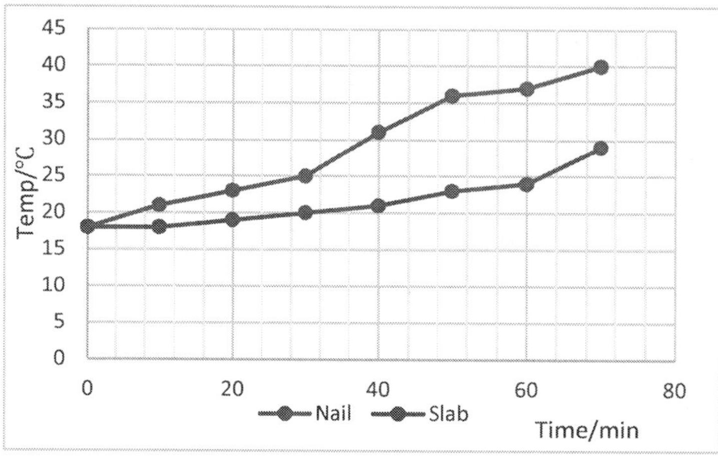

Figure 6: Surface temperature vs. time for mix containing 2% steel fibers

Analysis:

The heating rate test was carried out in order to explore the effect of passing current through the concrete. The test was carried out on one mix only, fibers 2%, which had the highest results for electrical conductivity. Readings for surface temperature were taken every 10 minutes for a period of 70 minutes

6 COST ANALYSIS

Table 5: Cost analysis

Mix	Cost of Material L.E/kg	Estimated Cost of Mix L.E/m3
Conventional Concrete		756
Fiber 1%	16	1781
Fibers 1.5%	16	2390
Fibers 2%	16	2908
Steel Shavings 10%	5	3379
Steel Shavings 15%	5	4511
Steel Shavings 20%	5	5687
Graphite 5%	11	1967
Graphite 15%	11	4226

7 CONCLUSIONS

In light of our research and after performing several tests on the mixes results were obtained that led to these conclusions:

- Electrical conductivity results show that mixes containing steel fibers are the most conductive followed by both the graphite and steel shavings which show relatively close conductivity results
- The limited feasibility study demonstrates that steel fibers are the most economic material in comparison with the other conductive materials tested
- The steel fibers have significantly increased the compressive strength and flexural strength of concrete
- The graphite powder decreases the unit weight of concrete
- The mixes containing graphite powder requires more water to account for the graphite absorption

- From the graphite mixes better finishability was obtained than the steel fibers and steel shavings mixes
- Graphite mixes are the least water permeable which is due to the high fineness of graphite powder

8 RECOMMENDATIONS

For further research:

- Larger scope of experimental work should be performed on reinforced concrete
- Conduct further tests such as rapid chloride permeability test (RCPT), chemical durability and thermal heating
- Use more reliable and higher accuracy set up for the heating rate test
- Test conductive concrete using voltages higher than 220 volts
- Use different dosages of materials together for further testing
- Acquire steel shaving waste from different sources to test different types of steel shavings considering that it is a waste material which can vary from one source to another

For Applicators:

- Adjust the dosages of the super plasticizer
- Mix the mixes containing steel fibers for an adequate time to reach a homogenous mix
- Increase darbying and vibration in steel fibers & shavings to ensure a smooth finishing surface
- Allocate more time for cleaning the steel shavings to avoid the reaction between any contaminants and the concrete components
- Allocate more time for sieving the steel shavings to have relatively close sizes throughout the mixes
- Coat steel shavings with zinc to prevent rusting
- Cover the mixer during the mixing process of graphite powder to limit the dispersion of the fine powder

Acknowledgements

Acknowledgment have to go to Eng. Hany Mosaad and Dr. Nageh Allam.

References

Tuan, Christopher Y. "Roca Spur Bridge: The Implementation of an Innovative Deicing Technology." Journal of Cold Regions Engineering, vol. 22, no. 1, 2008, pp. 1-15.

Yehia, S. A., and Tuan, C. Y., "Evaluation of Electrically Conductive Concrete Containing Carbon Products for Deicing," ACI Materials Journal, V. 101, No. 4, July-August 2004, pp. 287-293.

Ramme, Bruce W., and Miller, Steven H., "An Electric Freeway to the Future," Concrete International, V. 33, No. 5, May 2011, pp. 45-48.

Tuan, Christopher M. "Electrical Resistance Heating of Conductive Concrete Containing Steel Fibers and Shavings." ACI MATERIALS JOURNAL, 1 Jan. 2004.

Yehia, Sherif, and Christopher Y. Tuan. "Conductive Concrete Overlay for Bridge Deck Deicing." ACI MATERIALS JOURNAL

Yehia, Sherif, and Joshua Host. "Conductive Concrete for Cathodic Protection of Bridge Decks." ACI Materials Journal, vol. 107, no. 6, 2010, pp. 577.

Tuan, Christopher y, et al. "Patent WO2010059169A1 - Conductive Concrete for Heating and Elelctrical Safety." Google Patents, Google, 27 May 2010, www.google.com/patents/WO2010059169A1.

Building Tomorrow's Society
Bâtir la Société de Demain

Fredericton, Canada
June 13 to 16, 2018 / *13 au 16 juin, 2018*

EVALUATING THE EFFECTS OF MINERAL FILLER ON THE VOLUMETRIC PROPERTIES OF HMA MIXTURES BASED ON SUPERPAVE MIX DESIGN SPECIFICATIONS

Al-Bayati, Hanaa Kh[1,3], Tighe, Susan[2]
[1] PhD Candidate, Centre for Pavement and Transportation Technology (CPATT), Department of Civil and Environmental Engineering, Faculty of Engineering, University of Waterloo, Waterloo, Canada
[2] PhD, PEng, Deputy Provost and Associate Vice-President Integrated Planning and Budgeting, Canadian Society for Civil Engineering, Norman W. McLeod Professor in Sustainable Pavement Engineering, Department of Civil and Environmental Engineering, Faculty of Eng., University of Waterloo, Canada
[3] halwan@uwaterloo.ca

Abstract: Volumetric properties of hot mix asphalt (HMA) are necessary requirements to ensure good performance for asphalt mixtures. However, volumetric properties appear to be directly influenced by various factors including the mixture grading, aggregate surface characteristics and compaction energy. This research is conducted to evaluate the influence of mineral filler on the volumetric properties of HMA. Two types of mineral fillers with three filler percentages were chosen to examine the effect of filler on various volumetric properties including voids in mineral aggregates (VMA), voids filled with asphalt (VFA), dust to binder ratio (Dp), and others. The obtained results indicated that the optimum asphalt binder content, VMA, and VFA decrease as the filler content is increased. Compared to other volumetric properties that were reduced when filler content increased, Dp proportion behaved inversely for both filler types in which Dp proportion increases when the filler content is increased. HMA mixtures that includes dust plant filler had the higher values of VMA, VFA, and optimum asphalt content (OAC) compared to the hydrated lime. The addition of filler with 2.5% percentage is very successful for both filler types due to satisfying all Ministry of Transportation Ontario (MTO) requirements for volumetric properties of HMA. Based on MTO specifications, the addition of 2.0% filler seems to be unsuccessful for both filler types due to lowering the Dp ratio. Mix design with 3.0% filler was also unsuccessful because of the lower value of OAC meaning that the mix is dry and there is insufficient asphalt binder to coat the aggregate particles.

1 INTRODUCTION

It is generally accepted that asphalt concrete (AC) is a heterogeneous material that is fundamentally composed of various constituents: asphalt cement, natural or artificial aggregate, mineral filler, additives, and air voids. To achieve a desirable performance for the asphalt mixtures, the volumetric properties of pavement mixes are utilized. Currently, the volumetric properties of asphalt mixtures can be categorized into two main groups: primary and secondary volumetric parameters (Dos Santos et al. 2013). While the primary volumetric parameters are directly related to the relative volumes of the individual constituents of asphalt mixtures: aggregates volume (Vs), air voids (Va), and asphalt binder volume (Vb), secondary volumetric parameters commonly referred to as volumetric properties of mixtures are Void Volume (Vv), Voids in Mineral Aggregates (VMA), and Voids Filled with Asphalt (VFA). Depending on the primary volumetric parameters, secondary volumetric parameters can be determined. The terminology of volumetric properties for the components of a compacted asphalt mixture can be found in Figure 1.

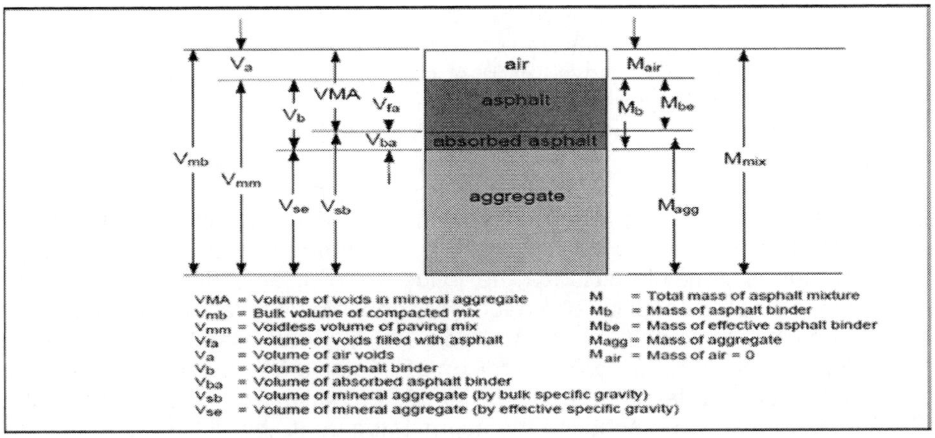

Figure 1: Components of a compacted HMA specimen (Hislop 2000, Huner & Brown 2001)

Marshall and Superpave methods are the common methods used in asphalt mixture design and incorporate volumetric criteria that mathematically calculated from the volumetric proportions of the constituent materials of the mixtures. In detail, the methods determine the optimum asphalt binder using HMA volumetric properties (Vv, VMA and VFA). Additionally, Superpave method can evaluate the filler content in the mixture and the percentages of initial and maximum compaction as a function of the number of gyrations in the Superpave Gyratory Compactor (SGC) (Dos Santos et al. 2013). In Superpave, a volumetric mix design protocol was established with limits on all three volumetric parameters (Hislop 2000). Depending on the mixture's nominal maximum size (NMS) and traffic volume, Superpave method establishes minimum values of VMA, and minimum and maximum values of the VFA as shown in Table 1.

Table 1: Superpave HMA volumetric properties (OPSS 1151 2007)

Traffic* Category (Note 1)	% of Theoretical Maximum Specific Gravity			Voids in Mineral Aggregate (VMA) % Minimum — Nominal Maximum Aggregate Size mm						Voids Filled with Asphalt (VFA)	Dust to Binder Ratio (Note 3)	Minimum Tensile Strength Ratio %
	$N_{initial}$	N_{design}	N_{max}	3.75	25	19	12.5	9.5	4.75			
A	91.5	96.0	98.0	11.0	12.0	13.0	14.0	15.0	16.0	70-80 (Note 4)	0.6-1.2	80.0
B	90.5									65-78		
C	89.0									65-75 (Note 5)		
D												
E												

Notes:
1. Traffic category as specified in the Contract Documents.
2. For Traffic Categories C, D, and E Superpave 9.5 mixes shall have a VFA range of 73 to 76%, while Superpave 4.75 mixes shall have a VFA range of 75 to 78%.
3. For Superpave 4.75 mixes, the dust-to-binder ratio shall be 0.9 to 2.0. Superpave mixes with gradation that pass beneath the PCS Control Point in Table 4, the dust-to-binder ratio shall be 0.8-1.6.
4. For Traffic Category A, Superpave 25.0 mixes shall have a VFA range of 67 to 80%
5. Superpave 37.5 mixes shall have a VFA range of 64 to 75%.

*The design equivalent to a single-axle load ranged (ESALS) for: A= Less than 0.3 million, B= 0.3 to 3 million, C= 3 to 10 million, D= 10 to 30 million, E= > 30 million.

2 LITERATURE REVIEW

It is generally agreed that a portion of aggregate that is suspended and freely discrete from aggregate particles in an asphalt binder can be known as a mineral filler. Based on this, it is reasonably postulated that mineral filler cannot be considered as aggregate or as a single component in a mixture; therefore, it is equitably counted as an integral component of mastic which is generally described as an actual binder for a mixture (Anderson & Bahia 1997, Remisova 2015). Additionally, due to its very small size, a filler is viewed as a fine material that has an ability to modify and enhance the properties of the asphalt-concrete mixture; and therefore, the filler appears to be a modifier and is not considered as a part of aggregate gradation (Zulkati et al. 2011).

Due to the diverse influence of filler addition on asphalt mixtures, numerous studies have investigated various aspects related to this influence and these aspects can be fundamentally classified into two main categories. The first category mainly includes the influence of filler addition on the asphalt mixture's performance. It was found that the addition of higher filler concentrations leads to strong mixtures. This was mainly attributed to better asphalt cohesivity and good stability that was obtained from a good packing distribution of the filler (Zulkati et al. 2011). The addition of lime as a filler can improve the resistance of hot mix asphalt mixtures to moisture damage, reduce oxidative aging, improve the mechanical properties and resistance to fatigue and rutting, and enhance the asphalt-aggregate bond (Dos Santos et al. 2013, Epps & Little 2001, National Lime Association 2006, Sutradhar et al. 2015). The second category is extensively focused on the influence of filler on the asphalt binder or mastic properties. It has been previously concluded that the presence of filler is highly related to a reduced optimum asphalt content (Tayebali et al. 1998, Muniandy et al. 2013, Brown et al. 1989, Kandhal et al. 1998). Mineral filler in hot mix asphalt plays a significant role in stiffening and toughening an asphalt binder. It seems to be responsible for improving adhesion of bitumen to aggregate (Sutradhar et al. 2015). In the perspective of volumetric properties, (Dos Santos et al. 2013) explored the influence of filler type and content on HMA volumetric parameters. The findings of the research revealed that VMA and VFA decrease as the filler content increases. It was concluded that lower values of VMA and VFA generally refer to the existence of thin asphalt film.

While comprehensively scanning the literature related to the filler addition into the asphalt mixtures, it should be noted that investigation of its impact on the volumetric properties of the asphalt mixtures is rarely found in the relevant literature. As many investigations were primarily limited in assessing the performance of asphalt mixtures that included mineral filler proportions, only one characteristic; namely, optimum asphalt content that can be counted as a volumetric property was evaluated (Muniandy et al. 2013, Patil et al. 2016). Additionally, as a part of the Marshall mix design, the relationship between asphalt binder content and some volumetric properties such as VMA and VFA were examined (Rahman et al. 2012, Kar et al. 2014, Sutradhar et al. 2015, Priyanka et al. 2015, Patil et al. 2016). To the best of the authors' knowledge, none of the literature studies mentioned above have ever explored the effect of filler on the volumetric properties according to the Superpave mix design. Therefore, this paper is a comprehensive study that mainly aims to investigate the influence of filler types with various proportions on various volumetric properties based on the Superpave mix design of HMA. Based on Superpave specification standards, asphalt mixtures were designed with 19 mm nominal maximum aggregate size (NMAS) and one binder type; namely, PG 64-28. To achieve this goal, two different types of mineral fillers that commonly utilized for preparing asphalt mixtures were used; namely, dust plant (DPt) and hydrated lime (HL) of various proportions (2%, 2.5%, and 3%) were chosen to examine the influence of mineral fillers on various volumetric characteristics including VMA, VFA, dust to binder ratio (Dp), and others.

3 MATERIALS AND METHODS

3.1 Materials

Natural aggregate (NA), and hydrated lime (HL) and dust plant (DPt) that are used as fillers in HMA were obtained from Miller Group company. The obtained sieve analysis of NA, targeted mix design, and MTO specifications are given in Figure 2. Table 2 shows some physical and mechanical properties of NA. The optical images of NA, DPt, and HL are shown in Figures 3-5, respectively.

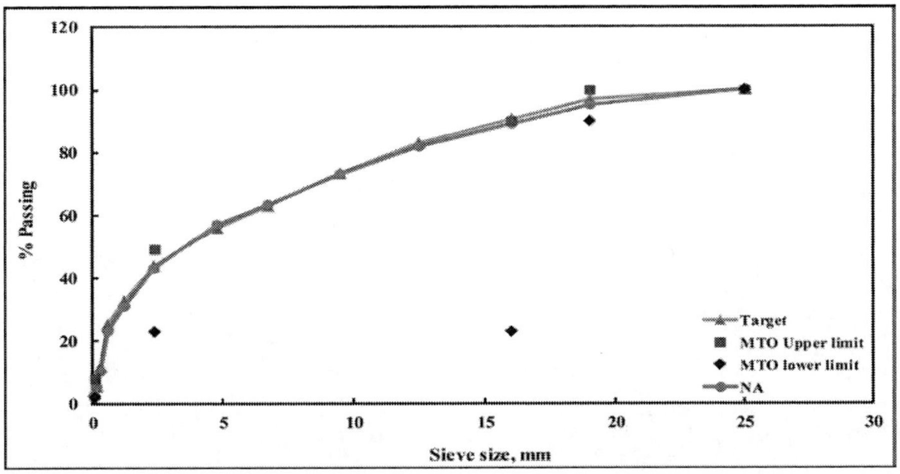

Figure 2: Particle size gradations of NA

Table 2: Physical and mechanical properties of NA

Aggregate type	Bulk relative density (BRD) (ASTM C 127)	Absorption wt.% (ASTM C 127)	Micro-deval abrasion loss, wt.% (ASTM D6928)	Fractured particles, wt.% (ASTM D5821)	Aggregate crushing value test (ACV) (BS 812-110)	Flat & elongated, wt.% (LS-608)	Freezing & thawing (LS-614)
NA	2.658	0.85	15.89	95.5	19.48	0.95	17.4

Figure 3: Optical image of NA

Figure 4: Optical image of HL

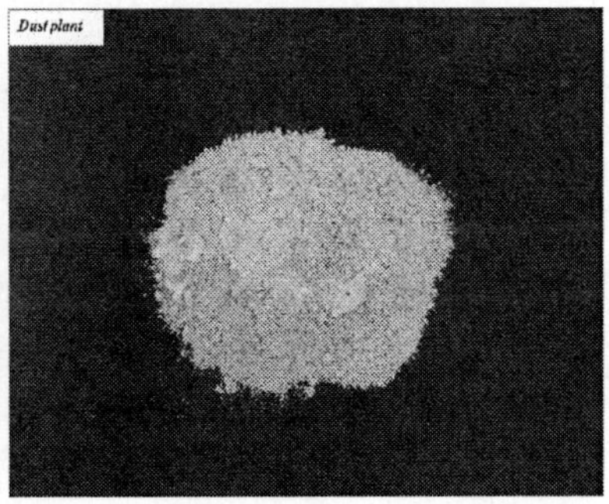

Figure 5: Optical image of DPt

3.2 Methods

3.2.1 HMA Superpave Mix Design

HMA Superpave mix design was performed according to AASHTO R 30-2 2010. A 19-mm Superpave® gradation designed for a road volume ranged between 10 and 30 million was utilized with an unmodified Performance Grade (PG) 64-28 binder. The main characteristics of the design procedure can be simply explained in the following brief descriptions. The design gyration level (N_{des}) was 100, whereas the maximum gyration level (N_{max}) was 160. The viscosity of 1.7 Poises and 2.8 Poises were established to define the mixing and compaction temperatures, respectively. The mixing temperature was 163°C, whereas compaction temperature was 150°C. Compaction was performed by using a Superpave gyratory compactor as shown in Figure 6. An optical image for the samples after compaction is presented in Figure 7.

Figure 6: Superpave gyratory compactor

Figure 7: Samples after compaction

4 RESULTS AND DISCUSSION

4.1 Aggregate Blends with Filler

As mentioned earlier, Superpave mix design was performed according to AASHTO R 30-2 2010. For filler addition, three proportions (2%, 2.5%, and 3%) were used to determine the mix blend. The mix blend that included two filler types with various proportions is given in Table 3.

Table 3: Mix design blend with filler

Filler content, %	Aggregate type, %			
	CA#1, HL8 stone	CA#2, ¼ chip	FA#1, Manufactured sand	FA#2, Blend sand
2.0	40	12	36	10
2.5	40	12	35.5	10
3.0	40	12	35.0	10

CA = Coarse aggregate, FA = Fine aggregate

4.2 Volumetric Properties of Asphalt Mixtures Including Filler

The laboratory data on the volumetric characteristics for HMA mix design with filler addition is tabulated in Table 4. More details on the results of the volumetric properties are discussed in the following sections.

Table 4: Mix design volumetric for different filler percentages

Filler type / Property	Dust plant, %			Hydrated lime, %		
	2.0	2.5	3.0	2.0	2.5	3.0
OAC (%)	4.9	4.83	4.54	4.81	4.79	4.26
VMA (%)	14.65	14.5	13.4	14.5	14.2	13.2
VFA (%)	72.69	72.50	70.10	72.30	72.00	69.50
Dp	0.43	0.6	0.8	0.47	0.6	0.8
Gmm N initial (%)	88.56	88.60	88.30	88.01	88.02	88.03
Gmm Ndes (%)	96.0	96.0	96.0	96.0	96.01	96.0
Gmm Nmax (%)	97.3	97.10	97.19	97.21	97.22	97.00
G_{mb}	2.398	2.40	2.425	2.401	2.407	2.423
G_{mm}	2.498	2.50	2.526	2.501	2.507	2.524

4.2.1 Optimum asphalt content

The optimum asphalt binder content in HMA with two different filler types is graphically analyzed in Figure 8. The obtained results revealed that a comparable performance that can be represented as an exponential equation for both filler types is registered. It is observed that the optimum asphalt binder content decreases when the filler percentage is increased. This behaviour can be mainly attributed to the fact that an increase in the filler amounts can lead to filling the voids that generally exist between aggregate particles, resulting in a reduction of the voids in the mineral aggregate. Therefore, the available space for asphalt binder is decreased (Mehari 2007). However, it is noteworthy that a considerable decrease in the optimum asphalt binder content is highly noticeable when filler content is increased from 2.5% to 3% for both filler types. Additionally, the results demonstrated that the HL with different percentages have a lower OAC than the DPt.

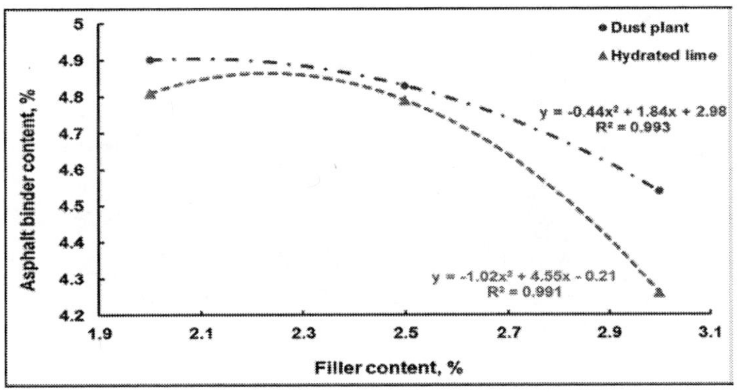

Figure 8: Optimum asphalt content

4.2.2 Voids in mineral aggregates (VMA)

Figure 9 illustrates the effect of filler types with different percentages on VMA of HMA mix design. It was demonstrated that an increase in filler content can lead to a decrease in VMA of the mixture for both filler types. The lowered VMA value could possibly indicate the existence of a thin asphalt film around the aggregate particles. It has been previously found that the utilization of high amounts of filler produces a thin asphalt film which could be more likely to be a detrimental factor with regards to mixture durability (Dos Santos et al. 2013). The findings also indicate a slight variation between two different filler types on VMA. This could be possibly attributed to the size of particles. During the compaction, the large particles of HL can work as rollers that provide less friction in the mastic, which leads to more compact packing and lowers

VMA (National Lime Association 2006). However, compared to the required percentage of the minimum VMA to MTO specification that is typically accounted as 13 for this category of the HMA mix design, the obtained outcomes were higher than the required percentage, resulting in a highly successful filler addition for both different types.

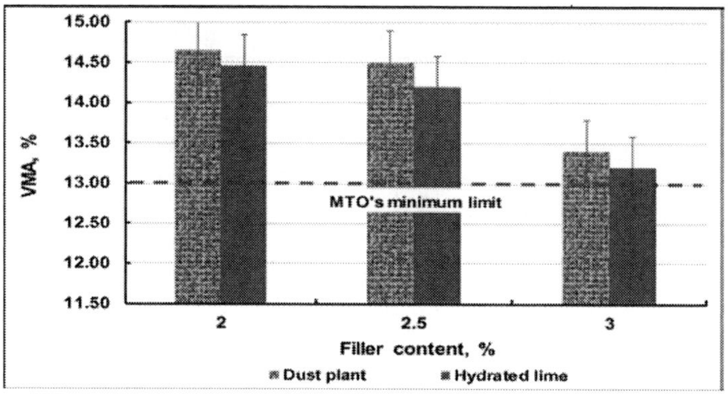

Figure 9: Voids in mineral aggregates (VMA)

4.2.3 Voids Filled with Asphalt (VFA)

The influence of various filler types with different proportions on VFA for HMA is presented in Figure 10. The findings showed that a similar behaviour for both filler types is obtained. It is noted that the VFA value proportionally decreases depending on an increase in the filler content. As explained in VMA property, the existence of a thin film appears to be the main reason behind the reduction in the VFA values. For the influence of different filler types on VFA, an insignificant variance between various fillers on this volumetric property is observed. The VFA values required for the MTO specification range between 65% and 75%. It is found that the research findings were higher than the lower limit percentage. This indicates that there is no impact of a filler type with various proportions on this volumetric property.

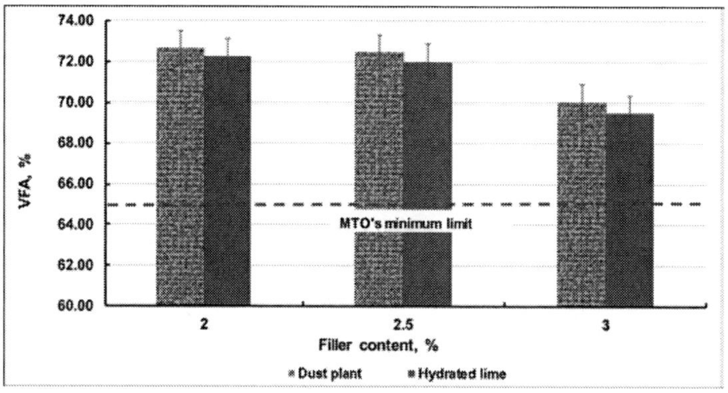

Figure 10: Voids filled with asphalt (VFA)

4.3 Dust to Binder Ratio (Dp)

Figure 11 demonstrates the Dp ratio (also known as the dust proportion) that represents another mixture design criterion. Dp can be defined as the ratio of aggregate fines expressed as a percent by weight that can pass through the 0.075 mm sieve (#200) to the effective asphalt content or optimum asphalt binder accounted as a percentage by weight. The findings of the investigation indicated that the Dp ratio seems to be directly proportional to filler content for both different filler types. The Dp proportion increases when the filler content is increased; therefore, it can be concluded that the Dp proportion behaves inversely compared to other volumetric properties for both filler types. The outcomes also revealed that the Dp values at 2% filler content are 0.43 and 0.47 for dust and lime filler types, respectively. It is interesting to note that the experimental values of Dp with filler content 2% were 0.6 for the dust and lime filler types, whereas the obtained values of Dp with filler content 3% were 0.8 for both different filler types. Depending on the MTO specification that requires an acceptable range for Dp between 0.6 and 1.2, it can be concluded that the obtained values of Dp for the mix design with a filler content of 2% are lower than the minimum required values, resulting in the mix design that is unacceptable for both filler types whereas the experimental values of Dp for mix design with filler content 2.5% and 3% were within the acceptable range of MTO specifications for both filler types. It has been previously reported that low Dp values generally indicate unstable mixtures, whereas high Dp values could possibly refer to a lack of sufficient durability (FHWA). An increase or decrease the air voids are affected by controlling the quantities of material passing the No. 200 sieve in the HMA. The increase addition of fine material leads to lower the air voids and vice versa. The durability of an asphalt mixture generally correlates with the air void. While the too high content of air void can provide passageways for the entrance of more air and water, resulting in a HMA damage, the too low percentage of air void results in a flushing and squeezing the excess binder out to the surface of HMA.

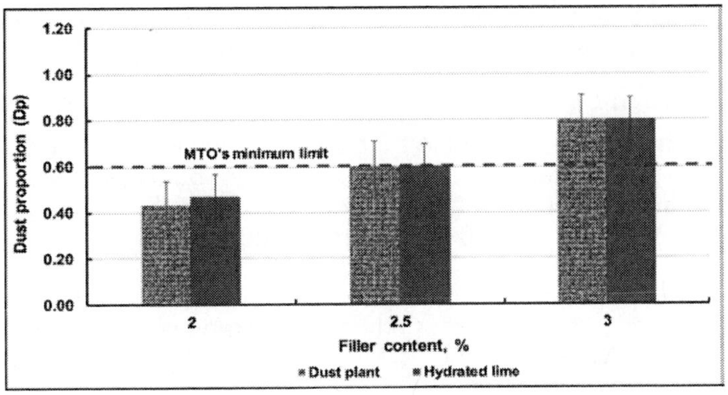

Figure 11: Dust proportion (Dp)

4.4 Maximum Theoretical Specific Gravity (Gmm) & Bulk Specific Gravity (Gmb)

It is generally known that Gmm simply refers to the ratio of the mass of the asphalt and aggregate mixture to the volume that does not include the air voids, whereas Gmb generally indicates the ratio of the mass of the asphalt and aggregate mixture to the volume that includes the air voids. The behaviour of the properties Gmm and Gmb for HMA using different filler types is displayed in Figure 12. It was observed that there is a comparable behaviour to Gmm and Gmb for different filler types with various percentages, indicating that there is no influence of filler type on these properties. When the filler content is increased, there is an increase in the Gmm and Gmb values for both filler types.

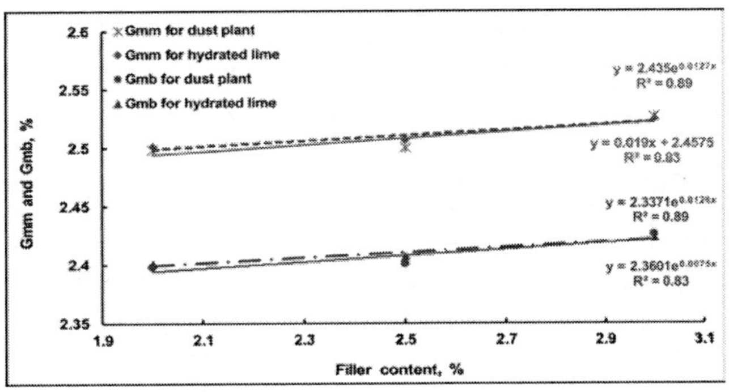

Figure 12: Gmm and Gmb of mix at different filler percentages and types

5 CONCLUSION

Based on the laboratory experiments that related to the application of various filler types in HMA, the following conclusions can be drawn:

- The optimum asphalt binder content for both filler types decreases when the filler percentage is increased. However, a significant decrease in the optimum asphalt binder content is observed when filler content is increased from 2.5% to 3% for both filler types.
- It is seen that an increase in filler content decreases VMA of the mixture for both filler types. A slight difference between two different filler types on the VMA property is noticeable. Compared to MTO specifications, the obtained results are higher than the required proportion.
- VFA values are proportionally reduced depending on an increase in the filler content. An insignificant variance between various filler effects on this volumetric property is found. Based on MTO requirements, it is indicated that the study findings are higher than the lower limit.
- Compared to other volumetric properties, Dp proportion behaves inversely for both filler types in which Dp proportion increases when the filler content is increased. The obtained values of Dp for the mix design with filler content 2% are unacceptable for both filler types due to lower values whereas the experimental values of Dp for mix design with filler content 2.5% and 3% were acceptable for both filler types based on the MTO specifications.
- Similar behaviour for the Gmm and Gmb properties with respect to different filler types with various proportions is observed. When the filler percentage increases, the values of Gmm and Gmb are increased for both filler types.
- The addition of filler of 2.5% is very successful for both filler types due to satisfying all MTO requirements for volumetric properties of HMA. Based on MTO specifications, the addition of 2.0% filler appears to be unsuccessful for both filler types due to lowering the dust to asphalt binder ratio. Mix design with 3.0% filler was also unsuccessful because of the lower value of OAC meaning that the mix is dry and there is insufficient asphalt binder to coat the aggregate particles.

Acknowledgements

The authors would like to gratefully acknowledge the funding provided by Ministry of Higher Education and Scientific Research/Iraq through Iraqi Scholarship/Doctoral Program. The authors would like to thank Miller Group Company for supplying natural aggregate and asphalt binder.

Reference

AASHTO R30-02, 2010. Standard Practice for Mixture Conditioning of Hot-Mix Asphalt (HMA). American Association of State Highway and Transportation Officials (AASHTO), Washington DC.

Anderson, R., & Bahia, H. 1997. Evaluation and selection of aggregate gradations for asphalt mixtures using Superpave. *Transportation Research Record: Journal of the Transportation Research Board*, **1583:** 91-97.

Brown, E. R., McRae, J. L., & Crawley, A. B. 1989. Effect of aggregates on performance of bituminous concrete. In *Implication of aggregates in the design, construction, and performance of flexible pavements.* ASTM International.

dos Santos Bardini, V. S., Júnior, J. L. F., & Klinsky, L. M. G. 2013. Influence of mineral filler on volumetric properties of hot mix asphalt. *International Journal of Pavements Conference.* IJPC Paper 156-1.

Epps, J. A., & Little, D. N. 2001. The benefits of Hydrated Lime in Hot Mix Asphalt. *Arlington (Virginia, USA): National Lime Association.*

Federal Highway Administration (FHWA). *Superpave Fundamentals Reference Manual*, U.S. Department of Transportation. NHI Course# 131053.
Hislop, W. P. 2000. Influence of aggregate type and gradation on critical voids in the mineral aggregate in asphalt paving mixtures. Doctoral dissertation. Iowa State University, Iowa, USA.
Huang, B., Shu, X., & Chen, X. 2007. Effects of mineral fillers on hot-mix asphalt laboratory-measured properties. *International Journal of Pavement Engineering*, **8**(1): 1-9.
Huner, M. H., & Brown, E. R. 2001. Effects of re-heating and compaction temperature on hot mix asphalt volumetrics (No. 01-04). NCAT report.
Kandhal, P., Lynn, C., & Parker, F. 1998. Characterization tests for mineral fillers related to performance of asphalt paving mixtures. *Transportation Research Record: Journal of the Transportation Research Board*, **1638**: 101-110.
Kar, D., Panda, M., & Giri, J. P. 2014. Influence of Fly Ash as a Filler in Bituminous Mixes. *ARPN Journal of Engineering and Applied Sciences*, **9**(6): 895-900.
Mehari, Z.B. 2007. Effect of different types of filler materials on characteristics of hot mix asphalt concrete. M.Sc. thesis. Addis Ababa Uinversity, Ethiopia.
Muniandy, R., Aburkaba, E., & Mahdi, L. M. 2013. Effect of Mineral Filler Type and Particle Size on Asphalt-Filler Matrix and Stone Matrix Asphalt Laboratory Measured Properties. *Australian Journal of Basic and Applied Sciences*, **7**(11): 475-487.
National Lime Association. 2006. Hydrated Lime-A Solution for High Performance Hot Mix Asphalt.
Ontario Provincial Standard Specification (OPSS 1151). 2007. Material specification for Superpave and stone mastic asphalt mixtures. Ontario Ministry of Transportation, Ontario.
Patil, S. B., Khot, A. A., Yadav, D. P., Bachal, R. B., & Patil, P. M. 2016. Effect of Filler and Aggregate on Bituminous Paving Mixes. *International Research Journal of Engineering and Technology (IRJET)* Volume: **03** Issue: 05
Priyanka, B. D., Kumar, P. A., Dedeepya, K., Shabuddin, A., & Rao, S. K. 2015. Use of fly ash as mineral filler for bituminous paving mixes. *International Journal of Research in Engineering and Technology*, **4**(1): 56-60.
Rahman, A., Ali, S. A., Adhikary, S. K., & Hossain, Q. S. 2012. Effect of Fillers on Bituminous Paving Mixes: An Experimental Study. *Journal of Engineering Science*, **3**(1): 121-127.
Remisova, E. 2015. Study of mineral filler effect on asphalt mixtures properties. *Bituminous Mixtures and Pavements VI*, **49**.
Sutradhar, D., Miah, M., Chowdhury, G. J., & Sobhan, M. A. 2015. Effect of Using Waste Material as Filler in Bituminous Mix Design. *American Journal of Civil Engineering*, **3**(3): 88-94.
Tayebali, A., Malpass, G., & Khosla, N. 1998. Effect of mineral filler type and amount on design and performance of asphalt concrete mixtures. *Transportation Research Record: Journal of the Transportation Research Board*, **1609**: 36-43.
Zulkati, A., Diew, W. Y., & Delai, D. S. 2011. Effects of fillers on properties of asphalt-concrete mixture. *Journal of transportation engineering*, **138**(7): 902-910.

Building Tomorrow's Society

Bâtir la Société de Demain

Fredericton, Canada

June 13 to 16, 2018 / *13 au 16 juin, 2018*

EVALUATION OF FLEXURAL BEHAVIOUR OF LOW DENSITY FIBRE REINFORCED CONCRETE USING DIGITAL IMAGE CORRELATION

Ibeawuchi, John[1,2], Moffatt, Edward[1] and Lloyd, Alan[1]
[1] University of New Brunswick, Canada
[2] John.ibeawuchi@unb.ca

Abstract: Fibre reinforced concrete over the years has gained popularity in structural applications through studies on its behaviour. Despite these studies there is the need to fully understand the flexural behaviour of fibre reinforced concrete (FRC). This paper presents the results of an experimental program to investigate the mechanical properties and post-cracking behaviour of FRC beams under four-point loading using digital image correlation (DIC) to monitor strain, displacement, and cracking. Expanded polystyrene low-density fibre reinforced concrete beams with steel and polypropylene fibre types at replacement levels of 0% and 0.5% by volume for each fibre type were cast and tested in flexure in accordance to ASTM C1609/C1609M-12. This experimental work detects the initiation and propagation of cracks on the specimens. It shows the failure mechanism, neutral axis shift and compression strain field of each beam. Results from the experiments showed that steel fibre produced a slight increase in compressive strength. However, flexural strength and post-peak ductility were significantly higher with the addition of either steel or plastic fibres. Steel fibre mixtures showed tension hardening behaviour while plastic fibre mixtures exhibited tension softening behaviour and a significant drop in peak load after first crack, which is a result of rapid crack propagation due to the low stiffness of the plastic fibres. DIC was used to measure displacement and strain fields and the values validate traditional LVDT measurements.

1 INTRODUCTION

Concrete is the most widely used building material in the world and is attributed to its availability and economy. The use of conventional plain concrete can sometimes be limited because of its brittle nature and low tensile properties. To improve the tensile properties, it has become common practice to use reinforcement to carry the tensile stresses that may develop in concrete. Concrete reinforcement affects crack development and propagation as it bridges the cracks and improves the behaviour of the concrete in flexure. Recent developments and research have given an alternative to the use of reinforcement bars in the form of fibre reinforcement, which involves the addition of discontinuous discrete reinforcing fibres of different types and geometry to the concrete matrix. In some cases, fibre reinforcement can be used to replace traditional reinforcement such as welded fabric, however, it can also be used in combination with steel reinforcement. Various forms and types of fibres are available, however, steel and plastic are the most common for a wide range of structural applications including but not limited to slabs on grade, mining, tunnelling and excavation support applications rock slope stabilization (ACI 544-2010).

Many researchers have studied the effects of test method, fibre geometry, aspect ratio, fibre type and volume fraction of the use of fibres in concrete and on the post-cracking strength of fibre reinforced concrete

(FRC). Carbon fibre has been found to give high compressive and splitting tensile strengths while steel fibre gives higher flexural strength and toughness (Wu et al., 2003). In hybrid construction containing both steel and plastic fibres, the concrete has been shown to improve in both strength and flexural toughness (Wu et al., 2003). Rizzutti and Bencardino (2014) analyzed the effects of fibre volume fraction on the mechanical properties of steel FRC. They observed that the addition of fibres does not significantly affect the compressive strength of concrete. Song and Hwang (2004) observed that the compressive strength of fibre reinforced concrete reached a maximum strength at a replacement of 1.5% by volume. They also observed that the flexural strength, splitting tensile strength and toughness index of fibre reinforced concrete improved with increasing fibre volume fraction. Others have attempted to model the pull-out behaviour of fibres in order to predict the flexural response of FRC. The property/quality of the concrete matrix, the fibre type, content and geometry determine the mechanical properties of FRC (Robins et al., 2002). These mechanical properties can be characterized from the analysis of stress-strain curves and load-deflection plots of the FRC. The primary reason for adding fibres in concrete is to improve the ductility and post peak behaviour.

The use of strain gauges, extensometers, linear variable differential transducer (LVDT) sensors and cross head movement have been well used in experimental research for measuring the displacements and strains of concrete. These techniques are not very accurate in the estimation of strain fields or early crack detection and local failure processes in structural members. An alternative technique is the use of digital image correlation (DIC) analysis.

DIC is one of the full field non-contact optical techniques used in measuring displacements and strains in experimental mechanics. Other techniques include photo-elasticity, geometric moiré, moiré interferometry, holographic interferometry, speckle interferometry (ESPI), and the grid method (Grédiac, 2004). These methods are however expensive, difficult to use and often cannot be used typically in outdoor conditions compared to DIC. DIC can be used in structural monitoring and inspections by capturing images and making comparisons/analysis to identify differences in the behaviour of a member under consideration. DIC is also a better alternative to methods like dye penetration in measuring crack opening (McCormick and Lord, 2010).

This paper presents an investigation into the behaviour of steel and polypropylene plastic FRC in flexure at a replacement level of 0.5% by volume. Behaviours were compared with a non-fibre low-density concrete mixture. This experimental work also validates the use of DIC in the measurement of displacement and strains as opposed to traditional methods of measurements. Low-density concrete containing expanded polystyrene (EPS) beads was used in the study due to its growth in popularity in the construction industry.

2 MATERIALS

All concrete mixtures were prepared with a general use (GU) ASTM C150 Type I Portland cement with a density of 3150 kg/m^3. Concrete specimens were cast using a water/cement ratio (W/CM) of 0.35 in addition to a normal density gravel as the coarse aggregate with a nominal maximum size of 9.5 mm and a standard lab grade sand. In order to maintain a low W/C and reach a suitable strength, a moderate dosage of a high-range naphthalene-based superplasticizer was used to achieve a workable slump. Two types of fibres were studied including hooked end steel fibres glued together in bundles by a water soluble glue and polypropylene plastic fibres (see Table 1). The density of all mixtures, expect the control, were reduced through the use of expanded polystyrene (EPS) beads with a nominal diameter of 6-10 mm.

Table 1: Properties of Fibres

	Steel	Plastic
Length l_f(mm)	30	30
Diameter d_f (mm)	0.5	1.0
l_f/d_f	60	30
Tensile Strength (MPa)	1500	800

3 MIX PROPORTIONS

Concrete mixtures were designed in accordance with ACI-211.2 and were modified by incorporating EPS beads. The expanded polystyrene beads were batched by volume as it was not accurate to batch by weight as a result of its low density. Two fibre volume contents of 0% (control) and 0.5% by volume were used. Table 2 presents the mix proportions of the various concrete mixes.

Table 2: Concrete Mix Proportions

Mix	Mix Designation	W/C	Coarse Aggregate (kg/m^3)	Fine Aggregate (kg/m^3)	EPS Beads (m^3)	Cement (kg/m^3)	HRWRA (mL/100kg)	Fibre Volume (%)
Control	NFRNWC*	0.35	875	693	-	550	975	-
Control	NFRLDC**	0.35	451	350	0.3	550	-	-
Steel	SFRLDC***	0.35	451	350	0.3	550	198	0.5
Plastic	PFRLDC****	0.35	451	350	0.3	550	-	0.5

*NFRNWC: Non-Fibre Reinforced Normal Weight Concrete

**NFRLDC: Non-Fibre Reinforced Low Density Concrete

***SFRLDC: Steel Fibre reinforced Low Density Concrete

****PFRLDC: Plastic Fibre reinforced Low Density Concrete

4 SPECIMEN PREPARATION

All concrete specimens were cast and cured in accordance with ASTM C192. A total of 5 specimens including three 100.6 x 203.2 mm cylinders and two 150 x 150 x 530 mm prisms were prepared for each mixture. The mixing procedure consisted of firstly mixing all aggregates (fine and coarse) for 30 seconds. This was followed by the addition of cement, EPS beads and fibres, which were then mixed for 30 seconds, 1 minute and 2 minutes, respectively. Water was then added and left to mix for another 3 minutes. The superplasticizer was finally added to achieve the desired workability. Concrete specimens were then cast after which the specimens were covered in wet burlap and plastic for 24 hours. The specimens were then demoulded and placed in a fog room (100% relative humidity) at room temperature (22±2°C) until they were tested.

5 TEST SETUP AND PROCEDURE

5.1 Compression test

The compressive strength test was performed on 100.6 x 203.2 mm cylinders in accordance with ASTM C39 at a loading rate of 35 psi/sec (0.24 MPa/sec). Compressive strength was measured at an age of 28 days.

5.2 Flexural test

The flexural strength (average of two specimens) was measured on the form finished side of 150 x 150 x 530 mm prisms under four-point loading configuration in accordance with ASTM C1609/C1609M-12. Two LVDTs were mounted on a jig at mid-span, one on each side in order to measure the displacement. The jig was clamped to the prism directly above the supports. The tests were carried out at a constant loading rate of 0.1mm/min in accordance with table 1 in ASTM C1609/C1609M-12 before and after a net deflection of L/900 (where L = 457.2 mm) had been reached. A photograph of the test setup is presented in Figure 1.

Figure 1: Test Setup for Flexural test.

5.3 Digital image correlation

DIC is a non-contact optical full-field deformation measurement system. A commercial, 3D acquisition and analysis software was used for this research (Correlated Solutions Vic-3D and Vic Snap). It was used to measure the displacement and strain on the specimens. The setup consists of two cameras with a resolution of 5 megapixels positioned 300 mm apart on a tripod and 1000 mm away from the specimen in such a way that the specimen fills the field of view of both cameras. An external light source was used to illuminate the field of view. Necessary adjustments to focus, aperture and exposure time were made to ensure higher image quality. Prior to testing, the specimen surface in view of the camera was painted white and speckled with random black dots. The system was then calibrated as specified by vic-3d testing guide. Images were captured at one frame per second. Applied force data was also recorded with a data acquisition system coupled to the DIC software during testing. Analysis and post-processing of data was done with image analysis software with a subset size of 35 pixels. Figure 2 and 3 presents the speckled specimens prior to testing and the DIC test setup, respectively.

Figure 2: Speckle pattern on specimen

Figure 3: DIC setup

6 RESULTS AND DISCUSSIONS.

6.1 Fresh concrete properties (Slump)

The workability of each mixture was determined by measuring the slump in accordance with CSA A23.2-5C. The results are presented in Table 3 and show that the workability decreased with addition of fibres and improved with higher content of superplasticizer.

6.2 Density

The density of each mixture was determined in accordance with CSA A23.2-11C and are presented in Table 3. The addition of fibres is shown to have negligible effect on the hardened concrete density, which is not surprising considering the volume replacement was less than 1%. The density of the low density mixes show a 20% decrease from the normal weight concrete mix.

6.3 Compressive strength

The compressive strength following 28 days of age is also presented in Table 3. The results show that the addition of steel fibres slightly increases the compressive strength. However, the addition of plastic fibres showed a 3% decrease in 28-day compressive strength. This decrease can be attributed to the tendency of plastic fibres to agglomerate in the mix, which can introduce air voids in the specimen. The addition of fibres also showed an improvement in post peak ductility capacity of the concrete specimens.

6.4 Flexural strength

The flexural strength was determined and calculated in accordance with ASTM C1609 and are presented in Table 3. It was observed that the addition of fibres increased the flexural strength. However, the use of steel fibres gave a significantly higher flexural strength when compared with the plastic fibre and non-fibre mixes as expected. Typical flexural load-deflection curves for the different mixes used in this study are presented in Figure 4-6, which compares the load-deflection plots of the DIC and that from LVDT. Increase in flexural strengths beyond a displacement of 1.5mm is attributed to strain hardening due to the effects of fibres. Both curves correspond with each other and proves the validity and reliability of the use of DIC to measure displacement and strain values. The DIC was used for measurement on one of the prisms while the LVDT was used for the other. Both methods were not combined as the jig holding the LVDT was in the view of the camera.

Table 3: Concrete Properties

Concrete	Density (kg/m^3)	Compressive Strength (MPa)	M.O.R (MPa)	Slump (mm)
NFRNWC	2408	65	3.79	45
NFRLDC	1930	31	2.71	65
SFRLDC	1933	33	3.87	70
PFRLDC	1929	30	3.14	25

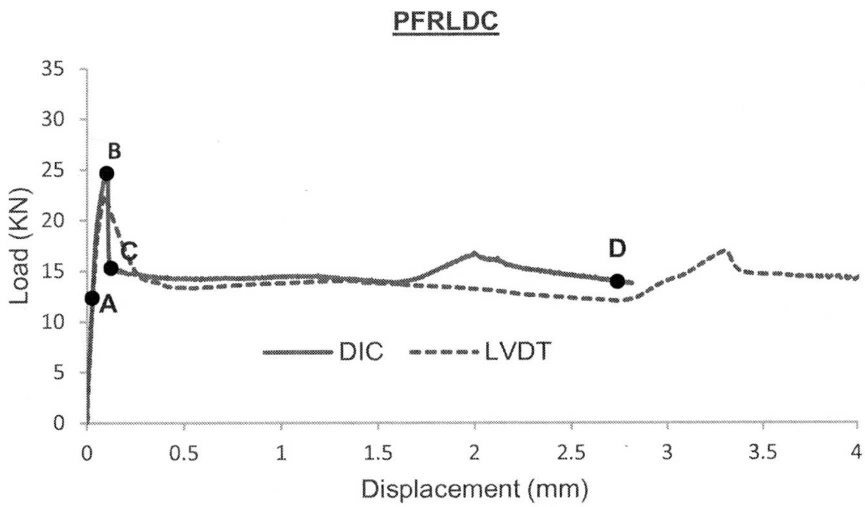

Figure 4: Load-Displacement curve for plastic fibre reinforced low-density concrete

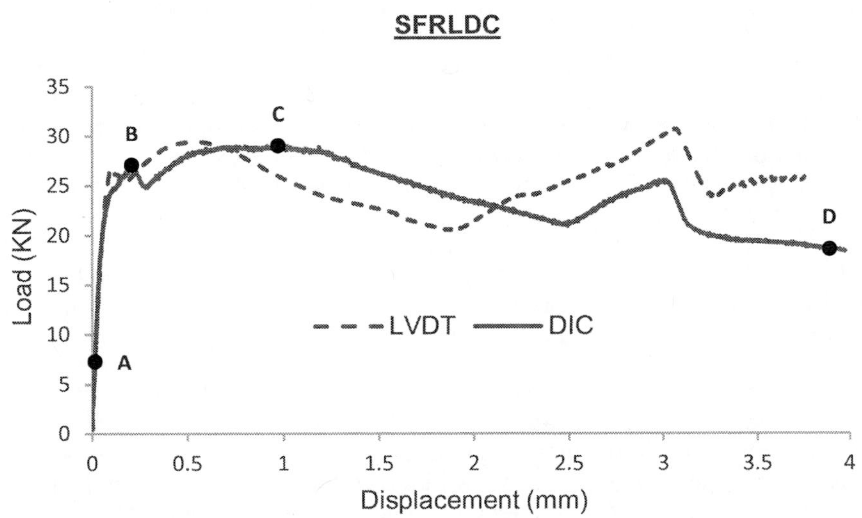

Figure 5: Load-Displacement curve for steel fibre reinforced low density concrete

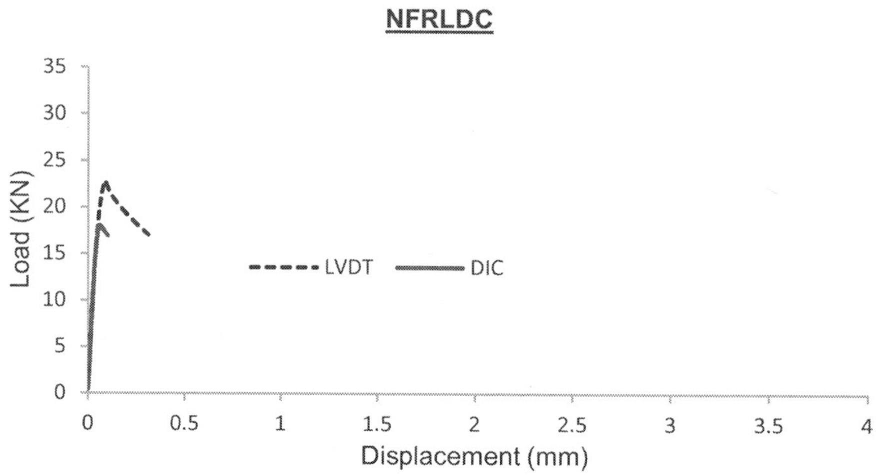

Figure 6: Load-Displacement curve for non-fibre low density concrete

6.5 Crack detection

Cracking in concrete is most times a sign of failure. This can be seen when a concrete specimen is subjected to constant loading. Due to these loads it is capable of reaching its capacity, which inevitably leads to cracking and in certain cases failure. In bending tests, as cracks propagate, there is a decrease in load and increase in crack width. It is assumed that the point of divergence from linearity of load-deflection curves is the point of first crack (Bensaid et al., 2014). Though this assumption holds value, it is still difficult to identify these cracks especially in cases where multiple cracks exist. In bending tests, these cracks develop in the tensile region of the specimen and propagate upwards towards the loading surface. With the aid of DIC, the cracks were immediately detected and their propagation was monitored until failure as presented in Figure 7. The figure shows the longitudinal (strains due to flexure) strain fields overlaid on the prisms at various times as the flexural test proceeded to specimen failure. Regions on the beam indicated by red show high strain concentrations and existence of cracks, whereas purple indicates low strain values (compression). From the image analysis, it was observed that SFRLWC (Figure 7 left) had more than one crack. The primary (first) crack corresponds with the deviation from linearity on the load deflection curve and was bridged by fibres. A secondary crack developed which led to the ultimate failure of the beam. Figure 7 shows the detection and propagation of cracks on the steel and plastic fibre concrete mixes studied. Figure 7A represents the strain distribution at the linear-elastic region. Figure 7B represents the first crack and initiation of cracks. Figure 7C shows the propagation of crack and Figure 7D shows the fracture of the specimens. These can also be seen in the load-deflection plots in figures 4 and 5.

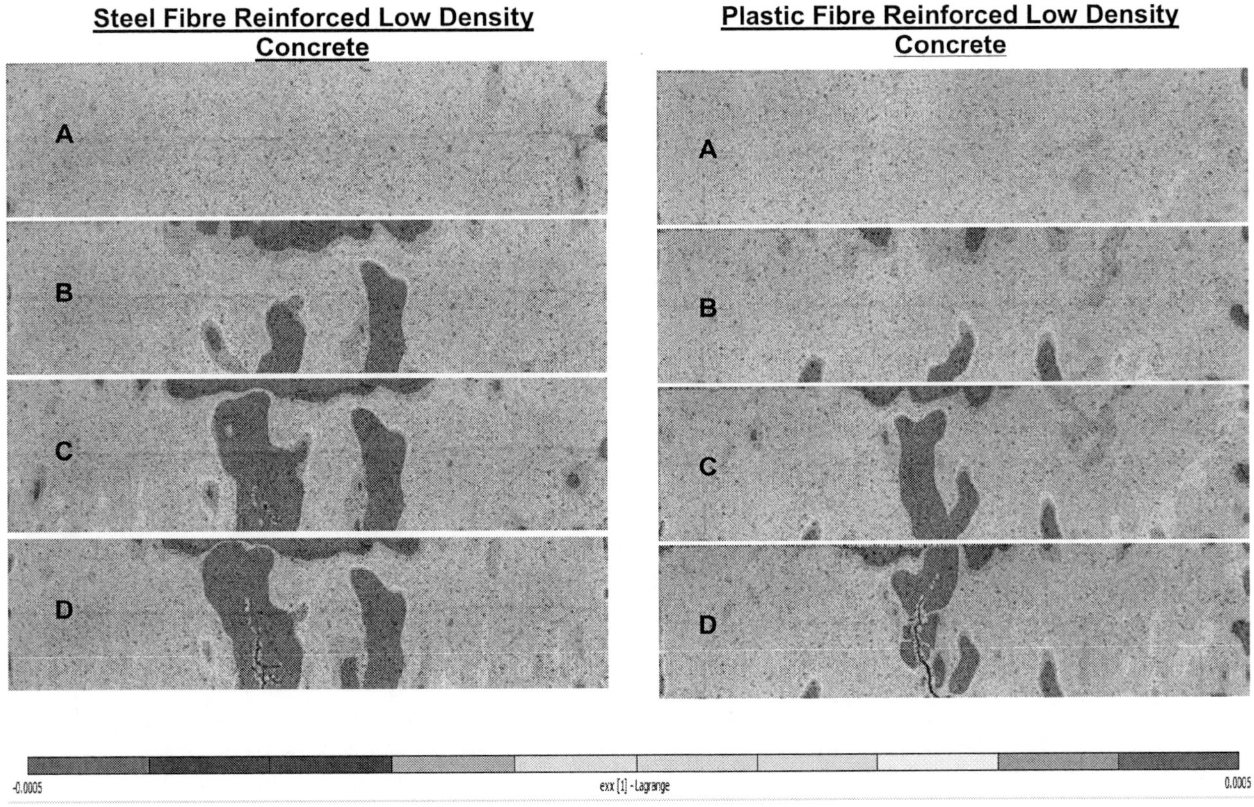

Figure 7: Contour diagram showing longitudinal strain distribution.

6.6 Neutral axis movement

The neutral axis is defined as an axis in the cross-section of a beam where stresses and strains are zero. If the beam cross-section is symmetric and the material is homogeneous and isotropic, the neutral axis would be at the centre. However, when concrete specimens are loaded, cracks develop causing the material to be anisotropic and the neutral axis shifts. By using image analysis, the shift of neutral axis of the concrete beams was observed. It was observed that the plastic fibre mix showed a high jump in the neutral axis position when cracks developed when the sectional strain was measured at the crack location. This can also be seen in the load-deflection curve (Figure 4) where there was a rapid drop in load after first crack. However, the movement of the neutral axis in the SFRLDC mix was gradual as a result of the fibres arresting the crack and the high flexural strength of the matrix. When cracks developed, the neutral axis moved up and was measured to be 49.4mm from the compression face of the beam as shown in Figure 9. This movement was gradual until failure and was measured to be 33.2 mm at fracture. The neutral axis position in the non-fibre mix is not presented in this work. From Figures 8 and 9, it can be seen that the strain distribution of the compression side remained largely linear. The strain distribution in the tension side is not representative of the real material strains due to the formation of cracks in the tension zone. From the movement of the neutral axes, strain diagrams were plotted considering only compression strain values with line of best fit extending into the tension region and are presented in Figure 10 and 11.

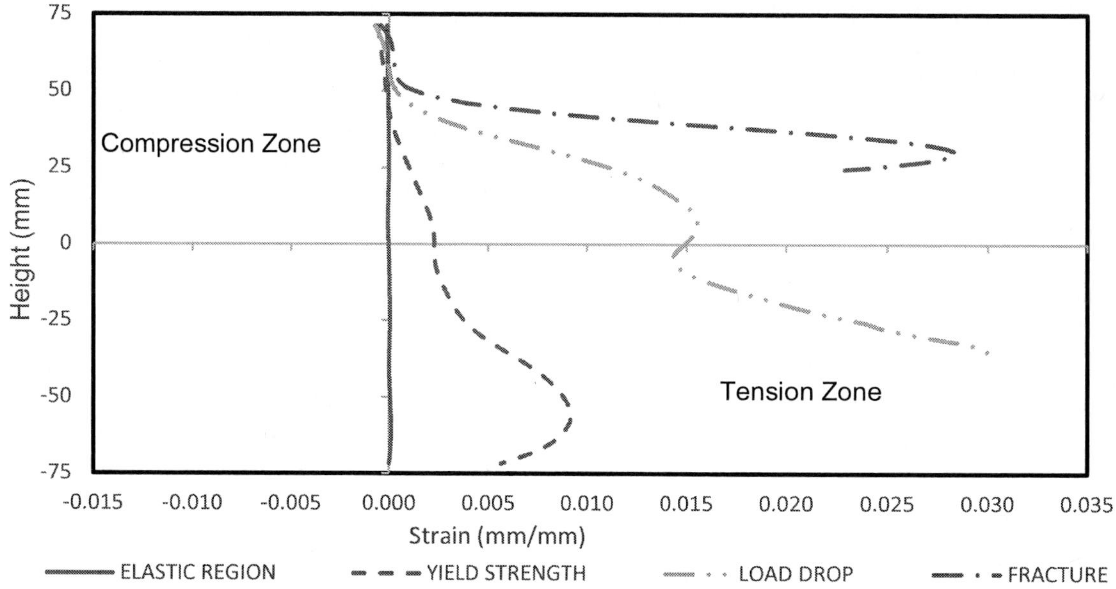

Figure 8: Neutral axis movement of PFRLDC shown from DIC

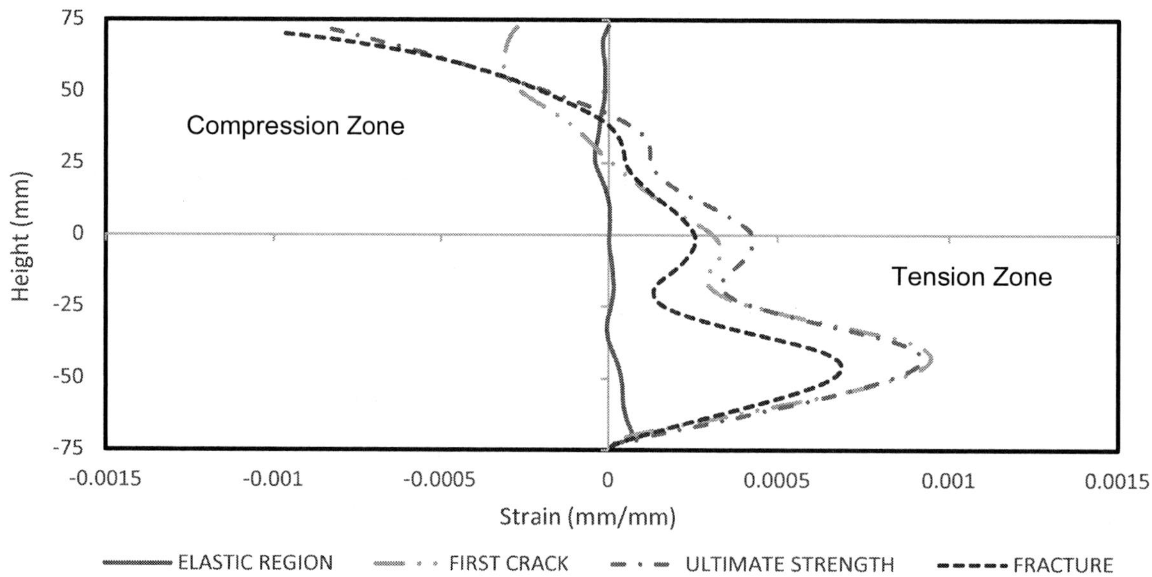

Figure 9: Neutral axis movement of SFRLDC shown from DIC

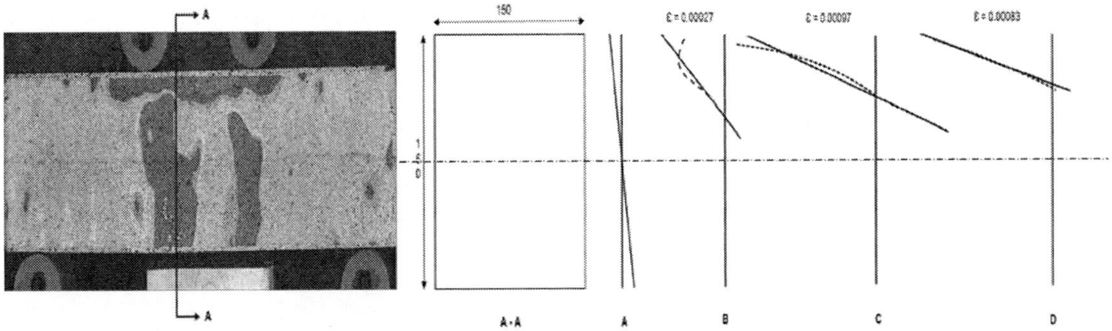

Figure 10: Neutral axis movement of SFRLDC plotted from DIC

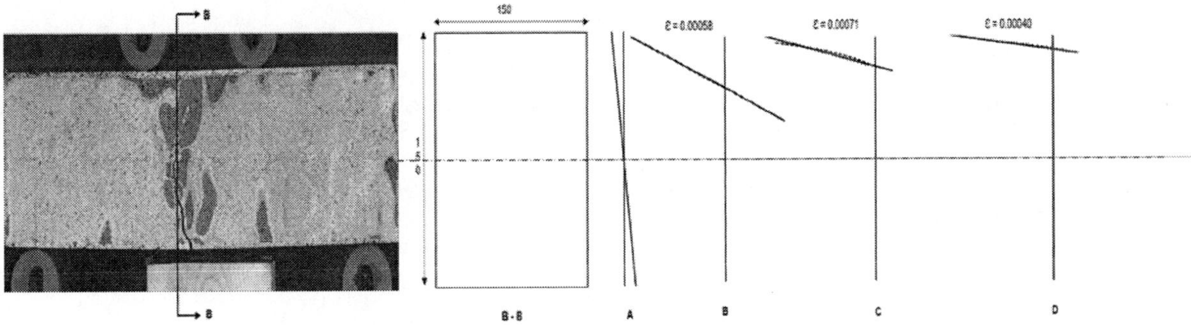

Figure 11: Neutral axis movement of PFRLDC plotted from DIC

Table 4a: Neutral axis position for SFRLDC

Point	Position	Distance* (mm)
A	Elastic region	75
B	Yield Strength	49.40
C	Ultimate Strength	36.85
D	Fracture	33.20

Table 4b: Neutral axis position for PFRLDC

Point	Position	Distance* (mm)
A	Elastic region	75
B	Yield Strength	31.08
C	Load Drop	18.14
D	Fracture	8.08

* Distance measured from the compression face of the specimen7.0. Conclusions

Based on the results presented within this paper, the following conclusions can be drawn:

- The effectiveness of steel fibres in arresting cracks is more pronounced than plastic fibres. This can be seen from the neutral axis plots where the steel fibre mix showed a gradual shift in neutral axis position compared to plastic fibre, which showed a more rapid movement.
- Steel fibre of low aspect ratio and volume fraction as low as 0.5% produces tensile hardening properties when used in low density concrete.
- The use of steel fibres slightly improved the compressive strength and greatly improved the flexural strength and ductility of the matrix.
- Digital image analysis is effective and a more accurate way of measuring deformation and fracture.

References

ACI 211.2. 2004. Standard practice for selecting proportions for structural lightweight concrete. ACI International. Moraga Dr, Los Angeles, United States.

ACI 544.1R. 2010. State of the art report on fibre reinforced concrete. ACI International. Moraga Dr, Los Angeles, United States.

ASTM C1609/C1609M. 2012. Standard test method for flexural performance of fiber-reinforced concrete (using beam with third-point loading), *ASTM International.* West Conshohocken, PA, United States.

Bensaid B., Mostefa H., Mohamed C. and Sofiane A. 2014. Failure mechanism of fibre reinforced concrete under splitting test using digital image correlation. *Materials and Structures,* no. 48: 2713–2726

Correlated Solutions. 2016. Vic-3D v7.2.6 Testing guide.

Correlated Solutions. 2016. Vic-3D v7.2.6 Reference Manual.

Grédiac. 2004. The use of full-field measurement methods in composite material characterization: Interest and limitations. *Applied Science and Manufacturing* 35(7-8):751-761.

McCormick, N. and Lord J. 2010. Digital Image Correlation. *Materials Today* 13, no. 12: 52–54.

Rizzuti, L. and Bencardino, F. 2014. Effects of Fibre Volume Fraction on the Compressive and Flexural Experimental Behaviour of SFRC. *Contemporary Engineering Sciences 7,* no. 8: 379-390.

Robins. P, Austin, S. and Jones, P. 2002. Pull-out Behaviour of Hooked Steel Fibres. *Materials and Structures* 35, no. 7: 434–42.

Song, P. S and Hwang, S. 2004. Mechanical Properties of High-Strength Steel Fiber-Reinforced Concrete. *Construction and Building Materials* 18, no. 9: 669–73.

Wu, Y., Jie, L., and Keru W. 2004. Mechanical Properties of Hybrid Fiber-Reinforced Concrete at Low Fiber Volume Fraction. *Cement and Concrete Research* 33, no. 1: 27–30.

Building Tomorrow's Society
Bâtir la Société de Demain

Fredericton, Canada
June 13 to 16, 2018 / *13 au 16 juin, 2018*

EVALUATION OF THE EFFECT OF FLY ASH ON THE CEMENT HEAT OF HYDRATION ACCORDING TO ASTM C1702

Abdel-Raheem, Mohamed[1,2], Hernandez, Luis Brando[1], Navarro, Linda[1], Silva, Grecia[1], Hernandez, Daniel[1], Gomez, Michael[1] and Garza, Rafael[1]
[1] Department of Civil Engineering, University of Texas Rio Grande Valley, USA
[2] mohamed.abdelraheem@utrgv.edu

Abstract: In the construction of mass concrete structures, special attention should be given to the cement heat of hydration (HH). The amount of HH released can be substantial that it might compromise the integrity of the structure. Uncontrolled HH can lead to the formation delayed ettringite formation and thermal stresses that may lead to the failure of the structure. One method to control the HH is through the integration of pozzolanic supplementary cementitious materials (PCSM) in the concrete mix that are known to lower the HH. Previous research shows that Fly Ash (FA) can be an effective replacement of cement especially in mass concreting as it not only enhances the compressive the strength but also it lowers the HH. However, one study shows that heat released using FA class-C can exceed the HH of plain cement. As such, there is a need to investigate this discrepancy and verify the results. This study presents a re-evaluation of the effect of FA on the HH of a cement HH using a high precision isothermal calorimeter. Eighteen paste mixtures were made for two different types of cements replaced at different percentages using FA classes C and F. The experiments were conducted in accordance with ASTM C1702.

1 INTRODUCTION

The hydration of cement is an exothermic reaction that can sometimes release a substantial amount of heat that can be problematic for the structure. In mass concreting, this amount of heat is considerable that it increases the temperature of the concrete. The problem occurs when a temperature gradient occurs due to the difference in the rate of cooling between the inner core and the outer surface of concrete. As the surface of concrete cools and shrinks faster than its core thermal cracking start to occur. Another problem that is caused by the rise in the temperature of the core of concrete is the formation of the delayed ettringite, which causes an expansion in the volume and cracking of concrete (Botang 2013, Patil 2015, Gajda and Vangeem 2002, Gajda 2016). As such the specifications usually limits the temperature difference between the outer and the inner of concrete to 20°C (36°F) with a maximum temperature of 70°C (158°F) (PCA 1997).

The methods used to treat the HH vary in nature. They can be classified into methods of constructions used and supplementary cementitious materials (SCM) incorporated in the concrete mix. The method of construction includes, precooling of concrete aggregates, post cooling of concrete, insulation of the concrete member, and pouring of concrete in thin lifts (PCA 1997, Botang 2013, Patil 2015, Gajda 2016). Some specifications require the incorporation of SCMS that are known to lower the HH in mass concreting (Alhozaimy el al. 2015). Previous studies addressed the effect of a variety of SCMs on the cement HH; the studies included fly ash (FA), silica fume, metakaolin, granulated slag, and others (Rojas 1996, Rojas and Frias 1995, Cheng-Yi et al. 1985, Frias 2000, Snelson et al. 2008, Persson 1997, Schindler 2003, Langan

2002, Ballim et al. 2009). Fly ash and granulated slag have been proven to be the most effective in lowering the HH of the concrete mix compared to other (Cengiz Duran Atis 2002, Snelson et al. 2008, Sokkary et al. 2004).

There are several factors that influence the amount of HH released, which are: the amounts of tricalcium silicate (C3S) tricalcium aluminate(C3A) in cement, water-cement(w/c) ratio, curing temperature, and cement fineness (Kirby et al. 2012, Orosz et al. 2017). The quantification of HH has been previously done using isothermal calorimetry because of its sensitivity and high precision (Pane et al. 2005, Mostafa et al. 2005, Xu et al. 2006). The sample used for the HH quantifications varied between cement pastes, mortar, and concrete. However, the quality and precision of the obtained raw data depends on the type of isothermal calorimeter used. Hence, it is expected to have variations in results not only because of the differing chemical composition of the material used but also because of the degree of precision of the instrument used. Another source of variation is the standard used for testing the material to quantify the HH. There are different standard tests that can be used to determine the HH. The Standard Test Method for Heat of Hydration of Hydraulic Cement (ASTM C186-05) is a more common test and has been used extensively throughout the literature. This test allows the determination of HH of hydraulic cement at any age, the temperature rise in mass concrete, and if the cement under test meets the HH requirement of the applicable hydraulic cement specification. Another high-precision standard that is specific to isothermal calorimetry is the Standard Test Method for Measurement of Heat of Hydration of Hydraulic Cementitious Materials Using Isothermal Conduction Calorimetry (ASTM C1702-17). However, it has not been very common in determining the HH of cementitious materials according to the literature reviewed.

2 PROBLEM STATEMENT

Although the literature contains several studies that focused on the evaluation of effect of FA on the HH of cement, the results are quite inconclusive. Most studies conclude that the incorporation of FA in the concrete mix can lower the HH significantly. However, one study shows the addition of FA class-C to the mix can increase the HH over time compared to plain Portland Cement (PC) (Schindler and Folliard 2003). This study is dated to 2003. The ability to measure and record the HH has drastically has changed since then as more accurate instruments and knowledge have become available.

3 OBJECTIVE

The purpose of this study is to re-analyze and re-evaluate the effects of different classes of FA on the HH of two different types of cements, namely type I, and III using a high precision isothermal calorimeter Calmetrix I-Cal2000. The study aims at drawing a conclusion on the performance of FA class-C since the results regarding its effect on the cement HH is quite contradictory. The study also serves as an alternative evaluation of the effect of FA class-F using ASTM C1702-17 standard, which is specific for isothermal calorimetry.

4 LITERATURE REVIEW

The effect of the FA on the cement HH has been studied throughout the years. The earliest study goes back to Meland (1983). This study investigated the effect of using FA on the HH when used as 10% and 20%. cement replacements. The study investigated the HH of a plain cement pasted, a mixture that incorporated FA as a cement replacement in 10% and 20%, and a mixture that included FA and plasticizer. The results showed that both mixtures that contained FA released lower HH compared to the control mix (Meland, 1983). Tokyay (1988) assessed the effect of three different types of Turkish fly ash on the cement HH. The FAs were used as 10%, 20%, and 40% cement replacements. The study concluded low-calcium FA is more effective at reducing the HH than the high-calcium types. In 1993, De Rojas et al. conducted a study to study to evaluate the effect of FA on the HH when used as a partial replacement of cement. The results show that increasing the percentage of FA reduces the temperature of the mixture. De Rojas and Frias (1995) conducted another study to compare the effect of FA versus Silica Fume (SF) on reducing the HH. The results showed that FA is more effective that SF. Several other studies concluded the same results about the effectiveness of FA in decreasing the peak temperature and the total HH released (Frias et al. 2002, Atis 2002, Maia et al. 2011, Alhozaimy et al. 2015).

Some other studies focused on the effect of the water/binder (w/b) ratio on the HH of mixtures that incorporated FA. As study by Langan et al. (2002) showed that increasing the w/b ratio delay the peak temperature and decrease the heat evolution for mixtures that contained FA class-C. On the contrary, the results obtained from Pane and Hansen (2005) show that decreasing the w/b ratio lowers the HH. The effect of FA combined with other SCMs, such as slag and metakaolin was tested. The results show that the FA when combined with slag becomes very effective in reducing the HH (Lee et al. 2014). However, the HH tends to increase in the presence of metakaolin (Snelson et al. 2008).

A study was conducted to compare the effect of different classes of Fly ash, namely class F and C, on the HH of a cement mixture (Schindler and Folliard 2003). The study sheds the light that the performance of FA varies based on the amount of the calcium compounds contents it includes. The study concluded that FA class-F is more effective in reducing the HH compared to class-C. In fact, FA class-C tends to increase the amount of HH released compared to a plain cement mixture due to the high contents of calcium (Schindler and Folliard 2003).

Class-C and class-F FA were both tested and compared to each other in mixtures that contained replacements of 15%, 25%, and 35% of PC (Schindler and Folliard 2003). Both class-C and class-F fly ashes with the highest replacement amount (35%) had the lowest adiabatic temperature rise when compared to the control sample (100% cement). However, after 40-50 hours, the heat release of the samples made with class-C FA showed an increase exceeding the HH released by the control sample. It was concluded that class-F FA is a better option for reducing the HH than FA class-C.

5 METHODOLOGY

A high-precision isothermal calorimeter was used to ensure the accuracy and consistency of the results, and the repeatability of the experiments in a controlled test environment. The isothermal calorimeter contains two separate channels that can accommodate two samples for testing at the same time. The tests were performed according to the ASTM C1702, which provides guidelines for measuring the heat of hydration of hydraulic cementitious materials specific to using an isothermal calorimeter. To conduct an ASTM C1702 test, the sample water is preconditioned for 2 hours at 23° C. After the water calibration is completed, the sample mix is made, and the samples are placed inside the isothermal calorimeter channels. With the software provided by the isothermal calorimeter manufacturer, the experiments are left running for 24 hours. As a result, raw data of the heat flow produced by the cement hydration reaction is generated and stored. The generated raw data can then be used to create thermal power and cumulative energy graphs for a more in-depth heat of hydration analysis.

According to ASTM C1702, a mixture sample is comprised of 50 grams of the cementitious binder mixed with water at a w/b ration of 0.5. In this research two different types of Portland Cement were considered, which are: Type I and III. FA class-F and class-C was used in different percentages as mass replacements of the different cement types. The replacement percentages used were 10%, 20%, 30% and 40% based on the percentages used in the literature. As such, eighteen mixtures representing the different combinations of cement type, FA class, and percentage replacements are needed to accomplish the goals of this study.

6 EXPERIMENTATION AND RESULTS

The heat flow of the different mixtures was measured using isothermal conduction calorimetry for 24 hours at constant temperature of 23° C According to ASTM C1702. Thirty-two different curves were generated reflecting the thermal power over time and the total heat released by the eighteen mixtures.

6.1 Cement Heat Curves

In order to evaluate the effect of different classes of FA on the HH, it is essential to generate the basic heat curves of the different types of cements that are used in the study as a medium for assessment for comparison purposes. The heat flow and heat of hydration curves for cement type I and III are given in Figures 1a and 1b, respectively. In agreement with the literature, the thermal power measured revealed that cement type III generates a rapid heat evolution in contrast with cement type I. According to the ASTM C 186 standards, cement type III reaches a sulfate depletion point at the time of 7 hours after hydration. The sulfate depletion point consists of a seizure formed by the action of aluminate activity due to the lack of a SCM in the mixture (Caldarone 2008).

6.2 Fly Ash Class-C

6.2.1 Cement Type I

The effect of FA class-C on the HH of PC Type I is shown in Figure 2. The results show that increasing the amount of FA class-C in the mix decreases the heat released and delays the peak temperature. The heat flow curves -see Figure 2 a- show that the increasing the replacement percentages extend the dormant period. This is due the prolonged reaction of Calcium Sulfate ($CaSO_4$) with Tricalcium aluminate (C_3A) that occurs during the dormant period. Also, the higher the percentage of FA class-C incorporated in the mix is, the longer the delay in the initial and final setting time will be.

To further analyze the effect of fly ash C on the HH of cement Type I, a cumulative energy (HH) graph was generated, as shown in Figure 2b. The results of this experiment confirm the results obtained from previous studies that assessed the impact of FA in cement mixtures on the HH, as in (Meland 1983, Tokyay 1988, De Rojas et al. 1993, De Rojas and Frias 1995, Frias et al. 2002, Atis 2002, Maia et al. 2011, Alhozaimy et al. 2015). However, it still contradicts the results reported by Schindler and Folliard in (2003), that indicated that using FA class C in cement mixtures increases the HH.

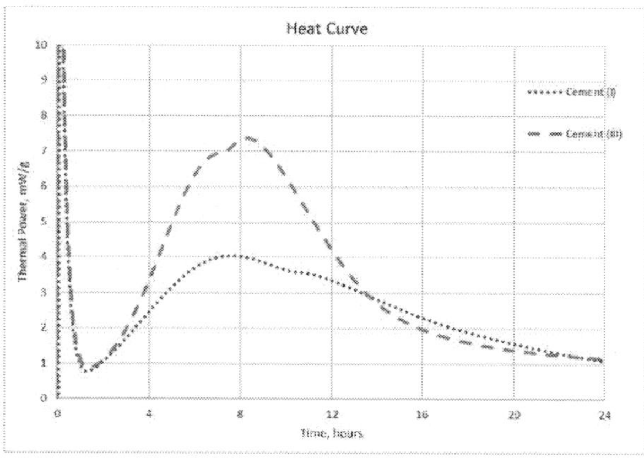

Figure 1a. Heat flow of cement type I and III

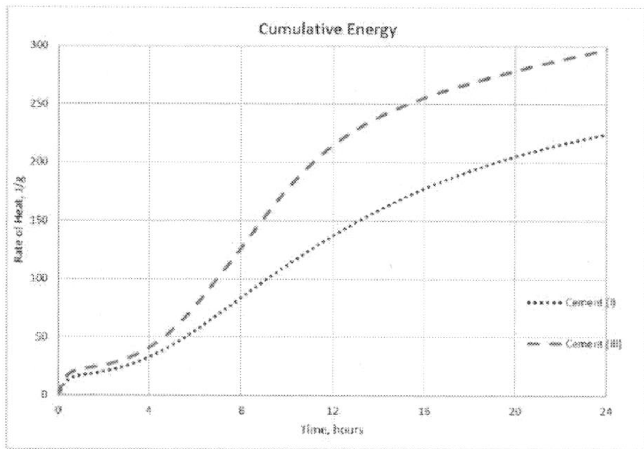

Figure 1b. Heat of hydration of cement type I and III

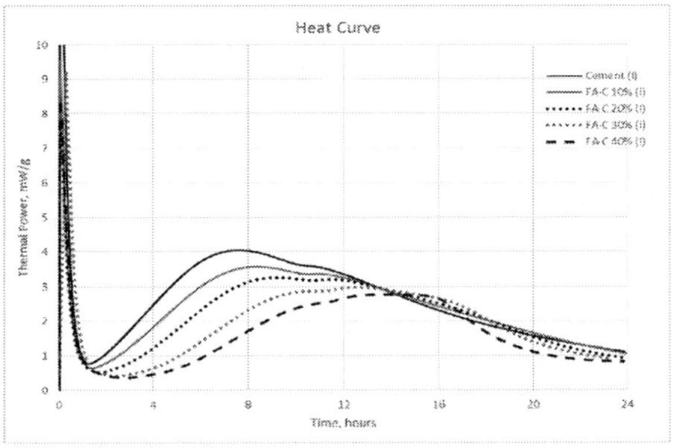

Figure 2a. Heat flow of cement Type I mixed with FA-class C

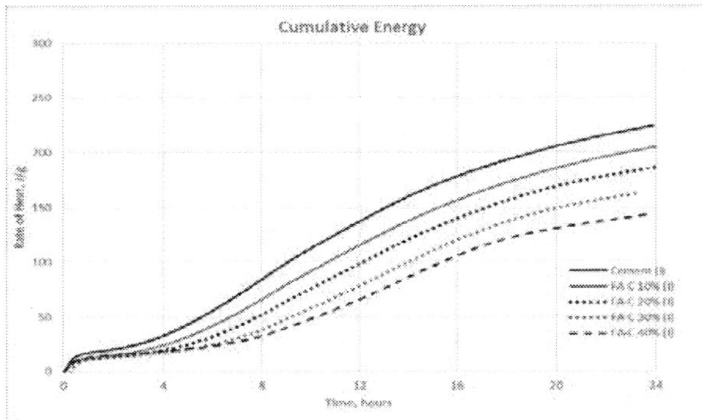

Figure 2b. Heat of hydration of cement Type I mixed with FA-class C

6.2.2 Cement Type III

The heat curves in Figure 3a show the contrast of heat release between different FA class-C replacement percentages of PC Type III. It is very clear that the heat evolution for PC Type III is higher than in Type I. Consistent effects are noticed on the HH of PC Type III with increasing the replacement levels of FA class-C. The highest heat index occurred at 10% FA class-C replacement level, which produced a thermal power of 6.7 mW/g. The increase in percentage of FA class-C replacement decreases the heat evolution; at 40 % FA class-C replacement, the heat curve lowers drastically. Additionally, the sulfate depletion and acceleration of aluminate activity are extended, which extend the dormant period. Similar to the effect of PC Type I, the setting time is increased with increasing the replacement level of FA class-C.

The cumulative energy (HH curve) collected from the isothermal calorimetry test is shown in Figure 3b. It can be observed that the OPC type III yield higher rate of HH, thus greater early compressive strength compared to PC Type I. Also, it should be noted that the FA class-C replacement maintained lower level of HH compared to OPC (Type I). In general, it can be deduced that the higher the percentage of FA class-C replacing PC Type III, the lower the HH will be.

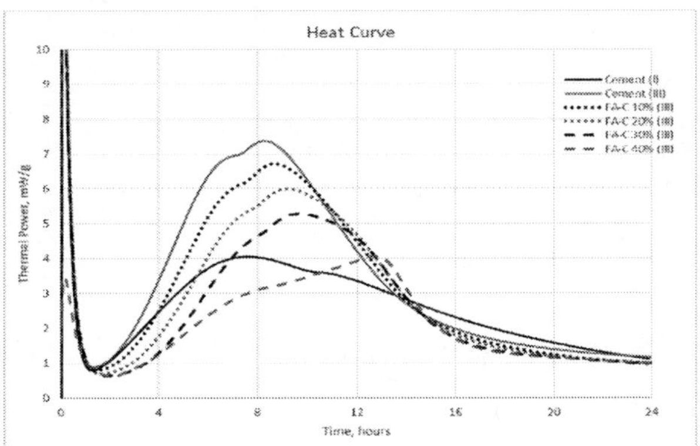

Figure 3a. Heat flow of cement Type III mixed with FA-class C

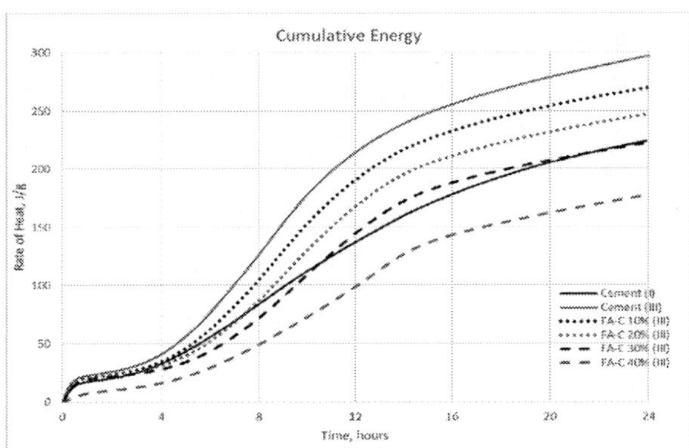

Figure 3b. Heat of hydration of cement Type III mixed with FA-class C

6.3 Fly Ash Class-F

6.3.1 Cement Type I

The results obtained from the isothermal tests for the replacement of Portland cement type I with FA class-F are shown in Figure 4. The effect of FA class-F on the HH of PC Type I is consistent with the results from the literature. The higher the amount of FA class-F used for replacing PC Type I, the lower the HH will be. The highest and the lowest heat peak recorded was at 3.6 mW/g and 2.8 mW/g using 10% and 40% FA class-F replacements, respectively. Also, the setting time is affected as well, the increase in replacement level slightly delays the final setting time. The main hydration peak span decreases in length drastically from 10% replacement level to 20%. This means that the dormant period lasts longer higher replacement level of FA class-F.

The cumulative HH data obtained from the isothermal test on the replacement of PC Type I with FA class-F, is shown in Figure 2b. The cumulative energy graph is formulated by integrating the thermal power graph. As shown, there is a significant drop of energy from 198 J/g at 10% replacement level to 158 J/g at 20% replacement.

6.3.2 Cement Type III

Portland cement type III has higher heat release than other types of PC. Nonetheless, FA class-F has similar effects on PC Type III compared to other type of cements. The HH of mixtures containing FA class-F as a replacement of PC Type III in the amounts 10%, 20%, 30%, and 40% were obtained using isothermal conduction calorimetry, as shown in Figure 5. The thermal power graph (Figure 5a) shows that increasing the amount of FA class-F decreases the amount of heat released. Also, the dormant period is prolonged by increasing the replacement percentages delaying the peak temperature. The delayed heat peaks (slightly shifted to the right) indicates that there is a small delay in the final setting time. of the cementitious materials when increasing the replacement of FA class-F. It should be noted that increasing the replacement percentage of FA class-F from 30% to 40% does not decrease the heat peak, but rather delay it.

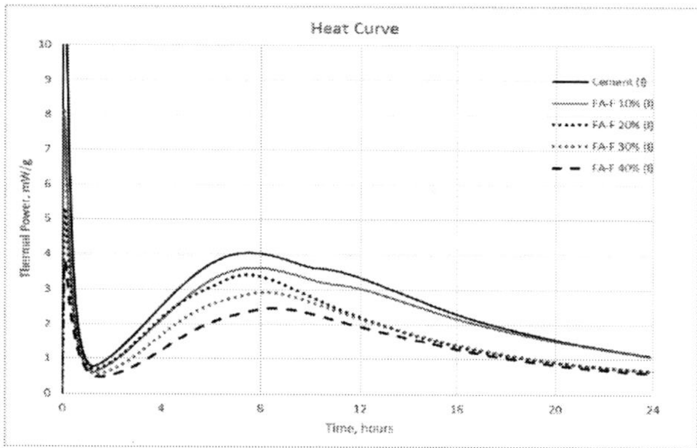

Figure 4a. Heat flow of cement Type I mixed with FA-class F

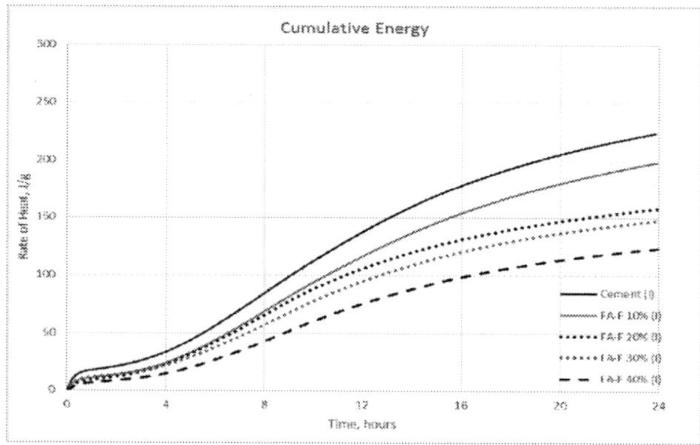

Figure 2b. Heat of hydration of cement Type I mixed with FA-class F

The cumulative HH curves for PC Type III mixtures made with FA class-F is given in Figure 5b. The energy released in these experiments is higher than with PC Type I. A pattern can be observed; when the replacement level of fly ash F increases, the HH decreases.

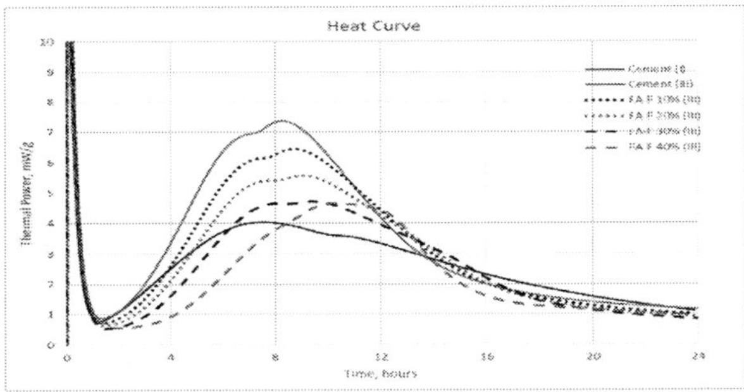

Figure 5a. Heat flow of cement Type III mixed with FA-class F

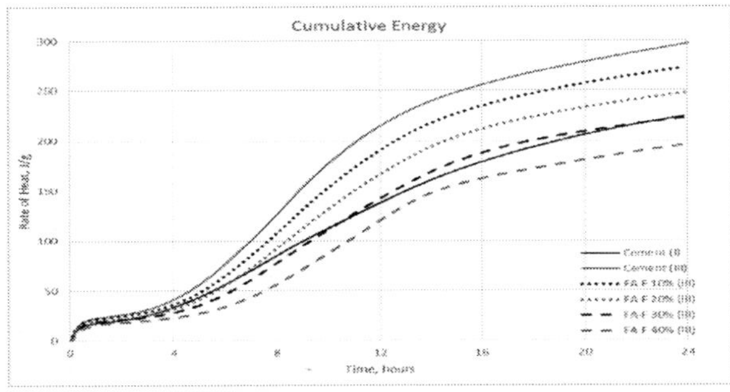

Figure 2b. Heat of hydration of cement Type III mixed with FA-class F

7 ANALYSIS AND FINDINGS

From the results it can be determined that PC Type I and Type III have similar patterns. However, when comparing the maximum heat flow from both cements, PC Type III recorded 7.4 mW/g and PC Type I reached to 4 mW/g, as shown in Figure 6. There is a substantial difference in heat peaks between these two cements. Nonetheless, both cements peaked at an indistinguishable amount of time. PC Type I peeked after 8.6 hours, where PC Type III peak occurred at 8.2 hours of testing. Among other similitudes, the dormant periods for both cement types lasted for about 6 hours. The total HH differs when comparing PC Type I and Type III. PC Type I had a total HH of 224 J/g, in contrast, Type III recorded 311 J/g, see Figure 7.

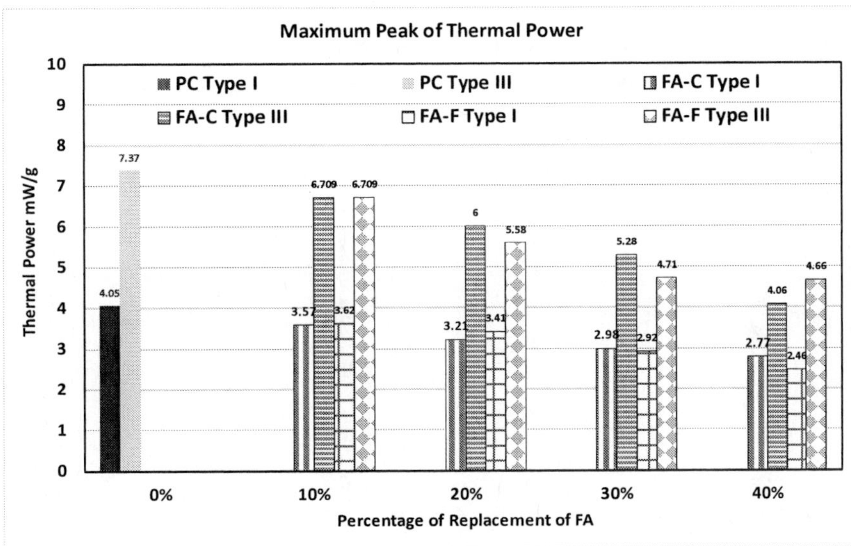

Figure 6. Comparison of maximum peaks of thermal power of PC Type I and III and samples containing different replacement levels of FA-C and FA-F

PC Type I with FA class-C recorded a maximum value of heat flow of 3.57 mW/g at 10% replacement level. The heat peak occurred at 10.8 hours of isothermal calorimetry testing. Also, the maximum total HH reached 229 J/g at 10 % replacement percentage. In addition, the duration of the dormant period begins to extend continuously with the increment of FA class-C replacement level. The durations of the dormant periods range from 6.4-10.5 hours. In contrast, PC Type I with FA class-F recorded a maximum thermal power of 3.62 mW/g at 10% replacement level. The maximum heat index was recorded after 8.7 hours of testing. Furthermore, the maximum total HH recorded for PC Type I with FA class-F was 198 J/g at 10%

replacement level. The dormant period slightly extends when the replacement level increases. The dormant period durations span from 6.3-7 hours, see Table 1.

Table 1: FA class C and F samples run with PC Type I

Parameters	PC Type I									
	FA class-C					FA class-F				
	0%	10%	20%	30%	40%	0%	10%	20%	30%	40%
Peak of Heat Flow (mW/g)	4.05	3.57	3.21	2.98	2.77	4.05	3.62	3.41	2.92	2.46
Time of Max Peak of Heat Flow (hours)	8.6	10.8	12.45	15.47	14.4	8.6	8.65	7.47	8.2	8.56
Max Total HH (J/g)	223.94	228.66	185.95	162.47	143.26	223.94	198.48	158.47	147.9	123.87
Dormant Period (hours)	6.35	7.17	9.65	10	10.46	6.25	6.35	6.16	6.82	7

PC Type III with FA class -C recorded a maximum value of heat flow at 6.709 mW/g at 10% replacement level. The maximum heat peak occurred at 8.2 hours of isothermal calorimetry testing, and it had the maximum total HH with 180.84 J/g. The dormant period, which is the stage before the hydration acceleration, range from 1.5-3.5 hours. On the other hand, PC Type III with FA class -F recorded a maximum thermal power of 6.46 mw/g at 10% replacement level, and it was recorded at 7.35 hours of testing. Furthermore, the maximum total HH was gotten from the 10% FA class-F replacement level with 272.52 J/g. The dormant period for most of the tested samples had a duration span ranging from 1and 3.3 hours, with the exception of the 40% replacement level that had an extended dormant period between 1.5 and 4.2 hours, see Table 2.

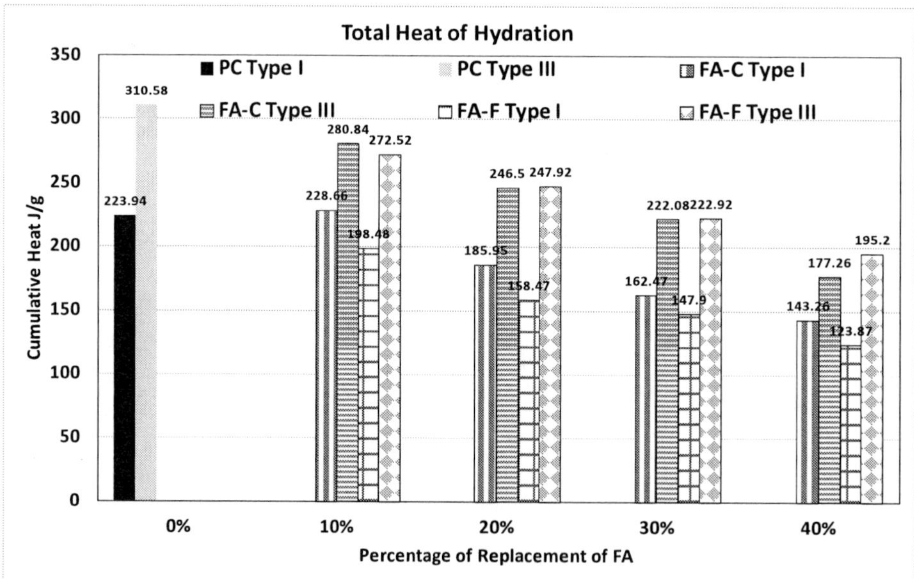

Figure 7. Comparison of the total HH comparison of PC Type I and III and samples containing different replacement levels of FA-C and FA-F

In comparison with PC Type I results, both follow the same pattern, as the percent replacement increases, the lower the maximum value of heat flow. Nevertheless, values for PC Type III are higher, because of its higher Tricalcium Aluminate (C3A) content, which is characterized for providing rapid strength at an early age by releasing large amounts of heat during the first days.

Table 2: FA class C and F samples run with PC Type III

Parameters	PC Type III									
	FA class-C					FA class-F				
	0%	10%	20%	30%	40%	0%	10%	20%	30%	40%
Peak of Heat Flow (mW/g)	7.37	6.709	6	5.28	4.06	7.37	6.46	5.58	4.71	4.66
Time of Max Peak of Heat Flow (hours)	8.2	8.65	9.32	9.73	12.85	8.2	8.75	9.1	9.5	10.4
Max Total HH (J/g)	310.58	280.84	246.5	222.08	177.26	310.58	272.52	247.92	222.98	195.2
Dormant Period (hours)	6.83	7.2	7.65	7.88	10.55	6.83	7.35	7.63	7.95	8.36

8 CONCLUSIONS

This study focusses on the evaluation of the effects of fly ash on the cement HH with the purpose of updating previous results using a high precision isothermal calorimeter. Based on the results and findings of this study, the following conclusions can be drawn:

1. The thermal power of PC constantly decreases when it is replaced with FA. The larger amount of PC replaced with FA, the lower the peak temperature rises.

2. PC with FA class-C has higher heat evolution than PC with FA class-F. However, there is a difference in maximum heat index between PC Type I and Type III. PC Type III has significantly higher heat evolution in comparison with Type I. Therefore, from the tests performed, PC Type III replaced with FA class-C have the higher heat outputs.

3. The maximum heat peak occurrence delays in time the higher the replacement level of FA on PC. Nevertheless, in PC with FA class-C the maximum heat peaks delays much more than in PC with FA class-F.

4. The total HH is decreased with the increase of replacement of PC with FA. The higher the replacement level, the lower the maximum HH.

5. The dormant period extends with the increment of replacement of FA in PC. FA class-C dormant period extends in larger intervals of time, where in FA class-F, the dormant period extends, but not

6. Significantly.

References

Alhozaimy, A, G Fares, O A. Alawad and A Al-Negheimish. "Heat of hydration of concrete containing powdered scoria rock as a natural pozzolanic material." Construction and Building Materials (2015): 113-119.

ASTM C1702-17, Standard Test Method for Measurement of Heat of Hydration of Hydraulic Cementitious Materials Using Isothermal Conduction Calorimetry, ASTM International, West Conshohocken, PA, 2017, www.astm.org

ASTM C1753 / C1753M-15e1, Standard Practice for Evaluating Early Hydration of Hydraulic Cementitious Mixtures Using Thermal Measurements, ASTM International, West Conshohocken, PA, 2015, www.astm.org

ASTM Standard C 1679 – 08, 2009, “Standard Practice for Measuring Hydration Kinetics of Hydraulic Cementitious Mixtures Using Isothermal Calorimetry”, ASTM International, West Conshohocken, PA (2009) www.astm.org

ASTM Standard C186-05, 2005, " Heat of Hydration of Hydraulic Cement," ASTM International, West Conshohocken, PA, (2005) www.astm.org.

Atiş, Cengiz Duran. "Heat evolution of high-volume fly ash concrete." Cement and Concrete Research 32, no. 5 (2002): 751-56.

Ballim, Yunus and Peter C. Graham. "The Effects of Supplementary Cementing Materials in Modifying the Heat of Hydration of Concrete." Materials and Structures 42, no. 6 (2008;2009;): 803.

Bofang, Z. (2013). Thermal stresses and temperature control of mass concrete. Butterworth-Heinemann.

Cheng-Yi, Huang, and R.f. Feldman. "Hydration reactions in portland cement-silica fume blends." Cement and Concrete Research 15, no. 4 (1985): 585-92.

Frias, M. "The effect that the pozzolanic reaction of metakaolin has on the heat evolution in metakaolin-cement mortars." Cement and Concrete Research. March 27, 2000.

Gajda, J. (n.d.). Mass Concrete—How Do You Handle the Heat? Retrieved May 12, 2016, from http://www.cement.org/for-concrete-books-learning/concrete-technology/concrete-design production/mass-concrete

Kirby, David M., and Joseph J. Biernacki. "The effect of water-to-cement ratio on the hydration kinetics of tricalcium silicate cements: Testing the two-step hydration hypothesis." Cement and Concrete Research 42, no. 8 (2012): 1147-156.

Langan, B. W., K. Weng, and M. A. Ward. "Effect of Silica Fume and Fly Ash on Heat of Hydration of Portland Cement." Cement and Concrete Research 32, no. 7 (2002): 1045-1051.

Mostafa, N.y., and P.w. Brown. "Heat of hydration of high reactive pozzolans in blended cements: Isothermal conduction calorimetry." Thermochimica Acta435, no. 2 (2005): 162-67.

Orosz, Katalin, Hans Hedlund, and Andrzej Cwirzen. "Effects of variable curing temperatures on autogenous deformation of blended cement concretes." Construction and Building Materials 149 (2017): 474-80.

Pane, Ivindra, and Will Hansen. "Investigation of blended cement hydration by isothermal calorimetry and thermal analysis." Cement and Concrete Research35, no. 6 (2005): 1155-164.

Patil, A. Heat of Hydration in the Placement of Mass Concrete. International Journal of Engineering and Advanced Technology (IJEAT) ISSN: 2249 – 8958, Volume-4 Issue-3, February 2015.

Persson, Bertil. "The effect of silica fume on the principal properties of concrete." Cement and Concrete Research27, no. 11 (1997).

Portland Cement, Concrete, and Heat of Hydration. (1997, July). Concrete Technology Today, Portland Cement Association (PCA). 18(2), 1-4. Retrieved July, 2014, from http://cement.org/tech/pdfs/pl972.pdf

Rojas, M S. and M Frias. "The influence of silica fume on the heat of hydration of Portland cement." Special Publication (1995):829-844

Rojas, M. "The influence of different additions on portland cement hydration heat." Cement and Concrete Research 23 (1993): 46-54.

Rojas, M. "The pozzolanic activity of different materials, its influence on the hydration heat in mortars. Cement and Concrete Research, 26(2), 203-213." (1996): 203-213

Schindler, A.K., and K.J. Folliard. "Influence of supplementary cementing materials on the heat of hydration of concrete." Advances in Cement and Concrete IX Conference, Copper Mountain Conference Resort in Colorado., August 2003.

Snelson, David G., Stan Wild, and Martin Ofarrell. "Heat of hydration of Portland Cement–Metakaolin–Fly ash (PC–MK–PFA) blends." Cement and Concrete Research38, no. 6 (2008): 832-40.

Sokkary, T.m, H.h., Assal, and A.m Kandeel. "Effect of silica fume or granulated slag on sulphate attack of ordinary portland and alumina cement blend." Ceramics International 30, no. 2 (2004): 133-38.

Xu, Q., Jiong H., Mauricio R., Keijin W., and Zhi G. "Isothermal calorimetry tests and modeling of cement hydration parameters." Developing a Simple and Rapid Test for Monitoring the Heat Evolution of Concrete Mixtures for both Laboratory and Field Applications, January 01, 2006.

Building Tomorrow's Society

Bâtir la Société de Demain

Fredericton, Canada

June 13 to 16, 2018 / *13 au 16 juin, 2018*

FIELD INSPECTION AND CLASSIFICATION OF PAVEMENT DISTRESSES OF ST JOHN'S CITY IN NEWFOUNDLAND CANADA

Ali, A.[1], Hossain, K.[1,4], Dhasmana, H.[2], Safiuddin, M[3], Bazan, C.[1] and Hussein, A.[1]
[1] Memorial University of Newfoundland, Canada
[2] University of Illinois at Urbana-Champaign, USA
[3] Ryerson University, Canada
[4] Corresponding Author: kamalh@mun.ca

Abstract: The City of St. John's (Newfoundland and Labrador, Canada) has a unique climate, with excessive precipitation year-round, moderate summers and extreme winters. The pavement foundation (subgrade) in this region consists mainly of strong metamorphic rocks. As the city is located on an island and has a small population size, roads of this city experience very low volume of heavy traffic. Despite having less traffic and strong foundation, most of the roads in St. John's suffer from serious structural and functional distresses. The distresses include but are not limited to deformation (structural, abrasion), moisture damages (pothole, ravelling), fatigue cracking (top down, bottom up), thermal cracking (longitudinal, transverse) etc. These distresses eventually lead to longer travel times, higher vehicular and pedestrian crashes and fatalities, excessive fuel consumption due to additional maneuvering actions as a result of poor road conditions, high CO_2 emissions and a sub-standard ride quality. To address these issues, a comprehensive understanding of these distresses under local conditions is necessary. As an initial effort, a small-scale field survey was conducted and the distresses were classified and ranked based on type and severity. In addition, summary of each distress mechanism has been presented, followed by some recommendations to mitigate and address these distresses in this region. It is expected that this study will not only contribute in enhancing the pavement service-life by developing an improved asphalt mixture design in future but also reduce deleterious emissions and improve traffic safety.

1 INTRODUCTION

More than 90% of the paved roads around the world are made from asphalt (EAPA 2015, WAPA 2010). Also known as flexible pavements, these roads generally consist of an asphalt surface course placed over the base course and subgrade of granular materials. Each layer plays a significant role in distributing the traffic loads received from the layer above and limiting the impact of environmental features such as freezing and thawing. Most of the distresses in pavements occur as a result of a large difference between high and low temperatures, water penetration and displacement in the subgrade, increase in traffic density, traffic loading, and inefficient compaction during construction. As a result, distresses in flexible pavements can be categorized as: (1) permanent deformation (at high service temperatures), (2) fatigue cracking (at intermediate service temperatures) and (3) thermal cracking (at low service temperatures).

St. John's has unique climatic conditions, i.e., a very high amount of precipitation throughout the year and temperate summers but extreme winter temperatures. With a population of just under a quarter million, the roads are exposed to a very low traffic volume. Nonetheless, the roads in St. John's show signs of serious deduction in structural as well as functional value. Instances of pavement deterioration due to traffic loads, tires, moisture, freezing and thawing, temperature variation, and other environmental factors are frequent in the city. These damages to the pavement not only require huge efforts in repair and maintenance activities but also result in general discomfort to the users if not taken care of in a timely manner. To address these issues, a field survey was conducted to identify the pavement distresses in St. John's. Also, an attempt has been made to classify the pavement distresses and to rank them based on their type and magnitude.

1.1 Literature Review

Researchers have classified pavement damages into three major categories: (1) crack type, (2) disintegration type dominated by potholes and (3) depression type (Sarie et al. 2015). Effects of traffic loads, materials and design parameters on pavement damage over time were also studied. It was revealed that pavement cracking is affected more by trucks with single and tandem axles than those with multiple axles (tridem and higher). Rutting (structural) occurs more due to heavier trucks with multiple axles than those with single and tandem axles (Salama et al. 2006). In addition, abrasion rutting (wearing of surface layer) occurs due to frictional impacts between tire studs and pavement surface that lead to loosening of mastic portion from pavement surface. Hicks (1999) provided a logical approach for determining the most effective preventive maintenance treatment for distresses in the flexible pavements at lower costs (Table 1). Te et al. (1995) developed a software program called "APDIS" which could evaluate pavement surface cracks using image processing algorithms.

Asphalt pavement materials play a vital role in determining the longevity of pavements. Aging of asphalt in the asphalt mix changes the structure, chemical composition and surface tension properties of asphalt. This causes the asphalt cement to become stiffer and more vulnerable to moisture damages (Hossain et al. 2018). Some other significant binder characteristics include: (1) temperature susceptibility, (2) adhesion/cohesion and (3) hardening and aging (VDOT 2012).

Table 1: Appropriate maintenance strategies for various pavement distress types (Hicks et al. 1999)

Distress\Maintenance	Crack sealing	Fog seal	Micro surfacing	Slurry seal	Cape seal	Chip seal	Thin overlay	Mill or grind
Rutting			#	#	#		#	
Fatigue cracking								
Longitudinal cracking	#	#	#	#	#	#	#	
Transverse cracking	#	#	#	#	#	#	#	
Bleeding			#			#		#
Ravelling		#		#	#	#		

2 DISTRESS IDENTIFICATION

1.2 Area of Study

This study focuses on the identification of major distress types and factors responsible for causing pavement deterioration in St. John's. The data collection process was entirely based on visual inspection. Roads included in this study are a few major roads including Newfoundland Road, New Cove Road, Elizabeth Avenue, Portugal Cove Road, Torbay Road and roads located inside the Memorial University of Newfoundland (MUN). Figure 1 presents the major road network in St. John's.

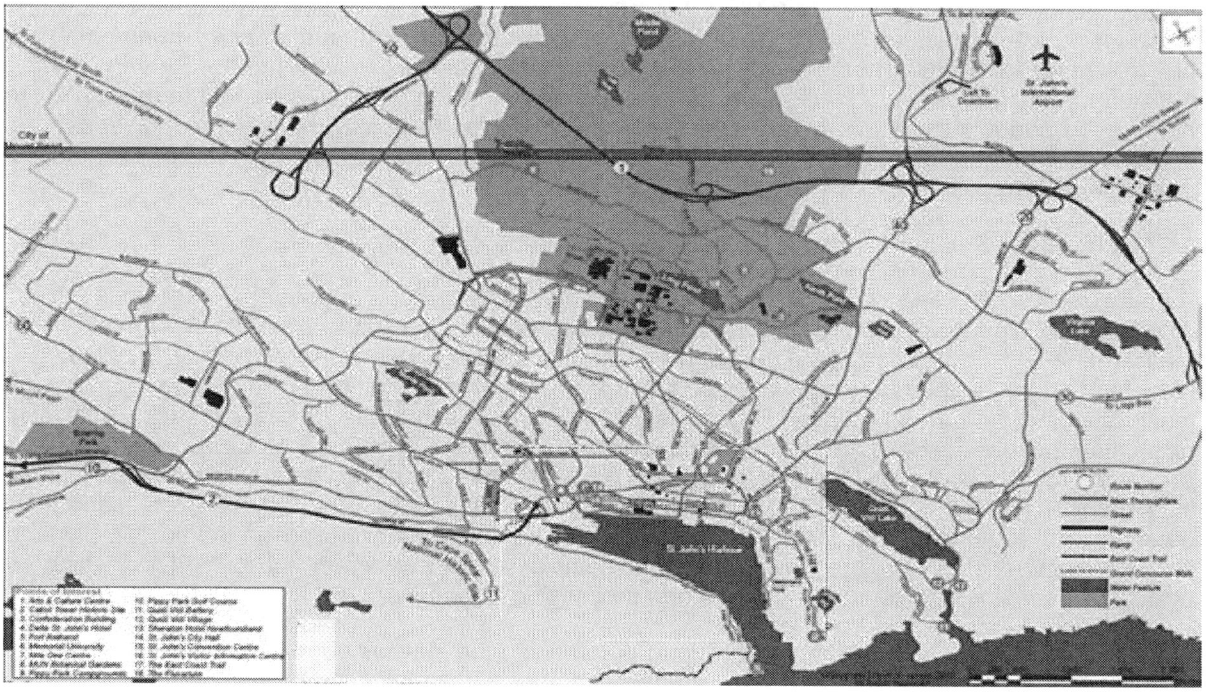

Figure 1: Map of the major road network of the City of St. John's (Map Viewer 2018)

1.3 Climate and Weather

Roads in St. John's are designed according to the historically available climate data. The average temperature in the city over the last 80 years ranged from -9.3 to 15.5 ºC. Air humidity is 79 to 83%, with an annual average of 81%. Average wind speed varies from 20 to 28 km/h.

Table 2: Weather in the City of St. John's (The Weather Network 2018)

Month	Average high temp (ºC)	Average low temp (ºC)	Record daily high with date (ºC)	Record daily low with date (ºC)	Wind speed (km/h)	Average precipitation (mm)	Humidity (%)
Jan	-0.9	-8.6	15.2 (Jan09/79)	-23.3 (Jan09/57)	28	150	82
Feb	-5.4	-9.3	16 (Feb4/96)	-23.8 (Feb18/90)	26	125	79
Mar	-2.5	-6.2	18.3 (Mar31/62)	-23.8 (Mar10/86)	26	131	82
Apr	1.6	-2	24.1 (Apr24/86)	-14.8 (Apr04/78)	22	122	83
May	6.2	1.5	25.6 (May30/72)	-6.7 (May05/64)	21	101	80
Jun	10.9	5.9	29.4 (Jun29/44)	-3.3 (Jun01/70)	21	102	79
Jul	15.4	10.5	31.5 (Jul 06/83)	-1.1 (Jul05/76)	21	89	79
Aug	15.5	11.1	31 (Aug07 /96)	0.5 (Aug24/83)	20	108	80
Sep	11.8	7.7	29.5 (Sep10/01)	-1.1 (Sep23/50)	21	131	81
Oct	6.9	3.3	24.6 (Oct04/87)	-5.6 (Oct31/46)	23	162	82
Nov	2.6	-0.7	19.4 (Nov06/50)	-13.4 (Nov25/83)	24	144	82
Dec	-2.2	-5.5	16.1 (Dec12/57)	-19.7 (Dec27/84)	27	149	83

3 DISTRESS IDENTIFICATION PROTOCOLS

Road maintenance is a critical step in the design of flexible pavements. In order to ensure the efficient functioning of pavements, maintenance procedures can be proposed based on the identification of pavement surface defects. Four steps are followed for this purpose: (1) Description: a short explanation of the substantial features of the distress together with some notes that might support if such a distress is identified; (2) Location: area of distress exposed to different loads and conditions is determined; (3) Causes of distress: factors causing the distress are identified; (4) Levels of severity: severity level is determined with respect to commonly defined qualitative judgement degrees (low, medium and high) and (5) Measurement and possible repairs.

1.4 Types of Pavement Damage

In this study, four major pavement distresses were identified, namely rutting, cracking (fatigue, block, longitudinal, transverse), potholes and delamination.

1.4.1 Rutting

Description: Rutting is normally defined as the vertical deformation of pavement materials that creates surface depression along the wheel path.

Location: Depressions can form over large areas of the road mostly along the wheel path. Figure 2 (a, b) shows examples of structural and abrasive rutting (due to studded tires) in the city of St. John's.

Identifying distress causes: Rutting is a prominent distress type in the flexible pavement. Pavement shearing can happen along the sides of the rut rendering the top surface clearly damaged. This heavily affects the ride quality. Some major theories explaining rutting behaviour in the literature are:

1. One-dimensional densification or vertical compression: Deformation under the wheel path without an accompanying hump on either side of the deformation is caused by material densification. It is predominantly induced by excessive air voids or insufficient compaction. It might result in a low to moderate or high severity rutting (Kandhal et al. 2003, Miljkovic et al. 2011).

2. Lateral flow or plastic movement (Visco-plastic deformation): Asphalt mix with inadequate shear strength or higher percentage of air voids can result in lateral displacement of materials with an accompanying hump. Air void network in the asphalt mix act as a lubricant and under heavy and repeated traffic loading a high percentage of air voids can accelerate this type of rutting. The damage intensifies at high temperatures and mostly standing loads. It might be quite difficult to predict this form of rutting on field as well as experimentally on account of the anisotropic behavior of asphalt.

3. Mechanical deformation (Structural rutting): Lateral movement of unbound material in the lower layers (subgrade) results in wider ruts in the pavement structure. It generally occurs due to irregularities in elastic modulus of the asphalt mix (too stiff or high elastic modulus).

4. Abrasive rutting (Wearing): This is caused by the loss of mastic phase or fine grains of aggregate in surface layer from the impact of studded tires (Sarie et al. 2015). This phenomenon, in particular, is heavily observed in the city of St. John's.

Levels of Severity: LOW (L): rut depth is ≥ 6mm and ≤ 12mm; MEDIUM (M): rut depth is ≥13mm and ≤ 25mm, HIGH (H): rut depth is ≥ 26mm (ODOT 2010).

Measurement and Possible Repairs: Special recording devices such as a profilometer or straightedge can be used to record rut depths. A Data Collection Vehicle (DCV) can also be employed to measure the extent of pavement rutting (ODOT 2010). Minor surface ruts can be filled and deeper ruts can be covered with an overlay. If surface asphalt is unstable, recycling can be performed. In the case of subgrade inconsistencies, reclamation or reconstruction is the best option (Adlinge et al. 2009).

(a) Structural rutting (b) Abrasive rutting

Figure 2: Rutting in the city of St. John's

1.4.2 Cracking

Four major types of cracking were identified as discussed in this section.

1.4.2.1 Fatigue Cracking (Alligator Cracking)

Description: Fatigue cracking comprises single crack or a series of interconnected cracks creating small, irregular regions on the pavement. Pavement deterioration due to fatigue mainly depends on the repetitive application of traffic loads, temperature of pavement and vehicle speed. According to Olard et al. (2005), it is possible to determine the fatigue behaviour of asphalt mixes by taking into account the level of strain imposed and the number of associated stress cycles. At low strain amplitudes ($<10^{-4}$), if the number of cycles imposed is greater than ten thousand, then the material damages due to fatigue (Perraton et al. 2003).

Location: Fatigue failure occurs in areas of the road subjected to repeated traffic loading (wheel paths) or wander within the lane.

Identifying distress causes: Failure of the surface layer or base due to repeated traffic loading results from the loss of mechanical properties (such as bearing capacity and stiffness modulus) of the pavement structure (Valdés et al. 2011) and consequently causes fatigue cracking. Other factors that contribute to fatigue cracking are low pavement thickness, aging pavement, poor quality of constituent materials and inadequate stiffness of the asphalt mix used in pavement (Adhikari et al. 2009).

Levels of Severity: LOW (L): area with none or only a few interconnecting cracks and the cracks are not sealed or spalled, average crack width is 6 mm and the crack pattern is no more apart than 32.8 mm (Figure 3a) (FHWA 2009, Salvatore et al. 2006, Scott et al. 2012); MEDIUM (M): interconnected cracks forming a complete pattern with slight spalling and sealing, average crack width is 6 to 19 mm and crack pattern is no more apart than 150 mm (Figure 3b) (Scott et al. 2012); HIGH (H): severely spalled interconnected cracks and average crack width is 19 mm (Figure 3c) (FHWA 2009).

Measurement and Possible Repairs: Linear distance of the affected wheel path or area of fatigue cracking in square meters is measured. Any zone is classified according to the level of severity. If fatigue cracking and rutting occur in the same zone, each distress should be treated separately (FHWA 2009). Changes in drainage pattern, base materials can be made. A full depth patch or spray patching can be performed for temporary repair (Ningyuan et al. 2001).

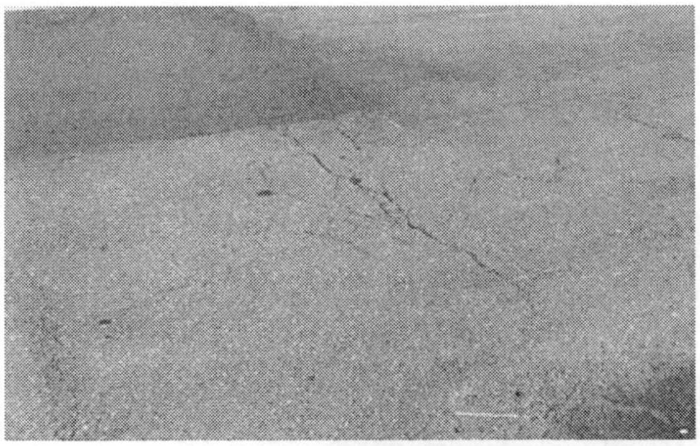

(a) Low Severity

(b) Medium Severity

(c) High Severity

FIGURE 3: Severity levels for fatigue (alligator) cracking

1.4.2.2 Block Cracking

Description: Interconnected cracks over an area not subjected to traffic load that divide the pavement into rectangular blocks ranging in size from approximately 30x30 cm to 300x300 cm (FHWA 2009, FDOT 2015).

Location: Block cracking occurs throughout the pavement width and not just in the wheel paths. They usually extend only a short distance on the hot-mix asphalt (HMA) surface. Figure 4 shows an example of block cracking in the city.

Identifying distress causes: This type of cracking can be caused by inadequate compaction, shrinkage and hardening of the asphalt mix (FHWA 2009, FDOT 2015).

Levels of Severity: LOW (L): tights cracks with negligible spalling, average width is 6 mm or less (Hall et al. 1993) and the width of sealed cracks can't be determined; MEDIUM (M): crack width is between 6 mm and 12 mm; HIGH (H): cracks with a mean width greater than 12 mm, severe spalling and moderate or serious random parallel cracking at the intersection of the cracks or near the crack exists (Hall et al. 1993).

Measurement and Possible Repairs: Affected area in square meters or the length of pavement at each severity level is measured. Thin wearing course can be constructed to repair a low severity block cracking. For more severe situations, overlays and recycling can be performed. For base problems, the reclamation and reconstruction of the pavement can be proposed (Adlinge et al. 2013).

Figure 4: Block cracking

1.4.2.3 Longitudinal Cracking

Description: Long cracks that are parallel to the pavement centerline. However, cracks that are located exactly in the center of the lane are classified as center-of-lane cracking.

Location: Longitudinal cracks can exist anywhere between the centreline to the outer edge of the outer wheel path (Figure 5 (a, b)). Their position within the lane (wheel path or non-wheel path) is more critical.

Identifying distress causes: Longitudinal cracks can be caused by frost heaving or construction failures (low compaction or poor construction technique) between the construction lanes. They can also be load-induced. Other reasons include low temperatures causing surface shrinkage in the asphalt mix and creation and propagation of cracks beneath the surface course (Scott et al. 2012).

Levels of Severity: LOW (L): non-wheel path cracks with negligible spalling, mean crack width is 6 mm or less; sealed crack width is difficult to determine (FHWA 2009); MEDIUM (M): moderate spalling with filled cracks of a mean width less than 6 mm and non-filled cracks of a mean width greater than 6 mm but less than 19 mm (Miller et al. 2003); HIGH (H): mean crack width greater than 19 mm.

Measurement and Possible Repairs: The linear distance in meters of the affected area is measured at each severity level. Crack treatment or spray patching can be performed to repair the damage.

(a)

(b)

Figure 5: Longitudinal cracking

1.4.2.4 Transverse Cracking

Description: Cracks that forms perpendicular to the centerline of the roadway. They are regularly spaced and start as thin, hairline cracks that widen later (Miller et al. 2003).

Location: Transverse cracks develop anywhere on the surface and propagate deeper with time.

Identifying distress causes: Transverse cracks are normally caused by shrinkage of the surface due to low temperatures, reflective cracks induced by the cracks beneath the surface, and a poorly constructed paving lane joint (Hall et al. 1993).

Levels of Severity: LOW (L): tight cracks with average width of 6 mm and slightly spalled or sealed cracks with a width that is difficult to determine; MEDIUM (M): cracks with average width between 6 to 19 mm, cracks less than 19 mm average width and low severity random cracking (FHWA 2009); HIGH (H): cracks with average width greater than 19 mm and cracks with average width less than 19 mm and medium to high severity random cracking (Miller et al. 2003). Severe spalling can also be noticed. Figure 6 (a, b) illustrates some examples of transverse cracking.

Measurement and Possible Repairs: Number and length in meters of cracks at each severity level are measured. Sealing can be performed for low severity cracks and an overlay can be placed for high severity cracks.

(a)

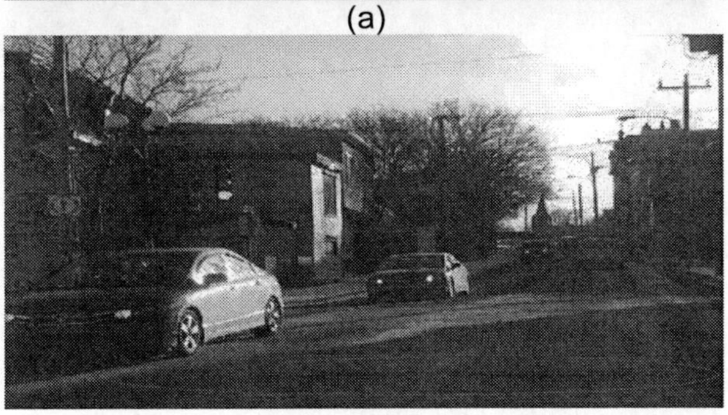

(b)

Figure 6: Transverse cracking

1.4.3 **Potholes**

Description: Potholes are localized distresses in the form of bowl-shaped holes of varying sizes.

Location: Potholes often form in the areas of poor drainage and high traffic of slow-moving vehicles. Figure 7 (a, b, c) shows some examples of potholes filled with water.

Identifying distress causes: Depressions are formed when pavement progressively deteriorates due to fatigue cracking or inadequate strength in one or more layers of the pavement structure. Small fragments of the pavement are removed and the distress then gradually propagates in the lower layers.

Levels of Severity: LOW (L): 25 mm deep; MEDIUM (M): 25 – 50 mm deep; HIGH (H): 50 mm deep

Measurement and Possible Repairs: Number of potholes and the affected area in square meters are measured for each severity level. They can be repaired by patching or excavation and rebuilding. Patching is a very common technique used to cover up the potholes on a regular basis. However, when done incorrectly patches can result in uneven roads causing discomfort to the drivers.

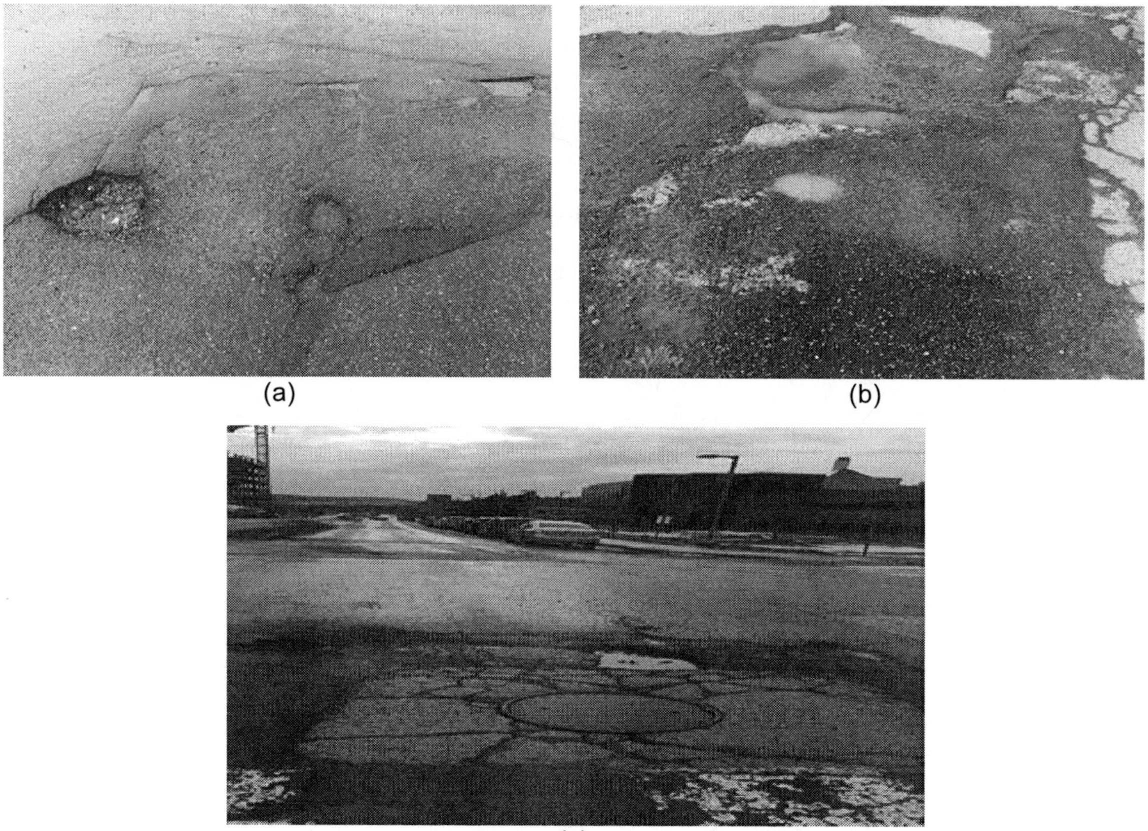

(a) (b) (c)

Figure 7: Potholes

1.4.4 Delamination

Description: Delamination is the gradual removal of an area of asphalt surface due to poor bonding with the underlying HMA layer. It reduces the serviceability of the pavement by causing subsequent slippage, cracking in the wheel paths, peeling and distortion in the layers.

Location: Mostly happens on the wheel path or shoulders of the pavement (Figure 8 (a, b, c, d)).

Identifying distress causes: Poor interfacial bonding between different layers, water percolation and heavy traffic loads are the major causes of delamination.

Levels of Severity: Delamination can extend from 30 square centimeters to tens of square meters. Even though it is undesirable at any depth, delamination deeper than the top two layers can cause surface distress during high traffic.

Measurement and Possible Repairs: Non-destructive test methods such as Ground Penetrating Radar or strain gages can be used to identify surface or subsurface delamination effectively. It can be treated by milling off the surface layer, replacing the wearing course or placing a thin asphalt overlay (Celaya et al. 2011).

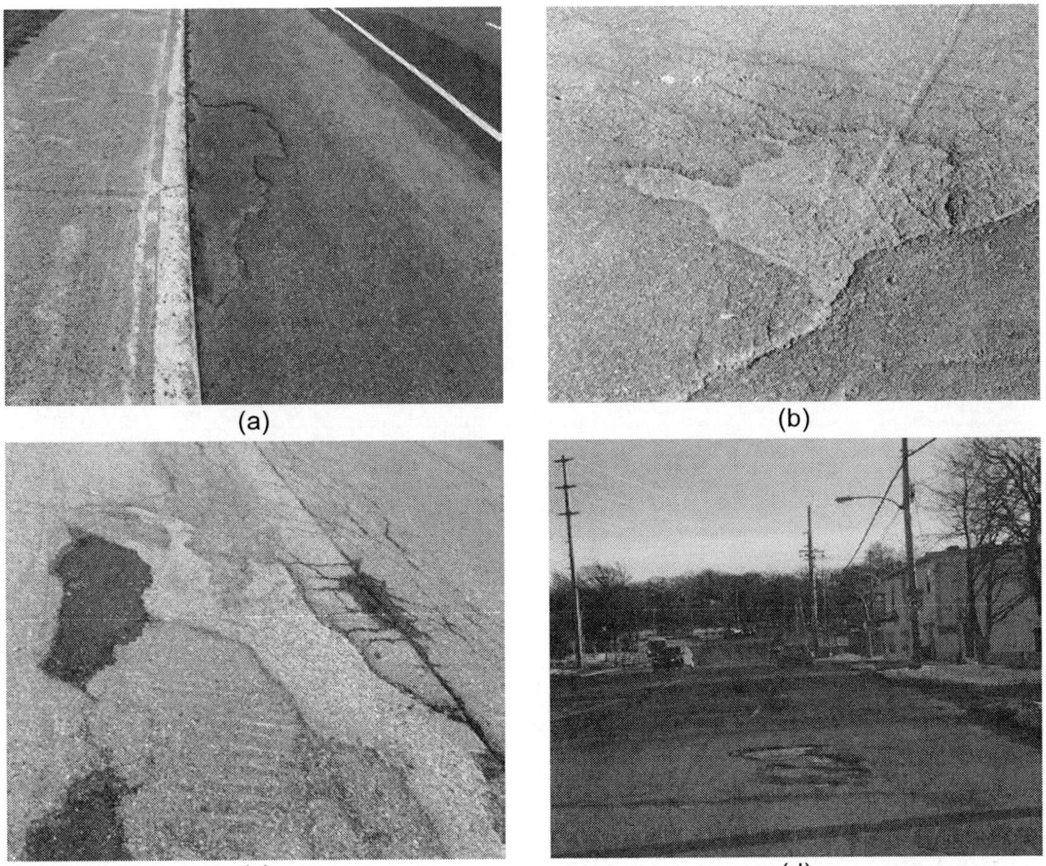

(a) (b) (c) (d)

Figure 8: Delamination

1.5 **Summary of Distress Survey**

A distress survey was conducted on 19 randomly selected road sections (without considering traffic volume and construction history) in St. John's city to identify different types of pavement distresses and document the level of severity for each distress type. Each road section of approximately 50 meters long was used for inspection and distress investigation. Major distresses such as rutting, fatigue cracking, block cracking, longitudinal cracking, transverse cracking, potholes, delamination and unevenness due to patching (patching related issues) were considered. Three levels of severity (low, moderate, high) were recorded to quantify the degree of distresses. If none of the distresses were found on the pavement, level of severity was indicated as "NO".

Tables 3 to 5 present the severity levels of each distress for different roads in the city. In the case of presence of more than one level of severity of distress, the highest severity was taken into account. The severity levels are designated as "LOW" for low severity, "MOD" for moderate severity and "HIGH" for high severity.

Table 3: Severity levels for Newfoundland road and Elizabeth Avenue road

Sl. No.	Type of distress	Newfoundland road				Elizabeth Avenue road			
		Section 1	Section 2	Section 3	Section 4	Section 1	Section 2	Section 3	Section 4
1	Rutting	LOW	MOD	MOD	MOD	HIGH	HIGH	LOW	LOW
2	Block cracking	NO	NO	HIGH	MOD	LOW	NO	NO	NO
3	Fatigue cracking	MOD	LOW	MOD	MOD	MOD	MOD	NO	NO
4	Longitudinal cracking	LOW	NO	MOD	MOD	MOD	LOW	NO	NO
5	Transverse cracking	MOD	NO	NO	NO	NO	NO	NO	NO
7	Delamination	NO	YES	NO	NO	NO	NO	NO	NO
8	Potholes	NO	LOW	MOD	NO	LOW	MOD	NO	NO
9	Patching related issues	HIGH	HIGH	HIGH	HIGH	MOD	HIGH	NO	NO

Table 4: Severity levels for New cove road, Portugal road and Internal road at MUN

Sl. No.	Type of distress	New Cove road		Portugal road		Internal road at MUN		
		Section 1	Section 2	Section 1	Section 2	Section 1	Section 2	Section 3
1	Rutting	LOW	HIGH	LOW	LOW	LOW	MOD	LOW
2	Block cracking	NO	NO	MOD	NO	HIGH	MOD	MOD
3	Fatigue cracking	LOW	LOW	MOD	LOW	NO	MOD	LOW
4	Longitudinal cracking	LOW	MOD	LOW	NO	NO	LOW	NO
5	Transverse cracking	NO	NO	NO	NO	NO	NO	NO
7	Delamination	NO	NO	YES	YES	NO	NO	NO
8	Potholes	NO	NO	LOW	NO	HIGH	MOD	MOD
9	Patching related issues	HIGH	HIGH	MOD	HIGH	MOD	LOW	LOW

Table 5: Severity levels for Torbay road

Sl. No.	Type of distress	Torbay road			
		Section 1	Section 2	Section 3	Section 4
1	Rutting	MOD	MOD	HIGH	MOD
2	Block cracking	LOW	MOD	MOD	LOW
3	Fatigue cracking	LOW	NO	MOD	LOW
4	Longitudinal cracking	LOW	LOW	LOW	MOD
5	Transverse cracking	NO	NO	NO	NO
7	Delamination	NO	NO	NO	NO
8	Potholes	LOW	NO	NO	MOD
9	Patching related issues	HIGH	MOD	LOW	HIGH

4 DISCUSSION

Table 6 shows the dominant distress types in the selected roads of St. John's. The first two sections on Elizabeth Avenue were taken between Freshwater Road intersection and Portugal Cove Road intersection and the other sections were considered between Portugal Road intersection and the north-east end of Elizabeth Avenue.

Table 6: High severity distress types in different roads of St. John's city

Sl. No.	Distress type	Road
1	Rutting	Elizabeth Avenue & Torbay roads
2	Block cracking	Newfoundland road & MUN roads
3	Fatigue cracking	Newfoundland & Elizabeth Avenue roads
4	Longitudinal cracking	Newfoundland road
5	Transverse cracking	Not noticed
7	Delamination	Present only in Newfoundland and Portugal roads
8	Potholes	Internal road at MUN
9	Patching related issues (e.g., unevenness, crack formation)	Everywhere

It is evident from Table 6 that Elizabeth Avenue is highly damaged from Freshwater Road intersection to Portugal Cove Road intersection where both rutting and fatigue related cracking can frequently be observed. After Portugal Cove Road intersection, the road is in quite a good condition without any damage. Delamination was observed on Portugal Cove Road which might have occurred mainly due to longitudinal gradient and loss of adhesion between the HMA and base layers. Delamination was also noticed in the second section of the Newfoundland Road. Potholes were mainly observed in the internal roads of Memorial University. Primary reason for potholes is the high traffic loads due to the ongoing construction. On Newfoundland Road, 9 to 14-meter-long longitudinal cracks were observed. Transverse cracking was not noticed anywhere.

5 CONCLUSIONS

Every year pavement failures increase the cost of operation of roads and vehicle maintenance. In addition, drivers and passengers face major discomfort due to the poor condition of roads.

In this study, 19 sections, each 50 m in length, were surveyed for the purpose of determining the various distress types on the roads of St. John's city. On the basis of visual survey and measurements conducted, major distresses noticed are rutting, fatigue cracking, longitudinal cracking, transverse cracking, potholes and delamination. Rutting and patching are the primary causes of unevenness on the roads. In general, other distress types are also frequently observed during the survey. The intent of this phase of survey was to obtain a preliminary understanding of the conditions of roads in St. John's. Results from this study will be utilized to design a more systematic survey with an advanced data collection system.

References

Adhikari, S., S. Shen, and Z. You. 2009. "Evaluation of Fatigue Models of Hot-Mix Asphalt through Laboratory Testing", *Transportation Research Record*, No. 2127, pp. 36-42.

Adlinge, S., and A. K. Gupta. 2009. "Pavement Deterioration and its Causes". *IOSR Journal of Mechanical and Civil Engineering*, ISSN: 2278 – 1684: 09 - 15.

Celaya, M., S. Nazarian, C. Rao, and H. V. Quintus. 2011. "Delamination Detection of Asphalt Pavements with Non-destructive Testing Devices". Transportation Research Board, No. 11 – 1487, Washington
European Asphalt Pavement Association. 2015. http://www.eapa.org/asphalt.php.

Federal Highway Administration. 2006 - 2009. *Pavement Distress Identification Manual For The NPS Road Inventory Program*, Cycle 4.

Florida Department of Transportation. 2015. *Airfield Pavement Inspection Reference Manual*.

Hall, K. T., C. E. Correa, S. H. Carpenter, S. H., and R. P. Elliott, R. P. 1993. "Pavement Distress Types and Causes. Distress Identification Manual for the Long-Term Pavement Performance Project". *NCHRP* W35 Appendix A - 1–31.

Hicks, R. G., J. S. Moulthtrop, and J. Daleiden. 1999. "Selecting a Preventive Maintenance Treatment for Flexible Pavements". *Journal of the Transportation Research Board*, 1680: 1–12.

Hossain, H., A. Karakas, P. Singhvi, H. Ozer, and I. L. Al-Qadi. 2018. "Effect of Aging and Rejuvenation on Surface Free Energy Measurements and Adhesive Property of Asphalt Mixtures". *Transportation Research Board 97th Annual Meeting*, 18-06052, Washington DC, USA.

Kandhal, P., and L. Cooley. 2003. "Accelerated Laboratory Rutting Tests: Evaluation Of The Asphalt Pavement Analyzer". *National Cooperative Highway Research Program*, NCHRP Report 508.

Map St. John's Newfoundland, Canada, 2018. http://map.stjohns.ca/mapcentre/.

Miljkovic, M., and M. Radenberg. 2011. "Rutting Mechanisms And Advanced Laboratory Testing Of Asphalt Mixtures Resistance Against Permanent Deformation". *Facta Universitatis - Series: Architecture and Civil Engineering*, 9(3): 407–417.

Miller, J. S., and W. Y. Bellinger. 2003. "Distress Identification Manual for the Long-Term Pavement Performance Program". *US Department of Transportation, Federal Highway Administration*, 129.

Ningyuan, L., T. Kazmierowski, S.Tighe, & R. Haas. 2001. "Integrating Dynamic Performance Prediction Models into Pavement Management Maintenance and Rehabilitation Programs". *5th International Conference on Managing Pavements*, (June).

Olard, F., H. Di Benedetto, A. Dony, & J. C. Vaniscote. 2005. "Properties of Bituminous Mixtures at Low Temperatures and Relations with Binder Characteristics". *Materials and Structures/Materiaux et Constructions*, 38(275): 121–126.

Oregon Department of Transportation. (2010). *Pavement Distress Survey Manual.*

Perraton, D., H. Baaj, H. Di Benedetto, and M. Paradis. 2003. "Évaluation De La Résistance À La Fatigue Des Enrobés Bitumineux Fondée Sur L'Évolution De L'Endommagement Du Matériau En Cours D'Essai : Aspects Fondamentaux Et Application À L'Enrobé À Matrice De Pierre". *Canadian Journal of Civil Engineering*, 30(5): 902–913.

Perraton, D., H. Di Benedetto, C. Sauzéat, C. De La Roche, W. Bankowski, M. Partl, and J. Grenfell. 2011. "Rutting Of Bituminous Mixtures: Wheel Tracking Tests Campaign Analysis". *Materials and Structures/Materiaux et Constructions*, 44(5): 969–986.

Salama, H. K., K. Chatti, and R. W. Lyles. 2006. "Effect of Heavy Multiple Axle Trucks on Flexible Pavement Damage Using In-Service Pavement Performance Data". *Journal of Transportation Engineering*, 132(10): 763–770.

Salvatore Cafiso, A. D. G., and S. Battiato. 2006. "Evaluation of Pavement Surface Distress Using Digital Image Collection and Analysis". *Seventh International Congress on Advances in Civil Engineering*. Istanbul, Turkey: 1–10.

Sarie, F., M. Bisri, A. Wicaksono, and R. Effendi. 2015. "Types of Road Pavement Damage for Road on Peatland, A Study Case in Palangka Raya, Central Kalimantan, Indonesia". 9(12): 53–59.

Scott, G., and T. A. Engineer. 2012. "Pavement Cracking : A Failure Indicator of Your Roads Changes to Professional Development Hour Requirements".

The Weather Network, 2018. https://www.theweathernetwork.com/ca/forecasts/statistics/newfoundland-and-labrador/st-johns.

Te, N., and H. Engineering. (1995). "The Information Obtained Through Accurate Condition Assessment of Pavement". 23–33.

Valdés, G., F. Pérez-Jiménez, R. Miró, A. Martínez, and R. Botella. 2011. "Experimental Study of Recycled Asphalt Mixtures with High Percentages of Reclaimed Asphalt Pavement (RAP)". *Construction and Building Materials*, 25(3): 1289–1297.

Virginia Department of Transportation. 2012. *Asphalt Plant Certification Study Guide.*

Washington Asphalt Pavement Association. 2010. http://www.asphaltwa.com/ pavement-types-pavement-types/.

Building Tomorrow's Society

Bâtir la Société de Demain

Fredericton, Canada

June 13 to 16, 2018 / *13 au 16 juin, 2018*

FINITE ELEMENT MODELLING AND ANALYSIS OF ASPHALT CONCRETE PAVEMENT

Berhanu Abesha Feyissa[1,2]
[1] Ingenieurbüro Abesha – Engineering Office, Freiburg, Baden - Württemberg, Germany
[2] info@ingbuero-abesha.de

Abstract: The stress-strain states of asphalt concrete pavement had been evaluated by employing the ABAQUS finite element model with varying loading, structural geometry and material properties. The analysis showed keeping structural geometry, loading and other material parameters constant, increasing the E-moduli of the binder twice yielded 4.9% and 14.3% reductions in vertical stresses developed at the mid-depth of the binder and base layers, respectively and increasing the same by forth fold yielded 11.1% and 27.6% reductions in these layers. However, there were slight increases in stresses in the wearing layer, in both cases. This held true as well in case of increasing the E-moduli of the base course, there were increase in stresses developed on wearing and binder courses. These implied that increasing E-moduli reduced stresses at lower depths and increased the same slightly in upper layers. On the other hand, the variations of the elasticity modulus of the wearing course were rather pronounced in change in the stress-strain states of the wearing course itself and did not have major influence on the stress-strain state of the lower layers. It had been observed that the magnitude of the vertical stresses in the wearing course were independent of the wheel configuration, however, the overlapping of stresses was gradually pronounced in the binder course. The analyses revealed that the subbase and subgrade layers suffer greatly due to the influence of the overlapping of stresses developed by dual/multiple wheels.

1 INTRODUCTION

Asphalt concrete (AC) pavement is one of the most dominant and important infrastructures that involve multiple layers of different materials' properties subjected to non-uniform traffic loadings and varying environmental conditions. The repetitive traffic loadings that the road experiences during its service life, combined with temperature fluctuations and other environmental actions, cause rutting, fatigue and other forms of deteriorations, which ultimately degrade the performance and durability of pavement structures. Both traffic volume and loads are increasing from year to year with rapid rate. The various categories of vehicles with different performance and operational characteristics contribute to pavement distresses. Estimations and predictions of the maximum distresses likely to happen within the pavement layer are vital for proper design of the pavement. Constitutive modeling of the deformation behavior of AC enables to evaluate the deterioration progress. On the other hand, better understanding of materials behaviors and structural responses of asphalt concrete in combination with of load-stress-strain computation tools (Vormwald, 2006) allow for optimization of asphalt pavement thicknesses and material choices.

1.1 Literature Review

The finite element (FE) method constitutes a particularly convenient tool for investigation of the influence of the material properties (e.g. strength, stiffness etc.) on the overall response of a structure. As stresses and strains are used to predict pavement distresses and thus the relative condition of the various layers in the pavement structure, the need for finite element analysis is significantly important. It is capable of simulating responses of pavements subjected to axle loads with different tire pressure and multiple wheel configurations. Several studies have been recently conducted to compare different analysis methods, namely FE pavement analyses versus the theory of multilayer system as well as isotropic versus anisotropic properties of pavement materials in order to examine the critical responses of pavement structures to traffic loads (Shayesteh et al. 2017) which had been found that the results of the FE analysis and those derived from the theory of multilayer systems had no significant difference among the mean values except the magnitude of tensile strain. Linear and non-linear analyses were preformed taking in account the effect of tyre pressure, axle load, asphalt layers rigidity and pavement layers properties (Abd Alla, 2006) to investigate the influence of the state of loading, tyre type and some boundary conditions as well as to evaluate the moving traffic loads, the effects of viscous damping and load frequency on the deflections and stresses. The author had concluded that the values of deflections and stresses in case of non-linear analyses were greater than those obtained from the linear ones. Besides, the moving wheel load and increasing speed had produced higher deflections. FE analysis and local empirical methods (Abed A. H. and Al-Azzawi A. A., 2012) were also used to evaluate rutting depth in flexible pavements by analyzing two dimensional linear elastic plane strain considering standard axle load with dual set of tires and contact pressure with an elastic multi-layers system and each layer being treated as an isotropic and homogeneous materials with specified resilient modulus and Poisson ratio. The FE models were developed for accelerated pavement testing to investigate the dimensionality effects of a moving load with a repeated load and an axisymmetric model (Wu et. al. 2011). The authors considered the load wander effect and the temperature change in pavement and calculated permanent deformations.

2 FINITE ELEMENT MODELING APPROACH

In this research, the pavement had been modeled from a single part where several regions were later created by partitioning the part in order to assign different sections' properties as well as to gain additional controls over the mesh generation processes. The material properties and layer thicknesses were selected based on the German's guidelines for standardizing the superstructure of pavement (RStO, 2012). Accordingly thicker and higher Elastic modulus of intermediate layers as well as minimum elastic modulus of subgrade layer of 45MPa, were adopted throughout the analysis.

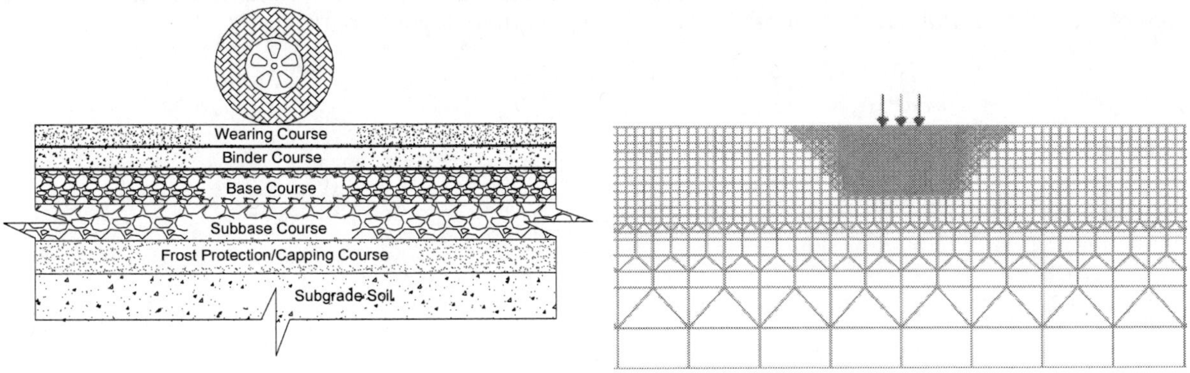

Figure 1: 2D Finite element modeling of asphalt concrete pavement

In both 2D and 3D simulations, as shown in Figure 1 above and 2 below, each model contained a single deformable solid part in a local coordinate system. The 2D continuum element was created by a shell feature in which thickness was considered small compared to the width and depth. The quadrilateral and triangular element types had been selected for smooth transition from courser to fine mesh. Different model types had also been used for single and multiple load cases. An attempt had also been made to reduce the boundary effect by applying load at the center of the model sufficiently far enough from the boundary. In addition, in order to get advantage of reduced computational time and memory space requirements, axis symmetry had also been used in some models. Different element types and sizes had been selected depending on the model type and according to the vicinity of the element with respect to the point of load application. The mesh density was controlled and managed by seeding the required area and/or edge of interest.

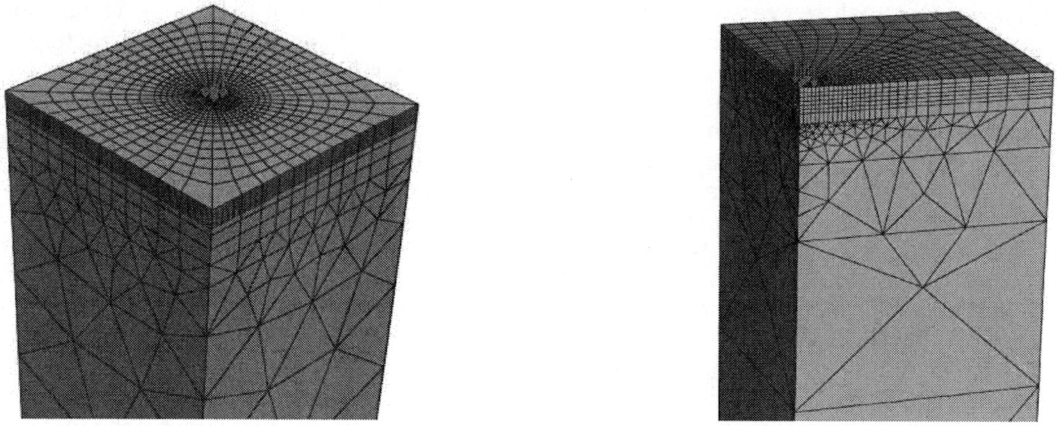

Figure 2: 3D FE Model with reduced boundary effect and axis symmetry

In the finite element modeling of asphalt concrete, the result output was extracted from the corresponding job.odb files and post processed to get the required pavement response. The influence of materials properties, variations of structural geometries and single and multiple wheel load assemblies over the various primary pavement responses had been evaluated. Result information from the output database had been analyzed from various angles.

3 RESULT ANALYSIS AND DISCUSSIONS

It is worth full here to discuss some important pavement responses obtained from FE analyses, as illustrated in Figure 3 below. In general, it had been found that, with relatively higher modulus of elasticity of the intermediate layers, the critical compressive, tensile and shear stresses were found at the top surface, at bottom of the base and within the binder courses of the pavement layers, respectively.

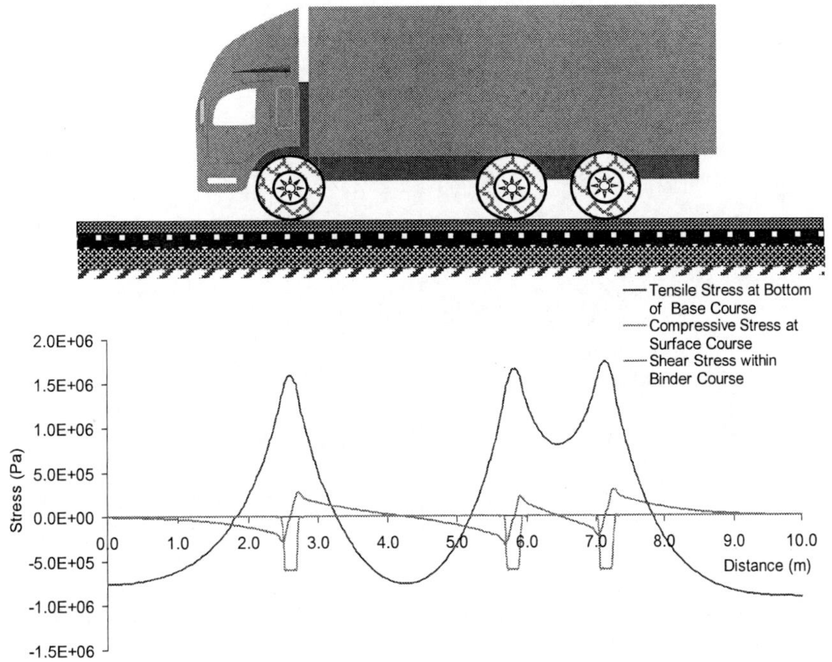

Figure 3: Illustration of critical stress states

The deflection behaviors as well as normal, shear and lateral strain-stress states have been evaluated at presumable critical positions in the model. Figure 4 showed the effect of thickness on the instantaneous deflection. The designations T-5-10-15-20 represented thickness of 5, 10, 15, and 20 cm of wearing, binder, base and subbase courses, respectively. The comparison was made with respect to contact pressure, change in E-Moduli and thicknesses of the layers. The analysis had shown that the instantaneous deflection obtained at top surface, contributed by all subsequent layers, had increased obviously with load. The deflection, in general, had decreased with increasing the E-modulus of the asphaltic layers and layer thicknesses. Doubling the total AC layer thicknesses from 150 to 300mm yields more than 10.9 % reduction in surface deflection. In addition, the deflection got smaller with increased the layer thicknesses, particularly increasing the base course thickness significantly reduced the surface deflection. Relatively, higher Elasticity modulus of binder and base courses had brought lower instantaneous deflections. The changes in the magnitude of surface deflections were rather sensitive to change in Young's modulus of the base courses than the wearing courses.

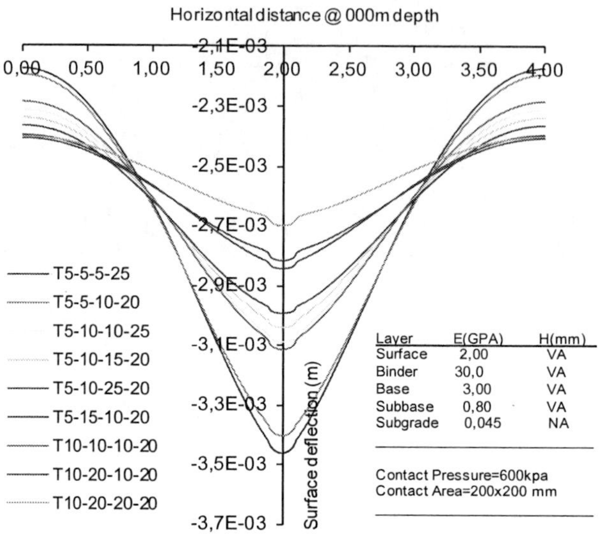

Figure 4: Instantaneous deflection behavior with change in AC thicknesses

The vertical stresses developed on the top layer were dependent neither the number of wheels nor their configurations. However, these stresses got overlapped some distance below the surface in case of multiple wheel loads. The magnitude and the area of influence depended on arrangement and the configuration of wheels. It was obvious that the magnitude of vertical stress developed at lower depths was by far greater than the one produced by a single wheel and dependent on the spacing of the wheels and axles.

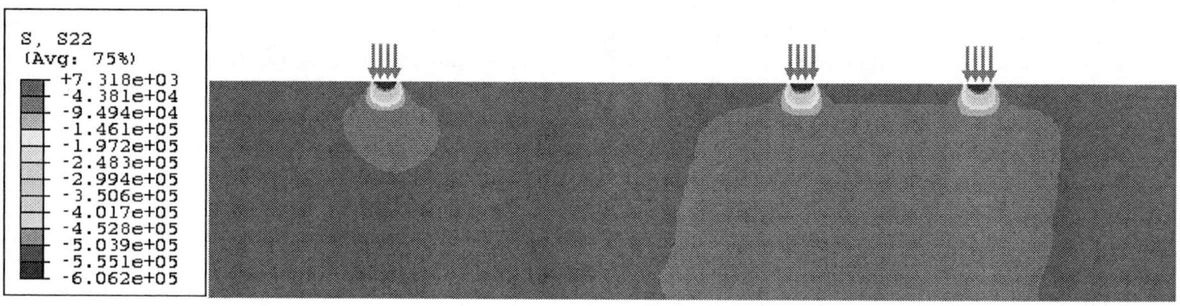

Figure 5: Stress state under dual wheel load application

As indicated in Figure 5 and 6, the magnitude of the stresses in the wearing course were higher than the other asphalt layers and independent of the wheel configuration. The overlapping of stresses was gradually pronounced in the base course.It had been revealed that the subbase and subgrade layers had suffered greatly with high stress-strain magnitudes due to the influence of the overlapping of stresses developed by dual and tandem wheel assemblies which had drawn serious concern on the stability of these layers.

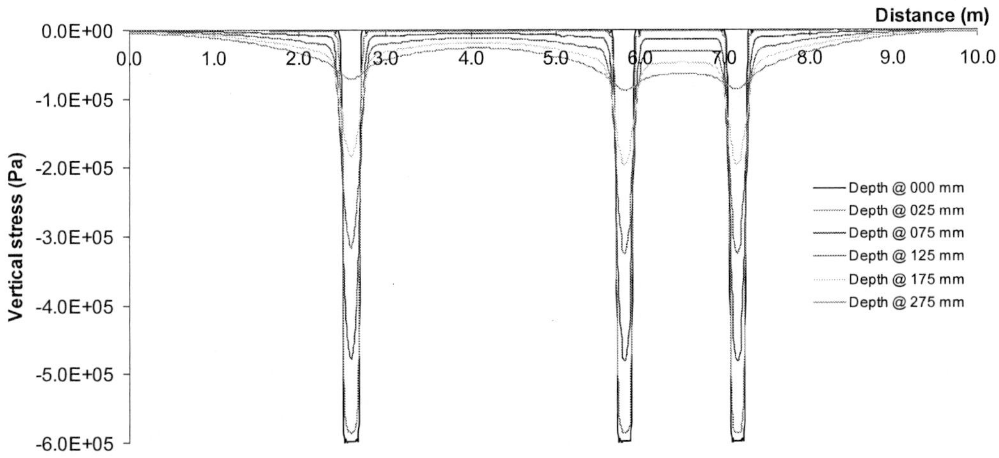

Figure 6: Influence of wheel configuration on the distribution of vertical stresses

The analysis had further revealed that increasing the binder E- modulus decreased in the magnitude of the vertical stress developed within each layers. Keeping structural geometry, loading condition and other material parameters constant, an increased the E-moduli of the binder twice yielded 4.9% and 14.3% reduction in vertical stresses developed at the mid-depth of the binder and base layers, respectively, just directly below the point of load applications. In addition, increasing the same by forth fold yields as well 11.1% and 27.6% reductions in these layers. However, there was a slight increase in stress in the wearing layer, in both cases. This held true as well in case of increasing the E-moduli of the base course. There were increased in stresses developed on both wearing and binder courses. These implied that an increased in E-moduli reduced stresses at lower depths and increased the same slightly in upper layers. On the other hand, the variations of the elasticity modulus of the wearing course were rather pronounced effect in the stress-strain states of the wearing course itself and did not have major influence on the stress-strain state of the lower layers.

The FE analysis indicated that the binder layer was subjected to remarkably high shear stress with maximum magnitude just outside of the projected contact area.

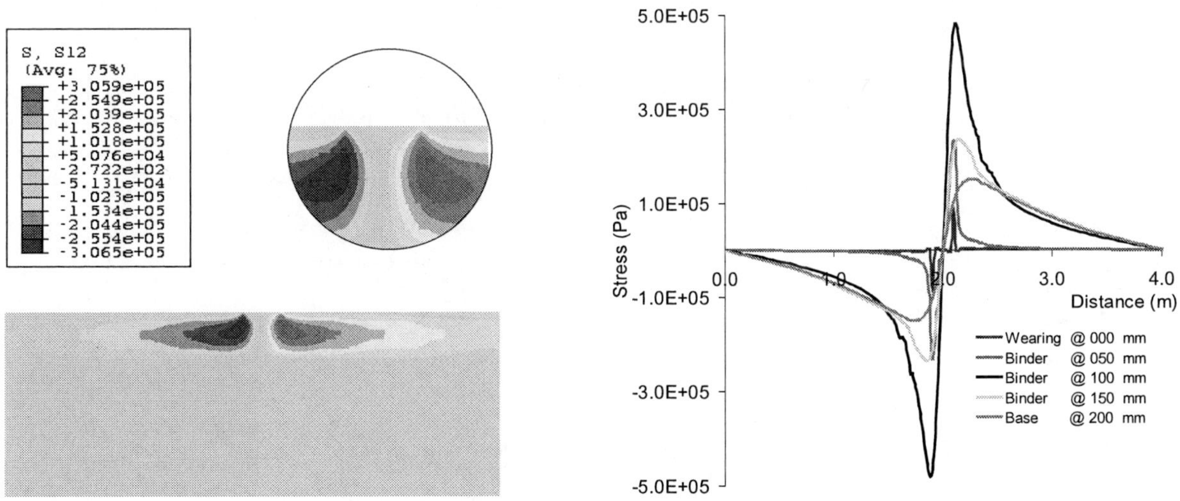

Figure 7: Shear stresses in the pavement section

In case of 3D analysis critical shear stresses were also observed close to the surface of the model but a few centimeters away from the wheel print. As shown in Figure 7 above, the maximum shear stresses had developed within the binder layer around the periphery of the projected area. The same was true in case of multiple wheel load configurations as shown in Figure 8 below. The magnitudes of shear stresses were higher on the binder course and the maximum magnitude of stresses were developed at the front and rear wheels. This was due to the fact that higher shear got developed some distance below the surface and outside of the loaded area. It was recommended that this layer was required to be stiffer and at the same time flexible so that the stress would remain within the capacity limit.

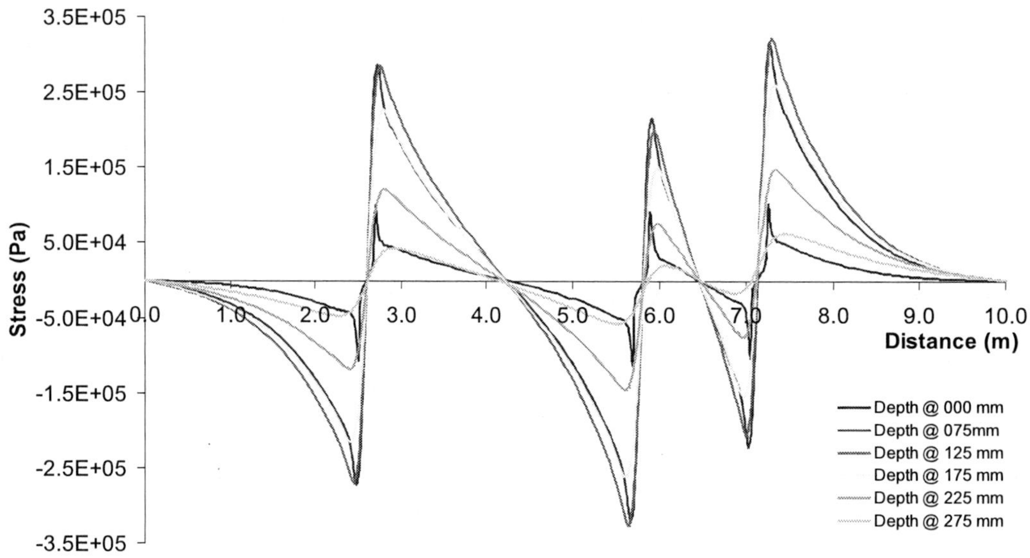

Figure 8: Shear stresses across the depth with dual wheel load configuration

Similar to the vertical stresses, the magnitude of such shear stresses depended on the magnitude of the wheel loads, contact pressures, layer thicknesses, material properties and other parameters of the layers.

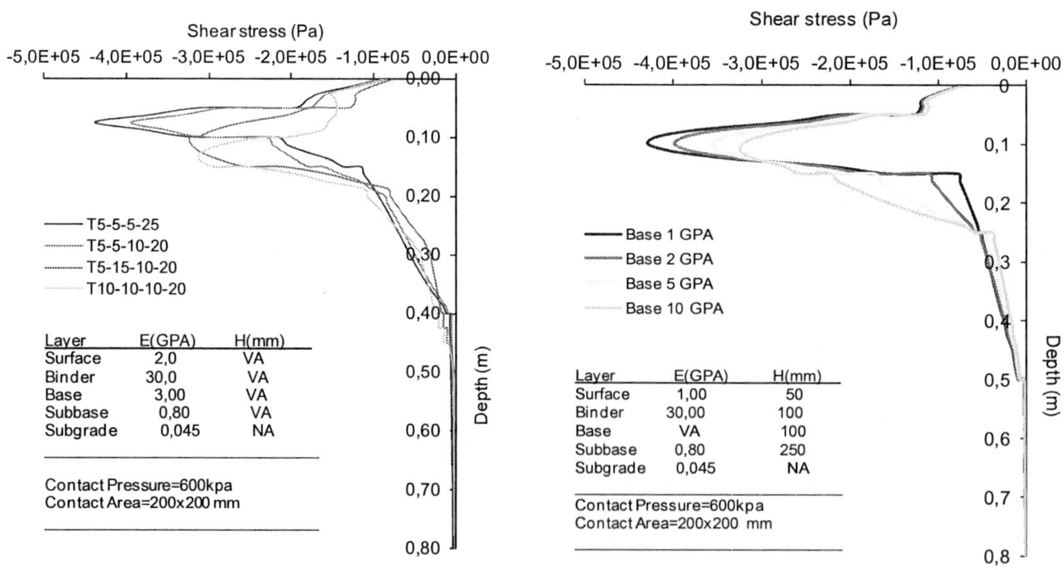

Figure 9: Shear stresses for different asphalt layer thicknesses

As shown in Figure 9 above, the shear stresses at the mid-depth of the binder layer decreased roughly by 25% when the asphalt layer thicknesses were increased from 150 to 300 mm. Higher elasticity modulus in the binder course rather increased the shear stresses developed within the binder layer itself and decreased the same within the base course and vice versa. The shear strains had reduced with increased in the E-moduli of the layers. The thicknesses of the layers had played constructive role in bringing down both shear stresses and strains in the AC layers.

The analysis had further shown the same results in case of tensile stresses as shown in Figure 10 and 11. Because of relatively higher elastic modulus at the AC layers, the maximum tensile stresses had occurred at the bottom of the base layer. The dual wheel assembly had affected the magnitude of the tensile stresses/strains produced at the base course.

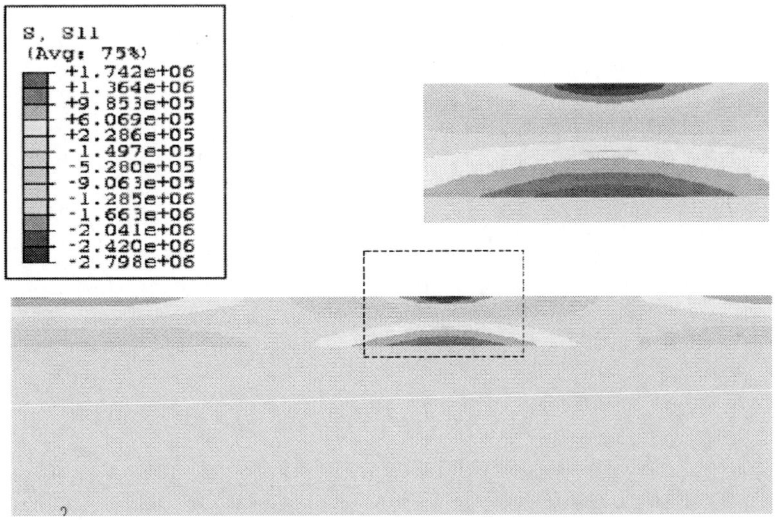

Figure 10: Lateral stress distributions

The area of influence due to tensile stress-strain at the bottom of the base course subjected to dual wheel was by far more than 1.75 times the area affected by a single wheel. This implied that the area of influence due to multiple wheel load configuration was very large which led to high stress/strain development for an extended base area in which the pavement would suffer greatly due to fatigue failure for multiple wheel load than the single wheel configurations.

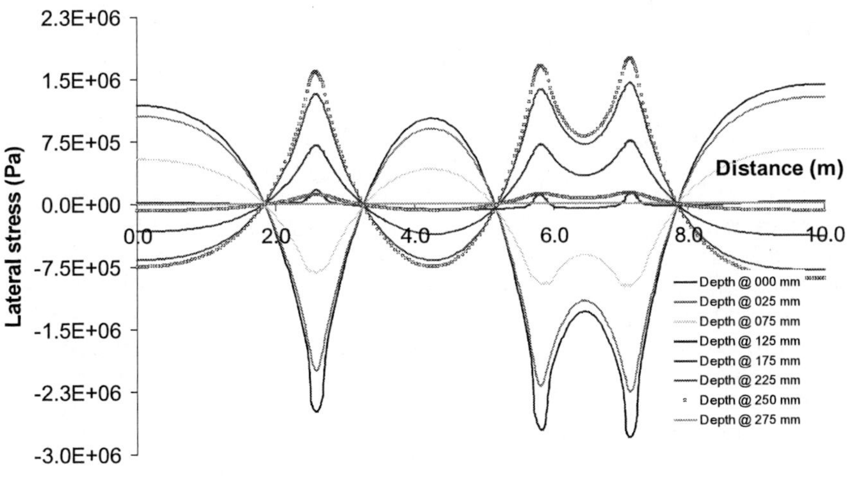

Figure 11: Horizontal stresses due to dual wheel loads

In addition, portion of the material at the bottom of base layer might have undergone relatively higher lateral deformations in case of multiple wheel load configurations.

The other important factor considered in the finite element analysis was the influence of the structural geometry on the stress-strain states of a pavement. In general, thicker asphaltic layers significantly reduced the magnitude of stresses and strains in the pavement structures, particularly for the substructure. More importantly the strain components are significantly decreased in thicker asphalt layers.

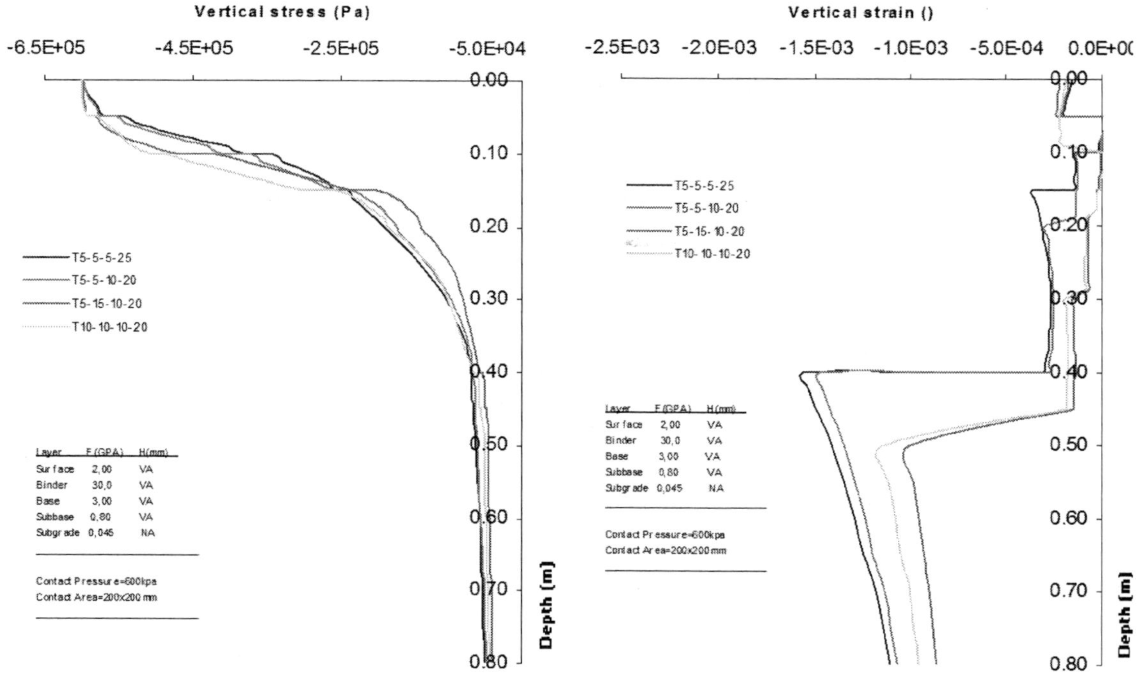

Figure 12: Vertical stresses and strains for different asphalt layer thicknesses

As can be seen from Figure 12 (left) above, at the mid depth of each layer, a reduced stresses were obtained when the thicknesses had been increased. The corresponding vertical strains were significantly reduced as shown in the Figure 12 (right) above.

4 CONCLUSIONS

The responses of the pavement for various conditions had been obtained from the finite element analysis. Different layers were created and assigned with sections' and materials' properties to evaluate the primary responses a pavement at presumable critical positions with varying loading characteristics, material properties and structural geometries. The FE analysis in general had shown that higher elasticity modulus of the asphaltic layers and/or layer thickness had reduced the instantaneous deflections as well as the stress-strain states at critical points. Critical vertical, shear and lateral/tensile/ stresses had been developed at the wearing, binder and base layers, respectively. It had been found that increasing the stiffness of any of the asphaltic layers resulted in increased stress states of the upper layers and reduction of the same in lower layers. It clearly indicated from the analysis that the strength of the asphaltic layers were in favor of the subbase and subgrade layers in decreasing the stress-strain states. In general, thicker asphaltic layers significantly reduced the magnitude of stresses and strains in the pavement structures, particularly for the substructure. More importantly the strain components were significantly decreased with thicker asphalt layers. From the analyses, it would be important to draw attentions that stiff materials' properties and thicker intermediate asphaltic layers would probably guaranty longer service life by reducing the stress-strain states within the sub-structural layers (subbase and subgrade layers), confining damage on top layers and/or maintenance works might instead be necessary or limited to the upper structural layers of the pavement.

Acknowledgements

I am very thankful for German Academic Exchange Service (DAAD) for the financial sponsorship throughout the research period. I would also like to thank the faculty staffs of Section of Road and Traffic Engineering and Material Mechanics of Department of Civil Engineering at Technische Univesität Darmstadt-Germany for their valuable assistance during the course of the research.

References

Abd Alla, E. M. 2006. The Rational Use of Finite Element Method in the Analysis of Flexible Pavements. *Journal of Engineering Sciences*, Vol. 34, No. 4, pp. 1185-1211.

Abed, A. H. and Al-Azzawi, A. A. 2012. Evaluation of Rutting Depth in Flexible Pavements by Using Finite Element Analysis and Local Empirical Model. *American Journal of Engineering and Applied Sciences*, Vol. 5, No. 2, pp. 163-169.

Dessouky, S., Masad, E., Little, D. and Zbib, H. 2006. Finite-Element Analysis of Hot Mix Asphalt Microstructure Using Effective Local Material Properties and Strain Gradient Elasticity. *Journal of Engineering Mechanics*, ASCE, Vol. 132, No. 2, pp. 158–171.

Erkens, S.M.J.G., Liu, X. and Scarpas, A. 2002. 3D Finite Element Model for Asphalt Concrete Response Simulation. *The International Journal of Geomechanics*, ASCE, Vol. 2, No. 3, pp 305–330.

Lee, H. J., Lee, J. H. and Park, H. M. 2007. Performance Evaluation of High Modulus Asphalt Mixtures for Long Life Asphalt Pavements. *Construction and Building Materials*, Science Direct, ELSEVIER, Vol. 21, pp 1079–1087.

Mun, S. 2003. *Nonlinear Finite Element Analysis of Pavements and Its Application to Performance Evaluation*. PhD Thesis, North Carolina State University.

RStO 12, FGSV, Ausgabe 2012. *Richtlinien für die Standardisierung des Oberbaues von Verkehrsflächen.*

Saad, B., Mitri, H. and Poorooshasb, H. 2006. 3D FE Analysis of Flexible Pavement with Geosynthetic Reinforcement. *The Journal of Transportation Engineering*, ASCE, Vol. 132, No. 5, pp. 402–415.

Shayesteh, A., Ghasemisalehabadi, E., Khordehbinan, M. W. and Rostami, T. 2017. Finite Element Method in Statistical Analysis of Flexible Pavement. *Journal of Marine Science and Technology*, Vol. 25, No. 2, pp. 142-152

Vormwald, M. 2006. *Materialmodellierung,* Vorlesungsmitschrift, TU-Darmstadt, Fachgebiet Werkstoffmechanik, Darmstadt.

Wayessa, S. G., Quezon, E. T. and Kumela, T. 2017. Analysis of Stress- Strain and Deflection of Flexible Pavements Using Finite Element Method Case Study on Bako-Nekemte Road. *Journal of Civil, Construction and Environmental Engineering*, Vol. 2, No. 4, 2017, pp. 100-111.

Wu, Z., Chen, X. and Yang, X. 2011. *Finite Element Simulation of Structural Performance on Flexible Pavements with Stabilized Base/Treated Subbase Materials under Accelerated Loading*, Louisiana Transportation Research Center.

Building Tomorrow's Society

Bâtir la Société de Demain

Fredericton, Canada

June 13 to 16, 2018 / *13 au 16 juin, 2018*

HEAT OF HYDRATION OF CALCIUM SULFOALUMINATE CEMENT – A LITERATURE REVIEW

Abdel-Raheem, Mohamed[1,2], Hernandez, Daniel[1], Navarro, Linda[1], Reyes, Manuel[1] Silva, Grecia[1] and Hernandez, Luis Brando[1,]

[1] Department of Civil Engineering, University of Texas Rio Grande Valley, USA

[2] mohamed.abdelraheem@utrgv.edu

Abstract: Calcium sulfoaluminate (CSA) cement is a sustainable alternative to Ordinary Portland Cement (OPC). The manufacturing of CSA has a lower carbon dioxide emission compared to OPC. This is attributed to the lower temperature needed to calcinate the raw materials and less energy consumed during the pulverization process, which reduces the carbon footprint. However, in many cases, the production process of CSA can be more expensive that OPC. The Heat of Hydration (HH) of cement can affect the properties of concrete. In mass concrete structure, high HH can lead to thermal stresses and delayed ettringite that may lead to the expansion, thermal cracking, and failure of the structure. Several factors affect the of heat released during the hydration of cement, such as: the cement mineralogical composition, curing temperature, water/cement (w/c) ratio, and cement fineness. This study presents a review of previous studies that focused on the assessment and quantification of the HH of CSA. This paper serves as a succinct guide for future researchers that aim to conduct advanced studies on the HH of CSA.

1 INTRODUCTION

Calcium sulfoaluminate (CSA) cement was mentioned for the first time in a French patent in the 1930's, but it was only in the 1970's and 1980's that its production started to become important (Garcia-Mate 2013). CSA cement can reach a compressive strength of 5000 psi in 24 hours, while Ordinary Portland Cement (OPC) achieves it in 28 days. The initial setting time of the CSA cement begins after approximately 20 minutes and its final setting time is then arrived to after ten more minutes. With the use of retarders, the processing time can be considerably extended. Therefore, even in low temperature environments, CSA cement maintains its reactivity unlike other cement types (Zhou et al. 2006). For these reasons, CSA cement is commonly used in airport start and runways, bridge reparations, and any other application where time is vital.

A key advantage of CSA cement is that it has a smaller impact on the environment in comparison to OPC; CSA can be produced at about 1200^{0}C, making it 200-250^{0}C lower than OPC, because it is relatively softer and more friable, making it easier to grind. As a result, the entire process will save more energy (Kasselouri et al. 1995).

When water makes contact with cement an exothermic reaction liberates heat, called heat of hydration. The release of excessive heat is a concern in mass concrete constructions, such as: dams, bridges, raft foundations, etc. Mass concrete elements generate substantial thermal gradients between the core and the surface that pose a considerable risk of thermal damage. This phenomenon is called Thermal Cracking (Bernardo et al. 2006).

Sufficient literature review exists regarding the heat of hydration and the early hydration phases of Calcium Sulfoaluminate cement. Common methods to quantify the heat of hydration- Isothermal Calorimetry, Semi-Adiabatic Calorimetry, Differential Calorimetry, nonisothermal-nonadiabatic differential BMR Calorimetry, which are discussed in this literature review. The objective of this study is to provide a comprehensive literature review on the Heat of Hydration of CSA cement that could serve as a guide for the scientific community for future references.

2 PROBLEM STATEMENT

Even though CSA is relatively new when compared to OPC, a great amount of research has been done. However, most previous studies are quite fragmented in the sense that they do not have continuity or fully cover CSA topics in depth. The fact that there is not a single paper that combines all this literature suggests that a comprehensive paper of all previous HH of CSA articles is needed.

3 OBJECTIVE

The main objective of this literature review to provide a tool to the scientific community that will facilitate future research on HH of CSA as it takes time and effort to know what has been done in the past. This literature review will compare results between different publications to look for discrepancies and similarities in their findings.

4 METHODOLOGY

The methodology adopted in this study is comprised of two steps. First, a thorough search was conducted to compile all previous research publications related to the HH of CSA. Second, the collected studies with classified according to the method used in assessing and quantifying the HH of CSA and organized chronologically for comprehensive review. Two categories were used to classify the studies based on the methods used to quantify the heat of hydration, which are: 1) Isothermal calorimetry, and 2) Semi-Adiabatic calorimetry.

5 ISOTHERMAL CALORIMETRY

The literature included eight studies that utilized isothermal calorimetry to assess and quantify the heat of hydrations of CSA. One of the first studies on the HH of CSA goes back to Dová et al. (2005). The authors presented a study on the heat evolution and mechanism of the hydration of CaO-Al_2O_3-SO_3 using an isothermal conduction calorimeter. The study addressed the hydration behavior of CaO-Al2O_3-SO at different temperatures range of 25, 40, 60 and 80°C, and different water-solid(w/s) ratios of 0.5, 0.7 and 1.0. For the calorimeter test, the authors made a combination of $CaCO_3$, $Al(OH)_3$ and $CaSO_4 \cdot 2H_2O$.Then, they calcined it twice for two hours at 1250°C to synthesize C_4A_3S. A first experiment was conducted to study the influence of the temperature on the rate of hydration. The results of the calorimeter test indicated that when the temperature increases the hydration intensity also increases and the duration of the induction period becomes shorter, as shown in Figure 1.

The second experiment focused on assessing the influence that w/s ratio has on the HH at different temperatures. The results showed that when w/s ratio increases at low temperatures (25°C and 40°C), the heat flow peaks increases slightly, and the induction period is shorter- see Figure 2a. However, at high temperatures (60°C and 80°C), the heat flow peaks and the length of induction period decrease by increasing the w/s ratio, see Figure 2b. As such, it is concluded that the hydration reaction of C_4A_3S is accelerated with increasing temperatures at w/s ratios of 0.5, 0.7, and 1.0. Also, the hydration accelerates if w/s ratio increases at 25°C and 40°C, but at higher temperatures the intensity of hydration decreases.

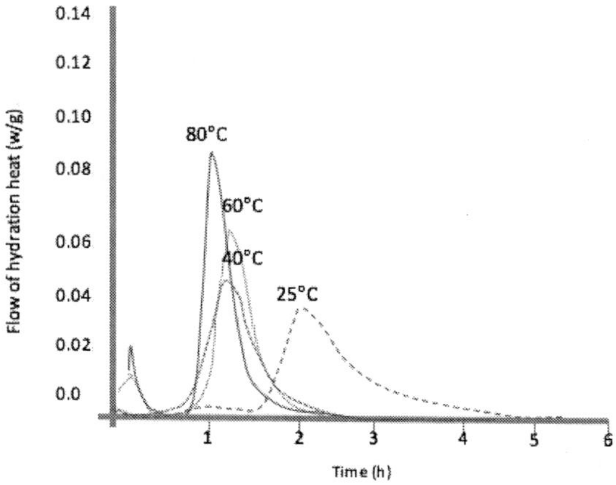

Figure 1a. Influence of temperature on the HH at w/s of 0.7 (Dová et al. 2005)

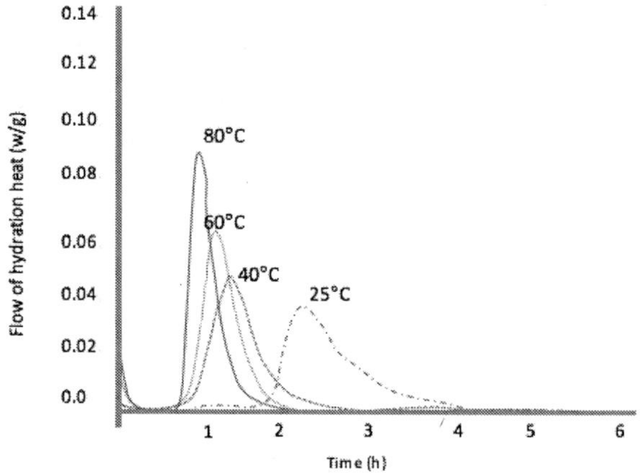

Figure 1b. Influence of temperature on the HH at w/s of 1 (Dová et al. 2005)

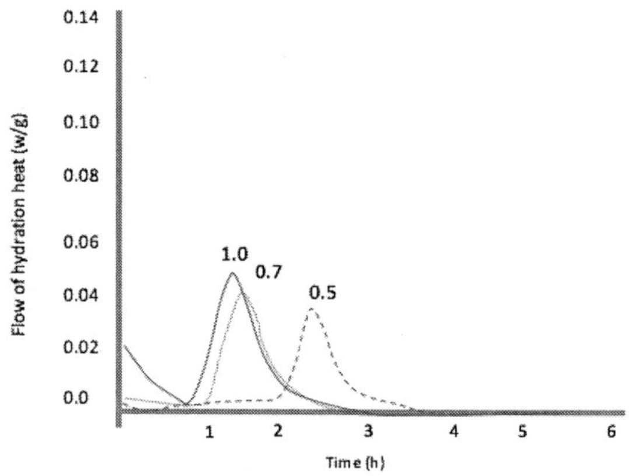

Figure 2a. Influence of w/s/ on the HH at W/S at 40°C (Dová et al. 2005)

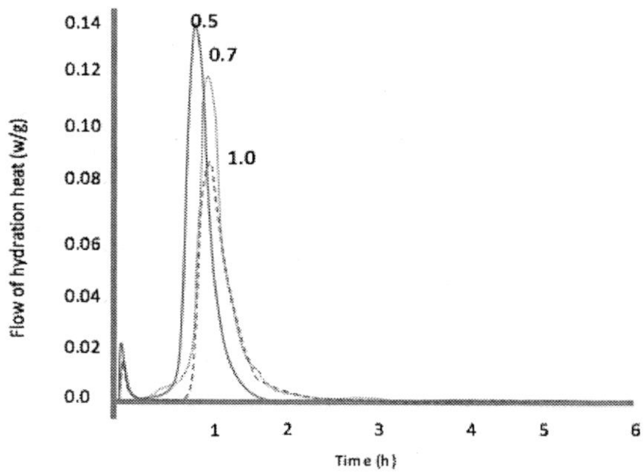

Figure 2b. Influence of w/s/ on the HH at W/S at 80°C (Dová et al. 2005)

Lura el al. (2010) conducted a set of experiments using isothermal calorimetry and chemical shrinkage measurements to assess the heat of hydration of Ordinary Portland Cement (OPC) and CSA. In this study, three experiments were conducted applying the two methods in order to determine the heat of hydration of PC samples and a mixture of CSA and anhydrite. The first setup is called a sole calorimetry, where a commercial conduction calorimeter with eight measuring cells was used. The time period used to get the heat flow was about 80 hours and a temperature of ±0.02 °C was set to keep the measuring cells in the calorimeter. The second experiment is a sole chemical shrinkage; a thin layer of demineralized water was placed on top of the three pastes. Regular intervals were chosen for three days to determine the corresponding measurements done by reading the oil level of each graduated pipettes.

Lastly, the third experiment is a combined calorimetric/chemical shrinkage, here several mixtures were conducted, and parallel measurements were used to measure the heat of hydration and chemical shrinkage. The results show that measuring the chemical shrinkage inside the calorimeter is as efficient as the sole measurement using the ASTM C-1608-07 procedures.

Gartner el al. (2011) studies the hydration of Belite-Calcium Sulfoaluminate-Ferrite cement paste using XRD-Rietveld, TGA, Differential Thermal Analysis (DTA), Isothermal calorimetry, and pycnometry. For isothermal calorimetry, the cement paste was studied at 20°C with a water-cement ratio of 0.5. The composition of the cement was as follows: 48% of belite (C2S), 28% of ye'elimite (C4A3S), 18% of ferrite solid solution (C2(AF)), and 6 % of anhydrite. The previous mentioned methods were implemented at the early hydration ages, which is between the first 28 days, and after 28 days (Gartner el al. 2011).

The specimens used in these procedures were prepared with different raw materials, such as: limestone, clay, bauxite, iron ore and anhydrite. All previous materials were ground together in the desired proportions and pelletized at 1280°C. The results from the hydration at very early ages for the calorimeter showed three peaks. The first one happened at 15 minutes, which is the beginning of the ye'elimite dissolution. The second one appeared at 2 hours, which is the main dissolution period of ye'elimite. Lastly, the third one appeared at 3.5 hours (Gartner el al. 2011).

In a different research, the hydration of two CSA cements (CSA-1 and CSA-2) was investigated at a range of 1 hour and 28 days with a w/c ratio of 0.72 (CSA-1) and 0.80 (CSA-2) (Winnefeld et al. 2010a). The chemical composition of CSA-1 was ye'elimite (50%), gehlenite (15%), calcium aluminate (8%), gypsum (22%) and others (5%). For CSA-2, its chemical composition consisted of ye'elimite (54%), belite (19%) and anhydrite (21%). All hydration experiments were carried out at 20 °C. The methods implemented were isothermal calorimetry, X-Ray Diffraction (XRD), Thermo-Gravimetric Analysis (TGA) and Scanning Electron Microscopy (SEM) (Winnefeld et al. 2010a).

Isothermal calorimeter was used to determine the rate of hydration heat liberated during the first 72 hours. The results showed that the main hydration peak of CSA-1 occurs at 7 hours and has a dormant period between 3-4 hours during hydration. The results showed that the main hydration peak of CSA-1 occurs at 7 hours with a dormant period between 3-4 hours during hydration. For CSA-2, the main hydration peak occurred after 7 hours and a second at 26 hours, with a dormant period similar to CSA-1 (Winnefeld et al. 2010a).

In a separate publication, the same authors studied the main phase of hydration in CSA, ye'elimite (calcium sulfoaluminate), with two calcium sulfates along with different reactivities (gypsum, and anhydrite) (Winnefeld et al. 2010b). The methods implemented were isothermal conduction calorimetry, TGA, and XRD. All the experiments were done at 20°C, and they used ye'elimite blended with different quantities of gypsum/anhydrite at different molar ratios (1:1, 1:2, 1:3 and 1:4) (Winnefeld et al. 2010b).

Isothermal calorimetry was used to determine the influence in the hydration heat flow of gypsum in ye'elimite. Without the addition of gypsum in ye'elimite, the results showed that there were two maxima peaks in the heat flow curve (Figure 3a). The first peak occurred after water was added, and the second one at 15 hours of hydration (Figure 3b). When gypsum was added to the mixtures, the graphs behaved similarly to the ones with pure ye'elimite. Nevertheless, it is important to notice that heat is more intense in the mixtures with the presence of gypsum, and its dormant period is shortened to about 3 hours. After the dormant period, the samples with lower amount of gypsum (1:1 and 1:2) showed two maximum peaks, the first one at 5 hours (1:1 and 1:2) and the second one was at 10.5 hours (1:1) and at 15.5 hours (1:2). The samples with higher amount of gypsum showed only one maximum peak (Winnefeld et al. 2010b).

When the anhydrite was added to the ye'elimite, the results showed that anhydrite has less impact on the very early reactions. The first peak shown in Figure 3a is caused by the addition of water. The results indicated that the addition of anhydrite in the samples accelerates the hydration of ye'elimite phase as gypsum does. In comparison to samples with gypsum, anhydrite's dormant period is shortened to about 4 hours, and its curves behave differently, see Figure 4b (Winnefeld et al. 2010b).

When the authors used potassium hydroxide (KOH) on the hydration kinetics of ye'elimite, the results showed that the initial peak increased significantly because there is an enhanced dissolution and precipitation of more hydrated phases in the first minute of hydration. The addition of KOH to the sample showed that the hydration of ye'elimite is accelerated with alkaline solutions. When gypsum and KOH were used in the samples, the results showed that the initial peak increases, and the hydration accelerates (Winnefeld et al. 2010b).

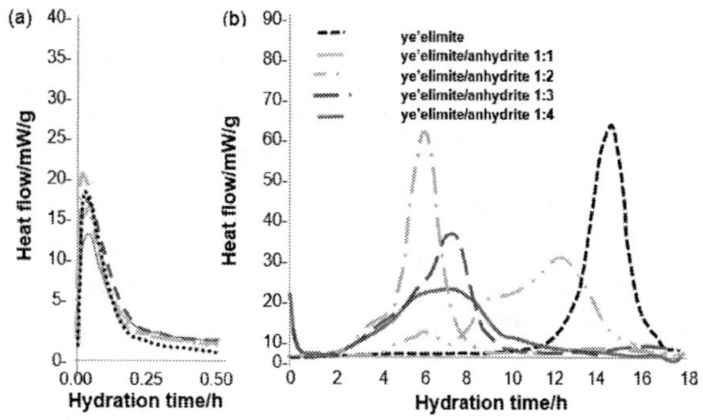

Figure 3a: Conduction calorimetry of ye'elimite pastes: influence of gypsum addition. q) first 30 minutes, b) from 0-18 hours (Winnefeld et al. 2010b).

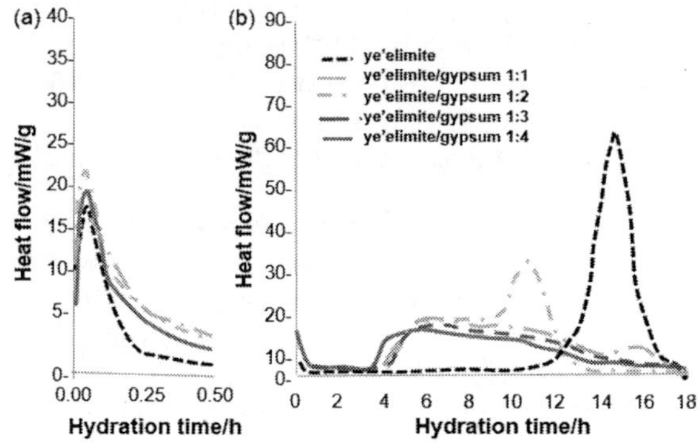

Figure 3b: Conduction calorimetry of ye'elimite pastes: influence of anhydrite addition. a) first 30 minutes, b) from 0-18 hours (Winnefeld et al. 2010b).

Another study presents a documentation for the synthesis of calcium sulfoaluminate-belite (CSB); three different CSAB clinkers were made using reagent-chemicals and different gypsum amounts (Chen and Juenger 2011). as study to analyze the dependency of temperature of calcium sulfoaluminate-belite cement hydration. A period of three days with a temperature of 23 °C was used to determine the hydration of pastes with a w/c of 0.45. Also, the phase composition was refined using a proposed proportioning method and an equation for calculating the minimum amount of gypsum needed for the synthesis of CSAB (Chen and Juenger 2011).

A specific rheometric cell was created to characterize hydration at early ages of CSA cement using demineralized water or a borate solution (Champenois 2013). One of the main advantages of this cell is the possibility to collect data simultaneously avoiding any bias due to slight temperature differences between two experiments. The objective of the study was the characterization of the early hydration ages of CSA using complementary techniques, such as: semi-adiabatic calorimetry and isothermal conduction calorimetry. The hydration of the cement was found using a micro-calorimeter under isothermal conditions at 24.0±0.5 ◦C. The temperature recorded in the samples for rheometry and conductometry were compared to the progress of the heat flux (Champenois 2013).

One study investigated the hydration phases of CSA (ye'elimite) mixed with two different forms of calcium sulphate, namely, gypsum and anhydrite. Two samples were added using the CSA clinkers with gypsum (CGy) and another one with anhydrite (CAn). There was a fast hydration on both samples within the first minutes. After that, the entire hydration for both samples decreased. The sample made of CSA and anhydrite had an initial slow stage than the one made with gypsum. Subsequently, there was an acceleration within the next two hours and a maximum heat flow was shown after 40 hours. The hydration rate for the CGy specimen is faster than the CAn. There were two peaks within the first 7 hours and the maximum heat flow was shown after 15 hours. This experiment showed that there is a higher reactivity in the system when the samples are mixed with gypsum after 8hours.The results also show that the hydration reaction is dependent on the availability of sulphates.

6 SEMI-ADIABATIC CALORIMETRY

The literature included only one study that used semi-adiabatic calorimetry to assess the HH of CSA. Berger el al. (2009) presented a study to investigate hydration of CSA cement at an early age with respect to gypsum content, the thermal history, and $ZnCl_2$ concentration. The CSA cements used in these tests were prepared by mixing a ground industrial CSA clinker with gypsum. The mortars and pastes used were made with a w/c ratio of 0.55, and with an increasing amount of gypsum from 0 to 35%. Also, in some of the solutions it was added $ZnCl_2$. The pastes were cured for 7 days at 20°C or submitted to a thermal cycle for 7 days Berger el al. (2009).

The heat curves were generated to assess the influence of gypsum on the HH of CSA – see Figure 4a. The results indicate that the cumulative heat decreases when the amount of gypsum in the mortars increases. Another interesting result is changing the amount of gypsum from 0 to3% of gypsum did not significantly change the total HH, but reduced the timing for the maximum heat fluxes. In the mortars with a 3-7% gypsum content, it was observed that the times for maximum heat fluxes decreased as the gypsum content increases. The mortars with a 7-10% gypsum content behave similarly by having the same minimum induction period, the time for maximum heat flux, and cumulated heat. In mortars with a gypsum content from 10-35% the results show that the induction period increases slightly when the gypsum increases Berger el al. (2009).

The results from the hydration of CSA with $ZnCL_2$ solution experiments were also determined by the semi-adiabatic calorimetry method, see Figure 4b. It was demonstrated that salt tends to retard the hydration process, but it becomes even more with the binders that did not contain gypsum. When the binders reach 20% of gypsum the retardation in hydration process is only observed at 0.5 Mol/L Berger el al. (2009).

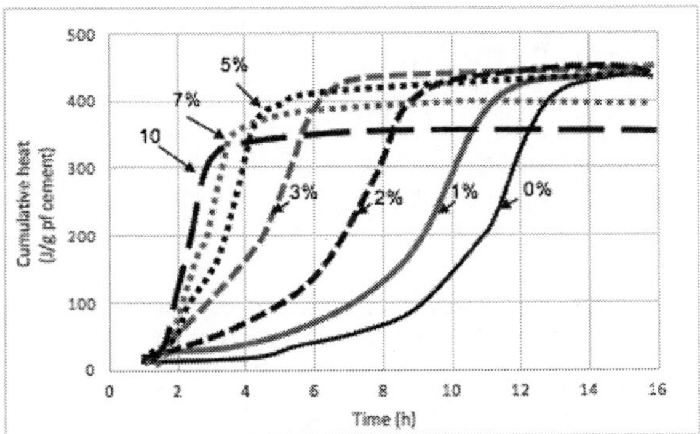

Figure 4a: Influence of the gypsum content on the heat produced by CSA mortars (Berger el al. 2009)

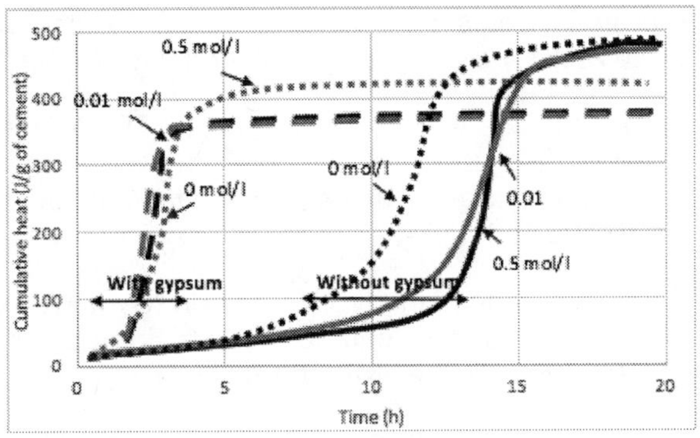

Figure 4b: Influence of the ZnCL2 concentration in on the heat produced by CSA mortars (Berger el al. 2009)

7 FINDINGS

The review of previous studies shows some interesting findings about the HH of CSA. It was found that the hydration of CSA increases by increasing the temperature in general. Also , increasing the w/s ratio at temperature ranging between 25°C-40°C will increase the hydration of CSA. However. increasing the w/s ratio at a temperature above this limit will slow the hydration of CSA (Doval et al. 2005).

Also, it was found that CSA made with gypsum behaves differently from the one made with anhydrite. It was found that gypsum accelerates the hydration reaction and shortens the dormant period more than the anhydrite (Winnefeld et al. 2010 a, b). One disadvantage of using excess gypsum in the CSA is that when the gypsum amount increases, the setting time in the cement decreases (Champenois 2013). Also, it is important to notice that the induction period is shorter when adding anhydrite to the CSA solution compared when gypsum is added, and the hydration is faster with gypsum compared with anhydrite (S. Allevi 2016).

In addition, it is found that when gypsum is not added to the CSA in the phase of ye'elimite, two maximum peaks are found in the heat flow curve. In the other hand, when gypsum is added to the solution of CSA at ratios of 1:1, 1:2, 1:3 and 1:4 the results show that the initial peaks increased. When anhydrite was used in the CSA solution, the results showed that anhydrite has less impact on the very early reactions compared with gypsum and the dormant period is shorter (Winnefeld 2010).

8 CONCLUSION

This study presents a summary review of the previous research that has been conducted to assess the HH of CSA and the factors that influence it. The review shows that the HH of CSA varies based on the chemical composition. Factors such as temperature and w/s ratios have different effects on the HH. The utilization of gypsum versus anhydrite in the synthesis of CSA affect the peak of the HH and the dormant period. As such, there is a need for future studies that standardize the different types of CSA based on the chemical composition and properties similar to the four main types of OPC.

References

Allevi, S., M. Marchi, F. Scotti, S. Bertini, and C. Cosentino. "Hydration of calcium sulphoaluminate clinker with additions of different calcium sulphate sources." Materials and Structures49, no. 1-2 (2016): 453-66. doi:10.1617/s11527-014-0510-5.

Berger, Stéphane, Céline Cau Dit Coumes, Patrick Le Bescop, and Denis Damidot. "Hydration of calcium sulfoaluminate cement by a ZnCl2 solution: Investigation at early age." Cement and Concrete Research 39, no. 12 (2009): 1180-187

Bernardo G., Telesca A., Valenti G. (2006) A porosimetric study of calcium sulfoaluminate cement pastes cured at early ages, Elsevier Cement and Concrete Research 36 (1042-1047)

Champenois, Jean-Baptiste, Celine Cau Dit Coumes, Arnaud Poulesquen, Patrick Le Bescop, and Denis Damidot. "Beneficial use of a cell coupling rheometry, conductimetry, and calorimetry to investigate the early age hydration of calcium sulfoaluminate cement." Rheologica Acta52, no. 2 (2013): 177-87. doi:10.1007/s00397-013-0675-9.

Chen, Irvin A., and Maria C. G. Juenger. "Synthesis and hydration of calcium sulfoaluminate-belite cements with varied phase compositions." Journal of Materials Science 46, no. 8 (2011): 2568-577. doi:10.1007/s10853-010-5109-9.

Doval, M., Martin Palou, and Vladimír Kovár. "Heat evolution and mechanism of hydration in CaO–Al2O3–SO3 system." Ceramics Silikaty49 (2005): 104-08.

Garcia-Mate, M., De la Torre A.G., Leon-Reyna L., Aranda M.A.G, Santacruz I. (2013) Hydration studies of calcium sulfoaluminate cements blended with fly ash, Science Direct Cement and concrete research (12-20)

Gartner, E. M., G. Walenta, V. Morin, P. Termkhajornkit, I. Baco, and J. M. Casabonne. "Hydration of a Belite-CalciumSulfoaluminate-Ferrite cement: AetherTM." 13th Int. Congr. Chem. Cem., Madrid (2011)

Kasselouri V., Tsakidridis P. (1995) A study on the Hydration products of a Non-expansive sulfoaluminate cement, Pergamon, Cement and Concrete research Vol 25 No 8 (1726-1736)

Winnefeld, Frank, and Barbara Lothenbach. "Hydration of calcium sulfoaluminate cements — Experimental findings and thermodynamic modelling." Cement and Concrete Research 40, no. 8 (2010a): 1239-247.

Winnefeld, Frank, and Stefan Barlag. "Calorimetric and thermogravimetric study on the influence of calcium sulfate on the hydration of ye'elimite." Journal of Thermal Analysis and Calorimetry101, no. 3 (2010 b): 949-57. doi:10.1007/s10973-009-0582-6.

Zhou Q., Milestone N.B., Hayes M., (2006) An alternative to Portland cement for waste encapsulation –the calcium sulfoaluminate cement system, Journal of Hazardous Materials 136 (120-129)

Building Tomorrow's Society

Bâtir la Société de Demain

Fredericton, Canada

June 13 to 16, 2018 / *13 au 16 juin, 2018*

HIGH STRENGTH REINFORCEMENT (HSR) - THE SMART SOLUTION FOR A SUSTAINABLE FUTURE

Hude, Florian[1,3], Silva, Jaime[2]
[1] MANA Ltd, Canada
[2] SAS Stressteel, Inc. USA
[3] f.hude@mana-barmill.com

Abstract: In Canada and in Germany the recent limitation of yield stress for reinforcement bars is f_{yk} = 500 N/mm². In the USA the code limit is 550 N/mm² and in some European countries reinforcement steel with higher yield strain (up to 600 N/mm²) is permitted. In the last decade many higher strength rebars were developed and the trend is towards this new reinforcement grades. The advantages of HSR are cost reduction (due to lower steel weight), size reduction of structural members (e.g. effective floor space increase as columns with high strength reinforcement have a cross section area compared to columns built with a common reinforcement), Increased flexibility for architects (due to higher load capacity with the same column dimensions or smaller dimensions with the same capacity) and reduced congestion issues result in in higher quality construction. A smart construction solution does not only give the designer one advantage (higher strength) but allows further enhancement: Changing the rib pattern in a way that the bar can be used as a threaded bar, there is no further manufacturing done to use the bar in a coupling system. This rebar-system allows further reduction of splice congestion, reduces development lengths and makes modular construction easy. In several countries (USA, Germany, Poland, South Korea, ...) high strength thread bars up to 670 N/mm² yield strength are used already. This paper will lay out the properties of HSR, describes the code compliance and will show several applications around the globe.

1 INTRODUCTION

Concrete is one of the oldest composite building materials. The Romans developed the "opus caementitium". The biggest deficiency ever was that concrete can only transfer very small tension loads. This was the reason why for old structures geometrical forms were chosen that the structure was mainly under compression (e.g. arches or vaults). The first person who used iron to increase the durability of concrete construction was Joseph Monier in the late 19th century. He used florist wire as he was a gardener. For the first applications, any kind of iron or steel was used to increase the tension strength of concrete. In the 1930s the first reinforcing steels were developed in the United States. (Russwurm, 1993)

In the 1950's, national codes for reinforcement were introduced in many countries. Since then, the steel quality was kept almost at the same level. In the German code (DIN 488, 2010) two steel grades are mentioned: reinforcing steel BSt 420 (f_y = 420 N/mm²) and BSt 500 (f_y = 500 N/mm²). The Canadian standards limit the yield strength also to 500 N/mm². According to ACI 318-08 (ACI 318, 2014) reinforcing steel with a yield strength exceeding grade 60, "the yield strength shall be taken as the stress corresponding to a strain of 0.35 %" but the maximum yield strength used in design calculations according to (ACI 318,

2014) is limited to 550 N/mm². The only code which permits higher maximum yield strength is the European Code (EC 2, 2011) (f_y < 600 N/mm²).

Meanwhile the concrete strength was consequently increased from normal strength concrete (NSC, f_c < 60 N/mm²) to high strength concrete (HSC, f_c < 100 N/mm²) and further to ultra high performance concrete (UHPC, f_c > 100 N/mm²). Although these changes allowed for increased concrete strengths, the strength of reinforcing steel was not increased for many decades. For this reason, it is argued in the literature that UHPC should be reinforced with high strength reinforcing bars (HSR) (see also (Reichel, 2010) and (Jungwirth, 2006)). One of the alternatives for reinforcing steel bar (a weldable rebar with a thread rib pattern and a yield strength f_y = 670 N/mm², further called "HSR" or "670W") is described below.

Furthermore, HSR allows for reduction in the quantity of reinforcement. This can be advantageous in different scenarios:

- Constant cross section: reduced reinforcement ratio leads to reduced congestion and saves installation time.
- Constant reinforcement ratio: reduced cross section leads to lower dead load and higher available floor space, which increases the sellable footprint.
- Change of construction method: composite construction can often be avoided by using HSR.

Another advantage of the 670W is the threadability of the bars. The ribs are in such a shape, that the bar is continuously threadable (see Fig. 1). Therefore, the bars can be used for preassembled rebar-cages (modules).

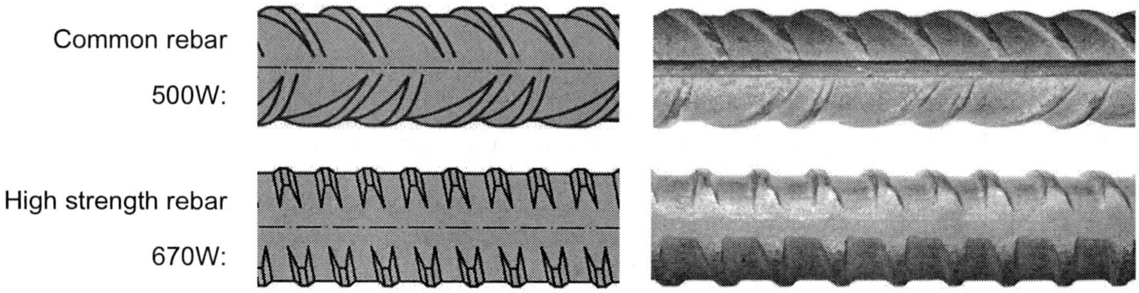

Fig. 1: Comparison of common and high strength reinforcement

2 HIGH STRENGTH REINFORCEMENT (HSR)

High strength reinforcement 670W was initially developed in Europe (SAH, 2018) to be used for geotechnical applications such as micro piles, soil nails and ground anchors. The idea was to use a high strength steel bar to decrease the necessary bar diameter and further to reduce the borehole diameter in deep foundation elements. A smaller borehole increases the drilling speed and reduces costs. Mainly there are two ways to increase the strength of steel: micro alloying or water tempering. The advantage of micro alloyed steel is a high ultimate strength to yield-strength ratio. Water tempcorised steel is much cheaper and has a sufficient ductility to be used as reinforcement for concrete structures.

The high strength thread bars are hot rolled and tempcorised. The bar is continuously threadable because of the shape and the geometry of the ribs. The bar diameters were chosen such that common reinforcement bar diameters used in Europe can be easily substituted. High strength reinforcement is available in a range of diameter $Ø_s$ = 18 to 75 mm (see Fig. 2).

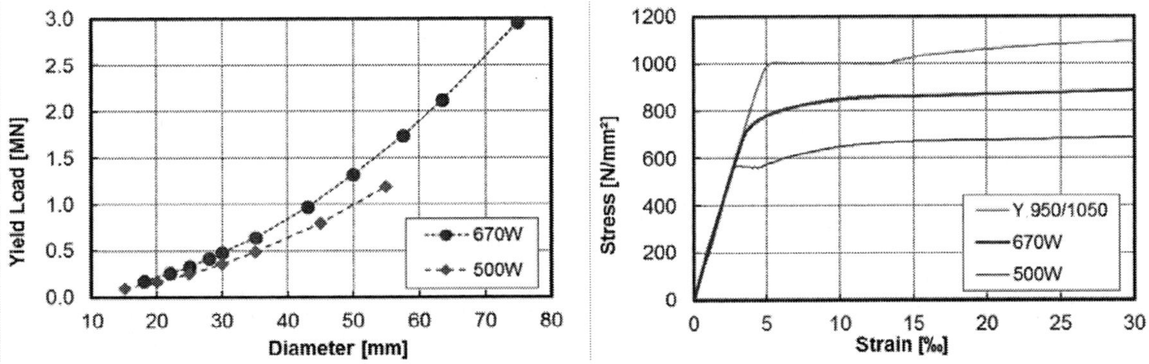

Fig. 2: Comparison 500W and 670W, Stress-strain-diagram for different steel grades

2.1 Material Parameters

HSR has similar mechanical properties as common reinforcement but provide additional advantages. The main mechanical properties are:

- nominal yield strength f_y = 670 N/mm²
- nominal ultimate strength f_u = 800 N/mm²
- high ductility (uniform elongation > 5%, elongation at fracture > 10%)
- no well defined yield point
- low relaxation
- low stress corrosion risk

HSR steel does not have a well defined yield point. For this reason, the yield strength is defined either as the 0.2 % proof-stress (see Fig. 2) or the 0.35 % offset method. The Young's modulus is slightly higher compared to common reinforcement (E = 200.000 N/mm²). The rib height is in comparison to common reinforcement bars is bigger, which leads to a higher relative rib area too. 670W was developed to be used with threaded accessories such as nuts and mechanical couplers in all available bar sizes.

2.2 Accessories

For HSR, there are several accessories and special parts available. These range from different nut types, mechanical couplers, transitional couplers for bar size reduction and end anchorages capable of developing the ultimate force of the bars (see (SAH, 2018) and Fig. 3).

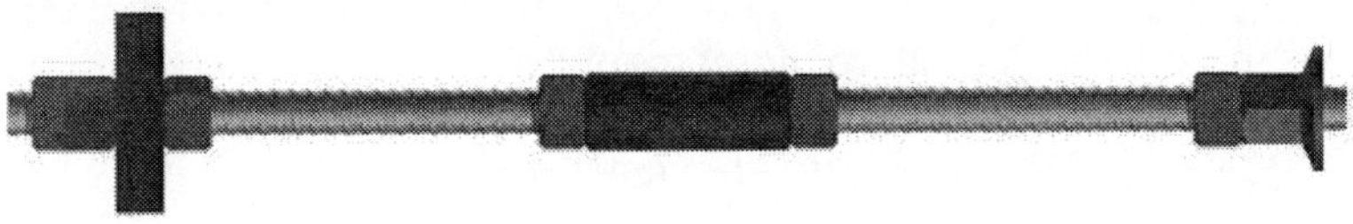

Fig. 3: Thread bar and accessories (plate end anchorage, coupler, end anchorage with anchor piece)

Lap splices can be avoided by using mechanical couplers. Development length can be reduced by using end anchorages. This helps to reduce reinforcement concentrations, discontinuity problems as well as bond loading in lap splice areas and avoid additional stirrups to strengthen the development length.

Using accessories is facultative though it is advisable as mentioned before. It helps to support the advantages of high strength reinforcement. Fig. 4 shows an example how reinforcement ratio or cross section can be reduced significantly. All three columns have the same load capacity.

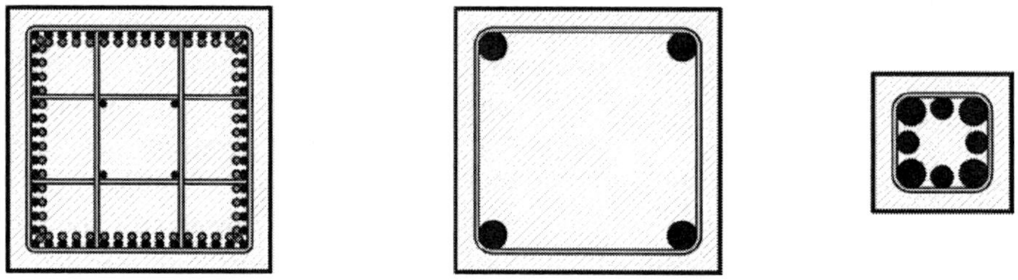

a) 500W, (ρ=8% at laps) b) 670W with the same bearing capacity and the same cross section area c) 670W with the same bearing capacity and ρ=20%

Fig. 4: Possible cross sections:

3 HSR AND CODE REGULATIONS

3.1 Eurocode

The range of yield strength in Eurocode 2 (EC 2, 2011) is between 400 N/mm² and 600 N/mm². High strength reinforcement exceeds this range by about 10 %. There is no upper limit for the bar diameter in the code. The rib geometry of continuously threaded bars is not covered by the EN 10080 but the relative rib area is larger than specified in the code.

In consideration of these divergences, some structural regulations need to be discussed. At first, due to the higher yield strength, it is not possible to keep the same limits of span to depth ratio to limit deflection. Higher stresses cause higher strains and higher strains cause higher deflections. Therefore, either stresses are limited or deflection needs to be calculated and if necessary constructional depth is increased. At second, the calculation of crack width and further the bond properties have to be considered. For a survey of differences in structural behaviour between 500W and 670W three beam tests with each grade were done. According to these tests, the higher relative rib area of 670W cause the same crack width but a lower crack spacing (Wechtitsch, 2006). A detailed research project confirmed these first results (Scheibe et al, 2013). Special points will be raised in the following paragraphs for the particular possible applications of HSR.

3.1.1 Columns, general

Serviceability limit state plays secondary role for columns. That is why columns are to be considered to be ideal for a first application for high strength reinforcement.

Following constraints for columns are given by (EC 2, 2011):

- Maximum reinforcement ratio: $\rho \leq 4\%$ ($\rho \leq 8\%$ at laps)
- Maximum concrete strain for concentric loading: $\varepsilon_c \leq -2.0‰$ (see Fig. 6a)

The aim for columns with high strength reinforcement is to provide a high bearing capacity and a maximum utilisation of the materials. For this reason, both constraints shall be opened.

3.1.2 Maximum reinforcement ratio

The restriction of reinforcement ratio is given by the possibility of casting. With small diameters it is not possible to place the concrete and provide a high quality for reinforcement ratios exceeding 8%. This is why (EC 2, 2011) limits the ratio. Columns armed with large diameters (up to 75 mm) can be casted even with reinforcement ratios up to 20% (see Fig. 4). For high strength columns the maximum ratio will therefore be set to ρ = 20% which is proven by experience (see also (Falkner et al, 2008)).

3.1.3 Maximum concrete strain

The maximum concrete strain for concentric loading (normal forces) is limited in (EC 2, 2011) with $\varepsilon_c \leq -2.0$ ‰. DIN 1045-1 which was the German equivalent code to (EC 2, 2011) allows $\varepsilon_c \leq -2.2$ due to effects concerning creep and shrinkage without calculation. This would cause a steel stress of σ_s = 2‰ · 200,000 N/mm² = 400 N/mm² which is much lower than the design yield strength of f_{yd} = 670 / 1.15 = 582 N/mm². For an economical design it is important to increase concrete strain up to about $\varepsilon_c = \varepsilon_{yd}$ = 582 / 205,000 = 2.84 ‰. This can be provided by calculating the effects due to creep and shrinkage (see Fig. 5, (EC 2, 2011) and (Falkner et al, 2008)).

Columns in high rise buildings are not loaded abruptly. They are loaded step by step. Months or years are needed to finish the building construction and for application of the full characteristic load. During this time, concrete redistributes the load to the steel (see Fig. 5).

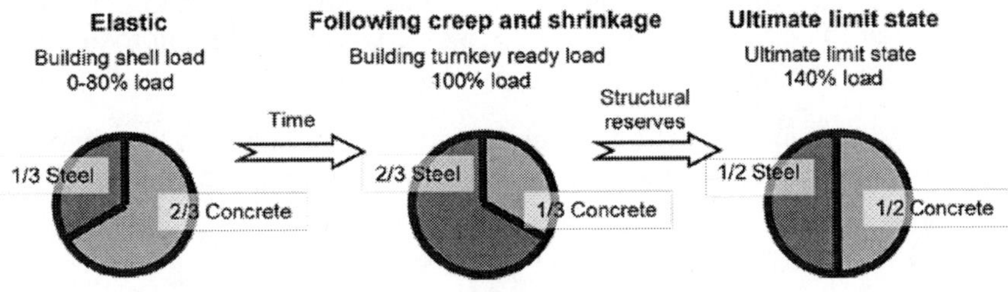

Fig. 5: Load redistribution of a column (Falkner et al, 2008)

Due to these effects concrete is unloaded and can be reloaded for ultimate limit state. The stress-strain-curve is modified and steel can be loaded to yield strength (see Fig. 6a). The strain difference $\Delta\varepsilon_{cs}$ includes effects due to load transfer from concrete to steel and unloading of the concrete.

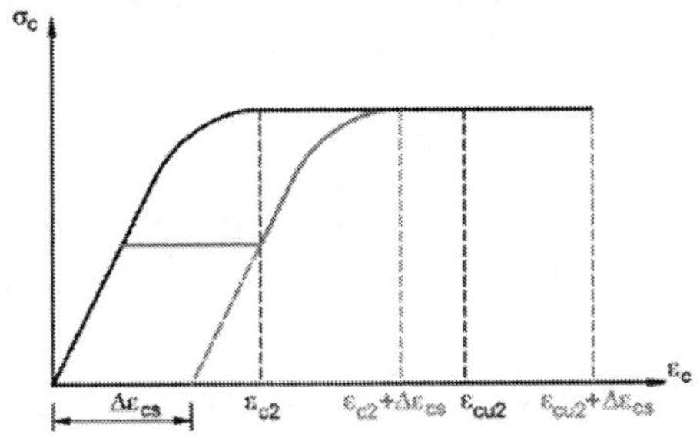

Fig. 6: a) Stress-strain relation for concrete with and without creep and shrinkage (Falkner et al, 2008)

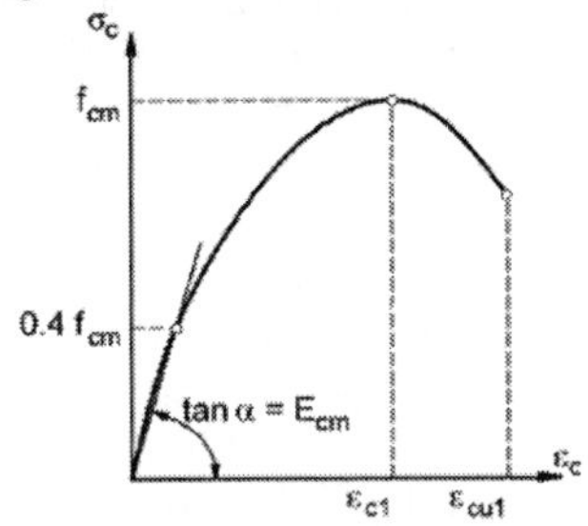

Fig. 6: b) nonlinear strength-stress relation for concrete according to (EC 2, 2011)

3.1.4 Further considerations

In addition to the consideration of creep and shrinkage, a nonlinear calculation of the bearing load of a cross section can be performed. The nonlinear stress-strain-relation is given in (EC 2, 2011) (see Fig. 6b).

It is obvious that the peak of the curve is at ε_{c1} (for example: C 30/37: ε_{c1} = 2.2‰, ε_{cu1} = 3.50‰). In Fig. 7b, the bearing capacity for different reinforcement ratios are shown. For low ratios ($\rho \approx 4$ %) the concrete carries more than 50%, for high ratios concrete carries only 10% to 20% of the ultimate load. Therefore, depending on the concrete strength, the bearing capacity will increase up to yield strain (ε_{yd} =2.84‰) of the steel even without any effects concerning creep and shrinkage (see Fig. 7a).

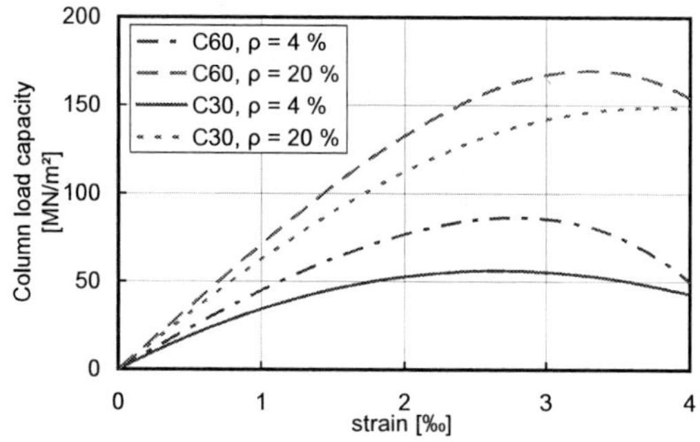

Fig. 7: a) Bearing capacity of 1m² column depending on the reinforcement ratio and concrete strength

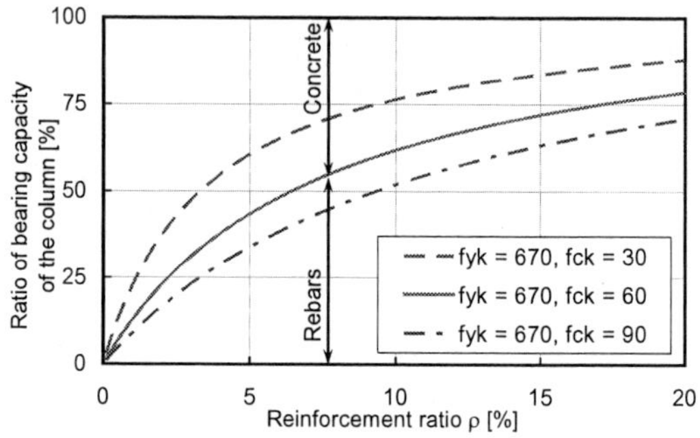

Fig. 6: b) ratio of bearing capacity of concrete and reinforcement of a column

This shows the advantage of using high strength reinforcement. For columns the absence of a well defined yield point does not provide any disadvantage. Only in the use of compression controlled flexural members a realistic stress strain curve has to be used for calculations as otherwise a brittle failure mode might occur.

3.2 ACI 318

ACI 318-14 (ACI 318, 2014) gives different restrictions depending on the field of application for the rebar (see Fig. 8). For longitudinal reinforcement, the maximum yield strength for design is f_y = 550 N/mm² (80 ksi). A further limitation is the maximum compression strain of concrete of ε_{cu} = 0.003. A main rule of design is the strain-compatibility of concrete and reinforcement. The yield strength of reinforcement can be found either by the 0.35% strain method or the 0.2% offset method. The theoretical maximum yield strength with the 0.35% strain method would be $f_{y,max.}$ = 0.0035 × 200.000 = 700 N/mm² (= 101 ksi). Using the strain limit of concrete, the maximal possible yield strength would be
$f_{y,max.}$ = 0.0030 × 200.000 = 600 N/mm² (= 87 ksi).

HSR is approved in the US by ICC-ES (ESR 1163, 2015) to be used for columns and shear walls with the full capacity. Different other approvals for similar steel grades (Grade 100 (ESR 3367, 2016)) would allow the use of fy = 690 N/mm² (= 100 ksi) for lateral support reinforcement. The limitation of f_y = 550 N/mm² (= 80 ksi) was driven by available reinforcement grades in the USA. ACI 318-14 (ACI 318, 2014) limits the concrete strain with 0.003. The commentary states that the actual test results "vary from 0.003 to higher than 0.008 under special conditions" (ACI 318, 2014).

Table 20.2.2.4a—Nonprestressed deformed reinforcement

Usage	Application	Maximum value of f_y or f_{yt} permitted for design calculations, psi	Applicable ASTM specification			
			Deformed bars	Deformed wires	Welded wire reinforcement	Welded deformed bar mats
Flexure; axial force; and shrinkage and temperature	Special seismic systems	60,000	Refer to 20.2.2.5	Not permitted	Not permitted	Not permitted
	Other	80,000	A615, A706, A955, A996	A1064, A1022	A1064, A1022	A184[1]
Lateral support of longitudinal bars; or concrete confinement	Special seismic systems	100,000	A615, A706, A955, A996, A1035	A1064, A1022	A1064[2], A1022[2]	Not permitted
	Spirals	100,000	A615, A706, A955, A996, A1035	A1064, A1022	Not permitted	Not permitted
	Other	80,000	A615, A706, A955, A996	A1064, A1022	A1064, A1022	Not permitted
Shear	Special seismic systems	60,000	A615, A706, A955, A996	A1064, A1022	A1064[2], A1022[2]	Not permitted
	Spirals	60,000	A615, A706, A955, A996	A1064, A1022	Not permitted	Not permitted
	Shear friction	60,000	A615, A706, A955, A996	A1064, A1022	A1064, A1022	Not permitted
	Stirrups, ties, hoops	60,000	A615, A706, A955, A996	A1064, A1022	A1064 and A1022 welded plain wire	Not permitted
		80,000	Not permitted	Not permitted	A1064 and A1022 welded deformed wire	Not permitted
Torsion	Longitudinal and transverse	60,000	A615, A706, A955, A996	A1064, A1022	A1064, A1022	Not permitted

[1] Welded deformed bar mats shall be permitted to be assembled using A615 or A706 deformed bars.

[2] ASTM A1064 and A1022 are not permitted in special seismic systems where the weld is required to resist stresses in response to confinement, lateral support of longitudinal bars, shear, or other actions.

Fig. 8: Maximum yield strength for reinforcement according to (ACI 318, 2014)

Also for the ACI based design, the same considerations as shown in chapter 3.1 for long term effects are valid. Again a higher strain than 0.003 can be achieved in calculating the stress-redistribution.

3.3 CSA A23.3

The regulations in CSA A23.3 are similar to ACI – especially the safety concept. But the deviation between CSA and ACI is, that CSA would allow a maximum compressive strain for concrete of ε_{cu} = 0.0035 (similar to Eurocode 2) and does define the limit for f_y to 400 N/mm² or the stress corresponding to a strain of 0.0035. The maximum yield strength of a bar could be $f_{y,max.}$ = 0.0035 × 200.000 = 700 N/mm² (= 101 ksi). This regulation is to some degree contrary to CSA A23.3 Clause 8.5.1. According to this clause, the yield strength shall not exceed 500 N/mm².

A general limitation of yield strength for reinforcement might be based on test data and empiric derived design formulas. In any event, for compression members neither deflection nor crack width limitation are important and hence there is no reason to limit the yield strength of the reinforcement lower than the stress corresponding to the crushing strain of concrete.

HSR further allows a less brittle failure mode for columns and give additional "hidden" safety as it can take additional load and allows a load redistribution close to failure from concrete to steel (see Hude, 2013; Mueller et al., 2012).

4 EXAMPLES OF USE

670W has been used successfully in multiple construction projects, changing the skyline landscape throughout major cities around the world. High rise buildings in Kuala Lumpur (Malaysia), Warsaw (Poland), Frankfurt (Germany), Miami and New York City (USA) among others were successfully built with HSR. In this paper only a few examples can be shown.

These systems are now present in multiple applications ranging from foundation to superstructure work. When combined with currently available methods of Building Information Management (BIM), using the threadability of HSR a modular system called "High Strength Reinforcement System" (HSRS®) was developed. This provides maximum efficiency in the construction process. The versatility and relative ease of installation of an HSRS® allows for the optimization of the reinforcement inside the section of structural concrete members. This provides advantages such as material savings, reduced labor times and improved quality of construction among others. 670W can also provide this advantages using stick building method as it will reduce the total steel quantity up to 40% compared to 400W.

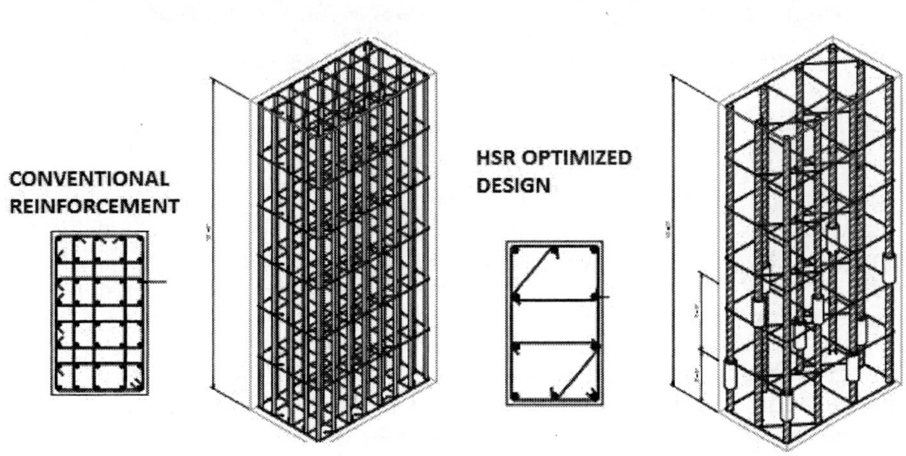

Fig. 9: Reinforcement conversion and layout optimization

4.1 Foundation Applications

Large bar diameters of high strength reinforcement can be a very cost effective option in foundation work. It can be used as the main reinforcing system or as an alternative to embedded steel members in concrete sections for deep foundations.

The use of threaded bars and mechanically coupled connections can eliminate complicated welded splices. Also, the bar threading has been proven to yield a high level of bond strength with the surrounding concrete. This application presents benefits in multiple aspects such as required labor, assembly times and inspection work for welding.

Very high reinforcement ratios can be accommodated and even two layers of reinforcement can be coupled (see Fig. 10).

Fig. 10: a) Multi-ring Caisson splice concept

Fig. 10: b) 55 Hudson Yards_ Caisson splicing

4.2 Super structures

670W is proving to be a cost efficient alternative on projects with a wide range of applications. From decongesting heavily reinforced sections, to allowing for fabrication off site on constricted jobsites using cages, the system aims to provide a feasible alternative for concrete construction. The combination of high strength concrete, HSR and innovative design tools has helped to bring this concept as an alternative to structural steel construction.

Fig. 11 shows the application of HSR in modules in New York City. One of this 63.5mm bars 670W (#20, Gr. 97) does replace 5 bars 35mm 400W (#11, Gr 60). It is obvious that common reinforcement would be impossible to be assembled in the same cross sectional area.

Fig. 11: a) HSR clusters and modular system

Fig. 11: b) Module pick up

A new high rise structure being built on 53 W53rd Street is a great example of the use of high strength systems. Commonly known as the MoMA Tower building, this structure is manly comprised of high strength concrete in combination with HSR. The complex geometry of the building and the challenge of reinforcing common framing joints identified as nodes, were best addressed by the use of HSR. Working together with the design team, a steel node concept using load bearing gusset plates, threaded bars and accessories was developed and implemented in critical joint connections. This approach reduced fabrication and erection times significantly. By using accessories designed for the HSR and threaded connections, the welding operations of the gusset plates within these joints were optimized and simplified benefiting the project significantly. (see Fig. 12)

In addition to the placement of reinforcement, coordination was also critical for the erection and installation of the big gusset plates at the core of the joints. Once all reinforcing steel entering the joints was modelled, the reinforcement layout dictated the shape of the plates and the size of the nodes.

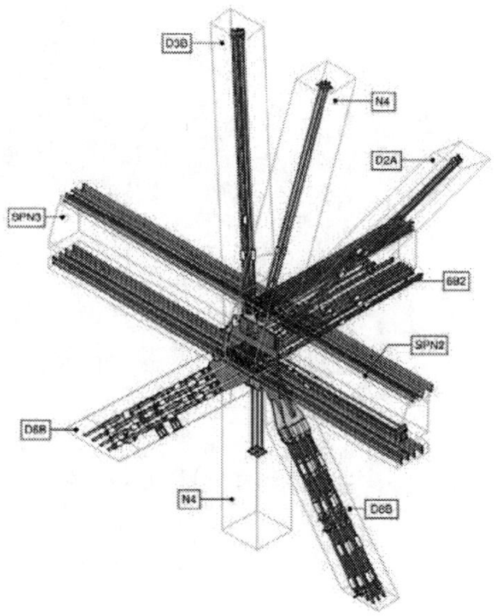

Fig. 12: a) Joint details at Node location

Fig. 12: b) Node erection and installation

5 CONCLUSION

Thread bars 670W have been used in geotechnical applications for over 10 years. Their versatility and improved mechanical properties led the way for use of the bars as concrete reinforcement in several other applications. High strength reinforcement fulfils all requirements to be used as reinforcement (high ductility, good bond behavior, low relaxation and low stress corrosion risk) and surpass the strength parameters of common reinforcement. Bond strength of high strength reinforcement is significantly higher compared to common reinforcement in normal strength concrete and it is slightly higher with UHPC.

The main advantage of using high strength reinforcement is high ductility and a remaining post failure load capacity.

In recent years, Architects and owners have increased their requirements for new buildings. These requirements concern structural safety, visually attractive design, high sustainability and a small ecological foot print. To fulfil all these requirements, it is important to use high performance products along with innovative methods aiming to optimize building members. Benefits related to production process, cross section decongestion and durability are achieved by the use of HSRS®.

As demonstrated by recently completed high rise structures and currently ongoing projects, the use of High strength reinforcement in combination with UHPC will help optimize the performance and construction of heavily reinforced concrete sections in new structures

References

Russwurm, D.: "Betonstaehle fuer den Stahlbetonbau", Institut fuer Stahlbetonbewehrung e. V., *Vieweg+Teubner Verlag*, Muenchen, 1993

DIN 488: "Reinforcing steels", *Deutsches Institut für Normung*, Berlin, 2010

ACI 318-14: "Building code Requirements for Structural Concrete", *American Concrete Institute*, Farmington Hills, 2014

EN 1992-1-1 (EC 2): "Design of concrete structures – Part 1-1: General rules and rules for buildings", *European committee for standardization*, Brussels, 2011

SAH: Homepage of Stahlwerk Annahuette: *http://www.annahuette.com*, reviewed on March 1st, 2018

Reichel M.: "Thin-walled precast segmental construction method in bridge construction made from Ultra High Performance Fiber Reinforced Concrete (UHPFRC): bearing behaviour, design and detailing", Doctoral Thesis, *Technische Universitaet Graz*, 2010

Jungwirth, J.: "Zum Tragverhalten von zugbeanspruchten Bauteilen aus Ultra-Hochleistungs-Faserbeton", Doctoral Thesis, *École polytechnique fédérale de Lausanne*, 2006

Wechtitsch, Michael: "GEWI-Steel SAS 670 as concrete reinforcement steel - Experimental and numerical investigations on the load carrying and deformation behaviour under bending", Diploma Thesis, *Technische Universitaet Graz*, 2006

Falkner, H. et al.: „Das neue Bewehrungssystem; Druckglieder mit hochfestem Betonstahl SAS 670/800", Teil I und II, *Beton- und Stahlbetonbau, Ernst & Sohn, Berlin*, May and August 2008

Scheibe, M., et al.: „Hochfeste Bewehrung im Stahlbetonbau - Identifizierung von Anwendungsmöglichkeiten, Entwicklung von Bemessungsgrundlagen und Ermittlung des wirtschaftlichen Einsparpotentials", BBSR-, Forschungsinitiative Zukunft Bau, Bonn (Foerderer); Stuttgart, *Fraunhofer IRB Verlag,* 2013

ICC-ES: ESR 1163 – Evaluation Subject: SAS Stressteel Grade 97 thread bar steel reinforcing bars and couplers, Reissued: 12/2015, *www.icc-es.org*

ICC-ES: ESR 3367 – Evaluation Subject: Dywidag grade 100 Threadbars® and couplers, Issued: 06/2016, *www.icc-es.org*

ICC-ES: ESR 2107 – Evaluation Subject: ChromeX® 9100 Grade 100 steel reinforcing bars, Reissued: 01/2016, *www.icc-es.org*

Hude, F.: High strength reinforcement – a solution against latent instability due to stress redistribution, conference proceeding, *CONSEC 13,* Nanjing 2013

Mueller, C., et al.: "Ultra-high Performance Spun Concrete Columns with High-strength Reinforcement", conference proceeding, 3rd International Symposium on Ultra-High Performance Concrete, Kassel, 2012

Building Tomorrow's Society
Bâtir la Société de Demain

Fredericton, Canada
June 13 to 16, 2018 / *13 au 16 juin, 2018*

INITIAL INVESTIGATION OF THE UTILIZATION OF STEEL CHIPS AS A PARTIAL REPLACEMENT OF COARSE AGGREGATES IN CONCRETE

Abdel-Raheem, Mohamed[1,2], Allam, Mohamed[1], Rodriguez, Arlynn [1], Beas, Moises[1], Gonzalez, Juan[1], and Tobias, Cesar[1]
[1] Department of Civil Engineering, University of Texas Rio Grande Valley, USA
[2] mohamed.abdelraheem@utrgv.edu

Abstract: The investigation of recycling different industrial wastes through its utilization in concrete has been lately on the rise. This is due to the fact that some of these wastes have proven to be useful in enhancing the properties of concrete. Previous studies have shown that the utilization of steel chips can be useful when used as a partial replacement of fine aggregates in concrete. This research study focuses on the utilization of steel chips as partial replacement of coarse aggregates. The study investigates the effect of steel chips on the compressive strength of concrete when replacing the coarse aggregates at different percentages. The results show that the compressive strength of concrete can significantly increase with the utilization of steel chips up to 25% replacement of the volume of coarse aggregates in the concrete mix.

1 INTRODUCTION

The continuous availability of Portland cement concrete is very critical to the construction industry. This has led to a dramatic increase in the production of Portland cement (Hanley, 2004). Concrete industry is one of the most demanding and most expanding industries because of urban expansion, rapid population growth, industrialization and the need to improve the standards of living. Unfortunately, Concrete production is consuming copious amounts of raw materials and requires expensive costs in mining for natural resources which lead to increase of greenhouse gas emissions, which impacts the environment negatively. As such, recycling and sustainability have become very critical aspects for the progress of the construction industry. This due to the limitations of natural resources, and the legislation of laws and regulations to protect the environment. Recycling industrial wastes by using them in concrete has several advantages. First, it minimizes the use or the need to process natural resources, such as sand and gravel, which comes with additional cost and energy consumption. Second, recycled industrial waste can be used as a substitute for portions of cement, which minimizes the construction cost and the greenhouse emissions associated with cement production. Incorporating the industrial waste in concrete serves as a legitimate venue of waste disposal provided that they do not possess any side effects on the health of the structure of the inhabitants. Finally, some of the industrial wastes investigated have been proven to enhance some of the properties of concrete. One of the promising industrial wastes that can be of great benefit to the construction industry is steel chips.

Steel chips are by-product of the mechanical cutting of steel cross sections. They vary in sizes and thickness. However, they can be very close to the sizes of the coarse aggregates. They are characterized by having cylindrical shape with smooth surfaces, as shown in Figure 1. Scale and steel chips had already been used multiple times for replacing sand as the fine aggregate in concrete, but never used to replace the coarse aggregate in concrete. Previous studies showed that steel chips and scale could be effective fine aggregate replacement as they enhance the compressive strength of the concrete, and they also could be used in shielding Gamma rays, which makes them very useful in manufacturing radiation shielding concrete mixtures (Alwaeli and Nadziakiewicz 2012).

There are limited ways to recycle steel chips, either by disposal in landfill or by melting. According to a study by Torkar el al. (2010), the melting of steel chips in Arc furnaces could be expensive, non-ecological and not effective. However, the study introduced a new complicated method for recycling steel chips using induction melting furnaces, but also it is very costly.

Figure 1: Steel chips used in this research as partial replacement of coarse aggregates

This research aims to use the steel chips as a raw waste material in concrete as a partial replacement of coarse aggregate. The objective is to determine the optimal replacements percentage that can be used without compromising the compressive strength of the concrete mix.

2 PROBLEM STATEMENT

Steel chips is a by-product that result from steel cutting and manufacturing. The majority of steel manufacturers tend to recycle only a few percentages of these wastes due to the prohibitive cost of re-melting and curing, and also due to the high emission of carbon dioxide gases that results from steel melting furnaces. In stead, these steel chips are dumped in landfills, which in turn affects the environment. A limited number of studies have discussed alternative approaches to reuse steel chips instead dumping into landfills; and even fewer studies discussed its reuse in concrete.

3 OBJECTIVE

This research focuses primarily on reusing steel chips as a partial replacement for the coarse aggregate in the concrete mix. The study focuses on the compressive strength of concrete as a performance measure of the approach. The study aims to assess the performance of the concrete at different percentages replacement of coarse aggregates with steel chips.

4 LITERATURE REVIEW

There have been very limited studies on incorporating steel chips as replacements for fine and coarse aggregate. The literature contained only two studies on the use of steel chips in concrete. The first study was conducted by Debieb and Kenai (2008) when they replaced fine aggregate with steel chips. The results show that the compressive strength of concrete decreased for the mixes that contained steel chips as a replacement of fine aggregates. Alternatively, Alwaeli and Nadziakiewicz (2012) replaced portions of the fine aggregate in concrete with scale and steel chips. In this study, sand was replaced in increments of 25%, 50%, 75%, 100%. The samples were used for testing the compressive strength and gamma radiation shielding. The results showed that the compressive strength of concrete containing steel chips is better than ordinary concrete of up to 25% (Alwaeli & Nadziakiewicz, 2012). At the same time, addition of scale and steel chips enhances the absorption of gamma radiation (Alwaeli & Nadziakiewicz, 2012).

On the other hand, several other studies focused on incorporating other waste materials as a replacement for coarse and fine aggregates in the concrete mix. One study investigated the effect of replacing the coarse aggregate with oil palm shells on the concrete compressive strength. The results showed that the compressive strength was about 41%-51% lower than the regular concrete but was still within the acceptable range for structural lightweight concrete (Basri et al. 1999). A study by Siddique (2003) investigated the effect of replacing the concrete fine aggregates with fly ash-class C. The results show that there is an increase in the compressive strength by increasing the percentage replacement of fly ash-class C in the mix. Aggarwal et al. (2007) studied the effect of the use of bottom ash and waste foundry sand in equal quantities as partial replacement of fine aggregates. The results showed an increase in the compressive, splitting tensile and flexural strength. Batayneh at al. (2007) used ground plastics and crushed glass as a substitute for sand. Crushed glass showed improvement of the compressive strength of concrete up to 20% replacement; however, the plastic was not as effective.

Ismail and Al-Hashmi (2008) conducted a study on the replacement of fine aggregate in concrete with waste iron and the results showed improvement in the compressive strength. In 2009, Ismail and Al-Hashmi replaced fine aggregates in concrete for waste glass. The results demonstrated the tendency of the slump to decrease as the waste glass ratio increased. The compressive strength increased by 4.23% when replacing 20% of fine aggregates with waste glass (Ismail and Al-Hashmi 2009). One study investigated the effects of ceramic industry waste on concrete. The end result for this coarse aggregate replacement showed good workability, and an increase in the compressive and flexural strength of the concrete, as well as the modulus of elasticity (Gobinath et al, 2011). Another study used recycled concrete aggregates as a coarse replacement. The results showed a decline in the quality of the concrete in terms of compressive, flexural strength and splitting tensile strength compared to the control concrete mix. In a different study, waste tire rubber was used in concrete as a coarse aggregate replacement. Although, the rubber chips had negative impacts on the compressive strength of concrete, the study found that the rubber chips' size had a direct impact on the compressive strength. Reducing the size of the rubber chips should result in an increase in its strength. (Huang et al, 2004). As such, many waste materials can ameliorate the properties and durability of concrete when used as replacements of fine and coarse aggregates.

5 METHODOLOGY

A plain concrete mix was used to assess the impact of steel chips on the compressive strength when used as a partial replacement of coarse aggregates. Purposely, the mix did not contain any admixtures or supplementary cementitious materials to exclude the effect of any possible interaction between the steel chips and the additives. Steel chips were used to replace coarse aggregates in concrete by volume at different percentages of 5%, 10%, 15%, 25%, 35% and 50%. Cylinders of 4 x 8 inches were made to test the compressive strength of concrete at 7 and 28 days according to the ASTM standards.

6 EXPERIMENTATION

A simple plain concrete mix with a target compressive strength of 25 MPa was used for comparative analysis and evaluation. The quantities needed for 1m^3 of concrete is given in Table 1. The water/cement ration was used as 0.45 to allow for good workability and easy compaction of the concrete specimens to attain high quality. Steel chips replaced the coarse aggregates at six different percentages (5%, 10%, 15%, 25%, 35% and 50%) by volume, as shown in Table 2. The unit weight of the steel chips used is 4559.81 kg/m^3, which is almost 3 times the unit weight of gravel (1641.08 kg/m^3). The compressive strength was tested at 7 and 28 days, and the results are shown in Figures 1 and 2 respectively.

Table 1: Concrete mix design used

Cement (Kg)	Sand (Kg)	Gravel (Kg)	W/C	Water (liters)
400.00	683.09	1394.25	0.45	180.00

Table 2: Concrete mix proportions for steel chips replacements at different percentages

Cylinder: 4" x 8"	Steel Chips/Gravel Replacement						
	Control	STC2a	STC3a	STC4a	STC5a	STC6a	STC7a
% Replacement	0%	5%	10%	15%	25%	35%	50%
Cement (kg)	4.61	4.61	4.61	4.61	4.61	4.61	4.61
Sand (kg)	7.78	7.78	7.78	7.78	7.78	7.78	7.78
Gravel (kg)	22.15	21.04	19.94	18.83	16.61	14.40	11.08
Water (kg)	2.10	2.10	2.10	2.10	2.10	2.10	2.10
Steel Chips (kg)	0.00	3.31	6.62	9.93	16.55	23.17	33.09

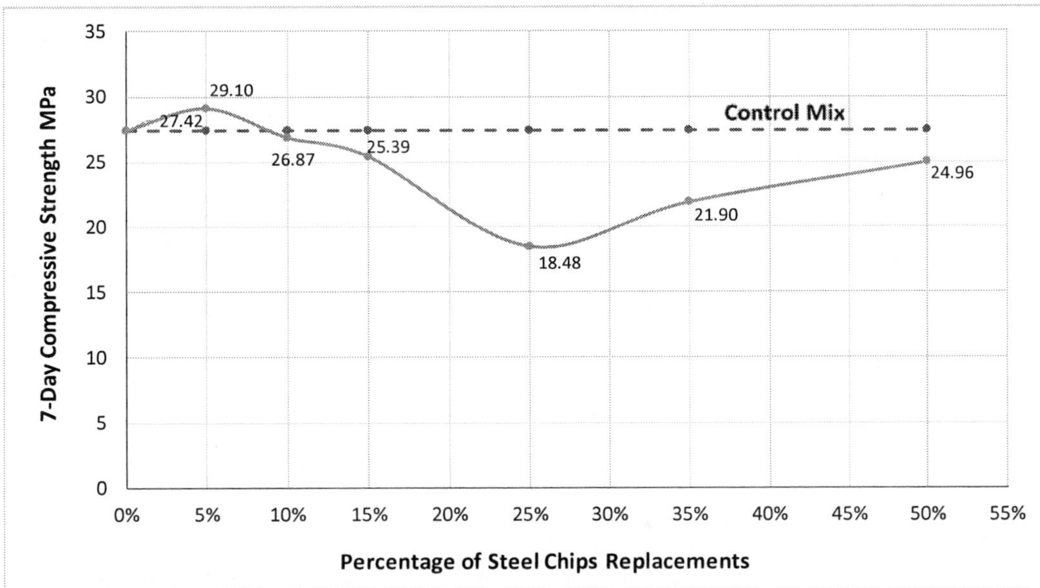

Figure 1: Compressive strength of concrete specimens containing different percentages of steel chips at 7 days

7 FINDINGS AND DISCUSSION

The results show steel chips will not affect the compressive strength of concrete if used as a partial replacement of gravel of up to 15%. However, the compressive strength of concrete can be increased by 18% by replacing 5% of the coarse aggregates with steel chips. These results along with the previous results obtained by Alwaeli and Nadziakiewicz (2012) proves the effectiveness of steel chips in enhancing the concrete compressive strength when used in the mix as a partial replacement of aggregates. However, the results obtained in this study should be further verified as the behavior of the compressive strength of steel-chips concrete does not follow a regular pattern. By investigating the compressive strength curve in Figure 2, it can be noticed that there is a drop in the compressive strength at 15% replacement of coarse aggregates, but it increases again when the replacements of coarse aggregates reach 35% and up. As such, it is recommended to only utilize steel chips as replacement of coarse aggregates in the concrete mix up to 15% until further investigation.

The water/cement ratio (0.45) used in this study was quite high to maintain goof workability; however, it affected the homogeneity of the mix. Since the unit weight of steel chips is almost 3 times the unit weight of gravel, it is expected that the steel chips will sink to the bottom of the concrete members in mixes with high water content. As such, it is recommended to repeat this study using different w/c ratios and investigate the changes in the compressive strength for concrete mixes containing steel chips as a replacement of coarse aggregates.

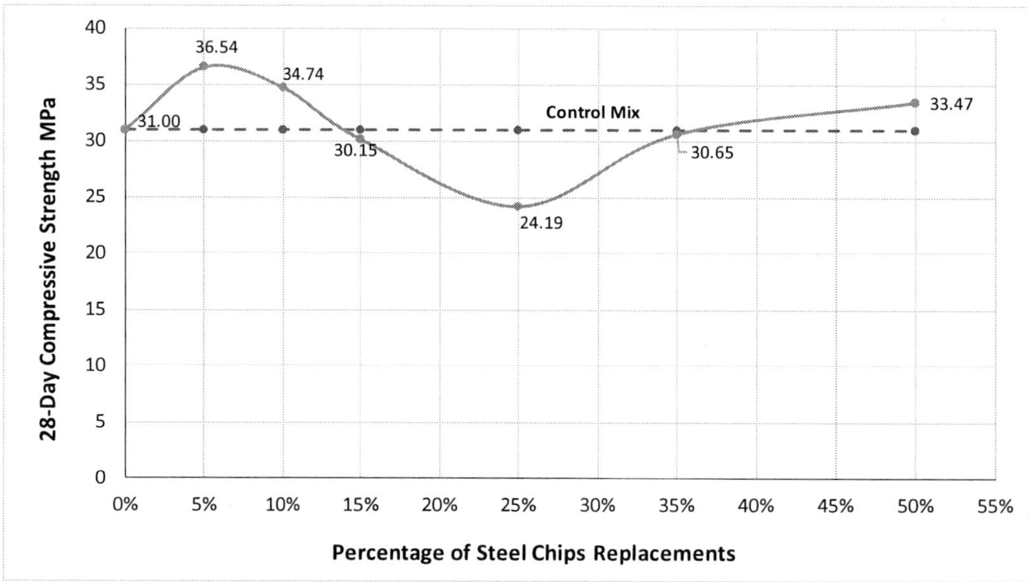

Figure 2: Compressive strength of concrete specimens containing different percentages of steel chips at 28 days

Although steel chips have proven to perform well at convenient percentages, there had been some considerations about the actual usage of the steel chips in the concrete mix that should be mentioned for the sake of clarity and integrity of this research.

- The concrete mixtures with steel chips had significant higher weights than the normal concrete, which could increase the dead loads in the construction members. This could cause limitation for its usage in the construction field. However; steel chips concrete can be efficient if used as heavy weight concrete.

- As a result of its heavy weight, the transportation of the steel chips concrete could be limited and expensive; special trucks have to be designed to carry such type, or if normal concrete truck mixtures are used, then it a smaller amount of concrete per truck can be transported.

- Theoretically, steel chips concrete could be used best as a pavement concrete or sidewalks due to its high durability and high density.

8 CONCLUSION

This research is a promising step forward for sustainability and recycling materials, which will save the environment from depleting its natural resources. Using a raw by-product in concrete without processing is very rare and limited; however, this research showed that steel chips could be a perfect by-product that replaces the coarse gravel in the concrete mix and behaves similarly and even improves the compressive strength of up to 15% if used in small percentages.

REFERENCES

Aggarwal, P., Aggarwal, Y., & Gupta, S. M. (2007). Effect of bottom ash as replacement of fine aggregates in concrete. Asian journal of civil engineering (building and housing), 8(1), 49-62.

Alwaeli, M., & Nadziakiewicz, J. (2012). Recycling of scale and steel chips waste as a partial replacement of sand in concrete. Construction and Building Materials, 28(1), 157-163.

Basri, H. B., Mannan, M. A., & Zain, M. F. M. (1999). Concrete using waste oil palm shells as aggregate. Cement and concrete Research, 29(4), 619-622.

Batayneh, M., Marie, I., & Asi, I. (2006). Use of selected waste materials in concrete mixes. Waste management, 27(12), 1870-1876.

Debieb, F., & Kenai, S. (2008). The use of coarse and fine crushed bricks as aggregate in concrete. Construction and building materials, 22(5), 886-893.

Hebhoub, H., Aoun, H., Belachia, M., Houari, H., & Ghorbel, E. (2011). Use of waste marble aggregates in concrete. Construction and Building Materials, 25(3), 1167-1171.

Huang, B., Li, G., Pang, S. S., & Eggers, J. (2004). Investigation into waste tire rubber-filled concrete. Journal of Materials in Civil Engineering, 16(3), 187-194.

Ismail, Z. Z., & Al-Hashmi, E. A. (2009). Recycling of waste glass as a partial replacement for fine aggregate in concrete. Waste management, 29(2), 655-659.

Ismail, Z. Z., & AL-Hashmi, E. A. (2008). Reuse of waste iron as a partial replacement of sand in concrete. Waste Management, 28(11), 2048-2053.

Senthamarai, R. M., Manoharan, P. D., & Gobinath, D. (2011). Concrete made from ceramic industry waste: durability properties. Construction and Building Materials, 25(5), 2413-2419.

Siddique, R. (2003). Effect of fine aggregate replacement with Class F fly ash on the mechanical properties of concrete. Cement and Concrete research, 33(4), 539-547.

Sri Ravindrarajah, R., & Tam, C. T. (1985). Properties of concrete made with crushed concrete as coarse aggregate. Magazine of concrete research, 37(130), 29-38.

Toutanji, H. A. (1996). The use of rubber tire particles in concrete to replace mineral aggregates. Cement and Concrete Composites, 18(2), 135-139.

Torkar, M., Lamut, M., & Millaku, A. (2010). Recycling of steel chips. Materiali In Tehnologije, 44(5), 289-292.

Building Tomorrow's Society

Bâtir la Société de Demain

Fredericton, Canada

June 13 to 16, 2018 / *13 au 16 juin, 2018*

INVESTIGATION OF TREATED CRCA IN HMA MIXTURES THROUGH EVALUATING LOW TEMPERATURE CRACKING

Al-Bayati, Hanaa Kh[1,3], Tighe, Susan[2]
1 PhD Candidate, Centre for Pavement and Transportation Technology (CPATT), Department of Civil and Environmental Engineering, Faculty of Engineering, University of Waterloo, Waterloo, Canada
2 Canadian Society for Civil Engineering, Norman W. McLeod Professor in Sustainable Pavement Engineering, Director of Centre for Pavement and Transportation Technology (CPATT), Department of Civil and Environmental Engineering, Faculty of Engineering, University of Waterloo, Canada
3 halwan@uwaterloo.ca

Abstract: One of the most important concerns for asphalt pavement in the cold regions of the world is the prospect of thermal cracks or low-temperature cracks. This is mainly attributed to the development of stresses, which exceed the pavement strength, leading to pavement cracking at extremely low temperatures. To evaluate the performance at low temperatures, the thermal stress restrained specimen test (TSRST) is performed to assess the influence of (CRCA) addition into asphalt mixtures based on Ontario Superpave specifications. The effect of different CRCA types with various proportions on the fracture stress and fracture temperature of asphalt mixtures is also examined. To achieve the objective, two different types of CRCA with various percentages (0%, 15%, and 30%) were chosen. The impact of treated CRCA on the behaviour of asphalt mixtures at low temperatures is further evaluated by applying combinations of various treatment methods. Heat treatment at 300°C followed by a short mechanical treatment is applied to enhance the physical and mechanical properties of CRCA. The application of the combination technique of heat treatment at 300 °C and short mechanical treatment method is very successful in improving various RCA properties. The addition of both CRCA types into the asphalt mixtures at various proportions is successful in achieving all MTO requirements for volumetric properties. While the application of the combination of various treatments leads to a significant reduction in the average fracture temperature, there is no impact for this technique on the fracture stress of the asphalt mixtures.

1 INTRODUCTION

Asphalt concrete is the basic material utilized in road construction worldwide (TAC 2013). Generally, it consists of approximately 95% aggregate and 5% asphalt material. According to the ministry of natural resources Ontario (MNRO), the average consumption of aggregate reached approximately 179 million tonnes per year in Ontario during the period 2000-2009, while this average is projected to amount to approximately 191 million tonnes between 2020 and 2029 (MNR 2010). Simultaneously, waste materials are increasing with the growth in the population due to rising demand for new highways, commercial buildings, housing developments, and infrastructure projects. This results in tremendous amounts of waste ending up in landfills every year. Due to considerable consumption and economic growth, the required natural resources will not only be depleted but will eventually be exhausted (Bolden 2013).

To reduce the exhaustion of natural materials, reduce waste materials in landfills and minimize environmental concerns, the application of waste materials in asphalt mixtures is becoming increasingly utilized worldwide. Among different waste materials, recycled concrete aggregate (RCA) is the main waste material that has been successfully applied in base aggregates and Portland cement concrete (PCC) aggregates. However, there were limited investigations into the utilization of RCA in the asphalt pavement mixtures (Mills-Beale & You 2010) due to the poor quality of RCA characteristics compared to natural aggregate (NA). Such properties include high water absorption, low density, and high porosity that result from the existence of adhered mortar on the RCA surface (Zhang et al. 2016). Recently, the use of RCA in the asphalt mixtures has become an attractive topic for many researchers worldwide (Shen & Du 2004, Paranavithana & Mohajerani 2006, Wong et al. 2007, Du & Shen 2007, Pérez et al. 2009, Wu et al. 2013, Pasandín & Pérez 2014, Radević et al. 2017). The successful use of RCA in the base and sub-base applications is the main reason behind examining the feasibility of using RCA in asphalt mixtures. It is believed that the problem correlated with RCA can be eliminated through the coating of RCA particles with a layer of asphalt cement (Zhang et al. 2016).

2 LITERATURE REVIEW

One of the most widespread concerns of asphalt pavements in the cold regions of the world is thermal cracks or low temperature cracks (Velasquez et al. 2008, Tan et al. 2012, Akentuna et al. 2016). At low temperature, the upper part of the pavement section is contracted, whereas the granular base holds the pavement's bottom section in place and prevents it from contraction ((http://dotapp7.dot.pdf). Then, when the developed stresses exceed the strength of the pavement (Tim 1995, Akentuna et al. 2016, Liu et al. 2017), the low-temperature cracks occur. Low-temperature cracks are the main cause of pavement degradation, increasing pavement roughness and reducing service life in the northern climates (http://dotapp7.dot.pdf). As previously mentioned, various studies have been conducted to investigate the use of RCA in hot mix asphalt (HMA) mixtures. Pérez et al. (2009) evaluated the resistance of asphalt mixtures that included RCA with NA aggregate with respect to fatigue cracking. Two different mixtures, a semi-dense mixture and a coarse mixture were used, and one type of RCA was used to prepare various HMA mixtures. Fatigue tests and dynamic modulus tests were performed to evaluate the asphalt mixtures. The results revealed that the fatigue behaviour for the coarse mixture with 50% RCA is similar to the behaviour of the same mixture without RCA. Additionally, the semi-dense mixture with 50% RCA had lower fatigue behaviour compared to the same mixture without RCA. The findings also indicated that the mixture that included RCA had a higher dynamic modulus than mixtures without RCA. The study concluded that the addition of RCA leads to an increase in the stiffness of a HMA mixture even if a large amount of bitumen is used. Zhu et al. (2012) concluded that the addition of CRCA without treatment causes poor moisture resistance and low temperature flexibility. The addition of treated CRCA, using pre-coating method by using liquid silicone resin, works to improve these properties. The addition of treated CRCA improves strength, absorption and adhesion with asphalt while it has a negative effect on permanent deformation at high temperature. However, mixture properties at high temperatures are still acceptable.

From the literature study, it can be stated that most of the previous research was mainly focused on the effect of untreated RCA on the mechanical properties of HMA mixtures such as stability, stiffness, moisture resistance, and dynamic modulus. Though a few studies were conducted for evaluating the thermal cracks in the HMA mixes, none of the above-mentioned investigations considered the effect of treated RCA on the thermal cracking resistance of asphalt mixes. In this study, experiments were set up to investigate the thermal crack susceptibility of the asphalt mixture samples containing CRCA using a thermal stress restrained specimen test (TSRST). The main objectives of this investigation are carried out in two phases. In the first phase, the effects of different untreated CRCA types with various proportions on the fracture stress and the fracture temperature of the typical Ontario HMA mixtures are investigated. Heat treatment at 300°C followed by a short mechanical treatment were performed to enhance the physical and mechanical properties of CRCA, then the influence of treated CRCA on the fractural temperature and the fractural stress of asphalt mixtures is examined in the second phase. A complementary objective for the second phase is that the impact of the combination of various treatment methods on the main RCA properties is further evaluated.

3 MATERIALS AND METHODS

3.1 Material

NA and one filler type that is commonly utilized for preparing asphalt mixtures; namely, dust plant, were obtained from the Miller Group and one type of asphalt binder, namely, PG 64-28 was used. In this research, two different RCA types were utilized, RCA#1 was provided from a ready-mix concrete plant through the crushing process of concrete that has unsatisfactory properties, performance, and age. Hence, RCA#1 can be categorized as fresh concrete that has not been used in civil engineering works. The second type, RCA#2, is classified as a granular A according to the Ontario provincial standard specifications (OPSS.MUNI 1010). RCA#2 was produced by Steed and Evans Limited in St. Jacobs, Ontario. In this study, CRCA is defined as the sieve fraction retained between 4.75 and 19 mm. The optical images of NA, RCA#1 & RCA#2 are shown in Figure 1-a, b, and c, respectively.

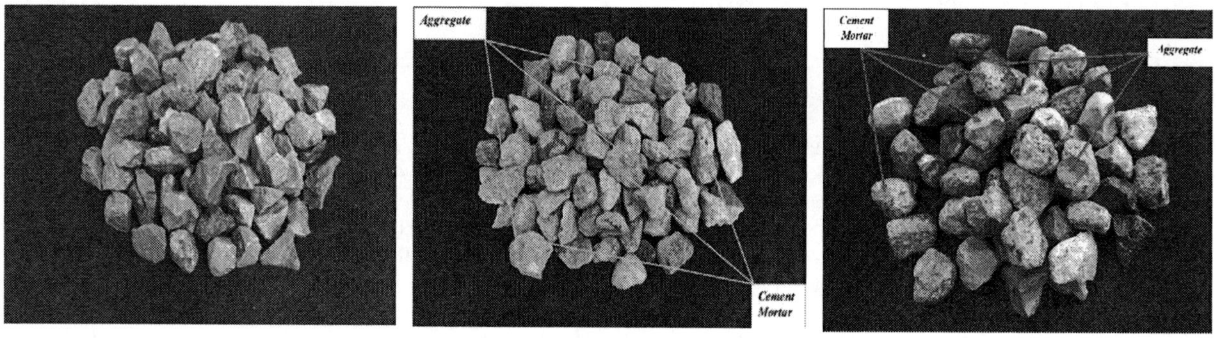

1-a. NA 1-b. RCA#1 1-c. RCA#2

Figure 1: Optical images of NA & RCA types

3.2 Methods

3.2.1 Preparation of treated and untreated CRCA

All CRCA types were washed thoroughly so that all noticeable impurities such as wood chips and others were removed. Then, the CRCA was dried in an oven at 105±5°C for 24 hr before undergoing various types of treatments and tests. The untreated CRCA tests were performed at room temperature (20°C), whereas the CRCA was heated at 300 °C for 1 hr to achieve the heat treatment. For the short mechanical treatment, the CRCA samples were placed in a Micro-Deval device for 15 minutes under the effect of metal balls. Then the samples were sieved with a 4.75 mm sieve to ensure that only the coarse aggregate was retained.

3.2.2 Superpave mix design

The Superpave mix design was performed according to AASHTO R 30-02 (2006). The design is equivalent to a single-axle load ranging between 10 and 30 million. Superpave mixture design procedure was carried out based on the nominal maximum aggregate size (NMAS), 19 mm. For the included untreated CRCA#1 and CRCA#2 mixtures, three proportions (0%, 15%, and 30%) and one percentage (30%), respectively, were added as a partial substitute for coarse NA. For the included treated CRCA, one portion of CRCA#1, (30%) was used. The gradation of virgin aggregate and various percentages of CRCA, targeted the mix design of Miller group, and ministry transportation Ontario (MTO) specifications are numerically tabulated in Table 1.

Table 1. Gradations with various CRCA proportions, targeted mix design, and MTO Specifications

Sieve Size	Passing (%) for Different CRCA Percentages				Target of	MTO
mm	0.0	15%CRCA#1	30%CRCA#1	30%CRCA#2	Mix Design	Limitation
25	100	100	100.0	100	100	100

Sieve Size	Passing (%) for Different CRCA Percentages				Target of	MTO
mm	0.0	15%CRCA#1	30%CRCA#1	30%CRCA#2	Mix Design	Limitation
19	95.2	95.2	95.3	96.6	96.8	100 - 90
16	89.0	88.7	88.5	86.6	90.6	90 - 23
12.5	81.8	81.1	80.5	73.9	83	
9.5	73.2	72.3	71.8	67.8	73.3	
6.7	63.3	63.0	63.1	61.5	63.3	
4.75	57.1	56.7	55.9	56.0	55.9	
2.36	42.8	42.8	41.3	41.8	43.5	49 - 23
1.18	30.7	30.7	30.5	31.2	32.5	
0.6	22.9	23.0	23.6	24.3	25.1	
0.3	10.2	10.3	10.3	10.3	11.8	
0.15	5.4	5.5	5.6	5.5	5.5	
0.075	2.1	2.2	2.2	2.1	3.8	8 - 2

3.2.3 Preparation of TSRST Specimens

Based on AASHTO R 30-02 (2006), the loose HMA mixtures were exposed to a short-term condition at 135°C for a period of four hrs before compaction to simulate the plant mixing and placement effects. An asphalt shearbox compactor (PReSBOX) was used to compact the beam specimens that were approximately 390mm x150mm x130mm in length, height, and width, respectively. Then, the beams were saw-cut into TSRST specimens with approximate dimensions 250mm× 50mm × 50mm at air voids of 7±1% that represent the typical values for compaction (AASHTO TP 10-93 1993, NCHRP 2007). The TSRST test was performed in accordance with AASHTO TP 10-93 (1993). The test specimens were conditioned at 5°C in an environmental test chamber for six hrs before starting the test. An initial tensile load is applied to the compacted beam specimens. Simultaneously, the specimen is exposed to a constant cooling rate of -10°C/hr and is restrained from the contraction by re-establishing the initial length of the specimen.

4 RESULTS AND DISCUSSION

4.1 Physical properties of NA and untreated CRCA

Table 2 reveals the obtained results of the physical and mechanical properties of NA and two different CRCA types. Generally, there is a considerable difference in the properties between the NA and both types of untreated CRCA. In terms of physical properties, namely, bulk relative density (BRD) and water absorption, a significant difference is registered between NA and untreated CRCA#1& 2. It is also indicated that a relative variation is observed in the physical properties between CRCA#1 and CRCA#2. It is also indicated that a relative variation is observed in the physical properties between CRCA#1 and CRCA#2. These findings confirm the outcomes of previous investigations which demonstrated that the absorption capacity of RCA is higher than NA (Ektas and Karacasu 2012, Butler et al. 2013a, Wu et al. 2013, Younis and Pilakoutas 2013, Pasandín and Pérez 2014, Singh et al. 2014). Adhered mortar, which has a higher porosity than NA, results in RCA being more susceptible to absorbing more water compared to NA (Tam et al. 2007). Hence, the presence of adhered mortar leads to increased water absorption, lowered density, and weaker bond strength (Wong et al. 2007). In the perspective of the mechanical properties, abrasion loss, adhered mortar loss, and aggregate crushing value, different observations can be indicated. Compared to CRCA#1, the outcomes indicate that the abrasion loss of CRCA#2 is significantly lower and closer to the NA value, indicating a strong type of CRCA. Such a strong type of CRCA has a lower quantity of attached mortar; therefore, the findings of the adhered mortar loss completely agree with the abrasion loss results. It is known that aggregate strength is an important factor in high-strength concrete and highly used pavements. Hence, the strength of Portland cement concrete aggregate and asphalt mixtures consider the importance of the influence of aggregate strength. According to British standards (BS) (882:1992), the aggregates could be possibly utilized in various applications depending on the maximum crushing values as in the following classifications: for a type with an aggregate crushing value (ACV) of less than 25%, this aggregate could be used in the production of heavy-duty concrete floor finishes. When the ACV value ranges between 25%-30%, this aggregate type could be utilized in the concrete used for pavement wearing surfaces. If the ACV value ranges between 30%-45%, this aggregate could be utilized in concrete used for other applications. Depending on the previous categories, the obtained findings indicate that NA and CRCA#2 can be used for heavy-duty concrete applications, whereas CRCA#1 can be utilized for pavement wearing layers.

Table 2: Physical and mechanical properties of NA and untreated CRCA types

Aggregate Type	Bulk Relative Density (BRD) ASTM (C 127)	Absorption, % ASTM (C 127)	Adhered Mortar, % Without Steel Ball	Micro-Deval Abrasion Loss, % ASTM (D6928)	Fractured Particles, % ASTM (D5821)	Agg. Crushing Value Test (ACV) (BS 812-110)
NA	2.658	0.8	-	15.89	95.5	19.48
CRCA#1	2.295	5.91	3.02	23.57	89.9	27.42
CRCA#2	2.421	3.74	1.08	16.03	95.72	23.28

4.2 Characteristics of treated CRCA

After applying the combination of heat treatment and short mechanical treatment, the main physical characteristics of CRCA are presented in Table 3. In general, it is obvious that a considerable improvement is registered for CRCA characteristics due to the application of a combination of different treatment methods. It is noted that the water absorption of CRCA significantly decreases from 5.91% to 4.13% due to the integration of heat treatment at 300 °C and mechanical method, indicating a successful treatment approach and recording a reduction of 30.12%. Simultaneously, a notable enhancement to the BRD characteristic is obtained. In terms of porosity, the laboratory results indicate that a significant reduction in the porosity is registered with an approximate reduction of 25.59%.

Table 3: Characteristics of treated and untreated CRCA

Type of Property	Untreated CRCA	CRCA after Heat & Sh.M.T.*
Bulk Relative Density (BRD)	2.295	2.447
Absorption, %	5.91	4.13
Porosity, %	13.56	10.092

Sh.M.T.* = Short mechanical treatment

4.3 Volumetric properties of asphalt mixture with CRCA addition

Table 4 shows the volumetric properties of different HMA mixtures that are prepared in this study. The laboratory results indicate that the mix design for various untreated CRCA#1 proportions (0%, 15%, 30%), untreated CRCA#2 (30%), and treated CRCA#1 (30%) successfully meet all the MTO requirements. As shown in Table 4 the optimum asphalt content (OAC) gradually increases with increasing untreated CRCA percentage. This is explained by untreated CRCA having a higher absorption capacity than NA due to the presence of adhered mortar. In contrast, there is a slight reduction in the bulk specific gravity (Gmb), voids in mineral aggregates (VMA), and voids filled with asphalt (VFA) of the mixes when increasing the CRCA proportion. In terms of the effect of the type of RCA on the volumetric properties of the asphalt mixtures, a considerable effect is registered for the type of RCA on these properties. For the same percentage of untreated CRCA addition, 30%, the asphalt mixture included CRCA#2 has a lower OAC than the mixture that included CRCA#1. In contrast, the asphalt mixture that included CRCA#2 has higher VMA and VFA values than the mixture that included CRCA#1. As a result, the volumetric properties of the asphalt mixture that included CRCA#2 are slightly better than the mixture that included CRCA#1. These outcomes correspond to the physical and mechanical property results. As previously shown in Table 2, CRCA#2 has a lower water absorption, higher BRD, and lower adhered mortar loss than CRCA#1. These findings have been confirmed by other researchers (Anderson & Bahia1997, Mills-Beale and You 2010, Bhusal & Wen 2011, Al-Bayati et al. 2018). The outcomes also demonstrate that there is an important decrease in the OAC for mixtures that included 30% treated CRCA compared to the mixture that included the same percentage of untreated CRCA. Additionally, a slight improvement is recorded for volumetric properties, VMA, VFA, and Gmb properties under the influence of the treatment approach for CRCA.

Table 4: Mix design volumetric properties for different CRCA percentages

Aggregate Type / Property	0 % CRCA	15% Untreated CRCA#1	30% Untreated CRCA#1	30% CRCA#1 with Heating+ Sh.M.T.*	30% Untreated CRCA#2	Acceptable Limitations of MTO Specification
OAC AC (%)	4.83	4.9	5.31	5.21	5.12	-
VMA (%)	14.5	13.6	13.66	13.88	14.0	13 min.
VFA (%)	72.5	70.8	70.7	70.9	71.4	65-75
Vv (%)	4.0	4.0	4.0	4.0	4.0	4.0
G_{mb}	2.4	2.395	2.373	2.393	2.384	-

Sh.M.T.* = Short mechanical treatment

4.4 TSRST of asphalt mixtures with various CRCA proportions

The TSRST results reveal a considerable reduction (i.e. more negative) in the fracture temperature, indicating a greater resistance of the mixture to low-temperature thermal cracks. In Table 5, the tabulated outcomes illustrate that the variation in the CRCA proportions has an impact on the fracture thermal stress and the failure temperature. The average values of fracture temperature are plotted in Figure 2. Compared to the control mix, 0% CRCA, it is noted that the average fracture temperature is reduced due to the CRCA addition. The reason behind this can be explained by the fact that RCA has more microcracks than NA. Due to the presence of microcracks, the strength of RCA is lower than NA. In addition, the cement mortar particles may become brittle under the impact of low temperatures, which leads to a higher probability of failure (Wu et al. 2013). The maximum reduction of the fracture temperature was up to -2.6 °C more than the corresponding low-temperature performance grade of the respective asphalt binder used. It is important to note that there is no significant impact for the RCA type on the thermal cracks at low temperatures. While the average fracture temperature of the mixture that included 30% CRCA#1 is -25.6 °C, the average fracture temperature of the mixture that included the same percentage of CRCA#2 is -25.4 °C.

To evaluate the influence of RCA treatment on the thermal cracks, the average fracture temperatures of the mixtures that included untreated and treated CRCA#1 with the same proportion are presented in Figure 3. It is interesting to note that the combination of the heat treatment at 300 °C and short mechanical treatment leads to a considerable decrease in the average fracture temperature. The average fracture temperatures are -25.6 °C and -28.2 °C for the mixtures that included 30% untreated CRCA#1 and treated CRCA#1 at the same percentage, respectively. Interestingly, this obtained result meets the corresponding low-temperature performance grade, -28.0 °C, of the respective asphalt binder used. Hence, this treatment type appears to be highly successful in the low- temperature regions.

Table 4: Results of TSRST for different asphalt mixtures including CRCA

Properties/Mixture Types	Control Mix, 0% CRCA	15% CRCA#1 Untreated	30% CRCA#1 Untreated	30% CRCA#2 Untreated	30% CRCA#1 with Heat Treatment
Binder PG	64-28	64-28	64-28	64-28	64-28
Fracture	-31.5	-26.21	-25.76	-25.95	-28.37
Temperature, °C	-31.1	-25.14	-25.70	-24.43	-27.95
	-28.1	-24.80	-25.41	-25.86	-28.27
Average	**-30.2**	**-25.4**	**-25.6**	**-25.4**	**-28.2**
Fracture Stress,	2.94	2.02	2.70	2.52	2.36
MPa	3.10	1.88	3.08	2.57	2.79
	1.57	1.65	2.76	2.63	3.18
Average	**2.54**	**1.85**	**2.85**	**2.58**	**2.78**

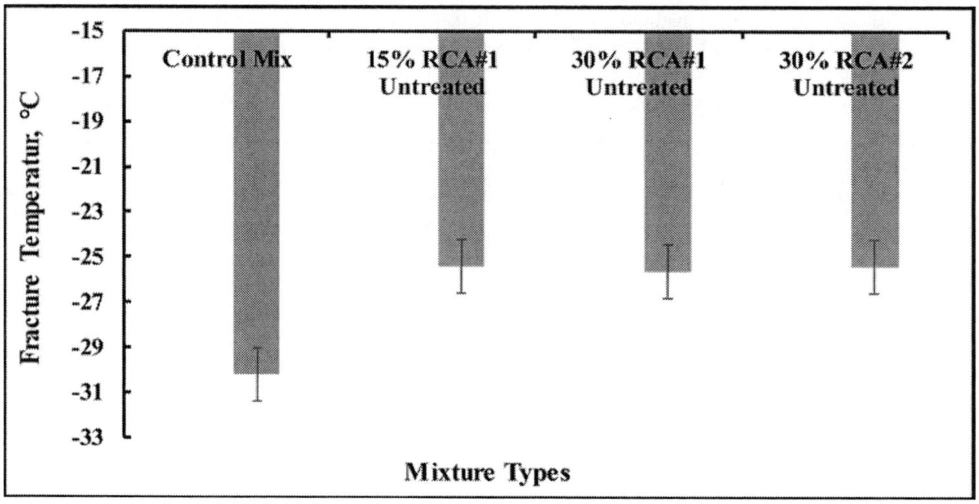

Figure 2: Fracture temperature of mixtures with different CRCA proportions and types

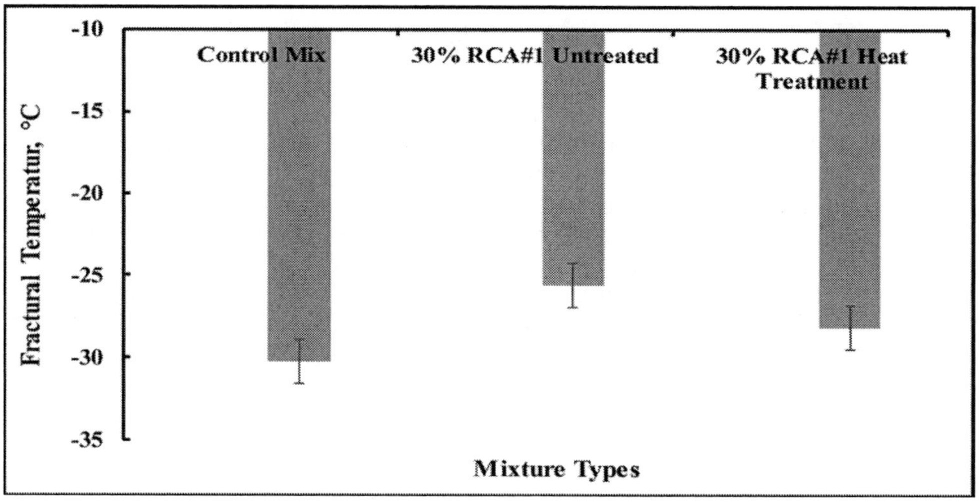

Figure 3: Fracture temperature of mixtures with treated and untreated CRCA#1

Figure 4 demonstrates the mean of the fracture stress for different asphalt mixtures. It is known that the higher value of the fracture stress refers to the largest resistance to the thermal cracks. The obtained results indicate that the fracture stress levels of the mixtures that included 30% untreated CRCA#1 & 2 are slightly higher than the fracture stress of the control mix. It is important to mention that the mixture that included 15% untreated CRCA#1 has a lower fracture stress compared to the control mix. Surprisingly, the values of the fracture stress for the mixtures that included treated and untreated CRCA#1 in the same proportions are approximately equal, indicating there is no impact of the RCA treatment technique on fracture stress. This obtained result is completely different than the average fracture temperature for the same mixture that included the same RCA type as earlier discussed. As a result, it can be concluded that the treatment approach for RCA has a significant effect on the average fracture temperature, whereas it has no impact on the fracture stress.

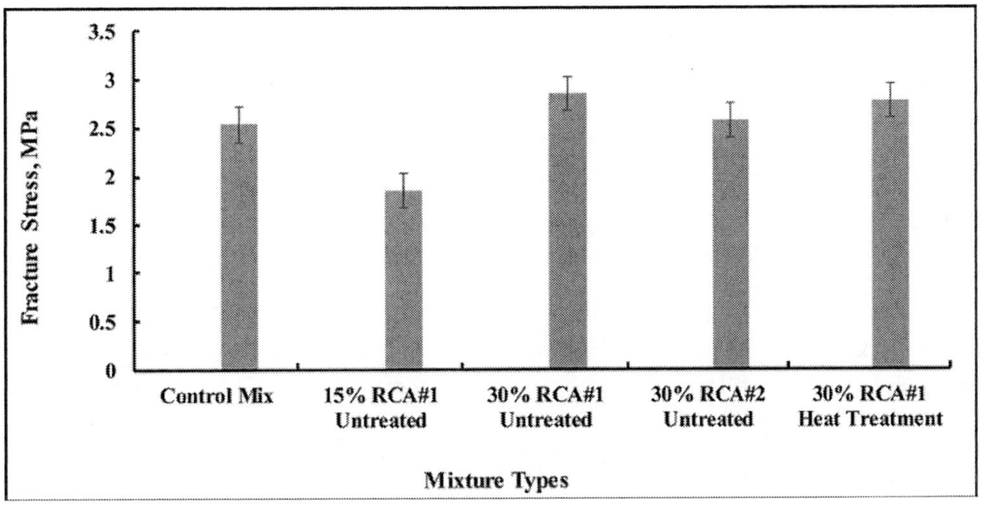

Figure 4: Fracture stress of mixtures with different CRCA types and proportions

5 CONCLUSION

Based on the obtained laboratory results, the following conclusions can be drawn:

- The Physical and mechanical characteristics of both CRCA#1 and CRCA#2 are inferior compared to NA. However, the properties of CRCA#2 appear to be better than those of CRCA#1, and closer to NA characteristics.
- The use of a combination technique of heat treatment at 300 °C and short mechanical treatment method is a highly successful method for improving various physical properties including water absorption, specific gravity, and porosity.
- The obtained results of volumetric properties demonstrate that the addition of both CRCA types into the asphalt mixtures at different percentages is very successful, evidenced by achieving all MTO requirements for asphalt mixtures.
- The findings indicate that the average fracture temperature is reduced due to CRCA addition compared to the control mix. However, there is no significant influence for the RCA type on thermal cracks at low temperatures. The application of the combination of various treatments results in a considerable reduction of the average fracture temperature.
- It is concluded that fracture stress of the mixtures that included CRCA at various proportions is generally higher, except for 30%, than the fracture stress of the control mix. In contrast to the average fracture temperature of the asphalt mixtures, there is no influence of the treatment technique of RCA on the fracture stress of the asphalt mixtures.

Acknowledgements

The authors would like to gratefully acknowledge the funding provided by Ministry of Higher Education and Scientific Research/Iraq through Iraqi Scholarship/Doctoral Program. The authors are also grateful to Steed and Evans Limited in St. Jacobs, Ontario for providing recycled concrete aggregate material. The authors would like to thank Miller Group in Toronto, Ontario for supplying natural aggregate.

References:

AASHTO R30-02, 2010. Standard Practice for Mixture Conditioning of Hot-Mix Asphalt (HMA). American Association of State Highway and Transportation Officials (AASHTO), Washington DC.

AASHTO TP10-93, "Standard Test Method for Thermal Stress Restrained Specimen Tensile Strength." American Association of State Highway and Transportation Officials, Washington DC, 1993.

Akentuna, M., Kim, S. S., Nazzal, M., Abbas, A. R., & Arefin, M. S. 2016. Study of the thermal stress development of asphalt mixtures using the Asphalt Concrete Cracking Device (ACCD). *Construction and Building Materials*, **114**: 416-422.

Al-Bayati, H. K. A., Tighe, S. L., & Achebe, J. 2018. Influence of recycled concrete aggregate on volumetric properties of hot mix asphalt. *Resources, Conservation and Recycling*, **130**: 200-214.

Anderson, R., & Bahia, H. 1997. Evaluation and selection of aggregate gradations for asphalt mixtures using Superpave. *Transportation Research Record: Journal of the Transportation Research Board*, (1583): 91-97.

Bhusal, S., Li, X., & Wen, H. 2011. Evaluation of effects of recycled concrete aggregate on volumetrics of hot-mix asphalt. *Transportation Research Record: Journal of the Transportation Research Board*, (2205): 36-39.

Bolden IV, J. J. 2013. Innovative uses of recycled and waste materials in construction application (Doctoral dissertation, North Carolina Agricultural and Technical State University).

Butler, L., Tighe, S., & West, J. 2013a. Guidelines for the Selection and Use of Recycled Concrete Aggregates in Structural Concrete. *Transportation Research Record: Journal of the Transportation Research Board*, (2335): 3-12.

Du, J. C., & Shen, D. H. 2007. Evaluation of Construction Material Recycling for Hot Mix Asphalt. *Journal of Environmental Engineering & Management*, **17**(4): 243-248.

Ektas, S., & Karacasu, M. 2012. Use of Recycled Concrete in Hot Mix Asphalt and an ANN model for Prediction of Resilient Modulus. *Ekoloji*, **21**(83): 53-60.

http://dotapp7.dot.state.mn.us/research/pdf/2009MRRDOC005.pdf).

Butler L., 2012. Evaluation of Recycled Concrete Aggregate Performance in Structural Concrete. Doctoral dissertation, University of Waterloo.

Liu, J., Zhao, S., Li, L., Li, P., & Saboundjian, S. 2017. Low temperature cracking analysis of asphalt binders and mixtures. *Cold Regions Science and Technology*, **141**: 78-85.

Mills-Beale, J., & You, Z. 2010. The mechanical properties of asphalt mixtures with recycled concrete aggregates. *Construction and Building Materials*, **24**(3): 230-235.

Ministry of Natural Resources (MNR). 2010. *State of the Aggregate Resource in Ontario Study.*

Retrieved May 2013, from Ontario Ministry of Natural Resources: http://www.mnr.gov.on.ca/stdprodconsume/groups/lr/@mnr/@aggregates/documents/document/286996.pdf

National Cooperative Highway Research Program (NCHRP), 2007. Improved Conditioning and Testing Procedures for HMA Moisture Susceptibility". Project Report No. 589, Transportation Research Board of the National Academies, Washington DC.

Paranavithana, S., & Mohajerani, A. 2006. Effects of recycled concrete aggregates on properties of asphalt concrete. *Resources, Conservation and Recycling,* **48**(1): 1-12.

Pasandín, A. R., & Pérez, I. 2014. Mechanical properties of hot-mix asphalt made with recycled concrete aggregates coated with bitumen emulsion. *Construction and Building Materials*, **55:** 350-358.

Pérez, I., Toledano, M., & Gallego, J. 2009. Construction and demolition debris in hot mix asphalt. In *Road Pavement Material Characterization and Rehabilitation: Selected Papers from the 2009 GeoHunan International Conference*: 141-146.

Radević, A., Đureković, A., Zakić, D., & Mladenović, G. 2017. Effects of recycled concrete aggregate on stiffness and rutting resistance of asphalt concrete. *Construction and Building Materials*, **136**: 386-393.

Rafi, M. M., Qadir, A., Ali, S., & Siddiqui, S. H. 2014. Performance of Hot Mix Asphalt Mixtures Made of Recycled Aggregates. *Journal of Testing and Evaluation*, **42**(2): 357-367.

Shen, D. H., & Du, J. C. 2004. Evaluation of building materials recycling on HMA permanent deformation. *Construction and building materials*, **18**(6): 391-397.

Singh, S.B., Mahalakshmi, N., & Thammishetti, N. 2014. Mechanical Properties of Recycled Course Aggregate (RCA) Concrete. *International Journal of Applied Engineering Research*, **9**(3): 363-372.

Tam, V. W., Tam, C. M., & Le, K. N. 2007. Removal of cement mortar remains from recycled aggregate using pre-soaking approaches. *Resources, Conservation and Recycling*, **50**(1): 82-101.

Tan, Y., Zhang, L., & Xu, H. 2012. Evaluation of low-temperature performance of asphalt paving mixtures. *Cold Regions Science and Technology*, **70**: 107-112.

Tim, A., 1995. Investigation of Low Temperature Thermal Cracking in Hot Mix Asphalt. Report No. CDOT-DTD-R-85-7, Colorado Department of Transportation.

Transportation Association of Canada (TAC), 2013. *Pavement Design and Management Guide*. Ottawa, Ontario.

Velasquez, R., Gibson, N., Clyne, T., Turos, M., & Marasteanu, M. 2008. Investigation of low temperature cracking in asphalt mixtures using TSRST. In *Proc., 6th RILEM Int. Conf. on Cracking in Pavements*: 405-414.

Wong, Y. D., Sun, D. D., & Lai, D. 2007. Value-added utilisation of recycled concrete in hot-mix asphalt. Waste Management, **27**(2): 294-301.

Wu, S., Zhong, J., Zhu, J., & Wang, D. 2013. Influence of demolition waste used as recycled aggregate on performance of asphalt mixture. *Road Materials and Pavement Design*, **14**(3): 679-688.

Younis, K. H., & Pilakoutas, K. 2013. Strength prediction model and methods for improving recycled aggregate concrete. *Construction and Building Materials*, **49**: 688-701.

Zhang, Z., Wang, K., Liu, H., & Deng, Z. 2016. Key performance properties of asphalt mixtures with recycled concrete aggregate from low strength concrete. *Construction and Building Materials*, **126**: 711-719.

Zhu, J., Wu, S., Zhong, J., & Wang, D. 2012. Investigation of asphalt mixture containing demolition waste obtained from earthquake-damaged buildings. *Construction and Building Materials*, **29**: 466-475.

Building Tomorrow's Society

Bâtir la Société de Demain

Fredericton, Canada

June 13 to 16, 2018 / *13 au 16 juin, 2018*

MECHANICAL PROPERTIES OF CONCRETE WITH RECYCLED SHREDDED RUBBER PARTICLES

Bandarage, Kumari[1], and Sadeghian, Pedram[1,2]
[1] Department of Civil and Resource Engineering, Dalhousie University, Canada
[2] pedram.sadeghian@dal.ca

Abstract: The waste tires from vehicles are continually increasing in the world and disposal of waste tires has become a global problem and required to find an adequate recycling system. As a method of recycling of waste tires, rubber particles of tire can be used in concrete and it has become a preferable solution for disposing waste tires. The goal of this research is to study the mechanical properties of recycled shredded rubber particles extracted from scrap tires in concrete to introduce a sustainable waste management system and subsequently to reduce the use of non-renewable resources in concrete. In this paper, fine aggregates of concrete were partially replaced with recycled shredded rubber particles less than 5 mm in size. Concrete cylinders with different proportion of rubber up to 50% of fine aggregates by volume were prepared and tested under compression. Also, specimens without rubber were prepared and tested as control specimens. The compressive strength and elastic modulus of the specimens are analyzed to characterize the behaviour of concrete. Overall, increasing the content of shredded rubber decreased the strength and elastic modulus, however increased the ductility of failure. It seems 40% shredded rubber content is the maximum to be recommended for more in-depth studies. This is an on-going research and more results will be presented at the time of the conference.

1 INTRODUCTION

Use of scrap tire particles in concrete is environmental friendly as it helps to reduce the use of natural resource for concrete and introduces the waste management of scrap tires. On the other hand, it will affect on the health of people and animals, if waste tire recycling process cannot be managed in a proper manner over the years. As a method of recycling of waste tire, rubber particles of tire can be used in concrete and it has become a preferable solution for disposing waste tires. Three different sizes of recycled tires can be normally taken into research as rubber powder (partially replace with cement), crumbed rubber less than 5 mm in size (partially replace with fine aggregates) and rubber chips larger than 5 mm in size (partially replace with course aggregates) in concrete.

Many researchers (e.g. Li et al., 2014, Bisht et al., 2017, Youssf et al., 2016, Atahan and Yucel 2012, and Blessen and Ramesh 2012) were involved to find the chemical and physical properties of concrete with recycled shredded rubber particles. Most of the studies have shown the reduction of compressive strength, tensile strength and the modulus of elasticity with an increment of shredded rubber in concrete mixture and on the other hand, increment of shredded rubber in concrete had improved the durability and freezing-thawing resistance of concrete. For example, Liu et al. (2016) experimented the mechanical and chemical properties of crumbed rubber concretes and found that 20% crumbed rubber in concrete were reduced the compressive strength by 3.9% and splitting tensile strength by 2.5% compared to the results of 0% crumbed rubber concrete sample but increased the durability, freezing-thawing resistance and sulfate resistance. However, the experimental results have shown that a 20% replacement of rubber could be the acceptable values of mechanical properties. Marie (2016) investigated the behavior of rubberized concrete when increasing the rubber content using the Ultrasonic Pulse Velocity Test (UPV) with shredded rubber particle sizes 4.75 to 0.15 mm and designed to target the compressive strength 25MPa at 28 days. The results showed that the UPV values decreased with the increase of rubber particle and it was concluded that up to 25% replacement of shredded rubber for the concrete could get the acceptable range of compressive strength. Mendis et al. (2017) experimented the effect of rubber particles on the flexural behaviour of steel-reinforced crumbed rubber concrete beams using shredded rubber mix contained less than 4 mm rubber particle sizes and it was concluded that the failure patterns of the beams were very similar to the normal concrete beams and lower tensile strength with compared to the normal concrete.

Most of the experimental studies concluded that 20 to 25% crumb rubber (less than 5 mm in size) replacements by volume of fine aggregates provided the acceptable range of mechanical properties to the concrete. But, those were based on their results and still couldn't find the exact recommendation with detailed material proportions and recommended code of practice. Also, most of the previous researches have based on crumb rubber (less than 5 mm in size with almost round particles) and could see very less interest in investigating of the properties of shredded rubber concrete less than 5mm in size replace with fine aggregate. Hence, the focus of this study is to analyse the strength, stiffness, and energy absorption of shredded rubber concrete using deferent proportions of 0 to 50% rubber less than 5 mm in size replaced with fine aggregate to comprehend the definite shredded rubber proportion to the concrete.

2 EXPERIMENTAL PROGRAM

2.1 Specimen Layout

A total of 24 concrete cylinders (100 mm diameter and 200 mm height) was prepared and tested under compression loading. The replacement of fine aggregates from shredded rubber by volume was used to prepare concrete mix at different percentages of 0, 10, 20, 30, 40, and 50% (by volume of fine aggregates) as shown in Table 1. Four identical specimens were prepared in each group. The specimens without shredded rubber were designed as control specimens.

Table 1: Test matrix and mix proportions of concrete specimens

Group #	Specimen group ID	Shredded rubber content (%)	Water/ cement ratio	Cement (Kg/m^3)	Water (kg/m^3)	Coarse aggregate (kg/m^3)	Fine aggregate (kg/m^3)	Shredded rubber (kg/m^3)
1	R0	0	0.33	656.8	221.1	949.8	567.5	0.0
2	R10	10	0.33	656.8	221.1	949.8	510.7	16.9
3	R20	20	0.33	656.8	221.1	949.8	454.0	33.9
4	R30	30	0.33	656.8	221.1	949.8	397.2	50.8
5	R40	40	0.33	656.8	221.1	949.8	340.5	67.8
6	R50	50	0.33	656.8	221.1	949.8	283.7	84.7

2.2 Material Properties

General use Portland cement was used for this study. Fine aggregate (natural sand) with bulk density 1791 kg/m^3 and coarse aggregate having a maximum size of 12.5 mm with bulk density 1666 kg/m^3 were used. Shredded rubber particles were used for this study with bulk density 535 kg/m^3. Figure 1 shows the shredded rubber particles used in this experimental program and Figure 2 shows the particle size distribution of shredded rubber, fine aggregates (sand), and coarse aggregates (gravel). The average particle size remained on the Sieve No. 04 was approximately 28 mm long. The figure shows that the shredded rubber used in this study had a gradation curve between the fine and coarse aggregates. A constant water cement ratio of 0.33 was considered for this experiment and the dosage of superplasticizer (PLASTOL 6400) 2 ml per 1 kg of cement was added to achieve the required workability for the mixture. The mix proportions of six groups are shown in Table 1.

Figure 1: Various sizes shredded rubber particles used in this study

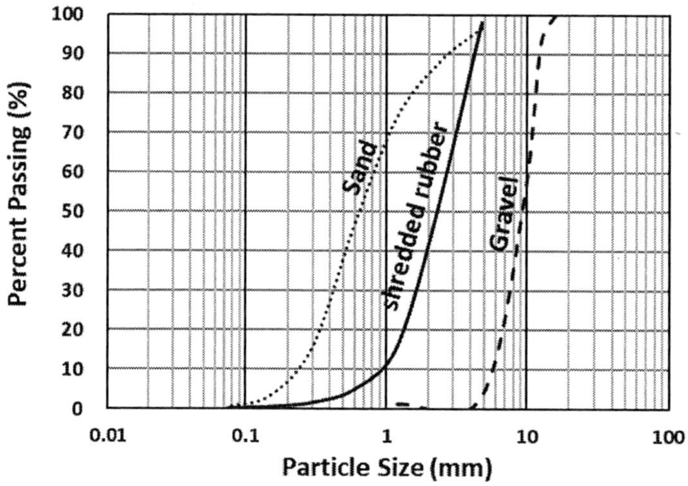

Figure 2: Particle size distribution of shredded rubber, fine aggregates, and coarse aggregates

2.3 Specimen Preparation

As shown in Figure 3, cylindrical plastic molds with a size of 100 mm diameter and height of 200 mm were used to cast all specimens. Since the two number of control specimens, 3.357 Kg coarse aggregates and 1.805 Kg fine aggregates (sand) added to the mortar mixing machine as shown in figure 3 and mixed around 2 minutes in dry condition. After that, 2.232 kg cement also added to the mixture and mixed another 1 to 2 minutes as required. Then, measured water 785ml added to the mixture regularly during operation of the mixing machine and 5ml superplasticizer also add to each mixture and continue another 2 to 3 minutes. The same procedure was followed for the all other mixtures with adding shredded rubber to the finer aggregate as specified in section 2.1 specimen layout. All concrete mixes were placed in cylinders which were already cleaned and applied mould oil, with three equal layers and compacted using 10 mm tamping rod each layer 25 times. Four specimens in each group were prepared. The samples were kept at curing room for 28 days. The density of each specimen was calculated by taken weight of each sample after 28 days dividing the volume of the specimen. specimens were weighed under dry condition after days. Calculated densities are tabulated in Table 2. Capping of the samples were done using sulfur mortar to give a smooth surface to concrete cylinders for applying compressive load.

Figure 3: Specimen preparation

With reference to the results taken from the average density, it could be clearly identified that the increment of shredded rubber in concrete mix had decreased the density of the concrete. The increment of the shredded rubber from 0% to 20% decreased the density 2530.73 kg/m^3 to 2441.76 kg/m^3 (3.52% decrease) also, the increment of the crumbed rubber from 0% to 50% decreased the density 2530.73 kg/m^3 to 2343.11 kg/m^3 (7.41% decrease). In addition to that, the results show, the density reduction rate has increased at 1.51%, 2.01% and 2.6% from R0 to R10, R10 to R20 and R20 to R30 respectively and from R30 to R50, the increment rate has decreased.

Table 2: Density of concrete specimens containing shredded rubber

Specimen ID	Average Density (kg/m^3)	Standard Deviation (kg/m^3)	Coefficient of Variation (%)	Density Reduction (%)
R0	2530.73	24.64	0.97	0.00
R10	2492.52	16.36	0.66	1.51
R20	2441.76	18.41	0.75	3.52
R30	2375.88	58.48	2.46	6.12
R40	2403.71	16.17	0.67	5.02
R50	2343.11	11.53	0.49	7.41

2.4 Test Setup and Instrumentation

The instrumentation setup for compression test was arranged as shown in figure 4. LP1 and LP 2 were fixed straight in the centre of the cylinder from 180^0 to measure the lateral stain during the test and LP3 and LP4 were fixed into the steel rings to measure the axial strain. These LPs were used to measure axial and lateral displacement during the test. The axial deformation referred to in this paper is the average value of the LP3 and LP4. The 100 mm gap between the two ring's center to center was maintained for each test. A steel cylinder was used to achieve the enough room for LP no. 03 and 04 during the test. The compressive strength test was done by minimizing the eccentricity as zero and the computer programmed to deform the specimens at a rate of 0.4 mm per minute. The specimens were compressed until it did not seem safe to deform the specimen any further.

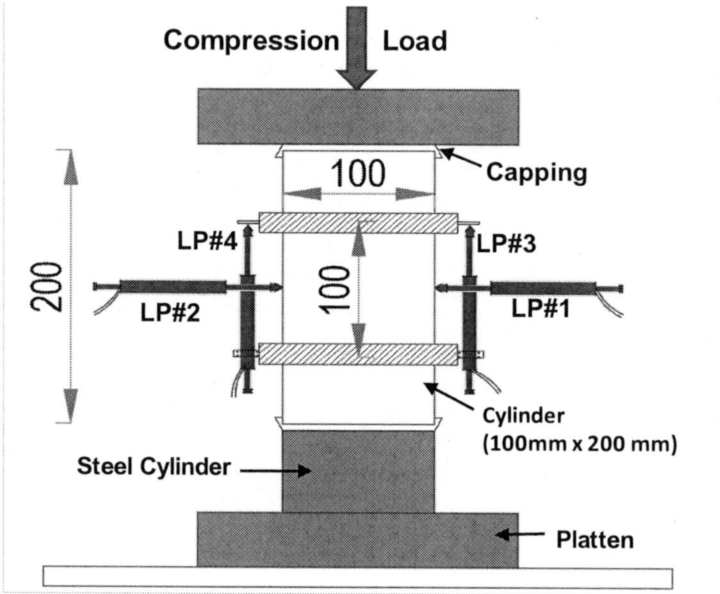

(a)

(b)

Figure 4: Test setup and instrumentation: (a) schematic of test setup and (b) photo of setup

3 RESULTS AND DISCUSSION

3.1 Failure Modes

Figure 5 shows the specimens failure modes after the tests. It should be noted that, the R0 cylinder (control) was failed as more brittle compared to the other specimens and behaviour of brittle failure was decreased with an increment of shredded rubber. The failure cracks were generated approximately in vertical direction for the specimens, but R50 specimen was shown the horizontal direction failure cracks too. Overall, increasing the shredded rubber content changed the mode of failure from typical diagonal cracks to a combination of longitudinal and transverse cracks. Three more specimens of each group will be tested to have a better understanding of the behavior of the specimens.

(a) (b) (c) (d) (e) (f)

Figure 5: Failure mode of the specimens: (a) R0; (b) R10; (c) R20; (d) R30; (e) R40; and (f) R50

3.2 Effect of Shredded Rubber on Concrete Strength

Figure 6 (a) illustrates the effect of shredded rubber on concrete strength. The results generally show that the increasing of shredded rubber content in concrete decreased the compressive strength. Maximum compressive strength (f'_c) of concrete was reduced by 26.0%, 30.2%, 42.0%, 43.1% and 70.0% for the R10-1 (57.5 MPa), R20-1 (54.2 MPa), R30-1 (45.1 MPa), R40-1 (44.2 MPa) and R-50-1 (23.3 MPa) respectively compared to the control specimen R0-1 (77.7 MPa). Also, the maximum strength reduction for the R10-1 to R20-1 calculated as 5.7% compared to the R10-1 compressive strength and R20-1 to R30-1 calculated as 16.8% compressive strength reduction compared to the R20-1. As shown in Figure 6 (a), it seems there is a significant strength, drop after 40% shredded rubber content. So, 40% shredded rubber content might be considered as a maximum shredded rubber content. More specimens need to be tested to make a reliable recommendation.

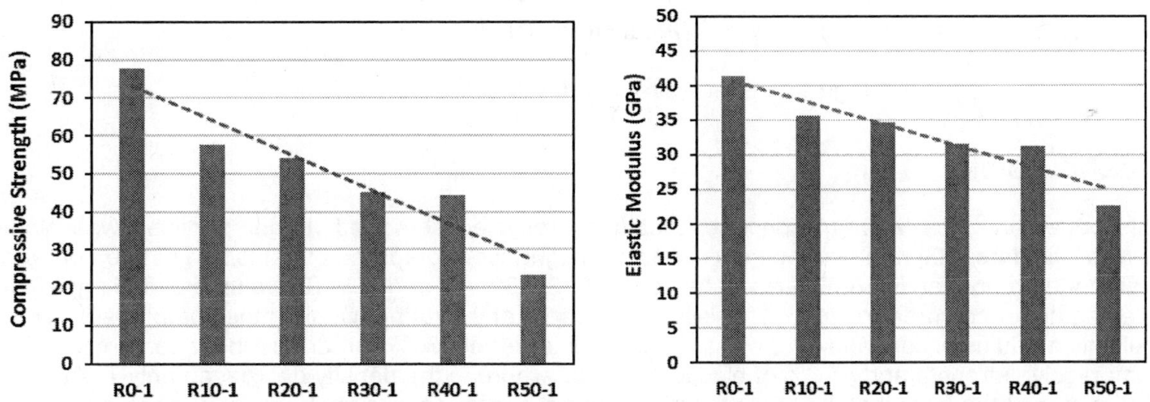

Figure 6: Effect of shredded rubber on: (a) concrete strength and (b) elastic modulus

3.3 Effect of Shredded Rubber on Concrete Elastic Modulus

The elastic modulus also has decreased due to the increment of the shredded rubber in concrete. The results in Figure 6 (b) show the elastic modulus of the test specimens obtained by fitting a straight line in the test data up to 45% of peak stress. The elastic modulus decreased by 14.0%, 16.4%, 23.7%, 24.6% and 45.2% for the R10-1 (35.6 GPa), R20-1 (34.6 GPa), R30-1 (31.6 GPa), R40-1 (31.2 GPa) and R-50-1 (22.7 GPa) respectively compared to the control specimen R0-1 (41.4 GPa). Similar to the compressive strength, it seems there is a significant elastic modulus drop after 40% shredded rubber content. More specimens need to be tested to make a reliable recommendation.

3.4 Effect of Shredded Rubber on Stress-Strain Curve

Figure 7 shows the stress-strain curves of the test specimens. The results show that the increasing of the shredded rubber content from 0 to 50% gives a lower slope (modulus) to the stress-strain curve with reducing the maximum axial stress (as explained above in section 3.2) and the strain at the peak stress. The specimen with 20% shredded rubber content showed a significant post peak behavior, indication the potential for higher energy absorption (area under the curve). Testing the rest of the specimens will provide a solid platform to evaluate the energy absorption capacity of the specimens. Moving forward in this research, concrete containing large quantity of shredded rubber can be used for cases under lateral confining pressure where the confinement can compensate the lack of strength due to shredded rubber. It has been shown that confinement is more effective on low strength concrete in comparison with high strength concrete (Sadeghian et al., 2008, Sadeghian and Fam 2015).

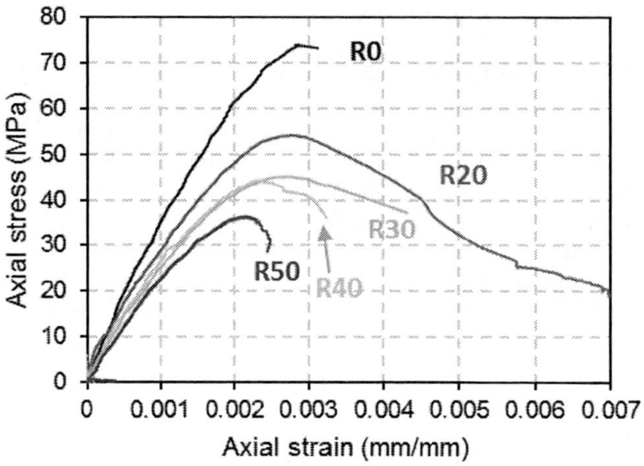

Figure 7: Stress-strain behavior of test specimens

4 CONCLUSIONS

In this paper, the behaviour of concrete containing recycled shredded rubber particles was studied. Concrete cylinders (100 mm diameter and 200 mm height) were prepared by replacing of fine aggregates with shredded rubber at different percentages of 0, 10, 20, 30, 40, and 50% (by volume of fine aggregates) and tested under compression load. The results showed that the rubber weaken the compressive strength and elastic modulus of concrete. For example, the cylinders with 50% shredded rubber content lost 70.0% of compressive strength and 45.2% of elastic modulus, whereas the density decreased only 7.4%. It was observed during the tests that the brittle failure of control specimen was changed to less brittle failure with increasing the rubber content. Also, the failure patterns of the concrete from typical diagonal cracks was changed to a combination of longitudinal and transverse cracks. Considering the results of this experiment, the shredded rubber used for the partial replacement for natural fine aggregates up to 40% might be recommended in an acceptable range due to control specimen mixture design was targeted to achieve the compressive strength of 44 MPa. As this is an ongoing project, more tests are continuing to get the appropriate results considering the performance of different types of mechanical properties due to the increment of the shredded rubber in concrete. More results will be presented at the time of the conference.

Acknowledgements

The authors are grateful for all the supports of the lab technicians (Brian Kennedy, Jesse Keane and Blair Nickerson) and students (Koosha Khorramian, Priyanka Garikapati and Mostafa Abyaneh) in conducting this study at Dalhousie University, Canada.

References

Atahan, A. O., and Yücel, A. Ö. 2012. Crumb rubber in concrete: static and dynamic evaluation. *Construction and Building Materials*, **36**: 617-622.

Bisht, K., and Ramana, P. V. 2017. Evaluation of mechanical and durability properties of crumb rubber concrete. *Construction and Building Materials*, **155**: 811-817.

Li, L., Ruan, S., and Zeng, L. 2014. Mechanical Properties and Constitutive Equations of Concrete Containing a Low Volume of Tire Rubber Particles. *Construction and Building Materials*, Elsevier, **70**: 291-308.

Liu, H., Wang, X., Jiao, Y., and Sha, T. 2016. Experimental investigation of the mechanical and durability properties of crumb rubber concrete. *Materials,* **9:** 3-172.

Marie, I. 2016. Zones of weakness of rubberized concrete behavior using the UPV. *Journal of Cleaner Production*, **116**: 217-222.

Mendis, A. S., Al-Deen, S., and Ashraf, M. 2017. Effect of rubber particles on the flexural behaviour of reinforced crumbed rubber concrete beams. *Construction and Building Materials*, **154**: 644-657.

Sadeghian, P. and Fam, A., 2015. Improved design-oriented confinement models for FRP-wrapped concrete cylinders based on statistical analyses. Engineering Structures, **87**: 162-182.

Sadeghian, P., Rahai, A.R. and Ehsani, M.R., 2008. Numerical modeling of concrete cylinders confined with CFRP composites. Journal of Reinforced Plastics and Composites, **27**: 1309-1321.

Thomas, B. S., and Gupta, R. C. 2015. Long term behaviour of cement concrete containing discarded tire rubber. *Journal of Cleaner Production*, **102**: 78-87.

Youssf, O., Mills, J. E., and Hassanli, R. 2016. Assessment of the mechanical performance of crumb rubber concrete. *Construction and Building Materials*, **125**: 175-183.

Building Tomorrow's Society

Bâtir la Société de Demain

Fredericton, Canada

June 13 to 16, 2018 / *13 au 16 juin, 2018*

NEW TEST METHOD FOR REINFORCED CONCRETE PIPE JOINT HYDROSTATIC INFILTRATION

Wong, Lui[1] and Nehdi, M.L.[2]
[1] VP Engineering & Quality, Con Cast Pipe, 299 Brock Road South, Puslinch, ON, Canada
[2] Professor, Department of Civil & Environmental Engineering, Western University, London, ON, Canada

Abstract: The hydrostatic performance of reinforced concrete pipe (RCP) depends on the gasket design and its materials properties, the joint design and its quality, the joint dimensional tolerance during pipe manufacturing, and more importantly, the quality of the concrete pipe installation. Existing specifications pertinent to testing RCP joint performance are limited to evaluating the exfiltration of the joint. It ignores the fact that, for a non-pressure gravity sewer, the true requirement for hydrostatic performance is to resist groundwater infiltration. Existing testing standard is only available in the field which is costly and difficult to perform. The test is limited by the groundwater condition. The repair cost is high in the case of failure. There is need to develop a simple yet effective method for gaging the infiltration capacity of the RCP joint at the manufacturing facilities. In the present study, a new testing method is developed to address this need. The proposed method was applied to conventional RCP diameters ranging from 600 to 1200 mm. The test results show that the proposed method provides a scientific and unbiased evaluation of both the joint capacity hydrostatic infiltration and the working external pressure of the joint. Recommendations for future development of a related standard test method are provided.

Keywords: Reinforced; Concrete; Pipe; Hydrostatic; Test; Joint; Infiltration.

1 INTRODUCTION

1.1 Background

For over a century, concrete pipe has been the primary solution for storm water management networks and wastewater collection systems. In such gravity flow, concrete pipe design has essentially focussed on the rigid pipe structure, soil-pipe interaction and joint structural integrity. The true hydrostatic performance of the reinforced concrete pipe (RCP) joint was not emphasized. Even though there are several specifications that require the manufacturer to conduct hydrostatic pressure testing of the pipe, such tests do not ensure confidence in the joint performance after in-situ installation.

Common hydrostatic tests performed by pipe manufacturers generally use internal pressure to detect exfiltration. However, actual hydrostatic conditions experienced in service by the pipe include not only internal pressure, but also external pressure when a pressure head from groundwater above the pipe level is present (ACPA 2009). The resistance to groundwater infiltration is one of the design criteria of RCP (MTO 2007). Most joint leakage issues reported in construction sites or via closed-circuit television (CCTV) inspections are a result of groundwater infiltration. Deep RCP installation with high groundwater table usually escalates the risk of leaky joints. Concerns of such infiltration have been reported as early as the 1970s (APWA 1971, Gottstein 1973). It was recognised that the problem of infiltration and inflow (I&I) would cause the sewer to have premature capacity issues and associated high water treatment costs. Such concerns continued to prevail in recent years (Pipeline 1999, Kesik, 2015, Hey et.al. 2016). Uncontrolled I&I in sanitary sewers causes detrimental financial and environmental effects to municipalities (FCM/NRC 2003). For instance, in a storm network, infiltration can carry fines from the embedment soil around the pipe into the sewer, causing settlement and misalignment problems (INDOT 2014, NYSDOT 2013).

The performance of the pipe joint is predominantly influenced by the joint quality, installation quality and external conditions. While, external conditions cannot be easily controlled, they need to be considered in design. The quality of the joint and installation can be evaluated, examined and inspected. Without an effective and simple qualifying test method for joint infiltration, designers resort to using more expensive pressure pipes in gravity flow applications to mitigate the I&I risks.

In recent decades, emerging alternative pipe materials such as PVC, high density polyethylene, polypropylene, fiber glass and other flexible pipe materials tend to promote better joint performance than that of concrete pipe, along with other advantages, such as chemical resistance, lightweight, longer lay length and ease of handling and installation. Hence, the concrete pipe industry is pursuing technological advancements to enhance the RCP sustainability, resilience and overall life-cycle value. Chief among the barriers that need to be overcome by modern RCP is mitigating challenges related to the hydrostatic joint performance.

1.2 Existing Standards

Twelve standard practices related to the joint hydrostatic performance were reviewed (Table 1). They can be categorized into three groups: joint quality, in-field quality, and pressure rating evaluations.

Internal hydrostatic tests are by far the most commonly used by RCP manufacturers to validate the joint quality. Single or multiple pipe sections can be pressurized internally to a certain pressure level and duration in either a horizontal or vertical orientation. However, these tests usually take place over a short period (typically 10 minutes or less) and conclude by visual inspection for exfiltration leakage. Standard practices such as CSA A257.2, ASTM C443, ASTM C1628 and BS EN 1916 require horizontal pipe sections to be tested not only in a straight alignment, but also in a deflected alignment. CSA A257.2 and BS EN 1916 require the joint performance under offset condition. While the intent of such tests is a proof of design, the results may not correlate to real field performance.

In-field tests allow the contractor to measure the amount of leakage using air or water after the installation of the pipeline. Some of these tests can have safety concerns, are difficult to perform, and time consuming. For an infiltration test such as the ASTM C1103, a certain level of groundwater is needed for the test. Such conditions may not be available at the time of the test. Other standard tests such as ASTM C969 and ASTM C1214 allow pressure reduction over a long section of pipe. If the total leakage exceeds the limit, it is hard to identify which joints are the source of the problem. In addition, these tests are usually difficult to execute during construction and only examine the pressure resistance over a short time period.

The last type of tests including AS/NZS 4058, AWWA C302 and ASTM C361 provides a pressure rating. Among these standards, AWWA C302 and ASTM C361 technically pertain to pressure pipe. The test method requires at least an additional 20% pressure beyond the required class to be met. These standards are more stringent and are commonly being specified in gravity sewer applications that only require an elevated joint performance. ASTM C361 requires the evaluation of the pressure rating under both aligned and offset positions. These tests only examine the internal pressure rating and do not reflect clear correlation to field performance.

1.3 Scope

The pipe joint performance expectation by municipalities is clearly a watertight joint. The pipeline needs to be designed to withstand hydrostatic pressure from the average groundwater level. Watertight joint is defined as a joint that provides zero leakage of water infiltration and exfiltration for a specific head or pressure application (AASHTO 2009). Existing tests conducted by pipe manufacturers evaluate exfiltration and do not address infiltration. Thus, it is very costly to discover performance issues during or after construction and the repair cost is far more expensive than the initial material and construction costs (Romer 2004). Many situations cause legal litigation between the contractor and the manufacturer. It is apparent that there is a misconception of the end user's expectation. Existing standards related to RCP joint performance do not provide sufficient consideration to certain field conditions, such as deep installation with a high groundwater condition. Current tests validate the resistance to exfiltration and assume that the joint performance to resist infiltration is interchangeable. In fact, the joint and the gasket are not symmetrical in nature, thus this assumption seems to be scientifically incorrect. There is no standard test method to validate the joint performance and capacity for infiltration. Leaky pipe joint is usually caused by the installation quality. However, it is challenging to quantify and validate quality during installation. End users generally demand RCP joints that can withstand infiltration.

Because of these gaps, it is important to develop an industry standard test method to validate RCP joints' hydrostatic performance, capacity and behaviour against infiltration. In order to simulate the sustaining hydrostatic pressure from ground water acting on the pipe joint, the method to pressurize the exterior of the joint is developed. The scope of the present research is to:

1. Provide a standard testing method that can be easily and safely conducted by any pipe manufacturer.
2. Recommend a standard testing procedure that can effectively qualify the RCP joint under a certain pressure rating.
3. Explore the factors influencing the joint hydrostatic capacity.
4. Establish a bench mark so that further study on joint quality or joint conditions, such as deflected and offset alignments, can be conducted.

Table 1 Summary of Joint Performance Standards

Standard	Ori.*	#. of Pipe	#. of Jt.	Med.**	Dir.***	Pressure Limit (kPa)$^{\#}$	Test Duration	Acceptance Criteria	Leak Exam.
						Joint Quality			
CSA A257	H	3	2	W	I	105 (A), 90 (D), 35 (O)	10 min	No leak	Visual
ASTM C433	X	2	1	W	I	90 (A), 70 (D)	10 min	No leak	Visual
ASTM C990	V	2	1	W	I	90 (A)	10 min	No leak	Visual
ASTM C1628	X	2	1	W	I	90 (A), 70 (D)	10 min	No leak	Visual
AS/NZS 4058	H	4	3	W	I	90 (A)	60 min	Limit leakage	Measure
	V / H	1	0	W	I	90	90sec/10mm wall	Limit leakage	Measure
					I	50, +20% of Opr.			Visual
BS EN 1916	H	2	1	W	I	50 (A), 50 (D), 50 (O)	15 min		Visual
GB/T 16752	V / H	1	0	W	I	60 CL1, 100 CL2,3	10 min	No leak	Visual
						Pressure Rated			
AWWA C302	H	2 Min.	1+	W	I	Operating	48 hours	No leak	Visual
					I	120 % of	20 min	No leak	Visual

						operating			
ASTM C361	X	2	1	W	I	120% of spec 75-300 375 (A), 375 (O)	Not specified	No leak	Visual
					In-field Quality				
ASTM C1214	H	Var	V	A	E	Initiate at 23.7	0.3 - 6 min/100ft based on size	Limit Leakage	Gauge
ASTM C969	H	<700ft	V	W	I / E	Min. 6 (2ft head)		Limit Leakage	Gauge
ASTM C1103	H	2	1	A / W	E	24 > GWT	5 sec	Limit Leakage	Gauge

*Ori.: H – Horizontal, V – Vertical, X – Not specified, **Media: W – Water, A- Air, ***I – Internal pressure, E – External Pressure, #A – Aligned, D – Deflected, O - Offset

2 TEST METHOD DEVELOPMENT

2.1 Testing Concept

To simulate infiltration, a small annular space is created outside the primary gasket of the RCP joint. A secondary gasket is introduced to seal the annular space and permit testing pressure to build up. Because the joint design will vary depending on the type of RCP, the secondary gasket needs to be properly selected based on the available space and the joint design. The gasket supplier is consulted for the adequate style based on the target testing pressure, material properties, and available space. Small inlet and outlet holes are drilled in opposite sides of the RCP joint, between the gaskets, to allow the water to be introduced into the annular space. Figure 1 shows the annular space of a conventional RCP joint.

The design of the test also maintains a safe operating environment. The equipment was designed to handle a maximum pressure of 685 kPa. To ensure a safe operating environment, the small amount of water in the enclosed annular space does not pose a flooding hazard of the testing area when failure occurs. Pressure is created by using compressed air to push the water into the annular space. The pressurization system scheme is shown in Figure 2.

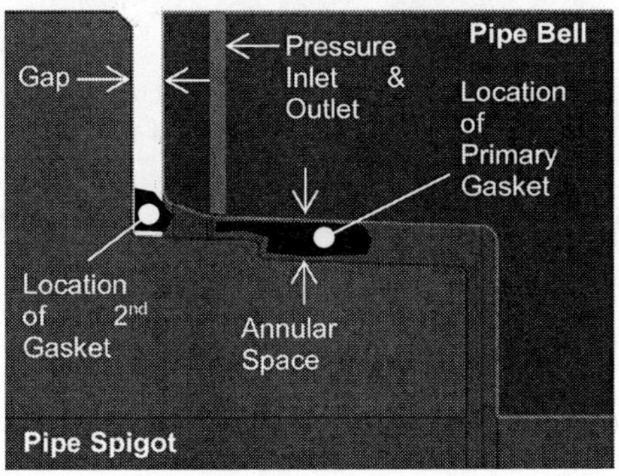

Figure 1 Typical RCP Joint

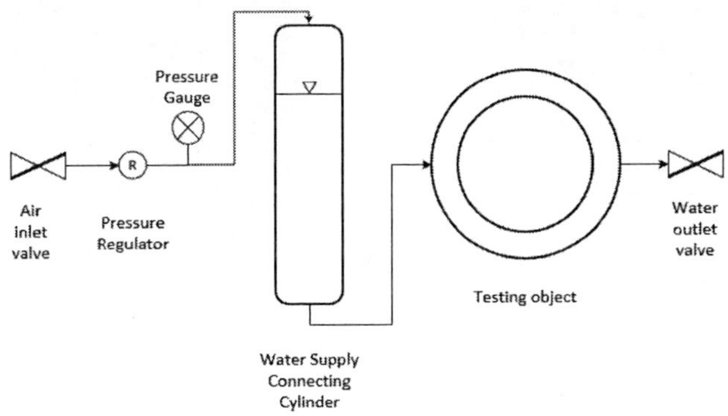

Figure 2 Pressurization System Scheme

2.2 Mechanical Apparatus and Setup

The mechanical apparatus includes an air inlet valve, pressure regulator, pressure gauge, water supply connecting cylinder (WSCC), connecting hosts, and water outlet valve. The air inlet valve is connected to a compressed air line where air pressure is introduced to the system. The pressure regulator allows the operator to control the target testing pressure. The pressure gauge displays the pressure level in the system with an accuracy of 2.5 kPa. WSCC made of a 4-inch diameter transparent fiber glass pressure cylindrical chamber contains sufficient water to supply the water demand in the annular space. The cylinder allows the operator to monitor water loss if a system leakage and/or joint leakage occur during the test. Water outlet valve allows the input of the water source during the setup process. The valve is kept closed during the test. Figure 3 and Figure 4 display the actual equipment and setup.

In the test setup, two pipe sections consisting of a bell end and spigot end are stacked vertically (Figure 4 and Figure 5). The total height of the combined pipe sections is between 700 and 1000 mm for ease of assessment and monitoring during the test. The test joint was created by homing two pipe sections sealed with a primary gasket and a secondary gasket. The secondary gasket (Figure 6) was placed between the shoulder of the pipe spigot and the bell end face to create a confined space for pressurization. Both homed pipe sections are tied down to the supporting steel base frame. Due to the internal pressure generated, the tremendous amount of uplift force is expected. The number of tie down straps and the strap rating are calculated based on the target testing pressure and the geometry of the annular space. The base frame is especially designed to carry the expected forces from the strap. Before the start of the test, the water shall be introduced from the outlet to the inlet and fill up the WSCC. This process is to ensure that the entire annular space is filled. It is suggested to pressurize the setup to a minimum of 50 kPa before the formal start of the test. The frame used in this research was designed for maximum pipe size with 1829 mm outer diameter tested to 685 kPa.

Figure 3 Testing apparatus

Figure 4 Base frame and test pipe setup

Figure 5 Inside view of test pipe

Figure 6 Testing and secondary gasket setup

2.3 Procedures

2.3.1 General

Considering that the groundwater pressure acting on the RCP pipeline is a sustained pressure, there is need to develop a process to evaluate both the ultimate and operation pressure. Using existing concepts from AWWA C302 and ASTM C361, the operating pressure is set not to exceed 80% of the ultimate pressure. None of the existing standards evaluates the ultimate capacity of the joint. The intent of the ultimate test is also to evaluate the capacity of the gasket and the joint design under the testing condition.

2.3.2 Ultimate Pressure

The first pressure stage is selected based on the expected ultimate pressure. Because there is no previous test results available for RCP joints, 100 kPa was selected based on the typical internal hydrostatic test from existing standards. The initial pressure is then incrementally increased by 25 kPa until failure. At each incremental attempt, the pressure is held for at least 10 minutes. Again, this duration is selected based on existing standard tests for exfiltration. Visual inspection for leakage, measurement of the water level in the WSCC and joint gap measurements are conducted in between each incremental attempt. Any leakage found is recorded. If the leak is found from the inside of the pipe joint, the pressure rating of the principle test gasket is determined to be the pressure value from the previous attempt. All other leakage, such as the outside joint, testing inlet and/or outlet or other locations will not be considered a failure. Adjustment of the strap shall be made to limit or eliminate the exterior leakage so the test can be continued. Time for each incremental attempt was recorded by the operator using a stop watch.

2.3.3 Operating Pressure

The operating pressure can only be conducted after the ultimate pressure is established. The operating pressure target shall be set to 80% of the ultimate pressure. Both the primary and secondary gaskets shall be replaced with new ones to avoid any possible damage incurred during the establishment of the ultimate pressure. To begin the test, the setup is recommended to be tested at a lower pressure level of 100 kPa followed by 70% of the ultimate pressure for 10 minutes. This is to ensure that the test setup is performing as expected. If no leakage is observed, pressure is increased to 80% of the ultimate. Visual inspection for leakage is completed and the target pressure is held for the next 20 hours with periodic visual inspection for leaks. Upon completion of the 20 hours, full inspection is carried out by the operator. Observation and measurement of the water level and joint gap shall be recorded. If no leakage is evident from the inside of the pipe, the test is successful, and the working pressure of the joint can be pressure rated at 80% of the ultimate pressure.

2.4 Measurements

The primary measurement in this test is the hydrostatic pressure inside the annular space. The analogue pressure gauge is used within the apparatus as the measurement device. Time can be measured using a digital stop watch of a smart phone. It is believed that the joint gap plays an important role in pressure capacity. Pipe joint movement is monitored using a predefined reference and measured by the operator manually using a caliper (Figure 8). The use of crack gauge and laser were attempted, but these methods did not provide the required precision.

Figure 7 Pressure Gauge and Time Measurement

Figure 8 Joint Gap Measurement During the Test

2.5 Sample Sections

Eight tests were performed using conventional RCP ranging having 600 mm, 900 mm and 1200 mm diameter. Single-offset pre-lubricated gaskets provided by two different suppliers were examined. In the routine quality hydrostatic performance test (CSA A257, 2014), the gasket is usually pressurized to 105 kPa, 90 kPa, and 35 kPa internally with straight, deflected and offset alignment, respectively for 10 minutes. The gaskets and pipes tested in this research are listed in Table 2.

Table 2 List of Gaskets Used in the Research

ID	Gasket Name / Model	Type	Supplier	Design Capacity	Remark
S01	7SOC	Wedge	1		Secondary
S03	158-4G	Wedge	2		Secondary
S04	210-4G	Wedge	2		Secondary
T02	TSS 135	Single Offset	1	105	Primary
T03	RFS 165	Single Offset	2	105	Primary
T04	RFS 185	Single Offset	2	105	Primary
T06	RFS 135	Single Offset	2	105	Primary
T07	TSS 185	Single Offset	1	105	Primary

3 RESULTS AND DISCUSSION

3.1 Joint Hydrostatic Capacity

Table 3 exhibits the summary of the infiltration tests conducted and their ultimate and operating pressure results. Total of eight tests were conducted using the equipment and procedure. Three tests were terminated before reaching the ultimate capacity of the testing setup. As part of testing the equipment, the maximum testing pressure was increased from 70% to 95% of the design capacity of the equipment. The other reason for test termination is the displacement of the secondary gasket. For the operating pressure test where the pressure was sustained at 80% of the ultimate pressure, four tests were completed for the minimum 20 hours of duration. Those tests failing to meet the 20-hour hold period were due to the failure of the primary or secondary gasket.

Table 3 Summary of Test Results

Pipe Size	Pri. Gskt.	2nd Gskt.	Infiltration						
			Gap	Straight Alignment (10 mins)			Straight Alignment (20 Hours)		
			mm	kPa	psi	Term. Reason	kPa	Dur.	Term. Reason
600mm RCP	T02	S01	8.6	225	32.6	T02 displaced	180	20 hr	Completed
600mm RCP	T02	S01	10.0	250	36.3	T02 displaced	175	8 min	T02 displaced
600mm RCP	T06	S03	5.5	650	94.3	95% of equip. cap.	525	20 hr	Completed
900mm RCP	T02	S01	9.0	425	61.6	T02 displaced, blow-out	300	0 min	T02 displaced
900mm RCP	T06	S03	8.5	550	79.8	S03 displaced	440	7 min	S03 Displaced
1200mm RCP	T07	S01	11.0	350	50.8	T07 displaced	280	2 min	T07 displaced
1200mm RCP	T04	S04	7.3	500	72.5	73% of equip. cap.	414	20 hr	Completed
1200mm RCP	T03	S03	8.0	480	69.6	70% of equip. cap.	414	22 hr	Completed

1.1 Joint Gap Monitoring

Joint gap was monitored by physical measurement during tests. The average joint gap measured at the time of actual primary gasket failure ranged between 8.6 and 11.0 mm. With less than a maximum 8.0 mm joint gap, the primary gasket did not reach actual failure. The joint gap exhibited an increase of 1 and 2 mm with respect to pressure during the test. This observation is due to the flexible straps and the certain degree of compression of the gasket.

1.2 Leakage Monitoring

Leakage was monitored during the test by visual inspection at each pressure stage and by inspection of the water level displayed in the WSCC. Figure 9 shows typical failure of the testing gasket when the ultimate pressure is determined. Infiltration leakage appears on the inner face of the pipe when the testing gasket is unable to withstand the pressure. Figure 10 illustrates typical failure of the secondary gasket essentially due to its displacement. Failure starts when a water mark appears in between the gasket through the gap. Water seeps through the gap and drips down the outside surface of the concrete. When the water drips down slowly, the tie down straps are adjusted to recompress the secondary gasket. If an improvement is made, the water mark on the surface will evaporate. If the water continues to seep out, pressure is held and the reduction in water level in the cylinder is in a steady pace, the test continues.

Figure 9. 600mm RCP with T02 at 250 kPa

Figure 10. 900 mm RCP Test at 600 kPa

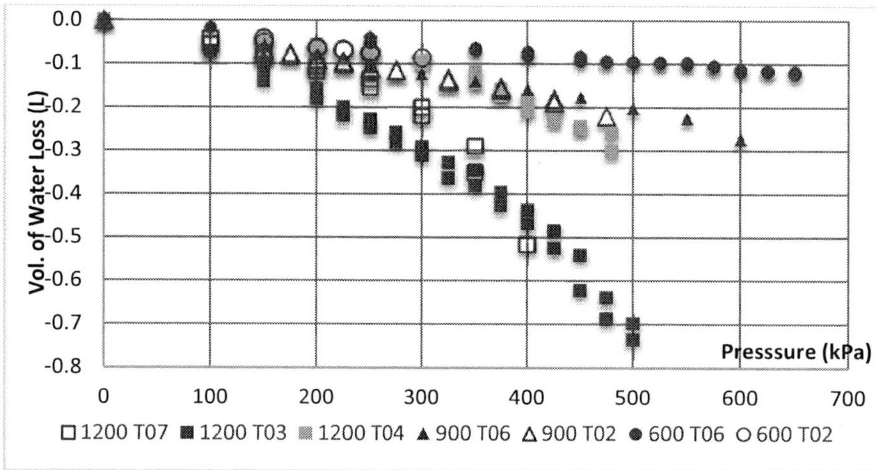

Figure 11 Volume of Water Reduction in the Water Supply Connecting Cylinder

The water level in the WSCC is also used in monitoring the leakage. The water level drops as the pressure increases. Figure 11 shows the volume of water loss calculated based on the water level change in the WSCC plotted against the pressure increase. The relationship is linearly correlated at different rates. The reduction of water volume is attributed to: (a) volume reduction of water under pressure, (b) absorption of concrete, (c) compression of rubber causing increase in annular space, (d) increase in joint gap causing increase in annular space, (e) displacement of the primary gasket, and (f) leakage. The differences in reduction rate reflected in the tests are likely related to displacement of the primary test gasket in its annular space. The single offset pre-lubricated gasket is designed to roll into the narrower gap from the step of the spigot during the homing process of the pipe. This leaves space behind the gasket from the inside of the pipe bell. The joint is more favorable to withstand the internal pressure but not favorable in resisting external pressure as the pressure pushes the gasket towards to the inside of the bell. With this observation, other than leakage, the change in water level in the WSCC due to (a), (b) and (c) is trivial. The joint gap movement during tests seems to consist of very small movement and is believed to be due to the primary gasket displacement causing an increase in the volume of the annular space. It is also observed in the 900 mm RCP test with T02 gasket that the joint failed when the joint gap is greater than 9 mm and the gasket has left the annular space. The amount of displacement is likely related to the rubber mechanical properties and size, but this needs further investigation. Leakage observed in the tests may be slow and controlled at the beginning, or sudden at the time of imminent failure. This can be easily observed and measured. Figure 11 displays the water level in WSCC and the mark of the corresponding pressure.

4 CONCLUSIONS AND RECOMMENDATIONS

4.1 Existing Standards

Existing standards related to RCP joint performance prescribe minimum specification, yet none of these standard tests evaluates the ultimate hydrostatic capacity of the joint design. Although some standards required the test to be performed under straight, deflected and offset alignments, none of the existing standards evaluates the joint's hydrostatic capacity in relation to the joint gap. Tested pipes tested are pressurized internally, therefore such tests do not validate the infiltration capacity of the joint due to the joints non-symmetrical nature. The internal pressure tests mostly limit to 10 minutes which do not reflect the sustaining ground water pressure.

4.2 Testing Procedure

The present research proposes a new RCP infiltration test set up and explores the capacity of existing RCP joint designs. Tests with higher-pressure range were conducted and multiple test were repeated to validate the set-up. It was concluded that the testing procedure is valid and reliable. It is recommended that given a target design ultimate pressure, the test can begin at 10% to 20% lower than the design ultimate. This will significantly reduce the number of test pressure increments and save time.

4.3 Existing Joint Design

The existing RCP joint design and combination of gaskets can withstand higher external pressure than the design intent of 105 kPa internal pressure according to CSA A257. The absolute ultimate pressure of some of the pipe gasket combinations were not reached. However, this test procedure is conducted by controlling the joint gap and does not reflect the actual installation condition of the pipe. Further investigation that involve simulating various levels of joint gap, deflected joint, and offset joint are required to quantify more realistic joint hydrostatic infiltration capacity.

4.4 Monitoring of Leakage

Traditional hydrostatic performance tests rely on visual observation of the leakage. In the proposed test, a WSCC is used to quantify the water volumetric loss during the test. In several cases during this research, the leak was first detected by observing the pace of water reduction exceeding the recorded pace. This eliminates guess work and helps the operator to focus on the weakest point of the setup.

4.5 Joint Gap

Joint gap impacts hydrostatic performance. In conventional hydrostatic testing, CSA A257, ASTM C433, ASTM C1628, BS EN 1916 require the joint to remain watertight with a 13 mm deflection on one side in the deflected alignment test. There is no maximum acceptable joint gap defined anywhere in the specifications. Accordingly, most engineers expect similar performance under 13 mm joint gap in the field. ACPA (2014) states that the manufacturer shall define the maximum allowable joint gap. However, there is no research in the open literature that relates actual hydrostatic performance to joint gap. It is believed that the joint capacity to withstand pressure reduces as the joint gap increases. Further investigation is needed to explore whether this reduction occurs gradually or rapidly.

4.6 Future Works

The proposed test method of RCP infiltration can be the foundation for a standard test procedure for the infiltration of RCP joints. Such a standard test method does not currently exist. This test procedure sets a benchmark for this standard test development. With the encouraging results observed for existing joint and gasket designs, this testing method can allow further development in understanding the relationship between joint hydrostatic performance and joint gap. Finally, the results can be developed into the installation guidance using a quantifiable measurement - joint gap. This can help the contractor to mitigate the infiltration risk and make sound decisions based on the actual field condition. The discovery of the time-dependent gasket movement in the annular space can lead to improvements in the existing single offset joint.

Acknowledgements

The authors acknowledge Con Cast Pipe for manufacturing the tested reinforced concrete pipes and for the financial support to this research project. The authors also acknowledge Hamilton Kent and Press Seal for their contribution in gasket samples.

References

American Concrete Pipe Association (2009). Concrete Pipe Joints Your Best Choice, ePipe. Resource #e-07-124 08/09

American Public Works Association (1971). Prevention and Correction of Excessive Infiltration and Inflow into Sewer System.

AS/NZS 4058. (2007). Precast concrete pipes (pressure and non-pressure). Australia/ New Zealand.

ASSHTO (2009). Pipe Joint Selection for Highway Culvert and Storm Drains, PP63-09

ASTM C361M (2014). Standard Specification for Reinforced Concrete Low-Head Pressure Pipe

ASTM C443. (2012). Standard Specification for Joints for Concrete Pipe and Manholes, Using Rubber Gasket.

ASTM C969 (2017). Standard Practice for Infiltration and Exfiltration Acceptance Testing of Installed Precast Concrete Pipe Sewer Lines

ASTM C990. (2009). Standard Specification for Joints for Concrete Pipe, Manholes, and Precast Box Sections Using Preformed Flexible Joint Sealants.

ASTM C1628. (2017). Standard Specification for Joints for Concrete Gravity Flow Sewer Pipe, Using Rubber Gaskets.

ASTM C1214 (2013). Standard Test Method for Concrete Pipe Sewerlines by Negative Air Pressure (Vacuum) Test Method

ASTM C1103 (2014). Standard Practice for Joint Acceptance Testing of Installed Precast Concrete Pipe Sewer Lines

AWWA C302 (2016). Reinforced Concrete Pressure Pipe, Non-cylinder Type

BS EN 1916. (2002). Concrete pipes and fittings, unreinforced, steel fibre and reinforced. UK.

CSA A257. (2014). Standards for concrete pipe and manhole sections. CSA.

Federation of Canadian Municipalities and National Research Council (2003). Infiltration / Inflow Control Reduction for Wastewater Collection Systems

GB/T 16752. (2006). Test methods of concrete and reinforced concrete drainage and sewer pipe. China PR.

Gottstein, L. E. (1973), Sewer System Evaluation for Infiltration / Inflow, American Consulting Service Inc.

Hey, G., Jonsson, K., Mattsson, A. (2016). The impact of infiltration and inflow on waste water treatment plants: a case study in Sweden.

Kesik, T. (2015) Best practices guide: Management of inflow and infiltration in new urban developments

Indiana Department of Transportation. (2014), Inspection Manual for Precast Concrete Pipe and Structure, Section 6-7

Ministry of Transportation Ontario. (2007). MTO Gravity Pipe Design Guideline. Section 2.4

New York State Department of Transportation. (2013), NYSDOT Geotechnical Design Manual, Section 21.2.11

Pipeline (1999) Infiltration and Inflow can be Costly for Communities, National Small Flows Clearinghouse Spring 1999 Vol. 10, No. 2

Romer, E., Kienow, K., (2004) Rubber gasket Concrete Pipe Joints Eliminating the Smoke and Mirrors, Pipeline Engineering and Construction

Building Tomorrow's Society

Bâtir la Société de Demain

Fredericton, Canada

June 13 to 16, 2018 / *13 au 16 juin, 2018*

NEW TESTING METHOD OF GFRP BARS IN COMPRESSION

Khorramian, Koosha[1] and Sadeghian, Pedram[1,2]
[1] Dalhousie University, Canada
[2] Pedram.Sadeghian@dal.ca

Abstract: Although Glass fiber-reinforced polymer (GFRP) bars have been recognized as an alternative for steel bars in concrete structures and there are standard methods for test them in tension, there is no standardized and convenient test method to figure out their compressive characteristics. Due to lack of information on compressive performance of GFRP bars, their contribution in compression is neglected in current design guidelines. This study introduces a new test method for testing GFRP bars in compression. This test method focuses on providing a circumstance under which the evaluation of compressive crushing strength and modulus of elasticity of GFRP bars would be possible without experiencing buckling of bars. A total of fifteen rebar coupons from two different manufacturers were prepared and tested under concentric compressive loading using the proposed test method. The coupons were divided into three groups including five similar specimens in each group. Different bar dimensions and material properties were considered. Two strain gauges were installed on each specimen to capture the strains at the middle height of the GFRP specimens. Moreover, two steel caps were attached to the end of each specimen to avoid premature failure as well as adjusting the alignment of the coupons. The results showed that the compressive modulus, strength, and crushing strains are consistently predicted using the proposed method. Also, the average ratio of compressive to tensile modulus of elasticity, strength, and strain for each group were between 1.02 to 1.09, 0.67 to 0.85, and 0.58 to 0.82, respectively. The results show the strength and modulus of GFRP bars in compression are close to those of in tension. Thus, GFRP bars can sustaining compression loads and ignoring their compressive contribution in concrete members is not reasonable.

1 INTRODUCTION

Glass fiber-reinforced polymer (GFRP) bars have been considered as an alternative to tensile steel reinforcing bars in construction industry. The major advantage of GFRPs over steel material is their corrosion resistance ability which makes it suitable for structural components susceptible to harsh environmental situations. Moreover, the electromagnetic transparency property of GFRPs make them appropriate for structures which operate magnetic resonance imaging (MRI) units. Because of the demand of GFRP bars in practice, there have been many researches on the flexural behavior of concrete beams reinforced with GFRP bars (Alsayed 1998, Ashour 2006, Benmokrane and Masmoudi 1996, Toutanji and Saafi 2000) as well as concrete slabs (El-Salakawy, Benmokrane and Desgagné 2006, Michaluk, et al. 1998), where the bars used as tensile reinforcement. However, there are a few researches on the capability of GFRP bars in compression (Khorramian and Sadeghian 2017, Tobbi et al. 2012, De Luca et al. 2010), mainly because of the doubts about the function of GFRP bars as well as lack of studies on their behavior in compression.

The reason why the demand for using GFRP bars in compression is not as high as in tension is many negative comments and guideline suggestions to neglect their contribution in the load-carrying capacity of structural members (ACI 440.1R. 2015, CAN/CSA S806-12 2012, Fib Bulletin 40 2007). This mainly arise from lack of research in studying the compressive behavior of FRP bars in compression. Another example is De Luca et al. (2010) that tested concentrically loaded concrete columns reinforced with GFRP bars and conclude that the contribution of FRP bars in compression can be conservatively.

However, the neglection of the contribution of FRP bars in compression is too conservative. Researchers showed that the contribution of GFRP bars in compression is comparable to steel (Tobbi et al. 2012) and experimentally observed considerable strains in compressive GFRP bars, which was more than concrete crushing strain (Mohamed et al. 2014, Khorramian and Sadeghian 2017). Therefore, GFRP bars under compressive stresses can be expected to be accepted and demanded for the solutions which requires corrosion resistance and electromagnetic transparency by considering their contribution to the member strength and stiffness.

The questions then would be evaluation of the characteristics of GFRP bars under compressive loads such as their stress-strain curve, ultimate compressive strength, and their crushing strain. The latter can be assessed using a test method for testing GFPR bars in compression. However, there is no standardized test method which evaluate the crushing strain, ultimate strength, and modulus of elasticity of GFRP bars in compressions. Khan et.al (2015) performed an experimental study on tension and compression testing of FRP bars by adapting the method recommended in standard test method for compressive properties of rigid plastics (ASTM D695 2015) by placing two hardened flat steel plates on top and bottom of specimens. However, their focus was mainly on the comparison of compressive characteristics of GFRP and CFRP specimens tested under compression, and it is noted that the mentioned ASTM standard was not designed specifically for FRP bars. Thus, are some gaps in finding a standardized test method to determine the compressive characteristic of GFRP bars. Therefore, this study is designed to propose a new test method for test of GFRP bars in compression.

2 PROPOSED TEST METHOD

Overall, this test method proposes GFRP bar, embedded in adhesive anchors and steel caps at the end, to be tested in a mechanical testing machine under monotonic compressive load up to failure while tracking the load and longitudinal strain. The purpose of designing this test method was to assist researchers and designers to assess compressive strength, modulus of elasticity, and crushing strain of GFRP bars as well as the stress-strain curve for their compression part. One of the features of this test method is the gripping and alignment is done during the process of preparation of test specimen instead of in the testing machine. The steel cap and adhesive anchors function as gripping method by confining the ends of GFRP bars to avoid premature failure and allow them to obtain their full compressive capacity. In addition, the alignment simply can be done when the adhesive anchors are installed using a level, which can avoid excessive bending and premature failure, and increase the accuracy of test results especially longitudinal strains. Testing five specimens, as shown in Figure 1, is recommended for each test condition to record consistent and accepted data sets.

The schematic illustration of the test specimen is presented in Figure 1. The components of test fixture include steel plates, steel rings, and adhesive anchors. The diameter of steel ring suggested to be considered twice the effective bar diameter while its length recommended to be the same as effective bar diameter, as shown in Figure 1. The steel ring must be thick enough not to be yielded or distorted by lateral pressure of adhesive anchors. The steel ring welds to a steel plate with an square cross section with a width equal 4 times the effective bar diameter with a thickness of at least five millimetres or thick enough not to be punched by the GFRP bars under compression loading. The GFRP bar should be 4 times as long as its diameter to give a free length equal to twice of its diameter.

The procedure begins with the preparation of specimens by building steel caps, the welded steel ring to the steel plates, followed by putting GFRP bars in place for first end of specimen while controlling the alignment, and concluded with doing the same for the other end. The ends of GFRP bar must be completely flat and perpendicular to its longitudinal axis. The alignment of specimen must be checked after putting GFRP bar at the center of steel cap and inserting adhesive into steel ring immediately using level the bar is at the center and completely perpendicular to the surface of the steel cap.

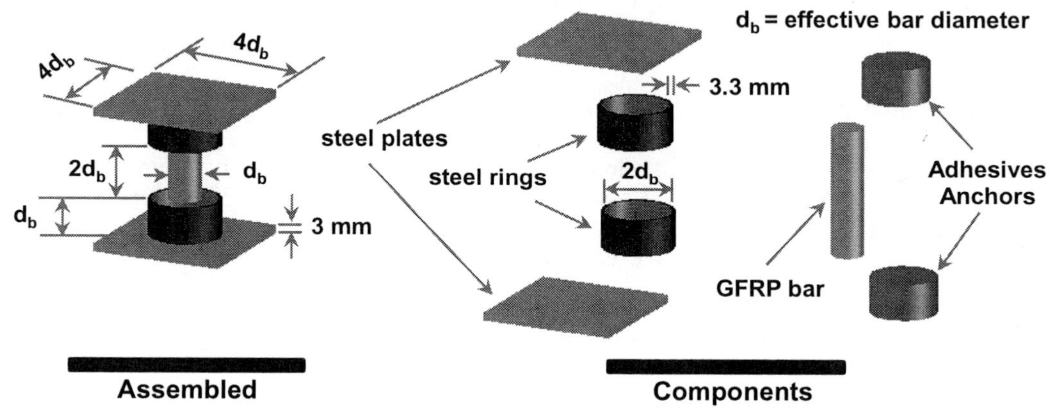

Figure 1: Proposed test fixture

For instrumentation, two strain gauges installed on two opposite sides at the middle of the bar is recommended to record the longitudinal strains. Strain gauges are proper measurement devices for this test set up due to the fact that the free length of bars is very limited, and, from another perspective their accuracy is reliable. The average of two strain gauges is an strong measure of alignment if properly installed. If the difference between the measured strains are not significant, the average value of two strain gauges is considered as longitudinal strain, otherwise, the existing misalignment leads to the creation of bending in bars and test results are not valid.

Once the preparation and instrumentation are done, the specimens can be tested by applying uniform and monotonic compression force. To distribute the load more uniformly to the steel caps, extra steel plates can be added to the ends of the specimens. It is recommended that tests perform using displacement control method with a testing rate that result in conclusion of test in five minutes. The test results are considered as acceptable if no premature failure in caps or buckling in the unbraced length observed. In other words, the tests are successfully performed if crushing of GFRP bars happens at the free length of the specimen.

3 EXPERIMENTAL PROGRAM

In this section, the application of the proposed test method is evaluated through testing three different groups of GFRP bars, differentiating by bar diameter, surface pattern, and manufacturer.

3.1 Test matrix

A total of fifteen GFRP bar specimens with different diameters and material properties were prepared and examined under compression. These specimens divided into three main group consisting five similar specimens. Each specimen is labeled with identification (ID) code as "Gx-y", where "G" stands for group, "x" represents the group number, and "y" shows the specimen number in each group. The test matrix is presented in Table 1. It is noted that the first group of specimens were tested by Fillmore and Sadeghian (2018), and the second group were tested by Khorramian and Sadeghian (2017) using the test method explained earlier in this paper. The last group was specifically tested for this paper.

Table 1: Test matrix

No.	Group	GFRP bar number	Nominal bar diameter (mm)	Specimen IDs	Reference
1	G1	#4	13	G1-1, G1-2, G1-3,	Fillmore and Sadeghian

No.	Group	GFRP bar number	Nominal bar diameter (mm)	Specimen IDs	Reference
				G1-4, G1-5	(2018)
2	G2	#5	16	G2-1, G2-2, G2-3, G2-4, G2-5	Khorramian and Sadeghian (2017)
3	G3	#6	19	G3-1, G3-2, G3-3, G3-4, G3-5	The current paper

3.2 Fabrication

Figure 3 presents the fabrication steps for the third group of testing specimens. The steel caps as well as the machined surface of all five specimens is shown in Figure 2(a). The specimens were put at the center of steel caps and a fast curing anchoring adhesive was applied and filled the empty space between steel cap and GFRP bar while the bar kept leveled and perpendicular to the cap [Figure 2(b)]. After curing of adhesive for both ends, two strain gauges were installed on the machined surface [Figure 2(c)]. The prepared specimens of G3 group are shown in Figure 2(d). It should be noted that the width of the steel square plate used at the ends of the steel caps were kept 50 mm and the same for all groups (instead of four times effective diameter as recommended earlier).

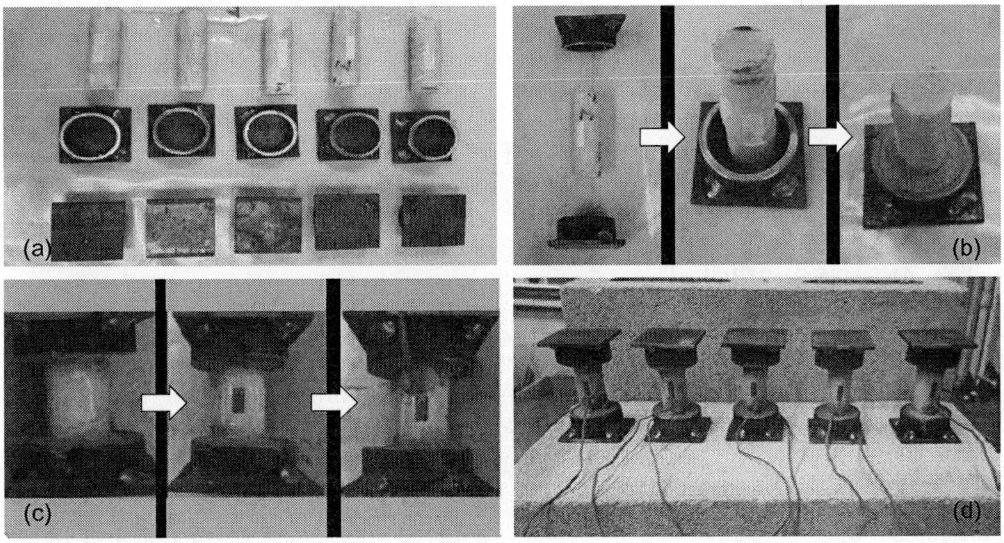

Figure 2: Fabrication of G3 group: (a) specimen components; (b) applying adhesives to bottom end; (c) installation of strain gauges; and (d) prepared specimens

3.3 Test set-up and instrumentation

It is noted that each specimen prepared according to the test preparation mentioned in the proposed test method section. The schematic test set-up and instrumentation is presented in Figure 3(a). To record the strains corresponding to each load step, two strain gauges (namely SG1 and SG2) were installed at the center of the GFRP bar as shown in Figure 3(a). For the sake of preparing the surface of bars for strain gauging, two different approach were used. The first one is to just machine the surface of bar and install the strain gauge directly to the GFRP bar while the second approach is to apply some resin around the center of the GFRP bars and apply the strain gauge on the surface of machined resin instead. In the second approach, it is believed that the cross-sectional area used to calculate the stresses is more accurate. Moreover, in the first approach, by grinding the surface of bar, the probability of hurting specimens by damaging the fibers in GFRP bar due to deeper surface preparation is increased. For group 1 and group 3 the second approach was used to prepare the surface for strain gauging. However, for the specimens in group 2 the first approach was used because of extra sand coat of the bars. As presented in Figure 3(a), the specimen with steel caps sits at the center of a steel plate where the load applies from the top using the loading machine which is presented in Figure 3(b).

To achieve the pure axial state of stress and avoid load eccentricities, a spherical platen was placed in the bottom of the specimens whose function was self centering the specimen in case of accidental eccentricities, as shown in Figure 3(b). In addition, to have more uniform stress in the specimen, two thick steel plates were put at top and bottom of the specimen to distribute the load uniformly at both ends of specimen [Figure 3(b)]. The tests were performed by a universal testing machine, capable of applying 2MN axial load. The loading method was selected to be in displacement increments to give a rate of 0.5 mm/min.

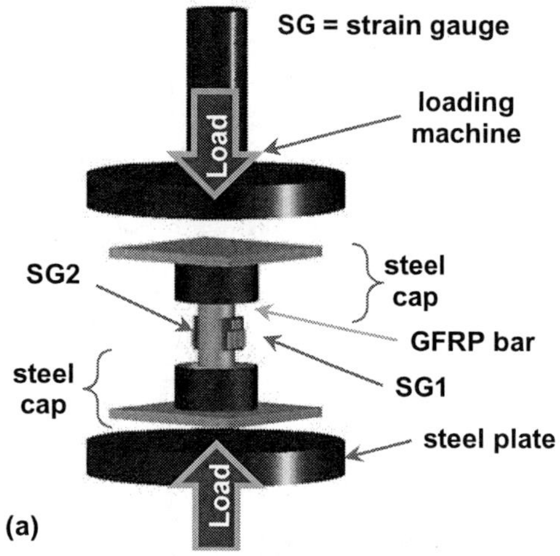

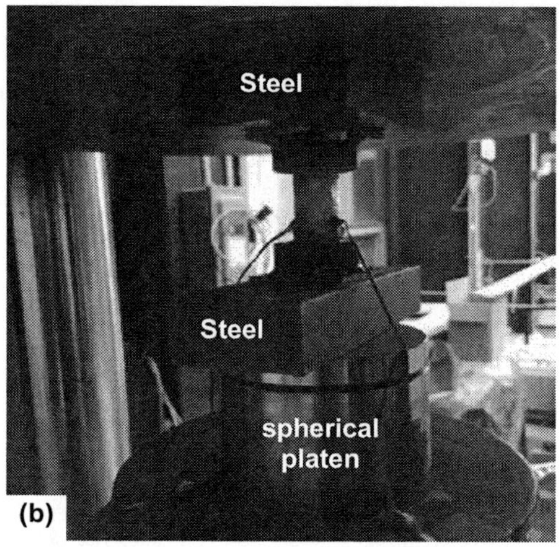

Figure 3: Test set-up and instrumentation: (a) schematic and (b) G3-1 specimen

4 RESULTS AND DISCUSSION

In this section, the results of experimental test data are presented. The modes of failure as well as the stress-strain behavior of GFRP bars tested under pure compression using the proposed test method are presented and a brief discussion and comparison of these compression test results and the tensile characteristics of the same material is presented.

4.1 Failure modes

Figure 4 presented the selected modes of failure of the compression coupons. It is noted that for all specimens no buckling happened before the peak load. The observations during the test showed a noise before reaching to the peak load followed by the crushing of some fibers which happened just before the final crushing of the whole bar and drop in the load. For specimens in group G1, the pattern of failure was like the one shown in Figure 4(a) in which an angled diagonal crushing pattern observed. The test specimens presented in Figure 4(b) except the one that is shown in the figure did not show any observable crushing pattern up to peak load, although their strains and peak loads were similar to the one in Figure 4(b). For specimens in group G3, the crushing pattern was still in the GFRP bar and not in the steel caps as shown in Figure 4(c).

Figure 4: Selected failure of test specimens: (a) G1; (b) G2; and (c) G3

4.2 Stress-Strain Behavior

The stress-strain of relationship of tested specimens for G1 and G2 groups are presented in Figure 5. All specimens experienced a linear stress-strain relationship up to some stage called "proportional limit" in this paper. In other words, if the specimens experience nonlinear behavior (i.e. G2 group), the proportional limit is defined as the point at the beginning of nonlinear part, as shown in Figure 5. For specimens in G1 group, no nonlinear part was observed, however, for G2 and G3 groups a nonlinear part derived by dividing the stroke displacement by a proper gauge length. The gauge length found by setting the slope of stress strain curve derived from stroke equal to the one obtained by the strain gauge [Figure 5(b) and Figure 5(c)]. The average compressive strength, modulus of elasticity, and crushing strain for G1 group was reported by Fillmore and Sadeghian (2018) as 559.03±35.54 MPa, 45.5±1.5 GPa, and 0.0122±0.0012 mm/mm, respectively, while for G2 group, these values were reported by Khorramian and Sadeghian (2017) as 534 MPa, 42.2±1.2 GPa, and 0.0133 mm/mm, respectively, for the graphs up to proportional limit. The average ultimate strength and strain of G2 groups were 738±74 MPa and 0.0190±0.0017 mm/mm, respectively. For the third group, due to some errors in data acquisition system, the first two specimens were failed, however, the rest of specimens showed a modulus of elasticity, proportional compressive strength, and proportional compressive strain of 49.3±0.84 GPa, 645 MPa, and 0.013 mm/mm, respectively, while the ultimate strength and ultimate crushing strain were obtained as 688.8±38.9 MPa and 0.0140±0.0010 mm/mm, respectively. The calculation of modulus of elasticity was done using the portion of the data between the strains of 0.001 mm/mm and 0.003 mm/mm to be compatible with the procedure used for defining the tensile modulus of elasticity of GFRP bars (ASTM D7205 / D7205M - 06 2016). It should be noted that the values of stress strain curves for both G1 and G2 groups are very similar which shows the tests were consistent in terms of main characteristics such as linearity of stress strain curve up to the proportional limit and the prediction of strength and strain corresponding to the proportional limit. For this study, the proportional limit is found using the average of linear parts which was coincidence with the break of strain gauges, as shown in Figure 5. The proportional limit can be studied in further investigations to find a proper criterion for defining this limit and proposing a method to find it.

The tensile strength, modulus of elasticity and rupture strain of GFRP bars for G1 group (Fillmore and Sadeghian, 2018) were reported as 839±49 MPa, 44.2±1.7 GPa, and 0.0209±0.0021 mm/mm, respectively, by performing tensile tests on five coupon specimens, and the same values for G2 group (Khorramian and Sadeghian, 2017) were 629±30 MPa, 38.7±1.5 GPa, and 0.0162±0.0011 mm/mm, respectively. The ratio of strength, modulus of elasticity, and strain at proportional limit in compression to the corresponding values in tension are 0.66, 1.02, and 0.58, respectively for specimens in G1 group while these values are 0.85, 10.9, and 0.82 for specimens in G2 groups. It should be noted the tensile test has not been performed for the third group of specimens due to time limitations.

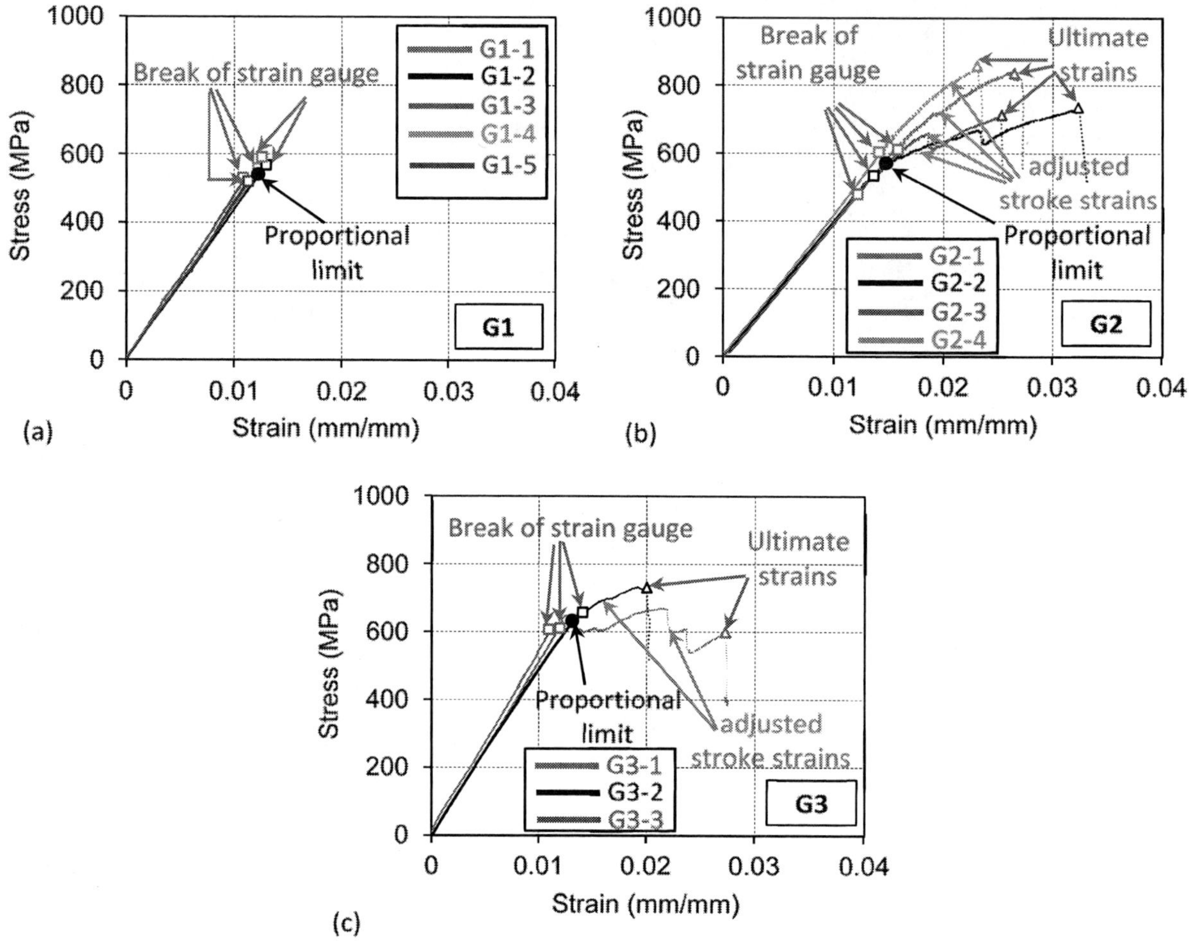

Figure 5: Stress-strain behavior of compressive test specimens: (a) G1; (b) G2; and (c) G3

Overall, these tests showed slightly higher modulus of elasticity in compression and tension, by comparing the results of compressive to tensile tests. However, if the guaranteed tensile characteristics of bars reported by the manufacturer is used for the sake of comparison, the ratio of modulus of elasticity obtained from compressive test of bars to the tensile guaranteed modulus of elasticity (α ratio) will be between 0.99 and 1.07, as presented in Figure 6(c). The compressive strength and strains at the proportional limit were comparable to the tensile corresponding values that emphasizes the potential demand of GFRP bars in compression and the required standardize test method to evaluate their compressive characteristics. The guaranteed tensile characteristics of the tested GFPR bars which were reported by the manufacturer are shown in Figure 6. It was observed that the ratio of compressive strength at the proportional limit to the tensile guaranteed strength (β ratio) is varied between 0.57 to 0.93, and the ratio of the ultimate compressive strength to tensile guaranteed strength (γ ratio) varies between 0.74 to 1.

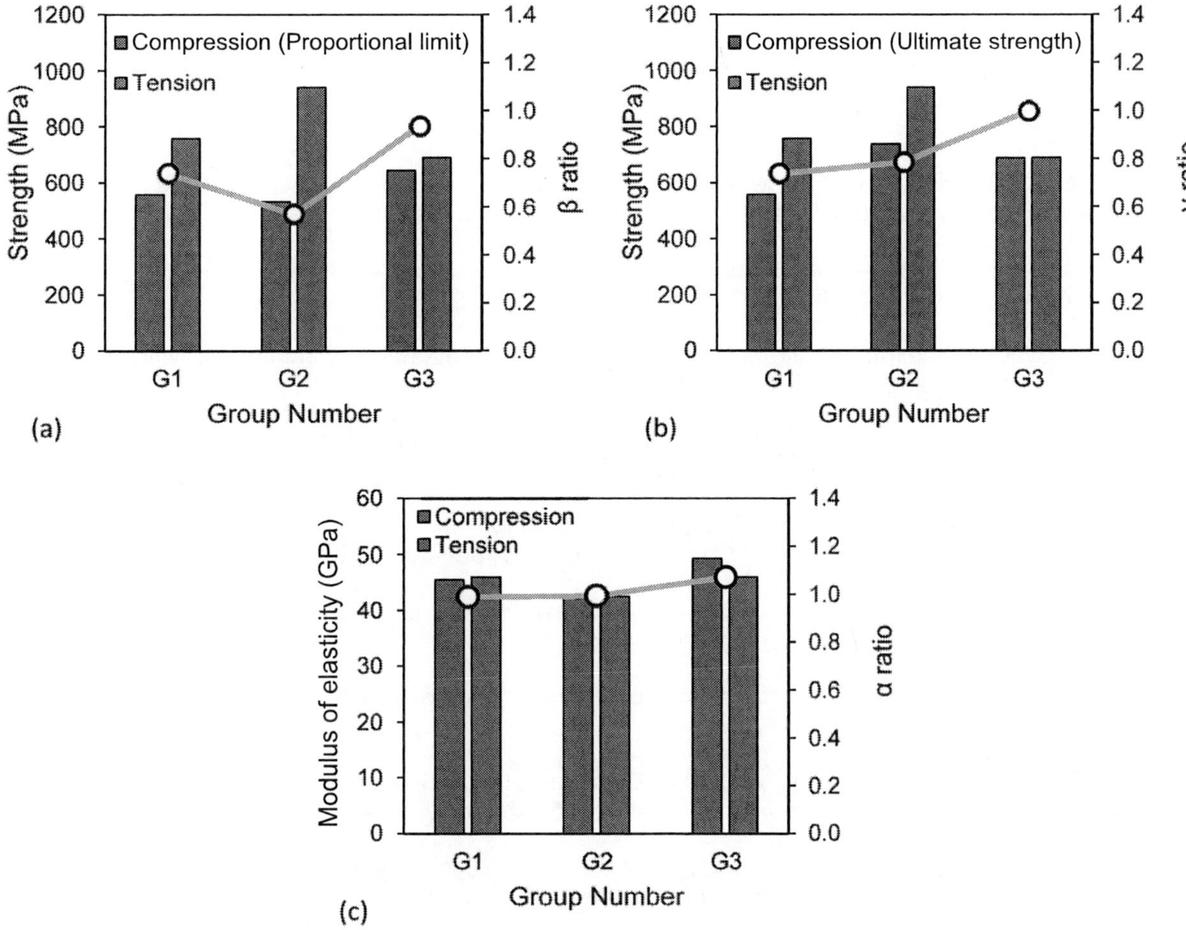

Figure 6: Tensile and compressive characteristics: (a) comparison of strength based on proportional limit; (b) comparison of strength based on ultimate strength; and (c) comparison of modulus of elasticity

5 CONCLUSION

This study proposed a test method for evaluation of the characteristics of GFRP bars tested in compression by providing a condition to avoid the buckling failure mode in bars and obtain crushing mode of failure. The test method suggested in this paper, applied to three set of specimens each of them including five identical specimens. The test groups enable the examination if the test method for GFRP bars with diverse diameters and produced by different manufacturers. The results showed a linear stress-strain relationship of the GFRP bars in compression up to a limit called the proportional limit in this study. The test results were consistent and determined the strength, modulus of elasticity, and the proportional limit or the effective compressive strain of the GFRP bars. The test method has the potential to be improved by introducing a procedure to find the proportional limit. Moreover, the ratio of compressive to tensile strength, modulus of elasticity, and strain for average of tested groups were between 0.66 to 0.85, 1.02 to 1.09, and 0.58 to 0.82, respectively, if the real tensile test results were considered. However, the ratio of compressive modulus of elasticity, proportional compressive strength, and ultimate compressive strength to the corresponding tensile values reported by the manufacturer were between 0.99 to 1.07, 0.57 to 0.93, and 0.74 to 1, respectively. Overall, test results showed that the compressive material properties for GFRP bars are comparable to the tensile properties which strengthened the need for an standardized test method for assessing these properties and showed the potential increase in demand for compressive GFRP bars. Thus, GFRP bars can sustaining compression loads and ignoring their compressive contribution in concrete members is not reasonable.

ACKNOWLEDGEMENT

The authors are grateful for the financial support of the Natural Sciences and Engineering Research Council of Canada (NSERC) and Dalhousie University in conducting this study.

REFERENCES

ACI 440.1R. 2015. Guide for the Design and Construction of Structural Concrete Reinforced Fiber-Reinforced Polymer (FRP) Bars. Farmington Hills, MI: American Concrete Institute.

Alsayed, S. H. 1998. Flexural behaviour of concrete beams reinforced with GFRP bars. Cement and Concrete Composites, **20** (1): 1-11.

Ashour, A. F. 2006. Flexural and shear capacities of concrete beams reinforced with GFRP bars. Construction and Building Materials, **20** (10): 1005-1015.

ASTM D695. 2015. "Standard Test Method for Compressive Properties of Rigid Plastics." American Society for Testing and Materials, West Conshohocken, PA.

A ASTM D7205 / D7205M - 06. 2016. Standard Test Method for Tensile Properties of Fiber Reinforced Polymer Matrix Composite Bars. West Conshohcken, PA: American Society for Testing and Materials.

Benmokrane, B. and Masmoudi, R. 1996. Flexural response of concrete beams reinforced with FRP reinforcing bars. ACI Structural Journal, **93** (1): 46-55.

CAN/CSA S806-12. 2012. Design and construction of building structures with fibre-reinforced polymers. Canadian Standards Association.

De Luca, A., Matta, F. and Nanni, A. 2010. Behavior of Full-Scale Glass Fiber-Reinforced Polymer Reinforced Concrete Columns under Axial Load. ACI Structural Journal, **107** (5): 589-596.

El-Salakawy, E., Benmokrane, B. and Desgagné, G. 2006. Designing and testing of concrete bridge decks reinforced with glass FRP bars. Journal of Bridge Engineering, **11** (2): 217-229.

Fib Bulletin 40. 2007. FRP Reinforcement in RC structures. Stuttgart: the International Federation for Structural Concrete.

Fillmore, B. and Sadeghian, P. 2018. Contribution of Longitudinal GFRP Bars in Concrete Cylinders under Axial Compression. Canadian Journal of Civil Engineering, NRC Research Press, In-press.

Khan, Q. S., Sheikh, M. N. and Hadi, M. N.S. 2015. Tension and compression testing of fibre reinforced polymer (FRP) bars. Joint Conference of the 12th International Symposium on Fiber Reinforced Polymers for Reinforced Concrete Structures (FRPRCS-12) & the 5th Asia Pacific Conference on Fiber Reinforced Polymers in Structures (APFIS-2015). Nanjing, China. 1-6.

Khorramian, K., and Sadeghian, P. 2017. Experimental and analytical behavior of short concrete columns reinforced with GFRP bars under eccentric loading. Engineering Structures, **151**: 761–773.

Khorramian, K. and Sadeghian, P. 2017. Short Concrete Columns Reinforced with GFRP Rebars Under Eccentric Loading. Canadian Society for Civil Engineering, CSCE, Vancouver, Canada.

Michaluk, C. R., Rizkalla, S. H., Tadros, G. and Benmokrane, B. 1998. Flexural behavior of one-way concrete slabs reinforced by fiber reinforced plastic reinforcements. ACI Structural Journal, **95** (3): 353-365.

Mohamed, H. M., Afifi, M. Z. and Benmokrane, B. 2014. Performance Evaluation of Concrete Columns Reinforced Longitudinally with FRP Bars and Confined with FRP Hoops and Spirals under Axial Load. Journal of Bridge Engineering, **19** (7): 04014020.

Tobbi, H., Farghaly, A. S. and Benmokrane, B. 2012. Concrete columns reinforced longitudinally and transversally with glass fiber-reinforced polymer bars. ACI Structural Journal, **109** (4): 551-558.

Toutanji, H. A., and Saafi, M. 2000. Flexural behavior of concrete beams reinforced with glass fiber-reinforced polymer (GFRP) bars. ACI structural journal, **97** (5): 712-719.

Building Tomorrow's Society

Bâtir la Société de Demain

Fredericton, Canada

June 13 to 16, 2018 / *13 au 16 juin, 2018*

PHOTOCATALYTIC NANOMATERIALS FOR WATER TREATMENT: IMPACTS OF WATER QUALITY ON TREATMENT EFFICACY, REUSABILITY, AND NANOMATERIAL RECOVERY

Stephanie Gora[1,3] and Susan Andrews[2]
[1]Department of Civil and Resource Engineering, Dalhousie University, Halifax, NS
[2]Department of Civil Engineering, University of Toronto, Toronto, ON
[3]stephanie.gora@dal.ca

Abstract: Many novel materials for water treatment are tested exclusively using indicator dyes (e.g. methylene blue), single contaminant solutions, or in synthetic water matrices. This includes photocatalytic materials such as titanium dioxide (TiO_2). This reliance on simplified bench-scale testing cannot accurately predict the behaviour of these materials in real water matrices and has led to frequent disappointment when nanomaterial-based treatment processes are scaled up. In this project, two TiO_2-based linear engineered nanomaterials (LENs) were synthesized and evaluated in terms of their ability to remove NOM and disinfection byproduct (DBP) precursors from four Canadian surface water matrices. Nanomaterials cannot feasibly be integrated into a water purification process unless they can be safely separated from the treated water before it is released to consumers and/or the environment, so the recovery of each material via settling and filtration was also explored. Finally, the used LENs were reused up to five times to establish whether they could be reused in a close-loop treatment process. The characteristics of the water matrices had strong impacts on the treatment efficacy, recoverability, and reusability of the nanomaterials. For example, nanomaterial settling was hindered in water matrices with higher concentrations of NOM and water matrices with higher levels of bicarbonate (i.e. alkalinity) were frequently less amenable to photocatalytic treatment. Other parameters such as pH and divalent ions also had strong effects on treatment efficacy and settling behaviour. The results of this project demonstrate that the characteristics of the water matrix being treated will have practical implications for the eventual design and energy efficiency of the UV/TiO_2 process and the projected costs of using TiO_2 photocatalysis for NOM removal in a drinking water context.

1 INTRODUCTION

Advanced oxidation processes (AOPs) couple two existing water treatment processes to exploit their individual capabilities as well as any synergies that might exist between them, most notably the formation of highly oxidative species such as hydroxyl radicals. UV/TiO_2 is a well researched but poorly adopted AOP that brings together TiO_2 nanomaterials, usually spherical nanoparticles with diameters below 50 nm, and UV light below 380 nm, which can be provided by conventional germicidal and "black" light lamps as well as by the sun. The UV light is absorbed by the nanoparticles and drives the release of electrons and electron holes, which can interact with water and oxygen to form reactive oxygen species (ROS), including hydroxyl radicals. The ROS then degrade contaminants that are adsorbed to the surface of the nanoparticle.

Disinfection byproducts (DBPs) are formed when species present in water interact with the chemicals added to drinking water for disinfection purposes. Previous research has demonstrated that UV/TiO_2 can degrade the precursors to organic DBPs but that full mineralization may not occur within a time frame amenable to full scale drinking water treatment system operation (Gora et al., 2018). This phenomenon, and its subsequent effects on overall DBP formation, are water matrix specific (Gora et al, 2018; Liu et al., 2010).

In this paper, we review the impacts of water quality on the removal of DBPs and DBP precursor surrogate compounds by commercially available nanoparticles and various LENs and their practical implications on the energy demand and cost of using UV/TiO_2 to treat different water matrices. We also briefly address the impacts of water quality on nanomaterial removal via settling and filtration.

2 METHODS AND MATERIALS

The sections below briefly describe the methods, materials, and calculations used in this project. Additional details regarding the materials, apparatus, and sample preparation methods can be found in earlier publications related to the project (Gora et al. 2018A; Gora et al., 2018B).

2.1 Materials

Raw water samples were gathered from the inlet of four surface water treatment plants in Ontario. Two of the plants use river water sources (Ottawa River and Otonabee River) and two use lake water sources (Lake Ontario and Lake Simcoe). The chemical characteristics of the four water matrices are summarized in Table 1.

Table 1: Raw water characteristics of four Ontario surface water matrices used in this study

Parameter	Units	Lake Ontario	Lake Simcoe	Otonabee River	Ottawa River
DOC	mg/L	1.6 – 2.0	3.4 – 3.6	3.8 - 5.2	4.8 – 6.8
UV254	cm^{-1}	0.02 – 0.03	0.05	0.09 – 0.15	0.16 – 0.24
SUVA	L/mg.m	0.8 – 1.5	1.4 – 1.5	2.0 – 3.3	2.9 – 4.0
Percent Humics[1]		47%	50 - 52%	56 - 58%	62 - 73%
pH		7.8 – 8.0	8.1 – 8.3	7.8 - 8.2	6.9 – 7.5
Alkalinity	mg as $CaCO_3$/L	89 – 92	115 – 122	84 – 86	26 – 28
Calcium[2]	mg/L	32 – 33	42 – 43	30 – 35	6 – 10
Chloride[2]	mg/L	23 – 26	42 – 45	10 – 13	2 – 5
Magnesium[2]	mg/L	8	7.6 – 7.8	3	2 – 3
Sodium[2]	mg/L	13 – 14	24 – 25	6 – 7	2 – 5
Conductivity[2]	µsiem/cm	300 - 315	378 – 408	188 – 244	61 – 112

[1]Percent humics based on LC-OCD fractionation results from Gora and Andrews, 2017; Zheng et al., 2015, and Taylor-Edmonds, personal communication, 2017

[2]Ontario Drinking Water Surveillance Program (2010-2012)

Evonik Degussa P25 nanoparticles were used without further modification as a treatment standard or as the base material for the synthesis of the LENs. The LENs were synthesized using a hydrothermal method as described in previous publications (Gora et al., 2018B). The temperatures used in the synthesis and calcination steps of the process were varied to manipulate the size and crystallinity (crystal phase structure) of the resulting nanomaterials. The morphology of the nanomaterials was characterized using SEM and TEM while their crystallinities were confirmed using HRTEM with SAED. The surface charge and surface area of each LEN was determined using zeta potential analysis and BET surface area analysis, respectively. The results of these characterization experiments are reported extensively elsewhere (Gora et al., 2018B).

2.2 Sample Preparation and Analysis

Adsorption experiments were conducted by spiking raw water with varying concentrations of TiO2 and allowing them to mix in an end-over-end mixing apparatus for four hours. All of the dye and NOM degradation experiments included triplicate or quadruplicate samples prepared by dosing samples of raw Otonabee River or Ottawa River water with P25 nanoparticles or one of the LENs and exposing them to 0, 5, 15, 30, 45, and 60 minutes of UVA irradiation. The samples were treated in a UV/TiO_2 batch reactor that consisted of four UVA lamps secured to a stand above a multiple location stir plate that was able to accommodate four beakers at once. The UVA LED bulbs (LZ1 UV 365 nm Gen2 Emitter, LED Engin Inc.) had a maximum irradiance at 365 nm. The average irradiance across the surface of the sample was calculated using a spreadsheet developed by Bolton and Linden (2003) and was determined to be 4.9 mW/cm^2. The irradiance of each lamp was confirmed before each test using a radiometer (International Light, ILT1400) equipped with a sensor optimized to measure light at 365 nm (International Light, XRL140B). Samples were analyzed for DOC and UV light absorbance at 254 nm and, as appropriate, for chlorine demand and disinfection byproduct formation potential. The concentration of methylene blue in the raw and treated water was determined by measuring the absorbance of the solution at 665 nm using a UV-Vis spectrophotometer.

2.3 Nanomaterial Removal

Bench-scale sedimentation and filtration tests were conducted to Aliquots of a 10 g/L TiO_2 stock solution were dispensed into an appropriate amount of MilliQ water to create triplicate samples for the settling tests. The samples were poured into the turbidimeter cuvette, sonicated for 5 minutes, and placed in the turbidimeter for a total of 2 or 3 hours. The turbidity at the midpoint of the cuvette was recorded at the beginning of the test and at ten minute intervals thereafter. The Time to Filter Test (Standard Methods 2710 H) was used to evaluate the filterability of P25 and the four LENs in different water matrices. Essentially, the time to filter a set volume of sample containing the nanomaterials was normalized to the amount of time required to filter pure MilliQ water through the filter under identical filtration conditions to yield a unitless value, the filtration index. A high filtration index implies that the suspension is resistant to filtration while a value close to 1 indicates that it has the same resistance to filtration as pure water.

2.4 Statistics and Calculations

One way ANOVA with Tukey's test was used to determine whether significant differences existed between different treatment levels. All analyses were conducted at the 95% confidence level. Error bars represent the confidence interval around the mean unless specified otherwise. The electrical energy per order (EEO) concept is currently listed as a "figure of merit" for the evaluation of advanced oxidation processes by IUPAC and is calculated as follows:

[1] $$EEO\left(\frac{\frac{kWh}{order}}{m^3}\right) = \frac{1000 \times Power\ Input\ (kW) \times Time\ (min)}{Volume\ Treated\ (L) \times \log\left(\frac{Initial\ Concentration\ (\frac{mg}{L})}{Final\ Concentration\ (\frac{mg}{L})}\right)}$$

An alternative way to compare the efficiency of different treatment systems is to calculate the cost of the energy required to remove a contaminant from a given volume of water:

[2] $$Cost\ per\ Volume\ (\$/m^3) = System\ Power\ Requirements\ (kWh/m^3) \times Cost\ of\ Power\ (\$/kWh)$$

3 RESULTS AND DISCUSSION

3.1 NOM Adsorption to TiO_2

Unlike other AOPs, the interaction between TiO_2 and organic contaminants includes two steps: adsorption, which can occur with or without light, and degradation upon exposure to light. Figure 2 shows the mass of DOC and amount of UV254 removed from four water matrices via adsorption alone. The P25 nanoparticles removed only modest amounts of DOC from the Lake Ontario, Lake Simcoe, and Otonabee River water matrices via adsorption and there were no significant differences (95% confidence level) in the mass of DOC removed from these matrices. Nearly twice as much DOC was adsorbed from the Ottawa River water matrix as from the other water matrices, indicating that the NOM in this matrix was more amenable to adsorption than that in the others. As discussed in Section 2.1, previous studies have demonstrated that Ottawa River contains more NOM, especially aromatic NOM, than the other water matrices used in this study. Aromatic NOM is more likely to be adsorbed onto TiO_2 than non-aromatic NOM (Gora et al., 2018).

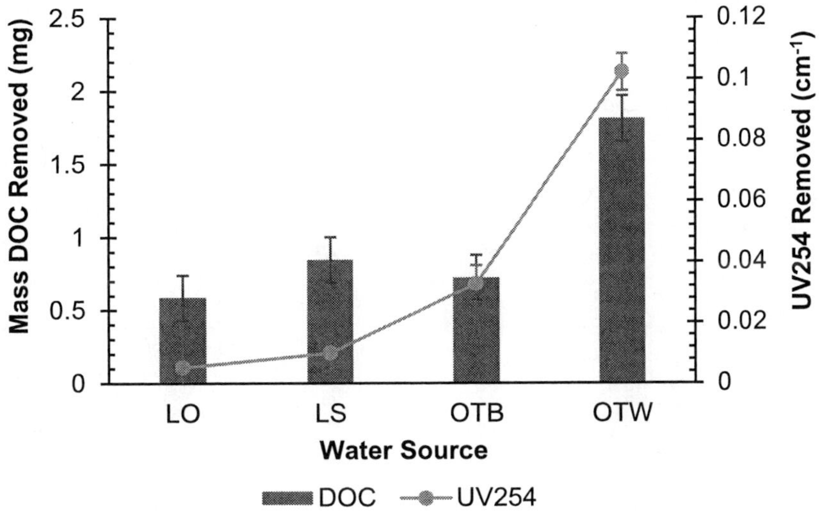

Figure 1: Removal of dissolved organic carbon (DOC) and UV254 from Lake Ontario water (LO), Lake Simcoe water (LS), Otonabee River water (OTB), and Ottawa River water (OTW) via adsorption onto P25 nanoparticles

In general, higher ionic content has been shown to increase nanomaterial agglomeration (Erhayem & Sohn, 2014). This, in turn, decreases the overall amount of surface area available and reduces the number of adsorption sites available for NOM to attach to. The agglomeration occurs because the ions depress the electrical double layer that surrounds the nanoparticles, decreasing the repulsive forces between individual nanoparticles and allowing attractive forces to dominate as they approach one another (Hotze et al., 2010). Not all ions affect adsorption in the same way, however, and there is evidence that some, notably calcium, can encourage NOM adsorption to TiO_2 (Liu et al, 2013). The Otonabee River water matrix and the two river water matrices contained more ions than the Ottawa River water matrix, and as such, the nanoparticles suspended in the Ottawa River water likely formed smaller agglomerates and had more surface area available for adsorption. pH values near the isoelectric point of a nanomaterial can also encourage agglomeration and subsequent decreases in available surface area and NOM adsorption. Changes in pH can also impact the configuration and charge of NOM molecules and the surface charge of the nanomaterials themselves. As a result (Mwaanga et al. 2014), NOM adsorption is most likely to occur at pH levels between the isoelectric point of the nanomaterial (pH 6.5 for P25) and the K_{PA} of NOM (approximately pH 3), and is particularly effective around pH 4 to 4.5, at which point the nanoparticles are well dispersed and positively charged, and thus most likely to attract the negatively charged NOM molecules to their surfaces (Gora & Andrews, 2017; Mwaanga et al., 2014). The four water matrices employed in this study all had a pH of approximately 8, which is above the isoelectric point of the nanoparticles and at a point where they would be partially dispersed and negatively charged, and thus less likely to adsorb NOM in general (Gora & Andrews, 2017).

The effects of LEN structure and crystallinity on adsorption behaviour have been discussed in previous publications (Gora et al., 2018A; Gora et al., 2018B). Basically, LENs with higher specific surface area adsorbed more NOM than those with less surface area. Surface area was a function of the synthesis conditions, and the LENs calcined at 550°C had higher surface area than those calcined at 700°C and thus adsorbed more NOM. Like P25, the LENs adsorbed more NOM from the Ottawa River water than from other water matrices.

3.2 NOM Degradation

Figure 2 shows the percent change in DOC, UV254, and THMfp when Otonabee River water (Figure 2A) and Ottawa River (Figure 2B) when these two water matrices were dosed with 0.25 g/L of TiO_2 in the form of P25 nanoparticles or the two LENs discussed previously and exposed to UV LED light at 365 nm for 60 minutes (UV dose = 17.6 J/cm^2).

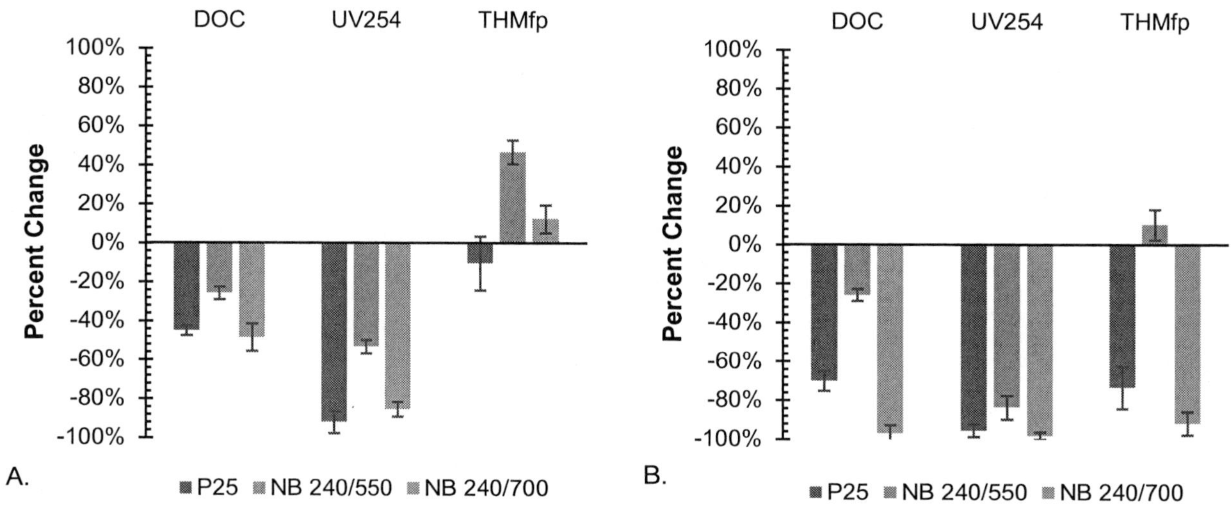

Figure 2: Percent change in DOC, UV254, and THMfp in Otonabee River water (A) and Ottawa River water (B) after 60 minutes of exposure to 0.25 g/L of TiO_2 and 60 minutes (17.6 J/cm^2) of UV LED light

All three nanomaterials were more effective for DOC and UV254 removal when used to treat Otonabee River water vs. Ottawa River water. UV/TiO_2 treatment actually increased the THMfp of the Otonabee River water relative to the initial raw water. The effects of UV/TiO_2 treatment on the THMfp of the Ottawa River water were material dependent, with NB 240/700 proving to be particularly. Indeed, throughout this study, NB 240/700 was strongly impacted by the water matrix used for the experiments. Results presented elsewhere demonstrated that this material was particularly effective for hydroxyl radical formation despite its low surface area, suggesting that, unlike with NB 240/550 and P25, UV/TiO_2 treatment with this nanomaterial proceeded mainly via hydroxyl radical mediated reactions. Ions such as bicarbonate can scavenge hydroxyl radicals, preventing them from interacting with target contaminants, in this case NOM. The Otonabee River water had a much higher alkalinity (a measure of bicarbonate content) than the Ottawa River water (84 to 86 mg as $CaCO_3$/L vs. 26 to 28 mg as $CaCO_3$/L), and this alkalinity likely had a particularly strong inhibitory effect on NB 240/700 relative to NB 240/550 and P25.

3.3 Methylene Blue Degradation and Comparison to NOM Removal

Methylene blue degradation has been used by many researchers as a measure of the effectiveness of photocatalytic systems (Mills, 2012). In this study, P25 and NB 240/700 were nearly equal in terms of their ability to degrade methylene blue, achieving 86 ± 4 % and 80 ± 2 % removal, respectively, after 30 minutes, suggesting that they were equally photocatalytically active under the experimental conditions employed in this study. The LENs that were calcined at 550°C were less photocatalytically active than P25 or the materials calcined at 700°C, achieving 40 ± 5% decolourization of methylene blue. In most cases, methylene blue degradation by the LENs was a good fit to a first order degradation model ($R^2 > 0.9$). This is in line with the findings of other researchers (Mills & McFarlane, 2007). The only material where this relationship did not have a R^2 above 0.90 was NB 240/700, which was a result of the very fast degradation observed between 0 and 5 minutes.

The Pearson correlation coefficients between methylene blue degradation and DOC degradation ranged from 0.74 for NB 240/700 in Ottawa River water to 0.97 for NB 240/550 in Otonabee River water. The correlation between methylene blue degradation and UV254 removal was much stronger than that with DOC for all three materials, with Pearson correlation coefficients ranging from 0.91 for NB 240/700 in Ottawa River water to 0.99 for NB 240/550 in Otonabee River water. There was no obvious relationship between correlation strength and water type. Methylene blue degradation was not a good predictor of the effects of UV/TiO_2 treatment DBPfp in either water source because DBPfp invariably increased relative to the raw water at short irradiation times. This finding has important implications for researchers evaluating the effectiveness of new photocatalytic materials because they suggest that surrogate compounds are not adequate indicators of a material's ability to improve the overall safety of drinking water.

3.4 Nanomaterial Separation

Previously published results (Gora et al., 2018B) showed that P25 and the four LENs settle out of purified water roughly in order of particle size. That is, P25, the smallest of the nanomaterials, settled the most slowly, while NB 240/700, one of the largest materials, settled the fastest. The trend was nearly reversed in the real water matrices. The proposed explanation for this was that the ions in the real water matrices compressed the electrical double layer around the nanomaterial particles, encouraging agglomeration (Liu et al., 2013). Some of the LENs, in particular the larger ones, settled more slowly in the real water matrices than in purified water. This was likely due to the presence of NOM in the river water as NOM has been shown to hinder the settling of nanomaterials (Liu et al., 2013). Indeed, the settling of the LENs was more hindered in the Ottawa River water than the Otonabee River water, likely because the former contained more NOM and fewer ions (Gora et al., 2018A). Although settling trends were not explored in depth in this study, the trends over time in Otonabee River water and Ottawa River water suggest that settling behaviour was complex, differed from one water matrix to the next, and that the dominant settling mechanism may have changed over time. In the Ottawa River water, for example, the LENs settled in a slow but steady manner, suggesting Type I settling (discrete particle settling). In contrast, the settling behaviour of P25 in Ottawa River water and all three materials in the Otonabee River water matrix was characterized by large initial drop in turbidity followed by slower, more gradual settling after approximately 15 minutes. This more complex settling pattern suggests that the particles, or more correctly, agglomerates, were not of uniform size in these suspensions.

Throughout this project, the water matrix impacted the filterability of industry standard P25 nanoparticles but not the lab synthesized LENs. In fact, filtration indexes for the latter were not statistically distinguishable from raw water containing no nanomaterials. This was likely because the LENs, whether present as single particles or agglomerates, were always larger in at least one dimension than the pores of the filter used for separation (0.45 µm). In contrast, the removal of P25 nanoparticles by filtration, or rather the resistance to filtration caused by the P25 nanomaterials during filtration, was strongly impacted by the water matrix. Specifically, P25 was more easily removed from the Otonabee River water than from purified water or Ottawa River water, likely because the higher concentration of ions in this water matrix drove the formation of larger, more filterable agglomerates.

3.5 Practical Implications

The practical implications of the bench scale findings presented in this paper and our previous work, for example the projected energy requirements and costs for systems operated with different TiO_2 nanomaterials, will determine the future feasibility of UV/TiO_2 for the treatment of different water matrices.

3.5.1 Process Design

Reaction rates determine reactor size and thus capital and operating costs. The Langmuir-Hinshelwood model is frequently used to model photocatalytic degradation:

[3] $$\frac{dC}{dt} = \frac{k_A K_L C_A C_{ROS}}{1+K_L C_A}$$

Where k_A is the observed reaction rate constant, K_L is the Langmuir adsorption constant, C_A is the concentration of the contaminant, and CROS is the concentration of ROS generated during UV/TiO_2 treatment. In reactions conforming to this model, an increase in the apparent rate constant should result in an increase in overall degradation, and seeing as the reaction rate constants for all parameters (DOC, UV254, THMfp, HAAfp) were higher in the Ottawa River water than in the other water matrices, dC_{NOM}/dt in the Ottawa River water should be higher than in the other water sources and thus the reactor volume required to treat this water matrix would be smaller than that required to treat the other water matrices. NOM is, however, heterogenous, and the parameters with which it was quantified in this study are all bulk parameters. Some, such as THMfp and HAAfp, are likely to increase at short reaction times as parent compounds are broken down into reactive intermediates. This does not fit the assumptions of the Langmuir-Hinshelwood model.

NOM removal is rarely the primary goal of UV/TiO_2 treatment, however, as the non-specific nature of the oxidative species produced during photocatalysis makes it an attractive option for the removal of otherwise recalcitrant chemicals such as taste and odour compounds. Brame et al. (2015) proposed a model for the UV/TiO_2 mediated degradation of organic contaminants that takes into account light attenuation (i.e. competitive light absorption by NOM or other inhibitory compounds), ROS scavenging by ions, and competitive adsorption by NOM and other species.

[4] $$\frac{dC_A}{dt} = \frac{-P_{ROS,0}}{1+\frac{k_N C_N (F+K_N S)}{k_A C_A (F+K_A S)}} 10^{-\mu l C_N}$$

C_A represents the concentration of the target contaminant, C_N represents the concentration of NOM (or other inhibitory compound), $P_{ROS,0}$ describes the ROS production rate without any light attenuation, k_N and k_A are the degradation rate constants for contaminant degradation and NOM degradation, respectively, while K_A and K_N are the adsorption constants for the contaminant and NOM, and F and S are terms that summarize other equations that describe the competitive adsorption and degradation of the contaminant and NOM (and other inhibitory compounds). A higher initial concentration of NOM would increase C_N, resulting in a decrease in the light attenuation factor ($10^{-\mu lCN}$), an increase in the denominator value, and a consequent decrease in dC_A/dt. Thus, all of the water matrices employed in this study would be expected to decrease the degradation rates of other contaminants, though this decrease may be related to different factors. For example, in the Ottawa River matrix, the high concentration of NOM would reduce dC_A/dt by increasing light attenuation and by competing with the target contaminant for adsorption sites and oxidative ROS species. In the Lake Ontario matrix, in contrast, degradation would be more likely to be inhibited by ROS scavenging by ions in the water matrix.

Brame's model does not explicitly account for nanomaterial agglomeration and does not differentiate between different ROS and/or photo-generated electron holes. The model also assumes that contaminant and NOM adsorption occur according to a multi-solute Langmuir adsorption isotherm model, but previous studies by our group and others have shown that NOM adsorption to TiO_2 is best described by a modified Freundlich model (Gora & Andrews, 2017). Nonetheless, the Brame model describes many of the important factors governing contaminant degradation via UV/TiO_2 and helps to elucidate the likely impact of different matrix components on system operation.

The overall process complexity and cost of any UV/TiO_2 system will also be impacted by the method used for nanomaterial separation and its effectiveness. In this study, filtration was more effective than sedimentation for nanomaterial removal and less likely to be impacted by the water matrix used as long as the nanomaterials, or their agglomerates, were large enough to avoid becoming stuck in the pores of the filter. As such, it will likely be the only option for nanomaterial recovery in water matrices containing NOM, an important implication for the space and energy costs associated with UV/TiO_2 systems.

3.5.2 Operating Costs

The sharply different DOC and, especially, THMfp degradation trends observed in the Otonabee River and Ottawa River matrices will direct impact the energy required to treat the different sources. The electrical energy per order (EEO), power per treated volume, and cost per treated volume to remove 90% of the initial DOC present in the raw Otonabee River water and Ottawa River are presented in Table 2.

Table 2: Electrical energy per order (EEO), power per volume, and cost per volume to reduce the DOC of two water matrices via UV/TiO_2 treatment with P25 nanoparticles and two linear engineered nanomaterials

Material	EEO		kWh / L		Cost / m^3	
	Otonabee River	*Ottawa River*	*Otonabee River*	*Ottawa River*	*Otonabee River*	*Ottawa River*
P25	225	113	0.23	0.11	$ 34	$ 17
NB 240/550	450	450	0.45	0.45	$ 68	$ 68
NB 240/700	180	53	0.18	0.05	$ 27	$ 8

A study by Yen and Yen (2015) explored the use of UV/H_2O_2 for DOC and THMfp removal from a synthetic water matrix made with commercial humic acids. Their experiments were conducted using a 9 W low pressure UV lamp (maximum irradiance at 254 nm) and three doses of H_2O_2. Under the experimental conditions used by Yen and Yen (2015) and in the current study, UV/H_2O_2 with was more efficient for DOC removal with P25 in Ottawa River water but comparable to UV/TiO_2 treatment with NB 240/700 in Ottawa River water.

The EEO concept is useful for comparing different types of light-driven systems and processes, however, it implicitly assumes first order degradation kinetics, and in the current study it was only possible to calculate EEO values for THMfp removal under two conditions, Ottawa River water with P25 and Ottawa River water with NB 240/700, because these were the only conditions under which first order degradation kinetics were observed. The EEO values for THMfp removal by NB 240/700 in Ottawa River water were comparable to those UV/H_2O_2 treatment with 10 mg/L of H_2O_2 (Yen & Yen, 2015), indicating that under some conditions and with highly photoactive materials, UV/TiO_2 may prove to be competitive with UV/H_2O_2. This should be confirmed by comparing the two processes in the same water matrices.

The cost to reduce the DOC of the Otonabee River and Ottawa River water matrices via UV/TiO_2 treatment was calculated assuming a rate of 15.7 cents per kilowatt hour, which was the on-peak cost of energy in Ontario in May 2017 when this analysis was initially conducted. The predicted cost to remove 90% of the DOC from the Otonabee River water was generally higher than that to remove DOC from the Ottawa River water, by a factor of 2 for P25 and 3 for NB 240/700. The cost to reduce the THMfp of the Ottawa River water by 90% ($7 to $15) and to within current Health Canada recommended values ($4 to $6) was considerably less because it was not necessary to achieve full mineralization of the parent NOM compounds to see a reduction in the overall THMfp of the water. In all cases, however, the cost per volume calculated in this study was orders of magnitude above that reported for full scale water treatment systems (Statistics Canada, 2011) and should only be used to compare between different TiO_2 nanomaterials and water matrices. Power costs and payment schemes in other Canadian jurisdictions vary in terms of magnitude and complexity, further complicating the analysis.

The cost estimates presented in this paper may overestimate the actual cost of operating a UV/TiO_2 treatment system because they were developed based on the bench-scale batch experimental set-up used for the experiments. In this configuration the light penetration into the samples was essentially limited to the top layer of water. In a real application, it is more likely that the system would exist as a flow through, serpentine reactor studded with UV LEDs. Depending on the width of the channel and the dose of TiO_2 added to the water, the amount of light penetration into the sample would likely be higher than that achieved in this project. This could substantially reduce the amount of power required for the irradiation step because each individual nanoparticle would be exposed to a higher dose of incident light and thus the time required to achieve the photocatalytic oxidation of DBP precursors would be shorter.

4 CONCLUSIONS

Throughout this study, the characteristics of the water matrix had important effects on the removal of DBP precursors via adsorption and degradation by TiO_2 nanomaterials. Degradation proceeded more slowly in water matrices that contained higher levels of ions, in particular bicarbonate (alkalinity) and calcium, possibly because these ions can act as ROS scavengers but also perhaps because higher concentrations of ions can compress the electrical double layer that surrounds the nanoparticles, encouraging agglomeration and an overall reduction in the surface area available for reaction (Liu et al., 2013). The reduced effectiveness of photocatalytic treatment in water sources with higher alkalinity has practical consequences in terms of energy use and the eventual operating costs of UV/TiO_2 treatment processes.

In general, the results of this study suggest that UV/TiO_2 treatment for DBP precursor removal will only be appropriate for low alkalinity waters with high NOM levels. Many regions of Canada, including parts of northern and eastern Ontario, the Maritime provinces, Newfoundland and Labrador, coastal British Columbia, eastern Manitoba, and various parts of Quebec use surface water sources that meet these criteria (Statistics Canada, 2011), and the single step treatment process might be competitive is at the very small scale. Small systems, which produce between 100 and 1,000 L of water each day, provide water for consumptive uses for small populations or full water services for residences or institutional facilities in areas where municipal water is not available for consumptive or non-consumptive uses (e.g. nursing stations). For example, the preliminary energy costs for the UV/TIO_2 treatment are comparable to the initial O&M cost estimates for 1,000 L/day potable water units used in some rural and remote communities in Newfoundland and Labrador (Picco & Chaulk, 2010).

Small communities often possess lower economic and operational capacity than larger ones and the water treatment systems in these communities sometimes employ minimal or non-traditional treatment processes (e.g. chlorination only), which can increase the risk of DBP formation (Guilherme & Rodriguez, 2014). Small systems are also often located in rural and remote areas, which can increase the cost of shipping of equipment and chemicals to the plant. UV/TiO_2 treatment is conceptually simple and, if the photocatalyst is recycled, has the potential to be relatively low energy and sustainable, which may make it particularly attractive for small communities in Canada.

Acknowledgements

Funding was provided through NSERC Strategic Project Grant (STPGP 430654-12), the NSERC Canada Graduate Scholarship, the Ontario Queen Elizabeth II Scholarship, and the Engineers Canada-Manulife Scholarship. The authors would like to acknowledge Leonardo Furtado, Michelli Park, Yijun (Jessie) Gai, Jenny Yue, Kennedy Santos, and Jim Wang for their help conducting laboratory work for this project.

References

Brame, J., Long, M., Li, Q., & Alvarez, P. (2015). Inhibitory effect of natural organic matter or other background constituents on photocatalytic advanced oxidation processes: Mechanistic model development and validation. *Water Research*, **84,** 362–371

Erhayem, M., & Sohn, M. (2014). Stability studies for titanium dioxide nanoparticles upon adsorption of Suwannee River humic and fulvic acids and natural organic matter. *Science of the Total Environment*, **468–469**, 249–257

Gora, S. L., & Andrews, S. A. (2017). Adsorption of natural organic matter and disinfection byproduct precursors from surface water on TiO_2 nanoparticles: pH effects, isotherm modelling and implications for using TiO^2 for drinking water treatment. *Chemosphere*, **174**, 363-370

Gora, S. L., Liang, R., Zhou, Y. N., & Andrews, S. A. (2018). Photocatalysis with easily recoverable linear engineered TiO_2 nanomaterials to prevent the formation of disinfection byproducts in drinking water. *Journal of Environmental Chemical Engineering*, **6** (1), 197–207

Gora, S., Liang, R., Zhou, Y. N., & Andrews, S. (2018). Settleable engineered titanium dioxide nanomaterials for the removal of natural organic matter from drinking water. *Chemical Engineering Journal*, **334,** 638–649

Guilherme, S., & Rodriguez, M. J. (2014). Occurrence of regulated and non-regulated disinfection by-products in small drinking water systems. *Chemosphere*, **117** (1), 425–432

Hotze, E. M., Phenrat, T., & Lowry, G. V. (2010). Nanoparticle Aggregation: Challenges to Understanding Transport and Reactivity in the Environment. *Journal of Environment Quality*, **39** (6), 1909-1924

Liu, S., Lim, M., Fabris, R., Chow, C. W. K., Drikas, M., Korshin, G., & Amal, R. (2010). Multi-wavelength spectroscopic and chromatography study on the photocatalytic oxidation of natural organic matter, *Water Research*, **44**, 2525-2532

Liu, W., Sun, W., Borthwick, A. G. L., & Ni, J. (2013). Comparison on aggregation and sedimentation of titanium dioxide, titanate nanotubes and titanate nanotubes-TiO_2: Influence of pH, ionic strength and natural organic matter. *Colloids and Surfaces A: Physicochemical and Engineering Aspects*, **434**, 319–328

Mills, A. (2012). An overview of the methylene blue ISO test for assessing the activities of photocatalytic films. *Applied Catalysis B: Environmental*, **128**, 144–149

Mills, A., & McFarlane, M. (2007). Current and possible future methods of assessing the activities of photocatalyst films. *Catalysis Today*, **129** (1–2 SPEC. ISS.), 22–28

Mwaanga, P., Carraway, E. R., & Schlautman, M. A. (2014). Preferential sorption of some natural organic matter fractions to titanium dioxide nanoparticles: influence of pH and ionic strength. *Environmental Monitoring and Assessment*, **186** (12), 8833–8844

Picco, B., & Chaulk, M. (2010). Clean and Safe Drinking Water Workshop. In *Clean and Safe Drinking Water Workshop*.

Statistics Canada. (2011). Survey of Drinking Water Plants, (16). Retrieved from www.statcan.gc.ca/pub/16.../16-403-x2013001-eng.pdf

Yen, H. Y., & Yen, L. S. (2015). Reducing THMFP by H_2O_2/UV oxidation for humic acid of small molecular weight. *Environmental Technology*, **36** (4), 417–423

Zheng, D., Andrews, R. C., Andrews, S. A., & Taylor-Edmonds, L. (2015). *Effects of coagulation on the removal of natural organic matter, genotoxicity, and precursors to halogenated furanones. Water Research*, **70**, 118-129

Leadership in Sustainable Infrastructure

Leadership en Infrastuctures Durables

Fredericton, Canada

June 13 to 16, 2018 / *13 au 16 juin, 2018*

PHYSICAL AND MECHANICAL PROPERTIES OF FLAX LIME CONCRETE

Garikapati, Krishna Priyanka[1], and Sadeghian, Pedram [1,2]
[1] Dalhousie University, Canada
[2] Pedram.Sadeghian@dal.ca

Abstract: Due to increase in production of construction materials, the emission of carbon dioxide levels and the requirement for production materials with a less environmental footprint is gaining more attention. One of the solutions for the problem is to replace high carbon dioxide emitting construction materials with plant-based materials, for example flax shives. Flax shives are obtained from the inner part of flax stem and produced as by-products of flax fiber production, from textile industry. So, as a solution to decrease amount of environmental footprint this paper involves the study of a special kind of concrete (without stone aggregates) made of flax shives and a lime binder. This product is called flax lime concrete hereafter. The study was mainly based on physical and mechanical properties of flax lime concrete like density, load capacity, failure mode, and load-deflection behavior. To study these properties three flax lime concrete beam specimens were prepared with a rectangular cross-section (152 mm x 51 mm) and length of 610 mm. The mix proportions were kept constant for all three specimens to check how loading and displacement were being varied for all three specimens of same mix proportions. A layer of jute, fiber-based fabric mesh was placed at the center of the specimens as reinforcement. Finally, specimens were tested under three-point loading to study the physical and mechanical properties of specimens. The flax lime concrete used in this study shows an average density of 559 kg/m^3 and flexural strength of 1.0 MPa. This paper is a part of an in-progress project focused on the application of flax lime concrete for prefabricated blocks for filling the cavity of building walls as an insulation.

1 INTRODUCTION

The manufacture of concrete raw materials and other building materials requires large amount of energy for their production also produces high levels of green house gases especially carbon dioxide (CO_2). A remedy for the problem is using plant-based materials like flax and hemp, which can absorb CO_2 as they grow also helps to replace non-renewable fibers such as fiberglass. Flax and hemp materials have a good flexural strength, so they can be used as reinforcement. These natural fibers are obtained from stalk (stem) of plants as short and long woody by-products. In this study short woody by-products known as shives and hurds were used, whereas long natural fibers have been used in the form of fiber-reinforced polymer composites (Hristozov et al. 2016, Sadeghian et al. 2018, and Betts et al. 2018). The main aim of this research is to study how short flax shives mixed with a lime-based binder work as a construction material, mainly as an insulation and/or filler material in the cavity of walls in buildings. The product can be cast-in-place for the wall cavities or it can be prefabricated as blocks or panels and then be placed in the wall cavities. This research mainly targets the prefabricated panels made of flax shives.

There are no or limited studies on flax shives, but there are a good number of studies on hemp shives. So, hemp lime concrete studies are taken as reference. Ip and Miller (2012) had investigated the levels of emission of greenhouse gases throughout the life cycle of a hemp-lime concrete wall (1.0 x 1.0 x 0.3 m^3). The mix proportions of binder used for hemp lime concrete wall, 75% of hydrated lime,15% of hydraulic lime, and 10% of pozzolans. The amount of CO_2 emission was 36.08 kg per wall from the time of construction till the end of the lifespan of the wall with the volume of 0.3 m^3. Nguyen et al. (2009) studied hemp lime concrete (a mixture of hemp shives plus a lime-based binder) also known as hempcrete. The hemp lime concrete was made of a binder containing 75% non-hydraulic lime of total volume of binder, 15% hydraulic lime, and 10% pozzolan. The ratio of binder to hemp was changed from 2.12 to 2.77 (for low compaction) whereas water to binder ratio was kept constant as 0.5. It was concluded that the hemp-lime concrete had a great deformation capacity prior to compression failure when heavy compaction was applied whereas, at low compaction, hemp lime concrete had low resistance to compression. Barnat-Hunek et al. (2015) investigated the mechanical and thermal properties of hemp-lime concrete. Six batches of hemp-lime concrete were prepared and several tests were conducted to see how properties were being varied with different mix proportions. Three cubic specimens (100 mm dimension) were prepared from each batch to test bulk density and absorptivity. It was concluded that if the amount of hemp shives increased, absorptivity decreased, and bulk density increased. In addition, there was reduction in the compressive strength with increase in hemp percentage.

There are several experiments conducted on the thermal conductivity of hemp-lime concrete. Thermal conductivity is mainly varied by the moisture content and density of hemp lime concrete. Generally, hemp-lime concrete is highly porous with open and interconnected porosity. A study by Amziane and Arnaud (2013) showed that hemp lime concrete with the density of 200 and 600 kg/m^3 had a low thermal conductivity of 100 W/m. k. They compared hemp lime concrete to other types of concrete of same density (e.g. cellular concrete). Another study made by Benfratello et al. (2013) concluded that thermal conductivity was a function of the amount of hemp shives used in hemp lime concrete. Also, when hemp was increased from 20 to 40% of total weight of the mixture, the density raised from 369 to 611 kg/m^3 and thermal conductivity decreased from 140.8 to 94.7 W/m.k.

Collet and Pretot (2014a) studied the hygrothermal behavior of hemp-lime concrete walls made of precast blocks under environmental conditions (temperature and relative humidity). The blocks were made with mix proportions of 72% of lime by weight, 28% hydraulic lime, hemp to binder ratio of 0.65, and water to binder ratio of 1.2. A test wall of dimensions 2.3 m long, 2.1 m high, and 0.3 m thick was made by using blocks. Holes were provided in the blocks to place wooden frame for networks. Thermocouples were used for temperature measurement whereas sensors were used for measurement of both temperature and relative humidity. After four months of drying, the wall was stabilized to the ambient conditions at 23 °C and 40% relative humidity. Then, measurements were performed on uncoated and coated wall. The wall was coated with the breathable commercial lime-based coating. The coated wall showed a reduction in vapor pressure through the wall and delayed vapor diffusion because of additional moisture resistance.

In another study, Collet and Pretot (2014b) performed an experiment to know how moisture content and density of hemp-lime concrete affects the thermal conductivity. For this purpose, they made a wall (2.3 m long, 2.1 m high, and 0.3 m thick) with a different binder to hemp ratio. The binder was made of 78% hydrated lime, 15% hydraulic binder, and 10% pozzolanic binder. From the experiment, it was concluded that if the density of hemp-lime concrete increased by 54% of its original value, the thermal conductivity increased by 66.66% of its original value. Also, when the moisture content of hemp-lime concrete increased from 0 to 90%, the thermal conductivity increased by only 17.5%. Therefore, the density of hemp-lime concrete had a major impact on the thermal conductivity of hemp-lime concrete.

Walker and Pavia (2016) performed tests on the moisture transfer and thermal properties of hemp-lime concrete by varying binder. Six mixes of various mix proportions of binder, hemp, and water proportions were used to cast wall with dimensions 1.0 x 0.9 x 0.3 m^3. Out of six mixes, the materials like cement and lime of two specimens were completely replaced by ground granulated blast furnace slag (GGBS), metakaolin and remaining four specimens were made as it is with lime and cement. Walls were tested for water permeability, thermal conductivity, and heat capacity. The test results proved that binder type did not had any notable influence on permeability, thermal conductivity, and heat capacity of the specimens. Kinnane et al. (2016) investigated regarding acoustic absorption nature of hemp-lime concrete. Hemp-lime concrete walls (1.0 x 1.0 x 0.3 m^3) were built using six mixes by varying proportions of the binder. They divided six mixes into two sets (with pozzolans and without pozzolans). Specimens were tested using an impedance tube testing. The walls with pozzolans showed higher sound absorption coefficient at different frequencies of sound.

The above studies proved that using by-products like hemp shives can have many environmental advantages to the construction industry. However, there are no or limited studies on flax shives. Therefore, the project was started by expecting the same results may also be possible with flax shives. The main aim of this project is to use flax shives to make flax lime concrete as an insulation material in wall cavities of buildings. As a result, the use of conventional synthetic materials can be avoided. This paper presents the first stage of the project focusing on limited beam specimens to get an idea of the basic physical and mechanical properties of flax lime concrete.

2 EXPERIMENTAL STUDY

2.1 Materials

For the production of flax lime concrete, flax shives (with the length less than 25 mm) were obtained from a local farm in Nova Scotia, Canada. The average bulk density of flax shives was measured as 141 kg/m^3 by following standard procedure. The average bulk density was obtained by filling a cylindrical container (100 mm diameter and 200 mm height) in three layers with dry shives and compaction was done with a tamping rod (25 strokes per layer). The mixture of binder was prepared by fast mixing autoclaved lime (92% hydrated) and General Use Portland cement. They were mixed in a ratio of 4:1 (i.e. for every four cups of lime, one cup of cement was used). Then flax shives, mixture of binder and water were mixed in proportion of 4:1:1 for preparation of flax lime concrete (i.e. for every four cups of flax, one cup of binder and one cup of water was used). Three specimens (Beam 1, Beam 2, and Beam 3) were prepared for testing with a rectangular cross-section (152 mm x 51 mm) and length of 610 mm. An all-purpose burlap (jute fabric mesh) was used at the center of each specimen, to act as reinforcement for the specimen. The openings of the mesh (approximately 5 mm x 5mm) helped to provide an integration between flax lime concrete and the reinforcement. All specimens were tested under three-point bending.

2.2 Specimen Preparation

A formwork shown in Figure 1 was used to cast the specimens. Grease was applied to the formwork. Ten cups of binder was prepared initially. To prepare binder, Lime (8 cups) and cement (2 cups) were well mixed until a uniform mixture of the binder was obtained. Then, 10 cups of flax were taken into a bowl and 2.5 cups of binder were added to it. Flax shives and binder were mixed thoroughly by hand until a uniform mixture was obtained. Then, 2.5 cups of water were added and again mixed thoroughly by hand. All the mixture was transferred into formwork and compaction was done for 2 to 3 minutes using a tamping rod with 25 mm x 25 mm cross-section that is 150 blows per layer. The procedure was repeated, and the mixture was transferred into the formwork. When formwork was filled to half (2 layers of mixture), all-purpose burlap mesh was placed. Then, the same process was repeated until formwork was filled. Compaction was done in 4 layers. For each layer using the tamping rod, blows were given along the whole layer in the same direction. Then, the specimen was covered with a plastic cover and transferred to curing room. The specimen was placed in curing room for 7 days and then transferred to the lab to be dried at room temperature. After at least 28 days after casting the average density of the specimens was obtained 559 kg/m^3.

2.3 Test Setup and Instrumentation

For testing the specimens, a hydraulic loading system was planned to be used. But due to the sensitivity of the specimens, olden days method of placing weights at the center was used to test the specimens. So, to test the actual specimen the setup shown in Figure 2 was made. Two roller supports, two dial gauges one plate and weights were used for the testing specimen. The span between rollers was maintained as 558.8 mm. A small plate was placed on a specimen which was of negligible weight. This plate was helpful to measure deflection in the specimen by using dial gauges. Once everything was set, markings on specimen were done and weights were placed at the center of a plate and the dial gauges were read. The weights were added to each specimen until it failed.

Figure 1: Specimen preparation: (a) formwork; (b) adding binder to flax; (c) mixing binder with flax; (d) adding water to mix; (e) mixing water with flax and binder; (f) finally obtained mix; (g) formwork filled to half to place burlap; and (h) casting and compaction

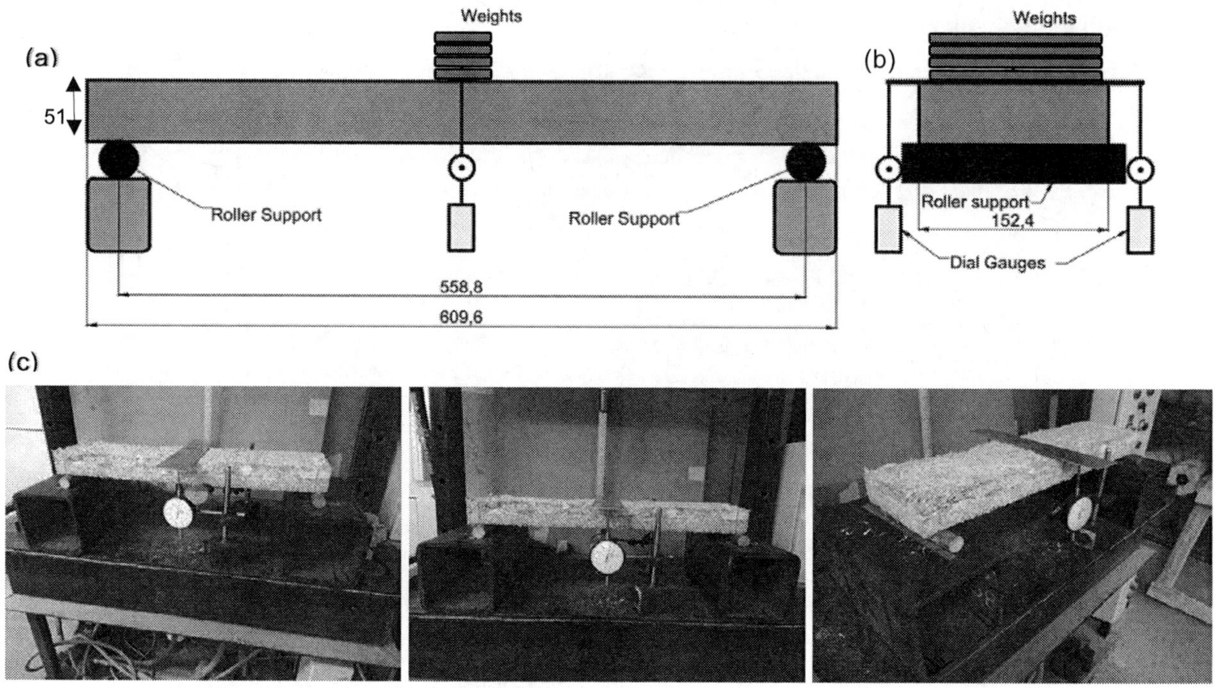

Figure 2: Test setup (dimensions in mm): (a) side view; (b) cross-section; and (c) photos of test setup

3 RESULTS AND DISCUSSION

3.1 Failure Modes

In all three specimens, flax shives were debonded on the application of weights. Due to debonding of flax shives, small flexure crack was initiated at bottom of the specimen, which was slowly propagated to the top by producing a cracking sound. The width of crack was observed to be increased on addition of weights. So, to provide support for the specimen all-purpose burlap was placed at half of the depth which can act as reinforcement. Burlap had shown a good effect on the specimen. Because it was observed that when the specimen was failed the crack can only propagate up to the depth where burlap was placed. So, the upper half of the specimen remained as it is without any damage. As shown in Figure 3, the cracks observed in all three specimens were similar to plain concrete specimen tested under bending, but the propagation of cracks was gradual.

Figure 3: Failure modes: (a) Beam 1; (b) Beam 2; and (c) Beam 3

3.2 3.2 Load-Deflection Behavior

Figure 4 shows the mid-span displacement of all three specimens. Since all the specimens had same mix proportions the mid-span displacement for all three specimens were nearly similar. In Figure 4(a), the point 'A' indicates the starting stage of cracking in the first the specimen. Once, after crack starts the first specimen produced about 2 mm of mid-span displacement before failing, where point 'B' indicates the failure of specimen. The Figure 4(a) also shows that displacement increased linearly with increase in load. In case of the second specimen, as shown in Figure 4(b), the total displacement produced was less than the first specimen. But, the displacement produced by second specimen from point 'A' to 'B' that is from starting stage of crack to failure was similar to first specimen. Figure 4(b) represented a linear relationship between load and displacement before Point "A". Figure 4(c) represents that total displacement produced by the third specimen was nearly equal to the first specimen and greater than the second specimen. All specimens produced nearly same displacement from cracking stage of Point "A" to failure at Point "B".

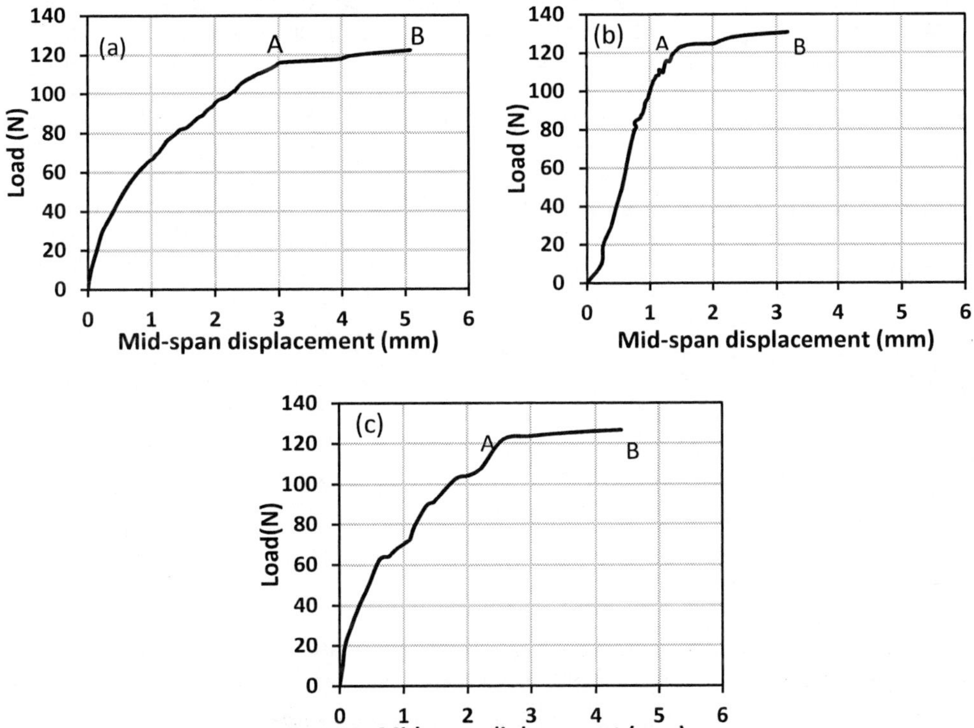

Figure 4: Load-deflection behavior: (a) Beam 1; (b) Beam 2; and (c) Beam 3

Table 1 shows the summary of the test results. The average loads at Point A and B were obtained 121 and 126 N, respectively. Also, the average displacements at Point A and B were obtained 2.3 and 4.2 mm, respectively. The displacement from Point A to B was almost 2 mm consistently. The average moment at mid-span at Point A (cracking) can be obtained as 17 N-m, which results in 1.0 MPa cracking strength of the flax lime concrete used in this study. The span/180 of the specimens provides a deflection of 3.1 mm, which is larger than deflection at cracking point. This means the blocks need to be stiffer. The solution can be placing the burlap mesh at a lower level closer to the flexural surface.

Table 1: Summary of test results

Specimen	Point A		Point B	
	Cracking Load (N)	Displacement (mm)	Failure Load (N)	Displacement (mm)
Beam 1	114	3.0	122	5.1
Beam 2	128	1.3	131	3.2
Beam 3	122	2.6	127	4.4
Average	121	2.3	126	4.2

4 CONCLUSIONS

In this paper, three beams specimens were prepared with a rectangular cross-section (152 mm x 51 mm) and length of 610 mm. The mix proportions were kept constant for all three specimens to check how loading and displacement were being varied for three specimens of same mix proportion. A layer of jute fiber-based fabric mesh (burlap) was also placed at the center of the specimens as reinforcement. After at least 28 days after casting, the average density of the specimens was obtained 559 kg/m^3. The specimens were tested under three-point loading up to failure. The following conclusions can be drawn:

- Failure crack in all three specimens was similar to flexure crack in the concrete. In all three specimens, the crack was started at center and propagated in an upward direction. Whereas the failure crack was not observed from center to the top surface since the burlap was placed at the center of depth. From the point of cracking to failure, the specimens did not take much load.
- From load-deflection curve, it can be observed that all the three specimens on average took about a load of 121 N at cracking, which corresponds to 1.0 MPa flexural strength. The total displacement produced by specimens on average was 4.2 mm. But, all the three specimens deflected by the same amount from cracking point to failure point.
- The usage of burlap at the center has provided good support to the specimen. But still, research is going on to change the position of burlap and then using flax lime concrete as insulation material. This paper is a part of an in-progress project focused on the application of flax lime concrete for prefabricated blocks for filling the cavity of building walls as insulation.

References

Amziane S, Arnaud L. 2013. *Bio-aggregate-based Building Materials: Applications to Hemp Concretes.* John Wiley & Sons, Inc., Hoboken, New Jersey, USA.

Barnat-Hunek D, Smarzewski P, Fic S. 2015. Mechanical and thermal properties of hemp-lime composites. *Composites Theory and Practice*, **15**(1): 21-27.

Benfratello S, Capitano C, Peri G, Rizzo G, Scaccianoce G, Sorrentino G. 2013.Thermal and structural properties of a hemp–lime biocomposite. *Construction and Building Materials*, **48**: 745-754.

Betts D, Sadeghian P, and Fam A., 2018. Experimental behavior and design-oriented analysis of sandwich beams with bio-based composite facings and foam cores. *Journal of Composite for Construction.* In-Press.

Collet F, Pretot S. 2014a. Experimental highlight of hygrothermal phenomena in hemp concrete wall. *Building and Environment*, **82**: 459-466.

Collet F, Pretot S. 2014b. Thermal conductivity of hemp concretes: Variation with formulation, density and water content. *Construction and Building Materials*, **65**: 612-619.

Hristozov D, Wroblewski L, and Sadeghian P. 2016. Long-term tensile properties of natural fibre-reinforced polymer composites: comparison of flax and glass fibres. *Composites Part B: Engineering*, **95**: 82-95.

Ip K, Miller A. 2012. Life cycle greenhouse gas emissions of hemp lime wall constructions in the UK. *Resources, Conservation and Recycling*, **69**: 1-9.

Kinnane O, Reilly A, Grimes J, Pavia S, Walker R. 2016. Acoustic absorption of hemp-lime construction. *Construction and Building Materials.* **122**: 674-682.

Nguyen TT, Picandet V, Amziane S, Baley C. 2009. Influence of compactness and hemp hurd characteristics on the mechanical properties of lime and hemp concrete. *European Journal of Environmental and Engineering*, **13**: 1039-1050.

Sadeghian P, Hristozov D, and Wroblewski L. 2018. Experimental and analytical behavior of sandwich composite beams: Comparison of natural and synthetic materials. *Journal of Sandwich Structures & Materials*, **20** (3): 287–307.

Walker R, Pavía S. Moisture transfer and thermal properties of hemp–lime concretes. 2014. *Construction and Building Materials*, **64**: 270-276.

Building Tomorrow's Society

Bâtir la Société de Demain

Fredericton, Canada

June 13 to 16, 2018 / *13 au 16 juin, 2018*

POTENTIAL APPLICATIONS OF SELF-CLEANSING NANO LOTUS LEAF BIOMIMICKED COATING IN DIFFERENT CONSTRUCTION SECTORS

Safiuddin, Md.[1,3], Hossain, K.[2], and Collins, C.M.[2]
[1] Angelo Del Zotto School of Construction Management, George Brown College, Toronto, Ontario, Canada; Department of Civil Engineering, Ryerson University, Toronto, Ontario, Canada
[2] Department of Civil Engineering, Faculty of Engineering and Applied Science, Memorial University of Newfoundland, Newfoundland and Labrador, Canada
[3] safiq@yahoo.com (corresponding author's email)

Abstract: Lotus leaf has a nanostructure with randomly oriented cone-like protrusions (nano-mountains) and nano-valleys between nano-mountains. The nanostructured lotus leave surface possesses an extremely high water-repelling capacity due to super hydrophobic property. This is well-known as "lotus effect", which has been mimicked by applying certain recently developed coating products. The "lotus effect" makes nano lotus leaf biomimicked coating suitable for applications in different categories of construction. The benefits of nano lotus leaf biomimicked coating are emphasized in this paper. It highlights the possible applications of nano lotus leaf biomimicked coating in various construction sectors, including buildings and infrastructure, such as roads, bridges, tunnels and sewers. In particular, the self-cleansing effect of nano lotus leaf biomimicked coating is focused to reveal its applicability in different types of construction. In addition, the self-cleansing mechanism of nano lotus leaf biomimicked coating for the protection of substrates is explained with illustrations. The present study also includes the different types of substrate, the techniques for the preparation of substrates, and the application methods for installing nano lotus leaf biomimicked coating on various substrates. Above all, some recommendations are given to improve the performance of nano lotus leaf biomimicked coating.

1 INTRODUCTION

Nanotechnology is a recent thrust in materials science although it has been in nature since long. It deals with the "boundary between atoms and molecules and the macro world, where ultimately the properties are dictated by the fundamental behaviour of atoms" (Ashby et al. 2009). Nanotechnology involves the behavior of materials at nanoscale. A radical change might occur in material performance and new applications in construction sectors would be possible if nanotechnology can be applied innovatively.

Current research shows that lotus leaf itself is using nanotechnology to obtain non-wetting, super water-repellent or super hydrophobic surface for self-cleansing (Kumar et al. 2011; Mozumder et al. 2011; Neinhuis and Barthlott 1997). The self-cleaning property of lotus leaf is linked with its special surface structure. Lotus leaf has a nanostructure with randomly oriented cone-like protrusions (Cheng et al. 2006; Li et al. 2017; Stratakis et al. 2009; Yamamoto et al. 2015), as shown in Figure 1. The nanostructured lotus leave surface possesses an extremely high water-repelling capacity, as illustrated in Figure 2 and Figure 3, due to super hydrophobic property (Button et al. 2004, Zorba et al. 2008). As a result, lotus leaf powerfully repels water, which carries away dirt as it rolls off the leaf, allowing the lotus to remain clean despite growing in turbid water (Robbins, 2001). This is widely known as "lotus effect" that has been mimicked on different substrates using some recently developed coating products. The "lotus effect" makes nano lotus leaf biomimicked coating suitable for manifold applications in different construction sectors such as buildings, bridges, pavements, and drainage infrastructure (Gonzalez et al. 2013, Gonzalez et al. 2016, Zhang et al. 2016).

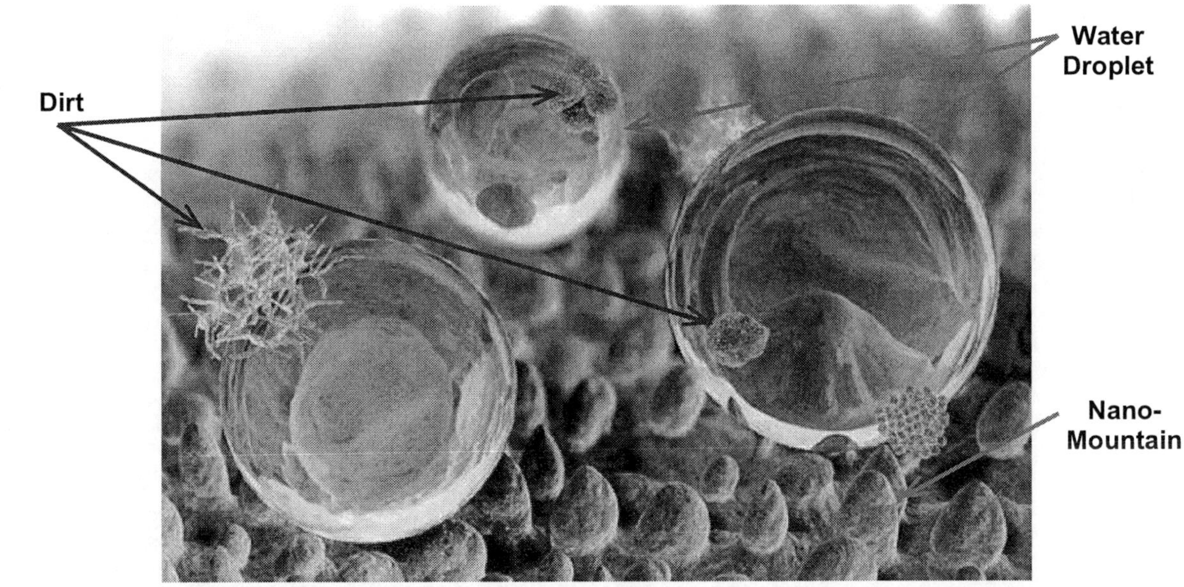

Figure 1: Water drops suspended on nanostructured lotus leaf (adapted from Rame-Hart 2015)

Figure 2: Natural cleansing of lotus leaf (adapted from Design and The Universe 2017)

This study focuses on the possible applications of nano lotus leaf biomimicked coating in various construction sectors. The working mechanisms and benefits of nano lotus leaf biomimicked coating are discussed in this paper. The self-cleansing effect of nano lotus leaf biomimicked coating is particularly highlighted to reveal its applicability on different structural systems. Furthermore, this paper imparts various substrate types, substrate preparation techniques, and application methods for applying nano lotus leaf biomimicked coating. Especially, some recommendations for improving the performance of nana lotus leaf biomimicked coating are given in this paper.

Figure 3: Cleaned lotus leaf in dirty water (adapted from Rame-Hart 2015)

2 SELF-CLEANSING MECHANISM

Lotus leaf has super hydrophobic surface. It remains clean and dry in spite of frequent exposure to dirt and rain (Forbes 2008). This is because lotus leaf contains nano waxy bumps (nano-mountains) and nano-valleys between them that prevent the accumulation of dirt and water. The nano-valleys between the nano-mountains are much smaller than the dirt particles. Therefore, the dirt particles stay on the tops of the nano-mountains since they cannot get into the nano-valleys. The nano-valleys between the nano-mountains are also too small for water droplets to get into, hence the water drops stay suspended on the tops of the nano-mountains creating very high surface tension.

Surface tension generally affects the top part of a water droplet – the bottom part sticks to the material with a flatter contact surface. In the case of lotus leaf, nanoscale bumps or nano-mountains prop the water drops up on the leaf surface, thus making them entirely surrounded by air (Cheng et al. 2006; MRSEC Interdisciplinary Education Group 2017). This induces surface tension in all sides of the water drops and makes them up even more on the leaf. As a result, the water drops attain a spherical shape with a minimum adherence to the leaf.

Water drops stay up with a contact angle in the super hydrophobic range of around 160^0 when rainfall occurs on lotus leaf (Cheng 2005; Cheng et al. 2006; Li et al. 2017; Yamamoto et al. 2015). In this position, the water drops are so loosely adhered to the leaf surface with a little contact area that any tiny movement or minor disturbance causes them to roll easily off. Consequently, the water drops swiftly roll off the lotus leaf gathering dirt along their way, thus making the leaf self-cleaning. A similar self-cleansing mechanism works in the case of nano lotus leaf biomimicked coating when it is applied on different structural elements or systems of bridges, pavements, and buildings. It has been illustrated in Figure 4 considering the exterior wall of a building. The wall remains clean because the water droplets cleanse the dirt when they roll down the vertical surface coated with nano lotus leaf biomimicked coating. Also, the penetration of water into the wall is substantially decreased because the water droplets swiftly move and cannot pass through the nanostructure of coating.

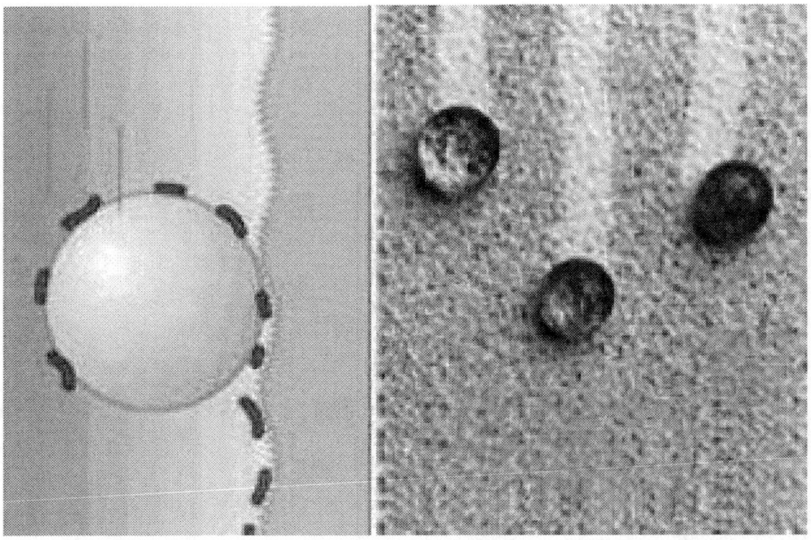

Figure 4: Self-cleansing mechanism of nano lotus leaf biomimicked coating (adapted from Design and The Universe 2017)

3 POTENTIAL APPLICATIONS AND BENEFITS

Super hydrophobic nano lotus leaf biomimicked coating has potentials for manifold technical applications such as self-cleaning window glasses or paints (Cheng et al. 2006; Li et al. 2017) and anti-corrosion metal surfaces or corrosion barriers (Vazirinasab et al. 2018). In fact, some such products, e.g. Lotusan (Sto Corporation 2017) and Hydrofoe (Lotus Leaf Coating 2018) are already commercially available in the market. A number of potential applications and associated benefits of using nano lotus leaf biomimicked coating are depicted hereafter.

- Nano lotus leaf biomimicked coating can be applied on the exterior walls or facades of buildings, as shown in Figure 5 and Figure 6, to minimize the dampness due to water sorption and to obtain a dust-free, self-cleansing wall surface. Water droplets will roll down swiftly with dirt and dust will not be collected on nano lotus leaf biomimicked coating because of its high water repellence. The buildings painted with a nano lotus leaf biomimicked coating product will self-cleanse during rain and stay dry due to the lotus effect. This one of a kind coating also resists the growth of mold, mildew and algae on the building surface.
- It can be applied on the exterior windows to keep them self-cleaned as shown in Figure 7. When water droplets will move down, they will carry the dust particles on their way, as they do on wall, thus keeping the windows self-cleaned.
- Nano lotus leaf biomimicked coating could be applied as a water-repelling coating on pavements for fast drainage, thus allowing them be dry quickly during wet season; this would contribute to safe driving and enhanced durability. For similar purposes, it can also be used on sidewalk for safe walk; however, it needs to be ensured that the friction property of the pavement surface is not affected in the presence of nano lotus leaf biomimicked coating.
- Nano lotus biomimicked coating can be used on the roof of buildings to accelerate the drainage of snow-melt water or rainwater; hence, it will minimize the algae or bacterial growth and thereby improve

the health of buildings for longer service life; also, the hydrostatic stress on the roof caused by water ponding will be reduced due to accelerated drainage in the presence of nano lotus leaf biomimicked coating.

- It can be used as a floor finishing material to produce water-repellent floor for enhanced service life. Water is involved with the damage mechanisms of many flooring materials. Water repellence will reduce the penetration of water into flooring materials and thus increase the service life of floor system with better durability.

Figure 5: Application of nano lotus leaf biomimicked coating and its effect on the facade of a building (adapted from Sto Corporation 2017)

Figure 6: Comparison of conventional and nano lotus leaf biomimicked coating (adapted from European Consumers Choice 2017)

- Nano lotus leaf biomimicked coating can be applied on the inner surface of building drainage structures, such as storm and sanitary sewers, to enhance the flow of storm water, waste water or sanitary waste by reducing surface friction.
- It can be applied on tunnel lining for keeping the tunnel structures dry and to resist the growth of mold and mildew in tunnels.
- Nano lotus leaf biomimicked coating can be considered for use in culverts to enhance the flow of overland water from upstream level to downstream direction. For similar reason, it can also be used in irrigation channels.
- It can be used on different components of bridge structures such as pier, deck, railing, divider, and abutment to enhance their service life with greater water resistance.
- Nano lotus leaf biomimicked coating can be applied on different metal and alloy surfaces as a corrosion barrier to improve their resistance against corrosion through super hydrophobicity.
- Due to super water-repellency of nano lotus leaf biomimicked coating, water droplets will not be retained on the glass surface; therefore superior optical clarity will be achieved in many applications such as windows, skylights and glazed curtain walls.

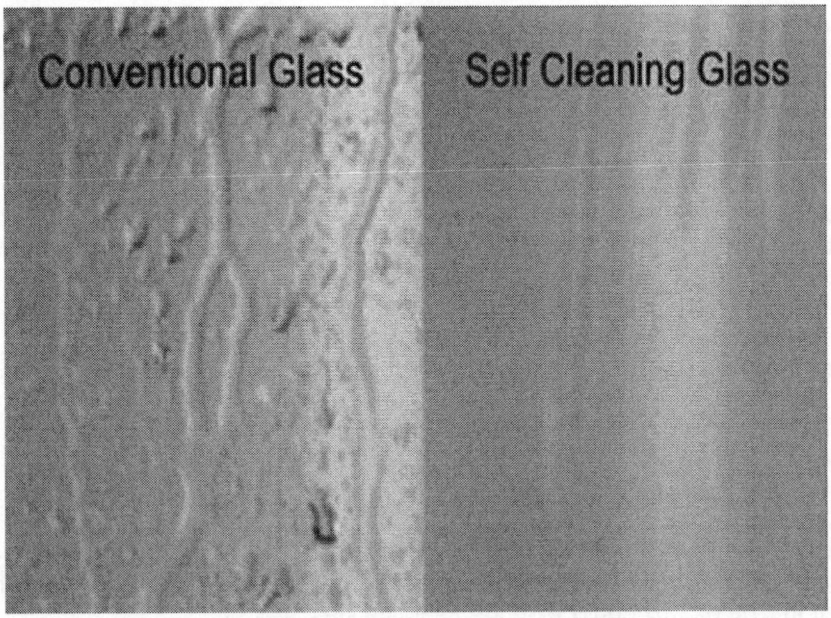

Figure 7: Comparison of conventional and self-cleaning glasses (adapted from Pro Clean Window 2018)
1

2 SUBSTRATE TYPES

Nano lotus leaf biomimicked coating can be applied on manifold substrate materials such as concrete, stucco, metal, glass, fiber cement board, masonry, exterior insulation and finish system (EIFS), and primed or pre-painted wood depending on the purpose of applications. In any kind of application, it must be ensured that nano lotus leaf biomimicked coating is well-adhered with the substrate to avoid premature blistering and de-lamination. A suitable bond coat as a primer can be used before applying nano lotus leaf biomimicked coating to guarantee good adhesion.

3 SUBSTRATE PREPARATION TECHNIQUES

Substrates must be cleaned to remove dirt, salt, and oil or grease. If the coating is applied on old surfaces, they must be free of mildew, algae, and efflorescence. The substrates must not have any contaminants on the surface that might affect the adhesion of nano lotus leaf biomimicked coating. Pitting, spalling, cracking, blistering, peeling, delamination, water damage, laitance, or any other surface defects should be repaired appropriately before applying the primer on substrate materials. Also, the substrates must have an adequately rough surface profile for the required level of coating adhesion (Safiuddin 2017). Abrasive-blasting, shot-blasting, high pressure water-blasting or at least sanding should be employed as per requirements for the preparation of substrates to achieve good adhesion of coating.

4 COATING APPLICATION TECHNIQUES

Nano lotus leaf biomimicked coating can be applied by dipping, brushing, rolling, spraying, and spinning as per applicability to the substrates. A variety of spraying equipment such as industrial sprayer, aerosol can, and non-aerosol pump sprayer can be used to spray-apply nano lotus leaf biomimicked coating. Acrylic-based compatible primer should be used on concrete, stucco, and masonry substrates before applying nano lotus leaf biomimicked coating. It can be applied onto the primed substrate in two coats. The first coat must be dried before applying the second coat to obtain the targeted surface structure and thickness of coating.

5 RECOMMENDATIONS

- An appropriate nano lotus leaf biomimicked coating product should be selected considering the type of substrate; it must be ensured that the coating and the substrate are compatible to provide better adhesion for improved performance.
- The substrate must be properly cleaned and made free of any contamination before applying any nano lotus leaf biomimicked coating product.
- The substrate must have an adequately rough surface profile for the required level of coating adhesion to obtain an extended service life of the treated surface.
- Nano lotus leaf biomimicked coating should be applied on the primed substrate surface for better performance and economy.
- A suitable applying technique such as dipping, brushing, rolling, spinning, or spraying as per the guidelines of the product's manufacturer should be used for the enhanced performance of nano lotus leaf biomimicked coating.
- The bond strength or adhesion of nano lotus leaf biomimicked coating after applying on the targeted substrates should be examined for extended service life.
- Lab-based experimental studies and monitoring of the field performance are required to validate the potential applications of nano lotus leaf biomimicked coating for guaranteed benefits in different construction sectors.

6 CONCLUSIONS

This study was carried out to explore the potential applications of nano lotus leaf biomimicked coating in various construction sectors. The self-cleansing mechanism of nano lotus leaf biomimicked coating has been explained in this study. The types of substrate, the preparation techniques for substrates, and the applying techniques of nano lotus biomimicked coating are also discussed in this study. Furthermore, some recommendations for improving the performance of nano lotus leaf biomimicked coating are given. Based on the present study, the following conclusions are drawn:

- Nano lotus leaf biomimicked coating significantly lessens the ingress of water into various substrates because of its super hydrophobic nature, and thus it minimizes the damaging effects of water involved in many deterioration mechanisms.

- Self-cleansing mechanism, associated with high surface tension and a contact angle of about 160^0, is achieved in the case of nano lotus leaf biomimicked coating due to its special surface texture with nano-mountains and nano-valleys.
- Nano lotus leaf biomimicked coating has good potentials for application in different structural elements or systems of buildings, pavements, bridges, and drainage infrastructure.
- Nano lotus leaf biomimicked coating products would perform better when they possess good bond strength or adhesion with the primed substrates.
- Appropriate selection of coating products, proper surface preparation, compatibility of coating products, and evaluation of coating adhesion before application will be conducive to improve the performance of nano lotus leaf biomimicked coating.

References

Ashby, M., Ferreira, P., and Schodek, D. 2009. Nanomaterials, Nanotechnologies and Design: An Introduction for Engineers and Architects. Elsevier Science Ltd., Massachusetts, USA.

Button, J., Emmanuel, F., and Middleton, R. 2004. Synthesis of Pavement Issues Related to High-Speed Corridors. Texas Transportation Institute, Texas, USA.

Cheng, Y.-T. 2005. Is the Lotus Leaf Superhydrophobic? *Applied Physics Letters*, **86**(14): Paper No. 144101.

Cheng, Y.-T., Rodak, D.E., Wong, C.A., and Hayden, C.A. 2006. Effects of Micro- and Nano-Structures on the Self-Cleaning Behaviour of Lotus Leaves. *Nanotechnology*, **17**(5): 1359-1362.

Design and The Universe. 2017. The Constantly Self-Cleaning Lotus. Retrieved from http://www.designanduniverse.com/articles/lotusan.php on October 12, 2017.

European Consumers Choice. 2017. StoLotusan Technology. Retrieved on February 26, 2018 from https://www.europeanconsumerschoice.org/house/sto-lotusan/

Forbes, P. 2008. Self-Cleaning Materials: Lotus Leaf-Inspired Nanotechnology. *Scientific American*, August Issue, pp. 1–5.

Gonzalez, M., Safiuddin, Md., Cao, J., and Tighe, S.L. 2013. Sound Absorption and Friction Responses of Nanoconcrete for Rigid Pavements. *Transportation Research Record: Journal of the Transportation Research Board*, **2369**: 87-94.

Gonzalez, M., Safiuddin, Md., Cao, J., and Tighe, S.L. 2016. Sound Absorption and Friction Properties of Nano-Lotus Leaf Coated Concrete for Rigid Pavement. *Materials Science*, **22**(3): 445-450.

Kumar, V.R., Bhuvaneshwari, B., Maheswaran, S., Palani, G.S., Ravisankar, K., and Iyer, N.R. 2011. An Overview of Techniques based on Biomimetics for Sustainable Development of Concrete. *Current Science*, **101**(6): 741–747.

Li, J., Zhou, L., Yang, N., Gao, C., and Zheng, Y. 2017. Robust Superhydrophobic Coatings with Micro and Nano-Composite Morphology. *RSC Advances*, **7**(70): 44234-44238.

Lotus Leaf Coatings, Inc. 2018. Hydrofoe: Super Hydrophobic Coating. Retrieved from http://lotusleafcoatings.com/coatings/ on February 24, 2018.

Mozumder, M.S., Zhang, H., and Zhu, J. 2011. Mimicking Lotus Leaf: Development of Micro-Nanostructured Biomimetic Superhydrophobic Polymeric Surfaces by Ultrafine Powder Coating Technology. *Macromolecular Materials and Engineering*, **296**(10): 929–936.

MRSEC. 2018. Lotus Leaf Effect. Materials Research Science and Engineering Center (MRSEC) Interdisciplinary Education Group, Nanoscale Informal Science Education Network, University of Wisconsin-Madison. Retrieved from www.nisenet.org/nano on October 13, 2017.

Neinhuis, C. and Barthlott, W. 1997. Characterization and Distribution of Water-Repellent, Self-Cleaning Plant Surface. *Annals of Botany*, **79**(6): 667-677.

Pro Clean Window. 2018. Window Care: Using Nanotechnology to Keep Windows Cleaner Longer. Retrieved from http://www.procleanwindow.com/nanotech/ on February 26, 2018.

Rame-Hart. 2015. Is Your Super Hydrophobic Surface the Real Deal or Is It a Fake? And What Is Pseudo Super Hydrophobicity Anyway? Rame-Hart Instrument Co Newsletter, October 2015.

Robbins, J. 2011. Engineers Ask Nature for Design Advice. New York Times, December 11.

Sto Corporation. 2017. StoColor Lotusan. Retrieved from http://www.stocorp.com/content/ on February 26, 2018.

Safiuddin, Md. 2017. Concrete Damage in Field Conditions and Protective Sealer and Coating Systems. *Coatings*, **7**: 22 pp.

Stratakis, E., Zorba, V., Barberoglou, M., Spanakis, E., Rhizopoulou, S., Pzanetakis, P., Anastasiadis, S., and Fotakis, C. 2009. Laser Structuring of Water-Repellent Biomimetic Surfaces. *SPIE Newsroom*, January 19, 2009; 3 pp.

Vazirinasab, E., Jafari, R., and Momen G. 2018. Applications of Superhydrophobic Coatings as a Corrosion Barrier: A Review. *Surface and Coatings Technology*, **341**: 40-56.

Yamamoto, M., Nishikawa, N., Mayama, H., Nonomura, Y., Yokojima, S., Nakamura, S., and Uchida, K. 2015. Theoretical Explanation of the Lotus Effect: Superhydrophobic Property Changes by Removal of Nanostructures from the Surface of a Lotus Leaf. *Langmuir*, **31**(26): 7355-7363.

Zhang, M., Feng, S., Wang, L., and Zheng, Y. 2016. Lotus Effect in Wetting and Self-Cleaning. *Biotribology*, **5**: 31-43.

Zorba, V., Stratakis, E., Barberoglou, M., Spanakis, E., Tzanetakis, P., Anastasiadis, S.H., and Fotakis, C. 2008. Biomimetic Artificial Surfaces Quantitatively Reproduce the Water Repellency of a Lotus Leaf. *Advanced Materials*, **20**(21): 4049–4054.

Building Tomorrow's Society

Bâtir la Société de Demain

Fredericton, Canada

June 13 to 16, 2018 / *13 au 16 juin, 2018*

PROPERTIES OF SPINNED STEEL AND BOND WITHIN CONCRETE

Abdel Moez, Mohamed[1*], Okasha, Hossam[1], Sabry, Hatem[1], Alaa, Omar[1], Dorgham, Hassan[1], Darwish, Mohamed[2], Sayed-Ahmed, Ezzeledin[3] and Abouzeid, Mohamed[3]

[1] Student, American University in Cairo, Egypt

[2] Lecturer, Ahram Canadian University; Adjunct Faculty, American University in Cairo, Egypt

[3] Professor, American University in Cairo, Egypt

[*] Corresponding Author: Mohamedalmoez@aucegypt.edu

Abstract: Ductility of reinforced concrete sections is of paramount importance in Construction engineering applications, especially in the areas associated with seismic and impact loading. The presented paper aims to evaluate the effectiveness of Spinned steel bars used as reinforcement bars. An experimental program was performed to evaluate the bond strength of these bars within concrete and their ductility, strengths and toughness of Spinned steel bars. Several tests were initially conducted on the Spinned bars (tension, bending and scanning electron microscopy) to assess their properties and present its potential to be used in reinforced concrete elements. The experimental investigation results reveal that Spinned steel bars have higher bond strength with concrete compared to regular steel reinforcement. Nevertheless, Spinned steel bars showed tendency for higher energy absorption, which can suggest its future use in design of structures subjected to dynamic loading.

1 INTRODUCTION

Reinforcement is an essential element for structures that are subjected to tensile stresses, where concrete have a strong resistance to compression, but weak resistance in tension; therefore, reinforced concrete is used as a composite material composed of both steel and concrete to resist both compressive and tensile stresses applied on structures. It is of paramount importance of reinforcement to have a high bond strength interlocking with concrete for the reinforcement to be fully subjected to tensile stresses without slippage from concrete; nevertheless, high resistance to tensile stresses for long lasting durable structures. As ductility is an essential parameter for structures subjected to dynamic actions such as seismic forces, wind and impact, design for such loads have been increasingly dependent on ductility enhancement (Tedesco et al, 1999). In the past years new methods of reinforcements, like prestressed reinforcement, began to arise through the process of twisting reinforced steel strands to obtain higher resistance to loads. As described in the patent presented by (Schmidt, 1940) who first published and presented the idea of twisting two reinforced steel bars together adopted the idea of twisting two bars together in a cold state from iron. Schmidt presented that the strength of rolled iron may be increased through the process of twisting in a cold state. Schmidt also claimed an increase in bonding capacity, accompanied by little or no increase of the elastic limit, and has no practical advantages. However, Schmidt did not mention a range for the number of pitches per unit length. If the number of pitches increased dramatically, it weakens the resistance of the twisted reinforcement to carry a higher load. Unlike what Schmidt stated, this paper presents a different perspective towards the modulus of toughness and ductility with different values that could lead to future benefits and usage. Another patent was presented by Dow and Ellis (2005) in which they discussed the idea of serpentines reinforcement, in which the shape of the reinforcements contained both peaks and troughs: this helped in developing the idea of obtaining pitches. However, they used woven metallic reinforcement not twisted steel bars. Utz (2007) presented the idea of rods crossing one another creating spaces between them; furthermore, he focused on the measurement of angles corresponding to the longitudinal axis from between 20 to 70 degrees, which is expected to grant both a higher resistance to load and shear as well as a higher bond interlocking with concrete. Then, Wright et al (2014) focused on using a spiral strip on top of a rod with flanges and twisting one bar right after the other, and mainly was concerned with the angles of the pitches created. On the other hand, Boguslavschi et al. (2016) focused on twisting one bar on itself as a single element claiming that it increases both bond with concrete as well as its increase in load resistance; however, the effect on ductility was not discussed within that patent which was also the case with Schmidt (1940). As both patents claimed that twisting reinforced steel bars enhanced its mechanical properties granting a higher resistance to load and bond interlocking with concrete due to the deformed shape that resulted from the process of twisting the bars. Meanwhile, these patents did not examine the effect of the process of twisting reinforcing bars on the overall ductility and the modulus of toughness of the twisted steel bars; the need to research these effects and their possible applications have emerged.

2 RESEARCH SIGNIFICANCE

On the other hand, enhancing ductility of steel has been subject to recent studies. One of the recent innovations was by Yamasaki (2017) who produced high-strength steel rods with high ductility through altering the chemical components of the steel alloy itself. A very similar approach was performed by Tian et al (2014) where in addition to changing the chemical composition of the steel, a re-crystalized ultra-fine grain structure was developed. Although this line of research is very common in literature, it is extremely expensive in practice as changing the chemical composition of steel at new temperatures needs innovations in the steel production industry itself, which is for sure not as economic as the regular readily available steels; hence, the efforts of these researchers cannot directly affect the construction industry at least on the short-run. On the other hand, the ductility enhancement of reinforced concrete members has been increasingly researched by members of the academic community in the field of structural engineering. Numerous researchers like (El Maghraby et al, 2010), (Meng and Elkhayat, 2016) and (Nguyen et al, 2013) have worked on using fiber reinforced polymers (FRP) within the reinforced concrete to enhance ductility. However, this approach significantly increases the cost as FRP is more expensive than steel. Hence, the ductility enhancement alternative that could be more feasible when compared to changing the steel microstructure and using FRP is spinning the steel; however, this expected effect of ductility enhancement has not been directly studied by researchers who worked on the spinned steel so far. That gap in research triggered the idea discussed in this paper of twisting two 10 mm reinforced steel bars on one another using the power lathe; through analyzing their characteristics, we could know their future potential use as a different method of reinforcement inside of concrete sections focusing on its effect on ductility and toughness. Twisting the two reinforced steel bars on one another creates pitches because of twisting. Three important concepts encouraged examining the concept of ductility of twisted reinforcement. The first concept is mainly concerned with a higher strain value. Since an additional length is required to spine the bar, the strain becomes larger since change in length of the bar increases; so, dividing the change in length by the original length of the twisted bar will result in obtaining a higher strain value. The second concept is concerned with the tendency of the twisted bar itself, when subjected to high tension, to deform to its original straight form, which will cause an additional strain inducing a pre-strain condition. Hence, the strain does not start from zero rather from a value before which increases the max strain value as well. Finally, the third and most important concept of ductility is mainly focused on the twisted bar being pre-stressed; the yielding point is expected to be achieved a bit earlier while the point at fracture is expected to occur at higher strains; so dividing the value of deformation at the point of fracture by the deformation value at yielding point, the ductility value will be higher. Accordingly, this paper focuses on studying the characteristics of spinned steel bars through performing several tests. First, microscopic tests to examine the microstructure of the spinned steel before loading were performed. Second, destructive tests were performed on the spinned steel bars themselves to study their mechanical properties and compare them to those of straight steel bars. Finally, bond test was performed to examine the interaction between the spinned steel and its surrounding concrete.

3 EXPERIMENTATION

3.1 Procedure of Twisting

The procedure is carried out by aligning two steel bars and welding their tips together from both ends to avoid their separation. Following on that, they are fixed from one end and the other end is inserted in the rotating shaft of the power lathe while the twisting process takes place as shown in Figure 1. The number of spines or pitches is controlled by the number of rotations performed by the power lathe and by stopwatch time to maintain the intended number of pitches per meter. The following number of pitches discussed in this study is 11 pitches per meter. To avoid the occurrence of cracks during the twisting process, a speed of approximately 120 rpm was obtained during the twisting process.

Figure 1: Twisting the bars using the power lathe

It is worth to note that to obtain 1 m of spinned steel, an additional length of 30 mm of each bar was acquired as the twisting process decreased some of the original length of the two bars taking a total of 60 mm additional length from the two bars twisted on one another. Knowing the exact length, volume of spinned steel samples was calculated by multiplying it to the unit weight of steel to obtain weights of steel. Then, equating the weight of spinned steel to the weight of the straight steel that is desired to obtain the cross sectional area, and solving for the diameter, which is 14 mm. That means that two 10 mm spinned steel bars are equivalent to one 14 mm straight bar in terms of weight.

3.2 Testing Program

Since the idea of using the spinned steel inside concrete section is still non- orthodox, so it was necessary to conduct some conceptual experimental work to ensure that the idea under study would be promising. Hence, two tests were conducted. The first test is the Scanning Electron Microscopy Test (SEM) and the second one is the direct tension test. The first test was conducted to make sure that during spinning the two steel bars together, there were no sizeable cracks developed and the steel was not damaged. Secondly, the tension test was conducted to examine the characteristics of the spinned steel and to determine how the new characteristics could be beneficial. Following on that bending test was performed to assure the ductility of the spinned steel and no visual cracks developed from bending. Finally, Spinned steel bond testing was performed to examine its behavior within the reinforced concrete section itself.

3.2.1 Scanning Electron Microscopy (SEM) Test

The SEM test enables viewing the internal stratification within a material at different microscopic scales. This test was performed according to ASTM E2142 – 08 (2015) to assure that no damageable cracks occurred during spinning process are of a size that constitute any danger to the reinforcement. Figure 2, Figure 3 and Figure 4 represent the surface within the spinned steel at different microscopic altitudes. Figure 2 shows the spinned steel at a microscopic altitude of 200 micrometers with no obvious cracks at all at that scale. Also, Figure 3 shows the microscopic altitude at 20 micrometers with no obvious cracks at all at that scale. On the other hand, Figure 4 shows the microscopic altitude at 10 micrometers in which two cracks have appeared that are circled in Figure 4. That means that one of these two circled cracks has a size of approximately 20 micrometers while the other one has a size of approximately 15 micrometers. Hence, it is safe to say that because these cracks are very small, their damage is minor and so are considered non-damageable and of a negligible size. Consequently, the spinning process could be considered as a non-crack-inducing process that does not induce any cracks in the steel.

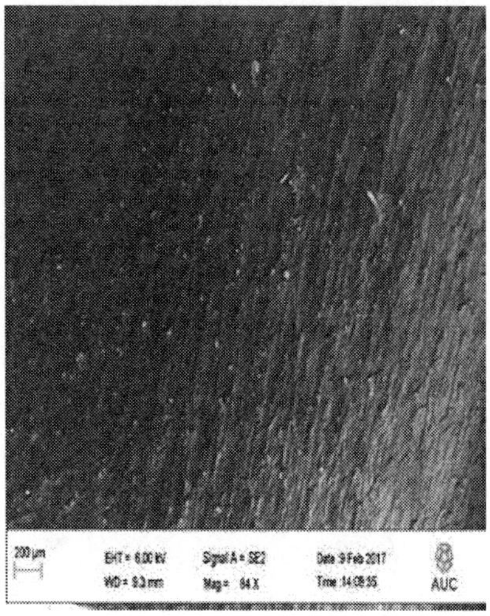

Figure 2: The spinned steel at a microscopic altitude of 200 micrometers.

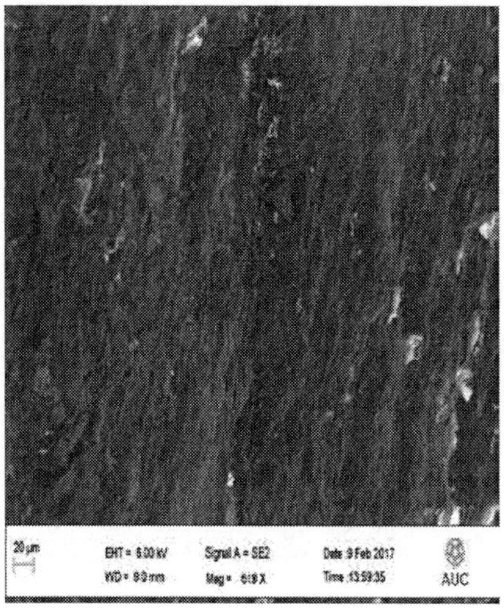

Figure 3: The spinned steel at a microscopic altitude of 20 micrometers.

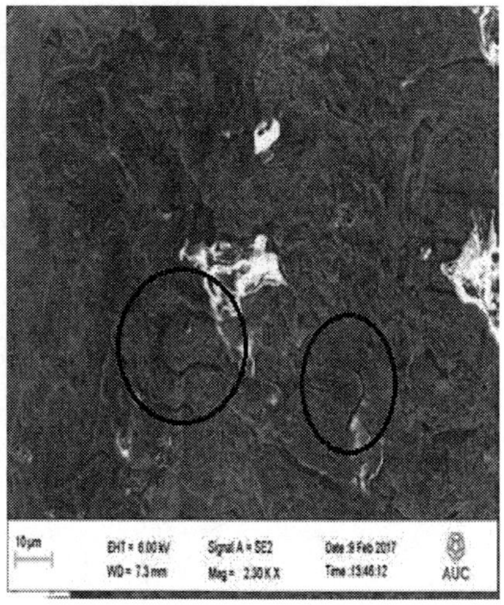

Figure 4: The spinned steel at a microscopic altitude of 10 micrometers.

3.2.2 Tension Test

Direct tension test was performed according to ASTM A370-17 on a control set of three specimens of straight bars of a 10 mm diameter and on another two sets of three spinned steel specimens of different spins per meter (11,17) within each of them two 10 mm bars were spinned together. As eleven spins per meter was to be better number of spins in terms of (load, ductility and energy absorption). The tension test results are presented in Figure 5 and Figure 6 and Figure 7

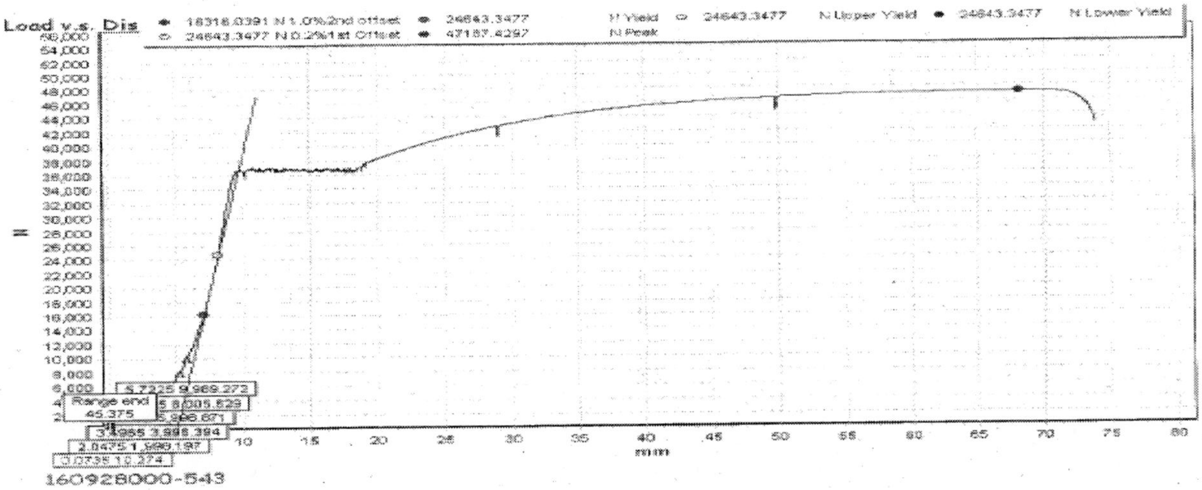

Figure 5 :Load – deflection curve for a straight bar specimen 10 mm

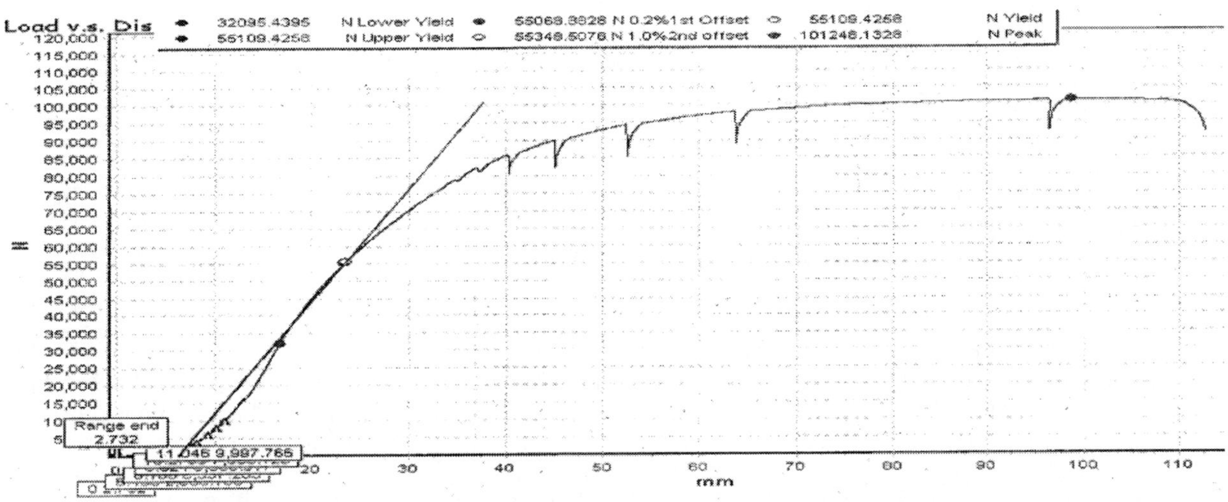

Figure 6: Load – deflection curve for a spinned specimen 11 spins per meter

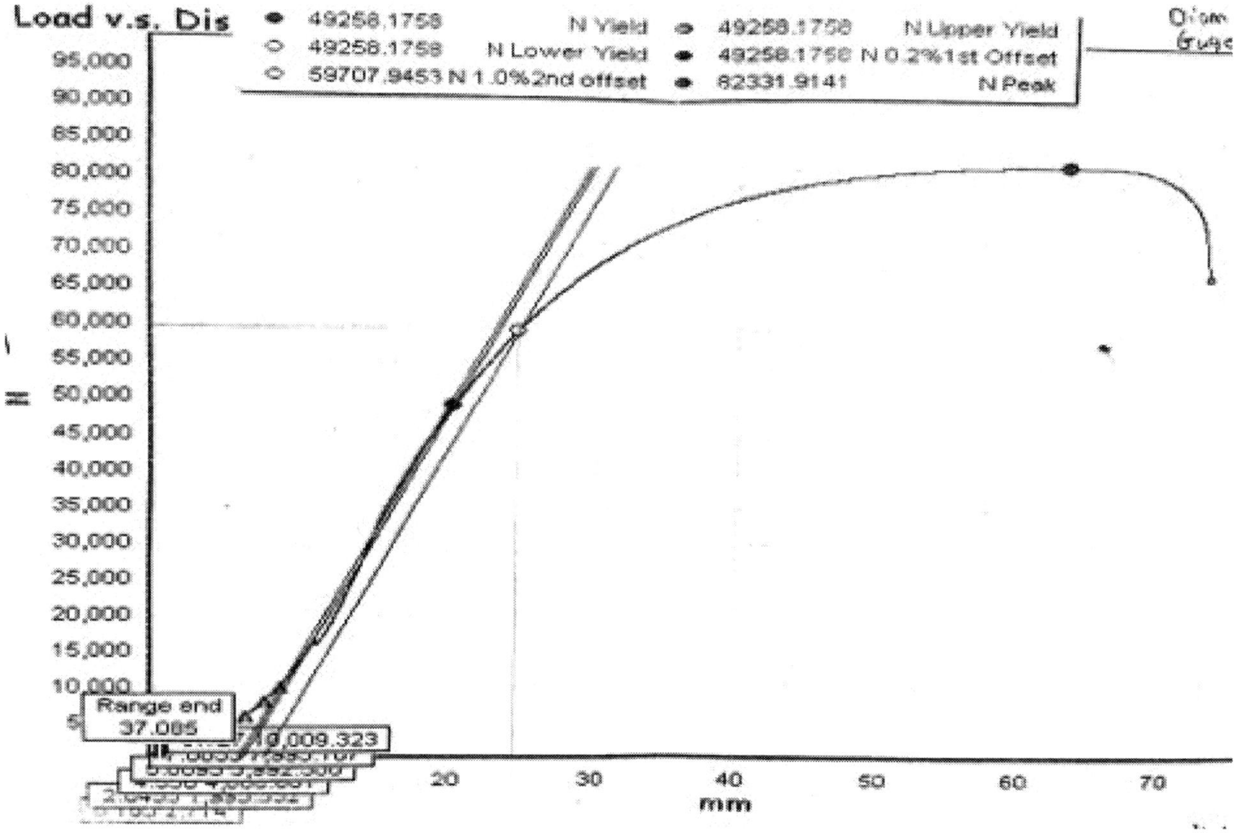

Figure 7: Load – deflection curve for a spinned specimen 17 spins per meter

The ultimate load that the straight bar specimen withstood was 47167 N while that of the a spinned bar specimen with 11 spins per meter was 101246 N. Hence, the ultimate load of the two spinned bars is higher than that of a straight bar, and the 11 spins per meter was higher than that of the 17 spins per meter; however, the real obvious change is in the elongation as the elongation if two straight conventional bars will still remain the same as that of one straight bar which was 74 mm while the elongation of the two spinned bars reached 112 mm which is significantly larger than what two straight bars would achieve together. That means that spinning the bars have significantly enhanced their ductility and the energy absorbed by the bars till failure (represented by the area beneath the curve) has been significantly increased, which can be explained more in Stress – Strain Curves of Figure 8 and Figure 9.

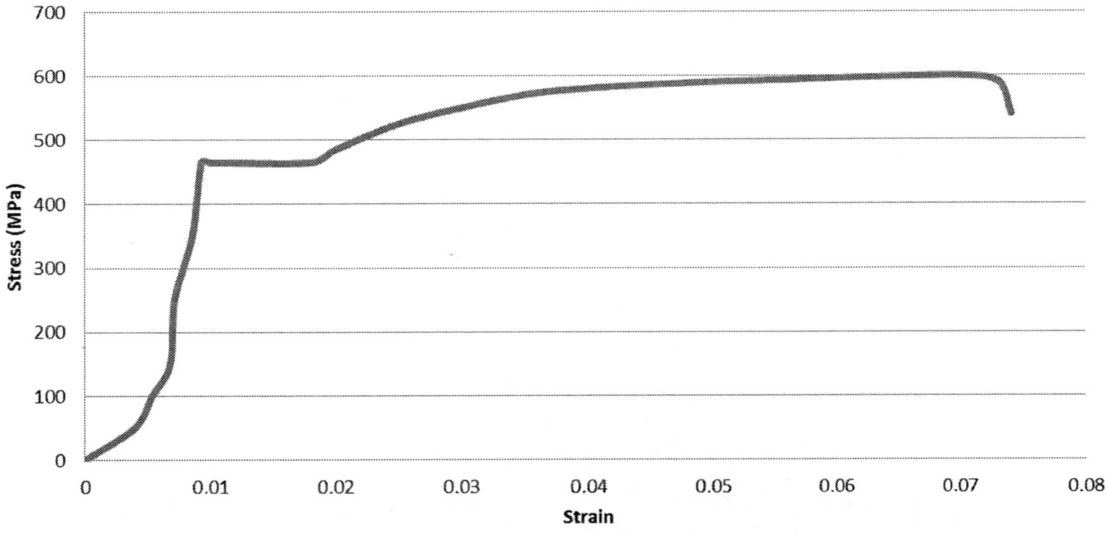

Figure 8: Stress – strain curve for a straight bar specimen

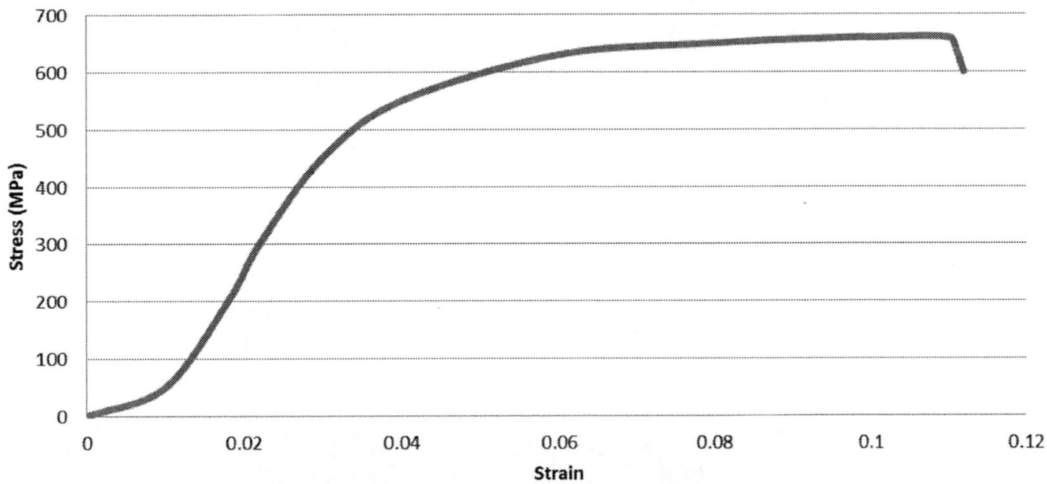

Figure 9: Stress – strain curve for a spinned specimen of 11 spins per meter

This enhancement in ductility can be also seen when comparing the stress – strain diagram of the spinned specimen shown in Figure 9 to that of the straight bar specimen shown in Figure 8. On the other hand, the ultimate strength has changed. This change in strength is attributed to the fact that the bars are not straight anymore which means that when calculating the stress, the load is not simply divided by the circular cross-sectional area of the bar as the area normal to the axial force in the case for the spinned steel became nearly oval. Hence, and although the ultimate load is still nearly the same as that carried by two straight bars, the change in the cross-sectional area due to the tilt in the bars is the main reason for the minor in ultimate strength. This tilted cross-sectional area could be obviously seen when examining the mode of failure shown in 10. However, this change in strength is still less than the increase in ductility which means that the area beneath the stress – strain curve of the spinned specimen shown in Figure 9 is significantly larger than that of the straight bar specimen shown in Figure 8. This area beneath the curve is the modulus of toughness which represents the energy absorbed per unit volume of the material till its failure. A comparison between the moduli of elasticity, resilience and toughness of the spinned and straight bars could be seen in Table 1. Although the modulus of elasticity has decreased due to spinning the bars, the moduli of resilience and toughness have significantly increased due to the significant increase in the strains that the spinned specimen experienced when compared to those experienced by the straight specimen. This could point out that the reduction in the modulus of elasticity is mainly due to the change in cross-sectional area coupled with the significant increase in elastic strain (which is inversely related to the modulus of elasticity)

Table 1: Comparison between the moduli of elasticity, toughness and resilience of the spinned and straight bars.

	Spinned Bars 11 Spins	Straight Bar
Modulus of Elasticity (MPa)	21,985	212,004
Modulus of Toughness (MPa)	57.16	36.36
Modulus of Resilience (MPa)	5.25	1.17

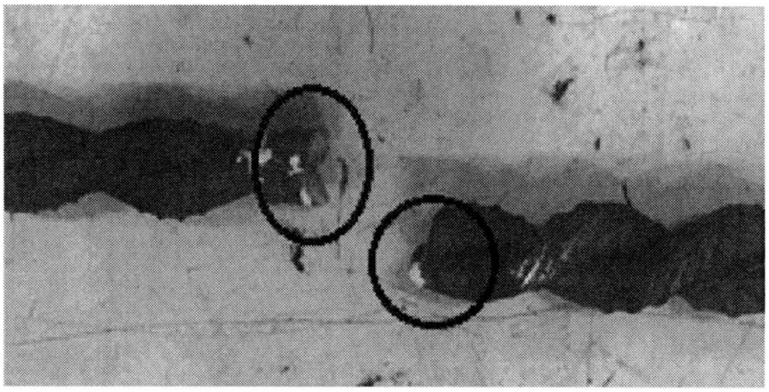

Figure 10: The mode of failure of the spinned steel specimen.

3.2.3 Bending Test

The cold-bending test was performed according to ASTM A370 - 17 (American Society for Testing Materials, 2017) to guarantee ductile behavior of the spinned bars when subject to bending. The test was performed on three spinned specimens. The three specimens exhibited the same deformed shape under the same load. As shown in Figure 11. The bent specimen behaved in a ductile manner with no brittle fracture experienced by the specimen during bending and no signs of brittleness in the bent spinned bars.

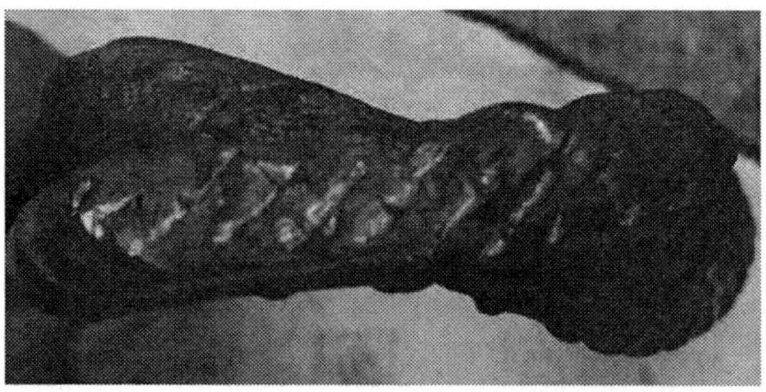

Figure 11: Side view of the cold-bent specimen

3.2.4 Bond Test

To assure that the bond between the spinned steel and the surrounding concrete is strong the bond test was performed according to ASTM C900-15. The specimen is a cylinder of a 150 mm diameter and a 300 mm height, embedded in it is the reinforcement to be tested. To compare the bond strength of the spinned steel to different sizes of reinforcing bars, three control groups were tested corresponding to one spinned group. Each group consisted of 2 specimens for consistency in results. In control group 1, a steel bar of a diameter of 12 mm is embedded in each cylinder specimen. In control group 2, two straight steel bars of diameter 10 mm are embedded in each cylinder specimen. In control group 3, a steel bar of diameter 16 mm is embedded in each cylinder. Finally, for the spinned group, two steel bars of diameter 10 mm of 11 spins per meter are spinned together and are embedded in each cylinder specimen. Hence, 3 cubes of dimension 150 mm × 150 mm × 150 mm were casted to ensure that the desired compressive strength of 55 MPa was reached.

Table 2: Results of Bond tests

Specimen	Type of Failure	Failure Force (N)
Ø12 Specimen 1	Steel slipped 20 mm out of cylinder	72000
Ø12 Specimen 2	Steel slipped 15 mm out of cylinder	73000
	Average	72500
2Ø10 straight Specimen 1	Concrete broke in 2 halves	92500
2Ø10 straight Specimen 2	Concrete broke in 3 parts	96500
	Average	94500
Ø16 Specimen 1	Slipping failure	115500
Ø16 Specimen 2	Slipping failure	129000
	Average	122250
2Ø10 11 spins Specimen 1	Steel broke without slipping or concrete fracture	91000
2Ø10 11 spins Specimen 2	Steel broke without slipping or concrete fracture	93000
	Average	92000

The results of the bond tests are shown in *Table 2*. These results when examined in correlation with the mode of failure of the spinned steel shown in Figure 12 prove that the bond between the spinned steel and the concrete is strong to the extent that the failure did not occur in the concrete and the steel did not slip off from the concrete; however, failure occurred in the steel itself which means that the bond between the concrete and the spinned steel was even stronger than the steel itself. On the other hand, when comparing the bond between the spinned steel to that of the other straight steel samples one could see that all the other samples using straight steel either failed due to slippage of the steel out of the concrete block or due to failure in the concrete block itself. This is mainly attributed to the fact that the contact surface area between the concrete and the spinned steel is significantly larger than that between the concrete and straight bars. This bond of two 10 mm spinned steel bars is even stronger than the bond of two straight 10 mm bars as the failure for these two straight bars occurred in the concrete block itself suggesting that the interlocking between the spinned steel and the concrete strengthens the bond between them. However, it is worth to note that bond strength is also a function of concrete strength; hence, these results may change in cases of different concrete strengths. The results obtained by Spinned steel in bond testing inside of concrete can be of paramount cost saving of additional lengths in development and splice length of reinforced concrete sections.

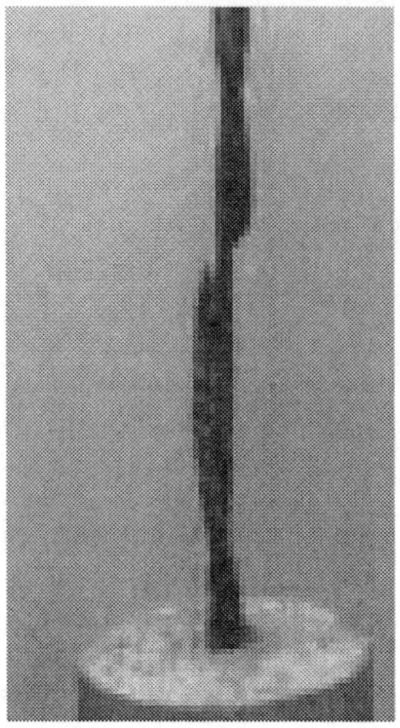

Figure 7: The Failure mode of spinned steel specimen

4 CONCLUSIONS AND RECOMMENDATIONS

Based on the scope of this work, the materials and technique incorporated, the following conclusions can be drawn:

- The spinning process of the steel did not induce significant cracks within the spinned steel bars.
- The spinning process of the steel bars significantly increased their ductility.
- The spinned steel bars have significantly higher moduli of toughness and resilience when compared to those of straight bars which could be attributed to the increase in ductility.
- The spinned steel bars behave in a ductile mode when subjected to cold bending.
- When examining the mode of failure that occurred in the spinned steel bars, it could be concluded that the bond between the concrete and the spinned steel bars is significantly higher than that of straight bars ; straight bars slip off the concrete block when subjected to the bond test, meanwhile spinned steel stays in place attached to the failed concrete section.

Based on these conclusions, an ongoing research is being performed to assess the use of spinned steel bars in reinforced concrete beams subject to static and impact loading. To validate the real-life application of the spinned steel, a pilot study needs to be performed on a larger scale involving spectrum of reinforcement and spinned bars at various pitches.

Acknowledgements

The authors would like to acknowledge the efforts of lab crew and the teaching assistants of the Construction Engineering Department at the American University in Cairo.

References

American Society for Testing Materials, "ASTM E2142 - 08(2015) Standard Test Methods for Rating and Classifying Inclusions in Steel Using the Scanning Electron Microscope," 2015. [Online]. Available: https://www.astm.org/Standards/E2142.htm. [Accessed 26 July 2017].

American Society for Testing Materials, "ASTM A370 - 17 Standard Test Methods and Definitions for Mechanical Testing of Steel Products," 2017. [Online]. Available:

https://www.astm.org/Standards/A370.htm. [Accessed 26 July 2017].

American Society for Testing Materials, "ASTM C900 - 15 Standard Test Method for Pullout Strength of Hardened Concrete," 2015. [Online]. Available: https://www.astm.org/Standards/C900.htm. [Accessed 26 July 2017].

Boguslavschi, N., Wright III, C. and Zalan, A., "Reinforcement for reinforced concrete". USA Patent US9243406B1, 26 January 2016.

Dow, M. and Ellis, Timothy W. Woven metallic reinforcement and method of fabricating same. US Patent US20050055933 A1, filed September 3, 2003, and issued March 17, 2005.

El Maghraby, Y. , Haroun, M. and Abou-Zeid, M., "Parameters Influencing Performance of Fiber-Reinforced Polymer Systems in Concrete," Transportation Research Record: Journal of the Transportation Research Board, vol. 2150, 2010.

Gao, G. , Zhang, H., Gui, X., Luo, P., Tan, Z. and Bai, B., "Enhanced ductility and toughness in an ultrahigh-strength Mn–Si–Cr–C steel: The great potential of ultrafine filmy retained austenite," ActaMaterialia, vol. 76, no. 1, pp. 425-433.

Meng, W. and Elkhayat, K. H., "Experimental and Numerical Studies on Flexural Behavior of Ultrahigh-Performance Concrete Panels Reinforced with Embedded Glass Fiber-Reinforced Polymer Grids," Transportation Research Record: Journal of the Transportation Research Board, vol. 2592, 2016.

Nguyen, H., Mutsuyoshi, H. and Zatar, W., "Flexural Behavior of Hybrid Composite Beams," Transportation Research Record: Journal of the Transportation Research Board, vol. 2332, 2013.

Utz, Mayer, Reinforcing member for structures made of reinforced or prestressed concrete or the same. US Patent EP1795667A2, filed November 22, 2006, and issued June 13, 2007.

Shingo Yamasaski, S. N., "Steel rod and high strength steel wire having superior ductility and methods of production of same". USA Patent US9689053 B2, 27 June 2017.

Schmidt, Rudolf. Twisting reinforcements for concrete. US Patent US2216758 A, filed October 20, 1936, and issued October 8, 1940.

Tedesco, Joseph W., William G. McDougal, and C. Allen. Ross. Structural dynamics: theory and applications. Menlo Park, CA: Addison Wesley Longman, 1999.

Tian, Y. Z., Bai, Y., Chen, M. C., Shibata, A., Terada, D. and Tsuji, N., "Enhanced Strength and Ductility in an Ultrafine-Grained Fe-22Mn-0.6C Austenitic Steel Having Fully Recrystallized Structure," vol. 45, no. 12, 2014.

Wright III, C., Boguslavschi, N. and Zalan, A., "Reinforcement for reinforced concrete and methods for manufacturing thereof". USA Patent US8915046B2, 23 December 2014.

Building Tomorrow's Society

Bâtir la Société de Demain

Fredericton, Canada

June 13 to 16, 2018 / *13 au 16 juin, 2018*

RECYCLED GROUND OYSTER SHELL FOR USE AS FILLER IN SELF-CONSOLIDATED GROUT

MacEachern, Daina[1] and Sadeghian, Pedram[2,3]
[1] Department of Civil and Resource Engineering, Dalhousie University, Canada
[2] Department of Civil and Resource Engineering, Dalhousie University, Canada
[3] Pedram.Sadeghian@Dal.ca

Abstract: In this paper, the mechanical properties of self consolidating grout with various levels of replacement of sand with ground oyster shell is investigated. The objective of this research is to determine the behaviour and structural capabilities of the alternate filler and determine its feasibility as an environmentally friendly and cost saving solution. Various levels of replacement of sand, namely 5, 10, 15, 20, 30 and 35% are investigated to find an optimal level. A base line grout without oyster shells was designed as a bench mark for strength and slump flow. One batch of mixes were designed with the same water content as the reference. The second batch was designed to have a similar slump flow as the bench mark by varying water content. Mixes consist of water, type N cement, mortar sand, superplasticizer, and ground oyster shells. Six grout cubes were prepared for each mix and were tested in compression at 38 days. Oyster shells were prepared by washing and drying followed by being passed through a jaw crusher twice and 4 minutes in a pulveriser. It was concluded that addition of all levels of ground oyster shells facilitated an increase in compressive strength. 20% replacement of sand with ground oyster shell provided the optimal compressive strength increase, with results 25% higher than the control.

1 INTRODUCTION

The production of concrete has a large negative impact on the environment. The cement production process emits significant amounts of carbon dioxide into the environment and the process consumes large amounts of water (Elyamanyet al. 2014; Pedersen 2004). Beyond simply adjusting the mixture proportions of a standard concrete mix, fillers can be added. Fillers are materials that can be added to the mixture in order to reduce the use or need for another more expensive or unsustainable material, or to vary other properties of the concrete or grout. The addition of fillers in concrete or grout mixes can help improve the sustainability, lessen the environmental impact as well as have an impact on physical properties. A very common use of fillers is in what is known as self-consolidating or flowable concretes and grouts in order to efficiently enhance the viscosity (Elyamany et al. 2014). Self consolidating concrete (SCC) and self consolidating grout or mortar (SCG or SCM) are a type of concrete and grout that has a high slump, eliminating the need for vibration to achieve good consolidation (Pedersen 2004). A few common environmentally friendly and recycled fillers include eggs shells, sea shells, lime stone rock, fly ash, stone dust, silicane fume and blast furnace slag (Elyamany, Abd Elmoaty, and Mohamed 2014; Pedersen 2004; Siffique 2008; Ye et al. 2007; Safi et al. 2015).

This paper will focus on the addition of waste sea shells, oyster shells in particular, to self consolidating grout. Safi et al. (2015) conducted a study to investigate the use of crushed sea shells as the fine aggregate in SCM and found that they could be used without negatively affecting the essential properties of mortar, causing only a small reduction in compressive strength. The study by Safi et al. made use of larger, sand sized crushed shells of various types but did not investigate the use of shells ground more finely than the sand. Yoon et al. (2003) conducted a study on the chemical-mechanical characteristics of crushed oyster shells. In this study they looked into the composition and characteristics of various sized crushed oyster shells as well as their effect when used in soil mortar, however; they did not specify which particle size distribution they used in their mixes. It was found that the compressive strength of their mortars decreased after 20% and 40% dosage ratio for cement/soil ratio of 0.2 and 0.1 respectively (Yoon et al. 2003). Olivia et al. (2015) investigated the mechanical properties of seashell concrete by partial replacement of the cement. The shells were filtered through a #200 sieve to ensure all particles were fine and conducted trial batches with 2,4,6 and 8% replacement of cement by weight. Olivia et al. concluded that the optimum compressive strength was 4% replacement, however; adding the seashells reduced all compressive strengths in comparison to the control with none.

2 EXPERIMENTAL PROGRAM

2.1 Test Matrix

The test matrix consisted of two main batches of grout with ground oyster shells replacing sand at various levels. A control mix with no ground oyster shell was used as a base line for both batches. The first batch was designed to have the same water to cement ratio as the control batch, only varying the amount of oyster shell at intervals of 5% replacement of the sand in comparison to the control. This allowed for a comparison and analysis of the effect of the filler on the workability and slump. The second batch was designed to have a similar slump as the control mixture, by varying the water to cement ratio along with the amount of oyster shell filler. This allowed for the effect on strength to be analyzed. As it is difficult to ensure the same slump every time, a range of ± 45 mm was allowed. Specimens were named based on the level of oyster shell in the batch (5-30) and whether they were water controlled (W) or slump controlled (S).

2.2 Material Properties

Both the masonry sand and the ground oyster shells were put through a sieve test to determine their gradation curves. The results of both tests are shown in Figure 1. Both the sand and ground oyster shells follow a similar gradation; however, the ground oyster shell is finer. This makes the ground oyster shell a good candidate to replace the sand in term of gradation.

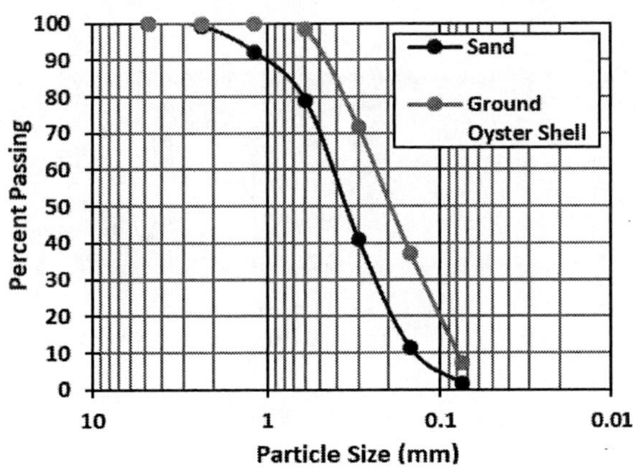

Figure 1: Gradation curve of sand and ground oyster shell

2.3 Specimen Fabrication

There were two main components to the fabrication of the test specimens. First the fresh oyster shells had to be prepared and finally the grout was mixed and cured. The oyster shells were collected from a beach in Brule Point, Nova Scotia. The shells were all empty and scoured by the tides and salt water. The oyster shells were then scoured and bathed in fresh water to remove as much salt and contaminants as possible. The shells were then placed in an oven at 70 degrees Celsius for approximately 24 hours. Once fully cleaned and dried, small 1 kg portions were sent through the jaw crusher twice each to break the full shells down into a size the pulverizer could manage. Next, in 1 kg size batches, the broken shells were placed in the pulverizer for 4 minutes until ground into a fine powder. This powder was collected and sieved in order to determine the particle size distribution. This process is shown in Figure 2.

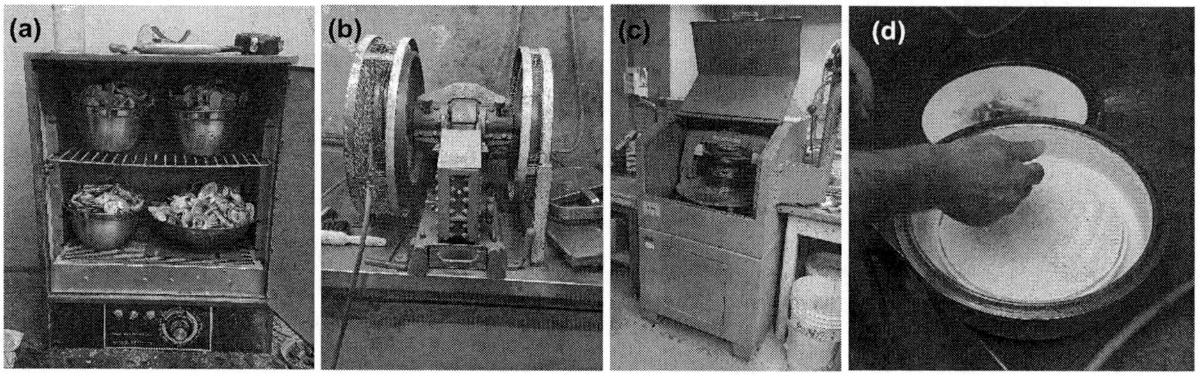

Figure 2: Ground oyster shell preparation: (a) shells are washed and dried; (b) shells are then put through jaw crusher twice; (c) shells are run through the pulverizer for 4 minutes; and (d) final oyster shell product

Once the ground oyster shell fines were prepared, the grout was mixed. All mixes contained water, superplasticizer, type N cement, masonry sand and the ground oyster shells. Depending on the batch of grout, there was a different variation of ground oyster shells, while maintaining the same quantity of fine aggregates, which includes the sand and oyster shells. To ensure a consistent mixture, the cement, fine aggregates and water were added in rotation and mixed with a power drill and paint mixer attachment. For each batch of grout, a modified slump test was conducted using a smaller tube (or cone) than the ASTM Standard for self-consolidating concrete to conserve material, ASTM 1611 (2009). A base line was set for this with the control mix. Three grout cubes were cast in 50x50x50 mm cubes for each batch of grout and cured for one day in lab before being removed from their molds and placed in the curing room. This process is shown in Figure 3.

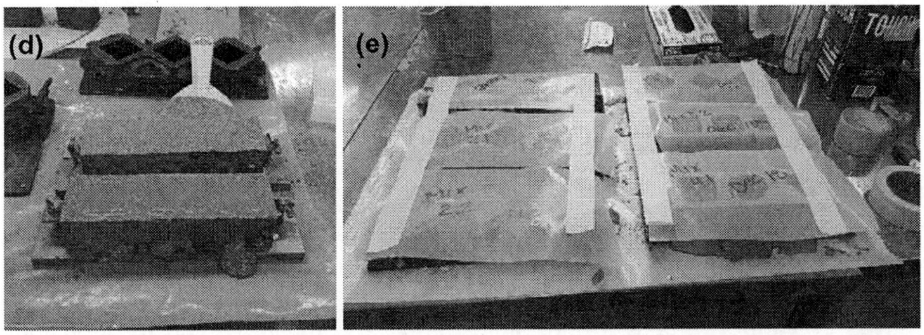

Figure 3: Specimen fabrication: (a) materials for the mix; (b) mixing materials; (c) modified slump test performed; (d) grout cubes prepared and; (e) grout cubes covered for one day of curing in lab

2.4 Test Setup

The test set up and procedure for the experiment was in accordance with ASTM C109 (2010). A spherical bottom fixture was used to minimize any accidental eccentricities. Final specimens were tested under compression using a universal testing machine with a constant strain loading rate of 1 mm/min. Load and stroke were recorded for all specimens until failure. Figure 4 below demonstrates the test setup.

Figure 4: Test setup

3 RESULTS AND DISCUSSION

All specimens were tested under compression at 38 days after casting and their results are provided in Table 2. It should be noted that the control specimen and GO5-W were tested in compression, however only the peak load was recorded.

Table 2: Summary of test results

Specimen group ID	Mini slump flow (mm)	Compressive strength		
		Average (MPa)	Standard Deviation (MPa)	Coefficient of Variation (%)
Control	250	40.85	1.85	4.52
GO5-W	330	38.50	1.07	2.79
GO5-S	295	42.32	3.14	7.42
GO10-W	348	44.41	0.81	1.81
GO10-S	235	49.32	1.94	3.94
GO15-W	363	44.92	0.72	1.60

GO15-S	235	47.84	0.51	1.07
GO20-W	363	46.09	0.51	1.11
GO20-S	238	51.15	1.86	3.64
GO25-W	310	43.88	0.98	2.23
GO25-S	220	46.14	3.11	6.74
GO30-W	295	44.13	1.37	3.10
GO30-S	223	46.40	0.46	0.98

3.1 Slump Flow Behaviour

The control mix was designed to have a mini slump flow of 250 mm, which through trial and error was determined to be a reasonable slump for a self-consolidating mix. It should be noted that the mini slump flow was obtained from small volume of grout filled in a 50x100 mm cylinder. Initially when determining a reasonable mix design, full scale slump tests were conducted as well as the modified smaller test to ensure a fair comparison. When adding the same amount of water to each batch, the slump increased, as expected. For 25 and 30% addition of ground oyster shell the slump decreased in comparison to the addition of 5% to 20%, but still increased with respect to the control. The constant slump batch was designed to maintain a slump of ± 45 mm of the control. The results of the mini slump test for the constant water samples are provided in Figure 6.

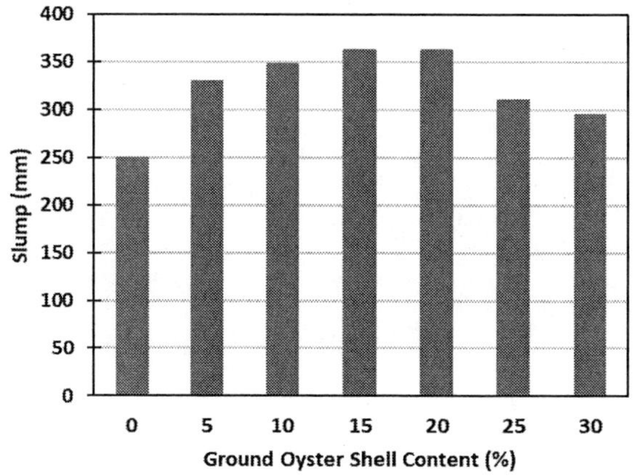

Figure 6: Variation of mini slump for constant water

3.2 Strength Behaviour

While maintaining constant slump, the results show an increase in strength in comparison to the control mix for all levels of oyster shell. The best strength results came from a 20% replacement of sand with oyster shell. The replacement of 20% of the sand corresponded to an increase in strength of almost 25% over the control. The 20% replacement was found to provide the highest strength increase for both constant slump and constant water, however; a lower strength increase was seen with constant water content. This is expected as more unnecessary water will lead to a weaker concrete; however, there is still an increase in the strength over the control mix in all but the 5% mix. The 5% ground oyster content mix, with constant water content appears to be off the trend when comparing it to the others. Figure 5 shows a compressive strength comparison for all ground oyster shell content levels tested.

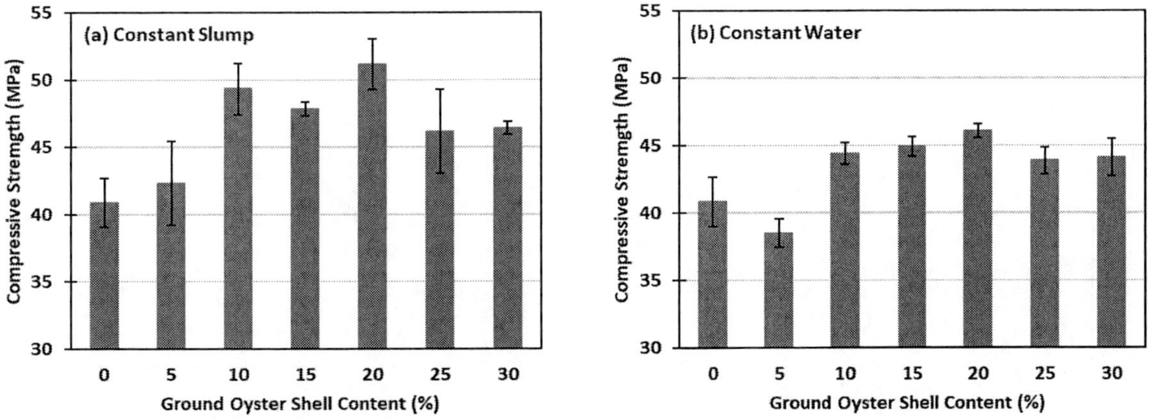

Figure 5: Variation of compressive strength: (a) constant slump mixes and (b) constant water mixes

Mix design proportions and details, for a cubic meter batch size, have been provided for the 20% ground oyster shell content mix with constant slump in Table 3.

Table 3: Mix proportions for 20% ground oyster content (by volume of sand)

Component	Quantity (1 m^3 batch)	
	Weight (kg)	Volume (m^3)
Water	170.5	0.170
Cement (Type N)	676.3	0.495
Masonry Sand	447.6	0.264
Ground Oyster Shell	60.0	0.066
Super Plasticizer	4.08	0.004
Mini Slump	238 mm	
28 Day Strength	51.15 MPa	

3.3 Load-Stroke Behaviour

Load and stroke were collected during the compression testing and the average curve for the three cubes of each constant slump batch is provided in Figure 7. Data from the 5% replacement batches has not been included as the discrepancies between the other data sets was too large. The cause of this discrepancy is unknown, however is likely due to an error in the mixing or curing stage. Further investigation is needed, including re casting and testing these specimens. Figure 7 shows that all variations followed a similar load-stroke shape, with the peak load and peak stroke varying slightly between specimens.

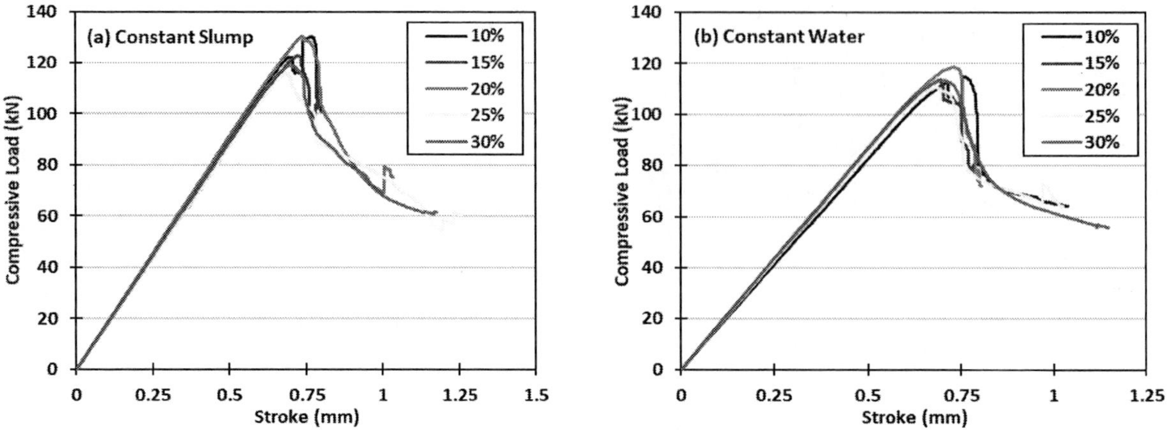

Figure 7: Comparison of load-stroke curves: (a) constant slump mixes and (b) constant water mixes

Table 4 gives the slope of the initial linear portions of these curves. It is seen that all constant slump specimens had a similar slope, with the average of all curves being 177.99 kN/mm. The constant water specimens had an average slope of 170.93 kN/mm.

Table 4: Initial slope of load-stroke curves

Ground Oyster Shell Content (%)	Average Initial Slope of Load-Stroke Curve (kN/mm)	
	Constant Slump	Constant Water
10	180.37	164.69
15	175.21	172.79
20	182.26	173.51
25	174.04	170.56
30	178.08	173.11
Average	177.99	170.93
Standard Deviation	3.07	3.28

4 CONCLUSION

Six variations of a control mix were designed to feature various levels of replacement of sand with ground oyster shell from 5-30%. Each level of replacement had two different batches, one with the same level of water as the control and the other with the same slump measurement as the control. All specimens were cast in grout cubes and tested in compression after 38 days of curing. It was found that for both controlled water and slump 20% replacement was the optimal design. 20% replacement of sand with ground oyster shell allowed for almost a 25% increase in strength over the control mix when slump was maintained. The next stage of this study will include mixing a larger batch of the 20% mix design to obtain the slump flow based on ASTM 1611 (2009). Further work on this topic will also include investigation into the particle size distribution of the ground oyster shell by conducting a hygrometer test on the fines. Various levels of replacement will be re-mixed in order to rule out any errors that may have occurred in the first round of mixing. Higher levels of replacement may also be considered, along with more levels of replacement between the 5-30% used in this study.

Acknowledgements

The authors acknowledge the financial support of Dalhousie University, the Nova Scotia Graduate Scholarship (NSGS) and the National Sciences and Engineering Research Council of Canada (NSERC). The aide of Dalhousie University's Mineral Engineering department with the crushing of the oyster shells was greatly appreciated. The authors also acknowledge the help of Jesse Keane for his help with testing and data acquisition.

References

ASTM. 2009. C1611/C1611M Standard Test Method for Slump Flow of Self-Consolidating Concrete. Annual Book of ASTM Standards Volume. ASTM International. West Conshohocken, PA, USA.

ASTM. 2010. C109/C109M - Standard Test Method for Compressive Strength of Hydraulic Cement Mortars (Using 2-in . or [50-Mm] Cube Specimens). Annual Book of ASTM Standards. ASTM International. West Conshohocken, PA, USA.

Elyamany, Hafez E., Abd Elmoaty M Abd Elmoaty, and Basma Mohamed. 2014. Effect of Filler Types on Physical, Mechanical and Microstructure of Self Compacting Concrete and Flow-Able Concrete. Alexandria Engineering Journal, Faculty of Engineering, Alexandria University, **53** (2):295–307.

Olivia, Monita, Annisa Arifandita Mifshella, and Lita Darmayanti. 2015. Mechanical Properties of Seashell Concrete. Procedia Engineering, Elsevier Ltd, **125**:760–764.

Pedersen, Bård M. 2004. "Alkali-Reactive and Inert Fillers in Concrete." Norwegian University of Science and Technology.

Safi, Brahim, Mohammed Saidi, Abdelhakim Daoui, Ahmed Bellal, Ali Mechekak, and Kamel Toumi. 2015. The Use of Seashells as a Fine Aggregate (by Sand Substitution) in Self-Compacting Mortar (SCM). Construction and Building Materials, Elsevier Ltd, **78**:430–438.

Siffique, Rafat. 2008. Waste Materials and By-Products in Concrete. Springer. India.

Ye, G., X. Liu, G. De Schutter, A. M. Poppe, and L. Taerwe. 2007. Influence of Limestone Powder Used as Filler in SCC on Hydration and Microstructure of Cement Pastes. Cement and Concrete Composites 29, Elsevier Ltd, **29**:94–102.

Yoon, Gil Lim, Byung Tak Kim, Baeck Oon Kim, and Sang Hun Han. 2003. Chemical-Mechanical Characteristics of Crushed Oyster-Shell. Waste Management, Esevier Ltd, **23** (9):825–34.

Building Tomorrow's Society
Bâtir la Société de Demain

Fredericton, Canada
June 13 to 16, 2018 / *13 au 16 juin, 2018*

CASE STUDY: REDUCING CEMENT USAGE OF STRUCTURAL CONCRETE BY OPTIMIZING AGGREGATE GRADATION IN FLORIDA STATE

Chung, Hung-Wen[1,3], Subgranon, Thanachart[1], Yang, Chun-Hao[2] and Tia, Mang[1]
[1] University of Florida, Department of Civil Engineering, United States
[2] University of Florida, Department of Statistics, United States
[3] chweying@ufl.edu

Abstract: The concrete industry in Florida is presently facing two major challenges, namely (1) the rising cost of cement and (2) the shortage of fly ash. The ascending price of cement has increased by 17.8% from 2012 to 2015. According to the forecast of fly ash utilization from American Road and Transportation Builders Association in 2015, the demand for fly ash will increase by at least 53% for the next 20 years. One of the possible solutions to these challenges is through a more effective design of concrete mixes for which the cementitious materials contents have been minimized.

Preliminary investigations have indicated that most FDOT concrete mixes have an excess of paste. The main reason for this is due to the mistaken assumption that concrete can always be made stronger and with quicker strength gain by increasing the content of cementitious material. In reality, 10 to 20% of the cement content in most FDOT concrete mixes can be removed without any adverse effects on the plastic or hardened concrete properties. This reduction in cement content can be maximized by using an intermediate-sized coarse aggregate along with the original aggregate (for example, using ¼" graded coarse aggregate with standard 1" graded coarse aggregate). By adjusting the gradation of the coarse aggregate blend, an optimum packing of aggregate (optimum aggregate gradation, OAG) can be obtained such that the aggregate volume content is maximized. The use of OAG along with the reduction of paste (cement) content can improve the properties of the concrete mix, including (1) improved workability of fresh concrete, (2) reduced shrinkage, and (3) reduced heat of hydration. Also, the reduction in the use of cementitious materials would significantly reduce the cost of concrete, and reduce the environmental impact by conserving natural resources, lowering energy consumption, and lowering carbon dioxide emission. According the results, reducing cement usage of concrete by OAG method is an approach to improve Florida structural concrete.

1 INTRODUCTION

The ascending price of cement and shortage of fly ash causes the difficulty and the higher initial cost of structural concrete. To solve the lack of raw cementitious materials' problem and save the budgeted cost, "Optimized Aggregates Gradation (OAG)" technique, which adjusts aggregates' gradation via adding medium aggregates (¼" stone) in original mixes, could be the proper prescription in Florida State. Using OAG methods, concrete could obtain higher workability. This research is concise and explicit as it would be able to practically applicate in the industrial field, which is more economic, environmental-friendly, valuable to the current Florida structural concrete mixes.

At a specific w/c ratio, increasing cement content does not improve strength, since the strength has relied on w/c ratio and individual cement content. (Yurdakul, 2010; Dhir et al., 2004). On the other hand, concrete with reduced cementitious content could extend structural durability (Moini et al., 2014) and prevent drying shrinkage cracks and strength; these cement-reduced mixes often performed poor workability. However, Optimized Aggregates Gradation (OAG) could prevent cement-reduced concrete from low slump and strength. On the contrary, overdesigned concrete with high cementitious content tends to crack, so it not only increases manufactured budgets but enhance maintenance costs; most importantly, deducting the service life of the structure. The research is to investigate the optimized aggregated blends (OAG) mixes with appropriate cementitious materials proportion in Florida structural concrete, and decrease materials usage, carbon dioxide emissions, and costs. In the preliminary result lab of OAG mixes conduct in the workability of OAG mixes are better than standard mixes. Besides, the strength of OAG also performed similar strength as SC mixes. This result displayed that decreasing paste content is not significantly influence the strength of concrete when aggregates gradation is optimized, and OAG improves the workability of cement-reduced concrete. Besides, according to this testing result and aggregate grading curves, we could be able to do quality control of different mixes concrete easily.

2 BACKGROUND OF OPTIMIZED AGGREGATES GRADATION

Optimizing aggregates gradation was rarely considered before the 1990s, but Shilstone (1990) develop a chart to adjust gradation (Fig. 1), which including two main factors, Workability Factor (WF) and Coarseness Factor (CF). The X-axis, CF, presents the amount of coarse aggregates (>3/8 in.) divided by the sum of the coarse and intermediate aggregates (> #8 and < 3/8 in.). The Y-axis, WF, means that the percent of the aggregates passed #8 and adjusted the sum of cementitious material in a mixture. (Lindquist et al., 2015) This chart includes different zones to estimate gradation of concrete, for instance, Zone I is gap gradation, likes convention concrete mix. Zone II is well-graduation which makes concrete good workability and dense. Therefore, Shilston Chart is clear to identify and estimate properties of fresh concrete.

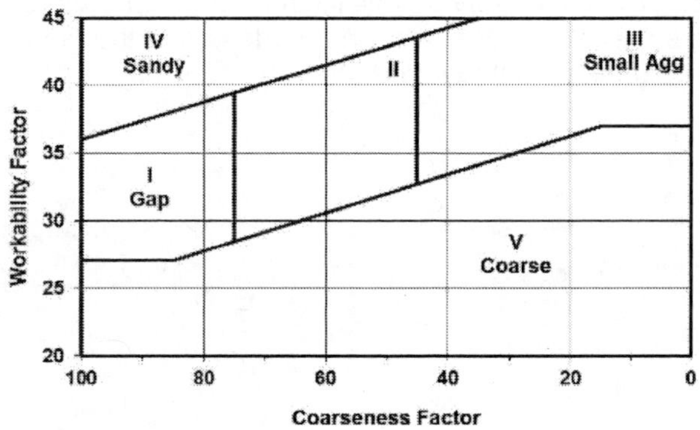

Figure 1– Shilstone Chart (Shilstone, 1990)

3 CONCRETE PROPERTIES

3.1 Plastic Properties

The results of workability, OAG concrete has higher workability than conventional concrete (around 1" higher). In this case, because of these OAG mixes controlled by WF and CF via shilstone chart (Fig. 1), the standard concrete (SC) mixes are in Sandy zone and OAG mixes are in Well-Graded zone. OAG mixes have higher workability than SC mixes and satisfying air requirement. The ability of decreasing paste content and amount of Type F-high range water reducing admixture are also an important topic of OAG tech, because the saving cost of concrete could be estimated in the future. As shown in Fig. 2, the graphs showed that SC mixes can decrease over 20% cement paste to get good slump in bridge deck mixes and OAG mixes can decrease 25% cement paste to get good slump in bridge deck mixes. (w/c_{max}=0.44, cement $content_{min}$=611 lb/CY, $Slump_{req}$= 2 to 4 and Air $content_{req}$= 1 to 6%, FDOT Standard Specifications for Road and Bridge Construction, 2017). However, to get good workability have to add more Type F when decreasing paste content. Because of adjusting CF and WF to Well-Graded zone, high range water reducing admixture could also be decreased in OAG mixes. Briefly, OAG mixes could enhance workability; as a result, it did save the usage of superplasticizer.

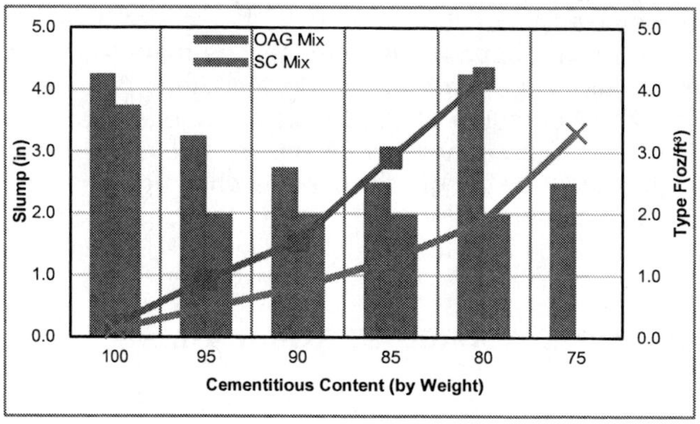

Figure 2– Slump and amount of superplasticizer (Type F) in different Mixes

3.2 Strength of Concrete

After compacting aggregates gradation which changes original gap gradation to well-graded gradation, OAG concrete still performs consistent strength as SC concrete after curing in 7 days and 28days. Fig. 3 and 4 are the strength of SC and OAG when descending paste content; although, the strength of OAG concrete did not enhance in this case, the workability is better than SC concrete as discussed in the previous section. Surprisingly, SC mixes have similar strength when decreased the paste content in SC concrete. This result means some of mix designs are overdesign in the field. Therefore, to save cost and raw materials, OAG mixes could be executed in the future.

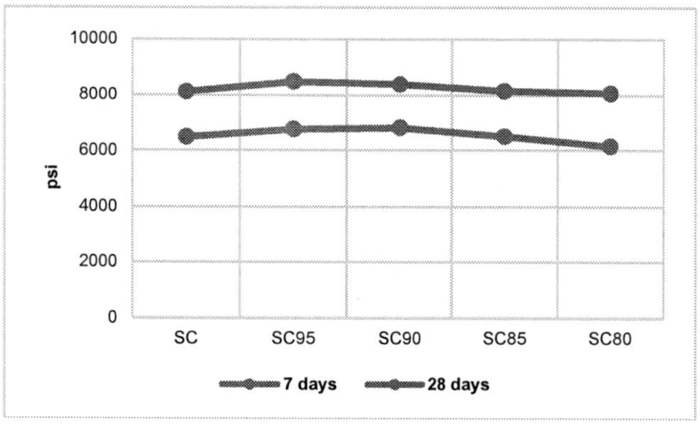

Figure 3– Strength of reduced paste content on SC

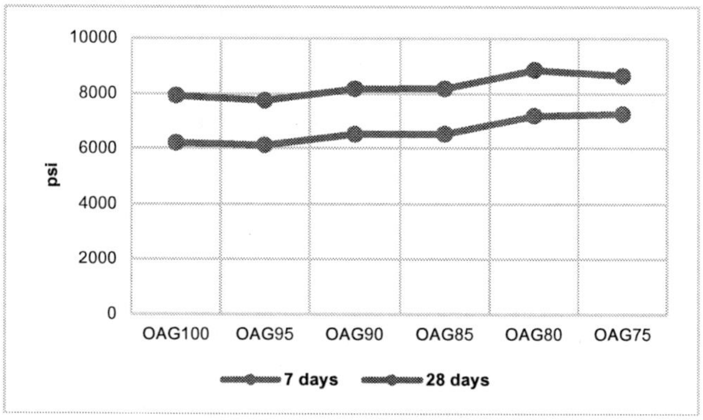

Figure 4– Strength of reduced paste content on OAG

3.3 Cost and Carbon Dioxide Prediction

Except for workability and strength, OAG tech without cementitious deduction also prevents concrete from crack, creep, and shrinkage, segregation, permeability. According to FDOT research (Project Number: BDV31-977-47), OAG concrete can decrease the opportunity of cracks, Due to capillary pressure and well-gradation aggregates distribution; OAG mixes also have lower permeability. Tia (2009) claimed that cement-reduced concrete reduces creep and shrinkage. (Project Number: BD-545-67)

Moreover, segregation could be avoided, because of controlling aggregates gradation. Based on preliminary testing results (Table 1), the cost of proper OAG(OAG85) mix reduces 12.9% total costs than SC mix. Moreover, the emission of carbon dioxide of proper OAG mixes(OAG85) also decrease around 4%. In the end, OAG concrete could be cost-saving, environmental-friendly and sustainable green concrete in a long term.

Table 1 The Cost and CO_2 Emission Evaluation

Mix Design	USD/yd^3	Compared with SC	CO_2 Emission Lb/yd^3	Compared with SC
SC	677.02	0%	446.24	0.00%
SC95	647.49	-4.4%	433.35	-2.89%
SC90	619.15	-8.5%	421.64	-5.51%
SC85	592.28	-12.5%	408.64	-8.42%
SC80	565.04	-16.5%	397.58	-10.90%
OAG100	678.71	0.2%	466.53	4.55%
OAG95	646.65	-4.5%	454.36	1.82%
OAG90	616.68	-8.9%	442.49	-0.84%
OAG85	589.59	-12.9%	429.82	-3.68%
OAG80	562.93	-16.9%	418.03	-6.32%
OAG75	531.05	-21.1%	405.63	-9.10%

4 CONCLUSION

- The OAG method could efficiently enhance the workability and maintain the air content of fresh concrete without adding admixture.
- The OAG method could reduce 25% cementitious content without losing strength, and the SC concrete only could take out 20% cementitious content without any sacrifice of strength.
- Decreasing the proper cementitious content in concrete will not influence strength, it can solve the problem of the shortage of fly ash and the rising cost of cement.
- The proper OAG concrete (OAG85) can reduce the cost 12.9% and the emission of CO_2 decrease 4% compared with Standard Concrete in bridge deck mix.

Acknowledgements

A compilation of this nature could not have been completed without the help and support of others. The Florida Department of Transportation(FDOT) is gratefully acknowledged for providing the financial support and equipment for this study.

References

American Road & Transportation Builders Association, 2015. Production and Use of CCP in the US. Market Forecast Through 333.

Dhir, R.K., McCarthy, M.J., Zhou, S., and Tittle, P. A. J.,2004. Role of cement content in specifications for concrete durability: cement type influences. Structures and Buildings, No. 157, Institution of Civil Engineers, pp. 113–127.

James M. Shilstone,1990. Concrete mixture optimization. Concr Int (10), pp. 33–39.

Moini, M.R., Sobolev, K., and Flores, I., 2014.The best aggregate blends for DOT Applications.American Concrete Institute Fall convention.

Lindquist W., Darwin, D., Browing, J., and McLeod, H.A.K., 2015. Implementation of concrete aggregate optimization. Construction and Building Materials, Vol. 74, pp. 49-56, ISSN: 0950-0618.
Tia, M., Liu, Y.J., and Brown D.,2009. Modulus of elasticity, creep, and shrinkage of concrete. State Contract Number: BD-545 -67, For the Florida Department of Transportation.
Yurdakul, E.,2010. Optimizing concrete mixtures with minimum cement content for performance and sustainability. Graduate Theses and Dissertations of Iowa State University.

Building Tomorrow's Society
Bâtir la Société de Demain

Fredericton, Canada
June 13 to 16, 2018 / *13 au 16 juin, 2018*

RESTRAINED SHRINKAGE TEST AND LAB SIMULATION OF MICRO-CRACKING TECHNOLOGY FOR CEMENT-STABILIZED SOILS

Wang, Shenglin[1,2]; Baaj, Hassan[1]
[1] Department of Civil and Environmental Engineering, University of Waterloo, Canada
[2] s545wang@uwaterloo.ca

Abstract: Pavement structural layers often present deterioration due to several factors such as mechanical loading, moisture and temperature. Cement-stabilized base and subgrade were documented to have higher stiffness and strength; however, this would come with other issues such as rapid hardening and early shrinkage cracks. Micro-cracking technique has recently been developed as a solution to shrinkage cracking. It consists of creating multiple hairline cracks in the stabilised layer in order to relieve the shrinkage and tension stress concentrations; therefore the development of wide cracks can be controlled. However, researches are still required investigating the ability if this technique to control shrinkage cracking for stabilized soils, as well as the lab simulation of the in-field pre-cracking process. In this research paper, two kinds of soils from southern Ontario (clay and silt) were stabilized with 6%, 9%, 12% cement by weight, respectively. The micro-cracking compaction was conducted at 3 days of curing. After 7 days of moist curing, both the pre-cracked and untreated specimens were retrained at both ends and underwent wetting and drying to initiate shrinkage crack. Meanwhile, flexural strength test for soil beams were conducted on 7 days, 28 days, and 56 days of the curing period. Results indicated that the stabilization effectively improved the durability properties and rupture modulus. With the increase of cement content and curing time, the values can be further improved. Moreover, Blenheim clay with 12% GU treated NM (not micro-cracked) had relatively higher shrinkage potential. The proper amount of cement content for Dresden silt and Blenheim clay accounted for around 6~9%, and at least 9%, respectively. Generally, the study presented a preliminary research for restrained soil shrinkage discussion and also attempted to investigate the pre-cracking process in laboratory circumstances.

1 INTRODUCTION

Fine grained soils for instance clay and silt are sensitive to moisture change. One day they are dry and hard, however another day they are wet and soft (Petry and Little 2002). For centuries, various methods have been adopted by engineers to improve the engineering properties of soils. Among those, cement-stabilization was recognized as an efficient method. The basic chemical mechanisms for cement stabilization include cementitious hydration, cation exchange, particle flocculation/agglomeration, as well as pozzolanic reactions (Prusinski and Bhattacharja 1999). The above reactions could help reduce the plasticity index, to enhance both strength and stiffness, and to improve the workability of clays and silts. The first soil-cement construction was carried out in South Carolina, United States, in 1935. Since then, substantial researches and constructions had been documented for soil-cement design and construction, providing promising results (Das 2015).

The main disadvantages of cement stabilization, on the other hand, include rapid hardening, hydration and drying-induced shrinkage cracking, sulphate attack, high cost, and CO2 emissions etc (George 1973). Silt and clay are more sensitive to wet-dry shrinkage due to the higher plasticity properties, capillary action, and larger specific surface area. Moreover, in order to provide sufficient strength and durability properties, the contents of stabilizers for fine soils are generally 2-5 % higher than those for coarse soils (Adaska and Luhr 2004), resulting in higher moisture requirement in soils. In pavements, shrinkage cracks from an underlying cement-stabilized base and subgrade can also reflect through the surface, allowing water penetration and exacerbating deterioration (Louw and Jones 2015). Micro-cracking technique is one of the pre-cracking technologies discussed recently. The general procedure in the field is to apply a vibratory roller (e.g. 12 ton) over the stabilized layer 24-72 hours after the final compaction. Afterwards, multiple hairline cracks will be created to relieve the shrinkage and tension stress concentrations; therefore the development of wide cracks can be controlled. Several trial micro-cracking sections had been constructed in the UK (Ellis and Dudgeon 2004) and in Texas, US (Sebesta and Scullion 2004). Field tests indicated a significant reduction of transverse cracks in micro-cracked sessions compared to controlled sessions after 20-30 months of service life. Also, a continuous growth of FWD (falling weight deflectometer) modulus was observed which could reach or exceed the values in contemporary controlled sessions. However, researches are still limited regarding the mechanism, effectiveness, and established procedure for such technology (Louw and Jones 2015), and there are very few studies which focus on its laboratory simulations.

In this research, two kinds of soils (clay and clayey silt) from a natural subgrade in southern Ontario were stabilized with 6%, 9%, 12% cement by weight, respectively. Restrained shrinkage test was applied to compacted soil beams after 7days of curing. Meanwhile, a micro-cracking process was simulated in lab for specimens with higher shrinkage potentials. On the other hand, flexural strength of both micro-cracked and untreated stabilized soil beams were tested at 7days, 28 days, and 56 days. Therefore, the modulus growth and self-healing properties of micro-cracked soils can be observed. Results of the tests will help to improve the knowledge of shrinkage potentials for cement stabilized soils, and it will introduce a method to simulate micro-cracking process in the lab.

2 MATERIALS AND METHODOLOGY

2.1 Characterisations of soils and cement

Preliminary, target soils were sampled from 2 different natural subgrades located in Chatham-Kent, Southern Ontario. Since the sites were originally used for crop cultivation, there is a considerable amount of organic matters; therefore, the top 10 cm of the soil was removed by a grader. Soils were then excavated and stripped to a depth of 30cm, preparing for mixing and compaction. Soil sampling was conducted when the soil was homogeneous and loose.

Based on the site location, the soils were named as Blenheim clay and Dresden silt. Their natural organic matters content, Atterberg limits, optimum moisture content, soil classification information as well as other geotechnical parameters were summarized below in Table 1.

Table 1: Soil geotechnical characteristics

	Blenheim clay	Dresden silt
Natural organic matter (%)	4.90	4.54
Plasticity index	21	0
Optimum moisture content (%)	17.3	15.9
pH value	4.84	6.51
Clay content (≤5μm) (%)	59	31
USCS* soil group	OH	OL
AASHTO soil group	A-7	A-5
Frost heaving susceptibility	Low	Moderate

* Unified Soil Classification System

Both of the soils had relatively high content of organic matters with approximately 4.5~4.9% by weight. Such matters containing soil humus or humic materials will lead to lower pH values and weaker alkaline environment; and will in turn, interfere with hydration and pozzolanic reactions of cement-treated soils (Harris et al. 2009). The Blenheim clay exhibited moderate plastic properties and had a significant percentage of clay particles; whereas the Dresden silt showed no plasticity. The optimum moisture content of Blenheim clay and Dresden silt under standard compaction effort accounted for 17.3% and 15.9%, respectively.

Cement used in this study is General Use (GU) cement conforming to Canadian Standards Association CAN/CSA-A3001-13. Some technical parameters of the cement were analyzed and presented below in Table 2.

Table 2: Physical and chemical parameters for cement used in research

Parameter	Value	Parameter	Value
w/c ratio by flow test	0.485	Vicat initial setting time (min)	133
Compressive strength 7d (MPa)	22.1	SO_3 (%)	3.9
Compressive strength 28d (MPa)	28.3	MgO (%)	2.5
Blaine fineness (m^2/kg)	383	Loss of ignition (%)	2.3

2.2 Sample preparation and test methodology

The general methodology of the study includes: soil mixing, specimen compaction, curing, micro-cracking compaction, restrained soil beams testing, and flexural strength testing. The schematic flowchart of the procedures is illustrated as follows:

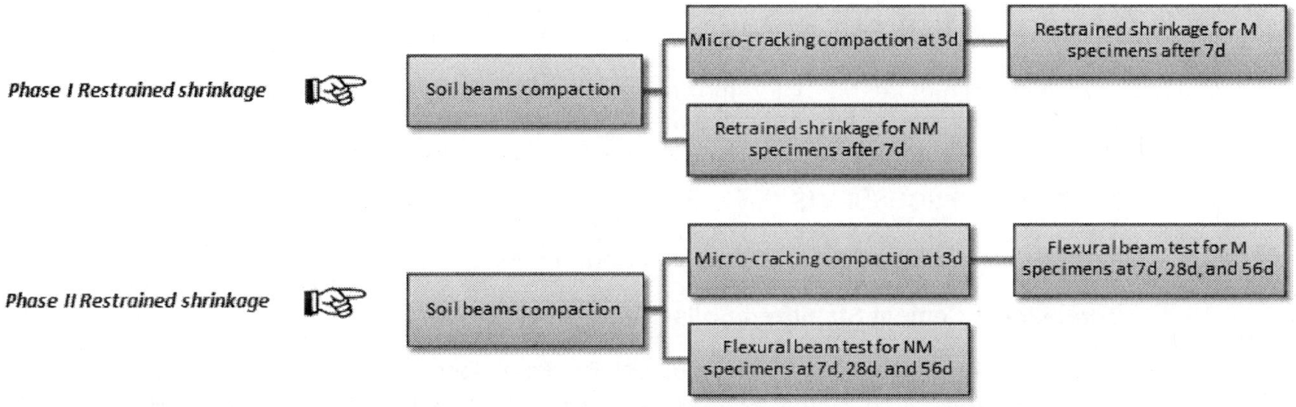

Figure 1: Schematic flowchart for lab tests

During the soil preparation, dry soil and water were initially mixed and mellowed for 24h to allow the moisture to distribute homogeneously. Then, the cement was mixed with wet soil by weight of 6%, 9%, and 12% respectively, and the mixture was compacted in moulds by hammer and tamper according to its optimum moisture content and the corresponding density. The test procedure was generally divided into two major phases: phase I, soil shrinkage potential test, and phase II, flexural beam strength test. The micro-cracking compaction was conducted after 3 days of moist curing, and restrained shrinkage tests started after 7 days of curing. On the other hand, flexural beam tests for micro-cracked (M) and non-micro-cracked (NM) specimens were carried out at 7 days, 28 days, and 56 days of curing respectively.

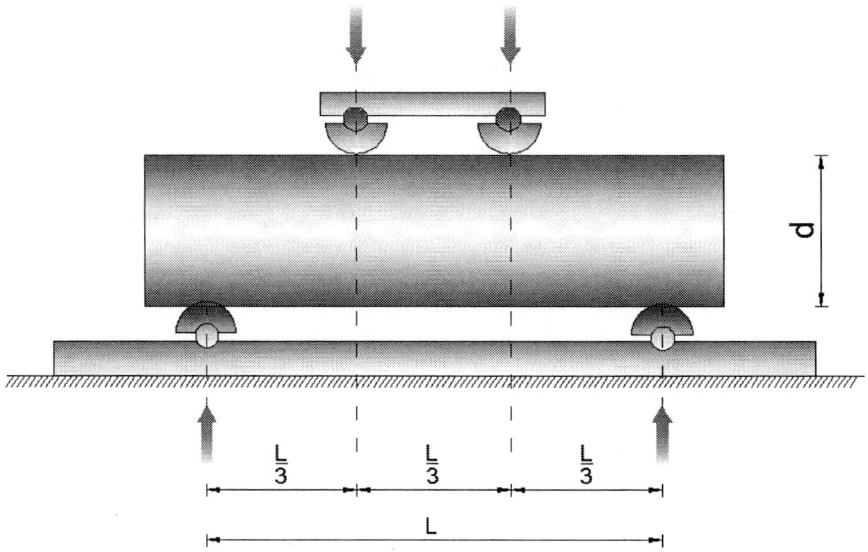

Figure 2: Flexure test of soil-cement by third-point loading method (ASTM D1635)

The flexural strength of the soil beams were then carried out with a third-point loading system (figure 2) as introduced in ASTM D1635, but using a smaller sized specimen (30mm by 30mm by 150mm). Vertical loadings were applied on one-third of the span length at a rate of 1.2mm/min until the specimen failed. The modulus of rupture can then be calculated from the following equation (ASTM D1635):

$$[1]\ \mathrm{R} = \frac{PL}{bd^2}$$

where R = modulus of rupture (MPa), P = maximum vertical load (N), L = span length = 120mm, and b = d equals to average width and depth of specimen = 30mm, respectively.

3 RESULTS AND DISCUSSIONS

3.1 Restrained Shrinkage Test for Stabilized Soil Beams

3.1.1 Durability of Cement Stabilized Soils

As it was illustrated previously, the dry mass of stabilized soil beams were measured at the end of each drying and wetting cycle. Figures 3 and 4 below present the variation of dry mass of the two soils compared to the initial weight, respectively. It should be noted that untreated soil beams were also subjected to wetting, however both of them collapsed during the first submerging period. Most of the stabilized soils, on the other hand, maintained their shape even when undergoing several wetting and drying cycles. In addition, higher cement content normally led to significant improvement of soil durability.

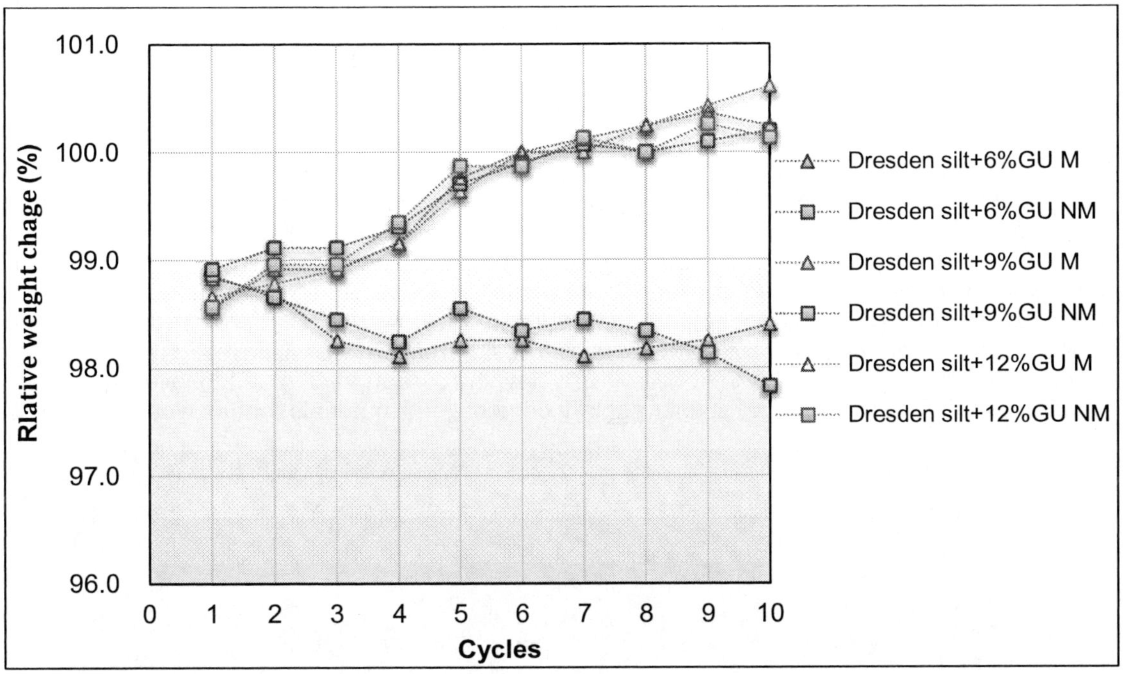

Figure 3: Relative weight change of Dresden silt after each wetting and drying cycle

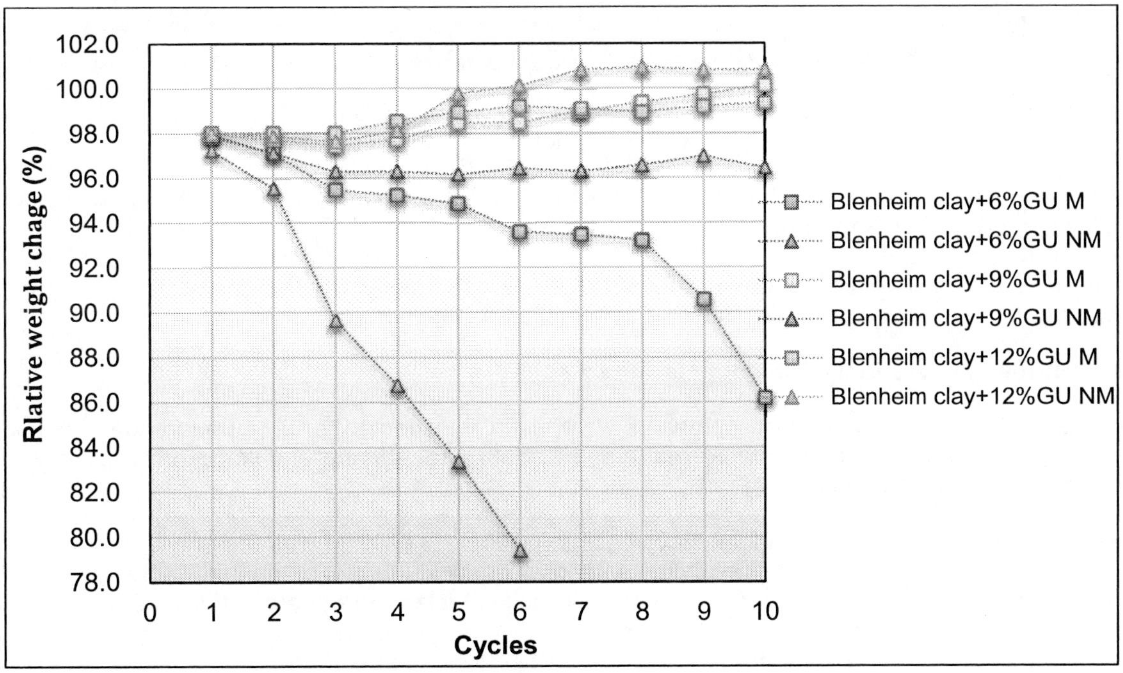

Figure 4: Relative weight change of Blenheim clay after each drying and wetting cycle

Dresden silt showed minimum weight loss after the wetting and drying deterioration; hardly any soil particles were seen peeled off from beam mass (figure 5). Silt with 6% GU cement showed a slight downward trend regarding the dry weight; nevertheless, the overall structure remained hard and stable during the wetting and drying. Meanwhile, silt with higher cement ratio (9% and 12%) had a continuous growth in the dry weight. The reason for this growth was due to the cement hydration: water which had reacted with cement was retained after drying, and the amount of retained water increased as the hydration continued. Empirically, the amount of retained water was estimated by 1/4 of cement percentage (PCA 1992).

On the other hand, the cement stabilized Blenheim clay showed much weaker durability properties than silt. Starting from the second cycle, non-micro-cracked Blenheim clay with 6% GU exhibited significant loss of weight; and it collapsed after 6 cycles, losing approximately 20% of soil particles compared to the initial state. Clay specimens mixed with 9% or more cement were more stable during the test; the weight of specimens remained consistent or slightly increased. According to PCA (1992) manual for soil-cement laboratory test, the permit loss of soil-cement sample after wetting and drying cycles shall not exceed 10% by dry weight of Dresten silt (A-5) and 7% of Bleinheim clay (A-7). From the perspective of wetting and drying, 6% GU stabilizing can be acceptable for Dresden silt. However for Blenheim clay, cement ratio around 9% or above can meet the criteria. Future mix design researches will investigate the Dresden silt with cement ratio between 6% and 9%. For Blenheim clay, the researchers intend to investigate the use of other hydraulic road binders including supplementary cementitious materials for this type of difficult soil.

Figure 5: State of soil beams after 6 cycles: Blenheim clay with 6% GU NM (left), Dresden silt with 9% GU M (right)

Specimens which had experienced micro-cracking showed improved performance of water resistance and durability properties. But the improvement was not considerable due to a short curing period (7d). Moreover, there were some difficulties on how to quantify the cracks produced by micro-cracking. The soil beams were visually seen to have a denser and stiffer structure. Consequently, whether the durability improvement was contributed by micro-cracking or "second compaction" is unclear.

3.1.2 Shrinkage potential of cement stabilized soils

After each drying period, restrained soil beams were visually inspected for the appearance of cracks and deformation. Since the soil beams were restrained at both ends, drying shrinkage of the soils will yield transverse cracks in the beams.

Among all the soil beams, the Blenheim clay specimen with 12% GU NM was the only specimen that had some extent of shrinkage potential. It exhibited shrinkage crack at the end of the third cycle; and the cracks became wider with the cycles ongoing. Subsequently, soil particles were peeled off from the beam surface during the submerging period. On the other hand, the same specimen that underwent micro-cracking showed no visible shrinkage cracking during the test. Therefore, the micro-cracking process enhanced the soil beam durability and/or relieved the tension potential. The other clay and silt specimens maintained their general integrity during the wetting and drying cycles.

Figure 6: State of soil beams Blenheim clay with 12% GU NM after 3 cycles (left), Dresden silt with 12% GU NM after 10 cycles (right)

Literature review indicates that the organic matter and soil mineral have crucial influences on the swelling and shrinkage properties. Expansive clay minerals, e.g. montmorillonite, had a large specific surface (SSA) and held a large cation exchange capacity (Luo 2007). High content of montmorillonite clay in the soil mass will result in a high water-hold capacity, high swelling and shrinkage potential, and high plastic property. However, according to the soil investigation, both candidate soils did not have high plastic property. The organic matters and clay particles in Blenheim clay brought relatively higher potential of shrinkage during cement stabilization, but generally the shrinkage and swelling potential for the two kinds of soils were not distinct. In the future, it is also necessary to introduce other soils which have different mineral components and higher plasticity properties for comparison. From the aspect of shrinkage potential investigation, 9% of cement dosage is optimum to both improve the durability as well as to prevent drastic hydration.

3.2 Flexural beam test

Results of rupture modulus were calculated based on equation [1] introduced previously. Samples were tested at the curing age of 7days, 28 days, and 56 days, respectively.

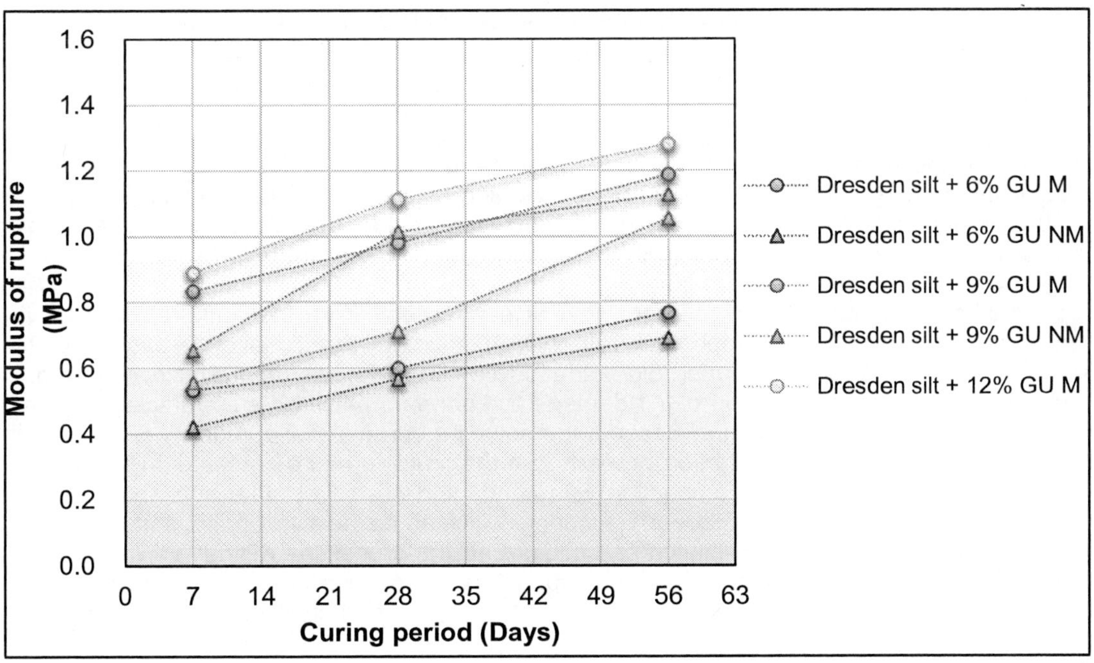

Figure 7: Modulus of rupture of Dresden silt

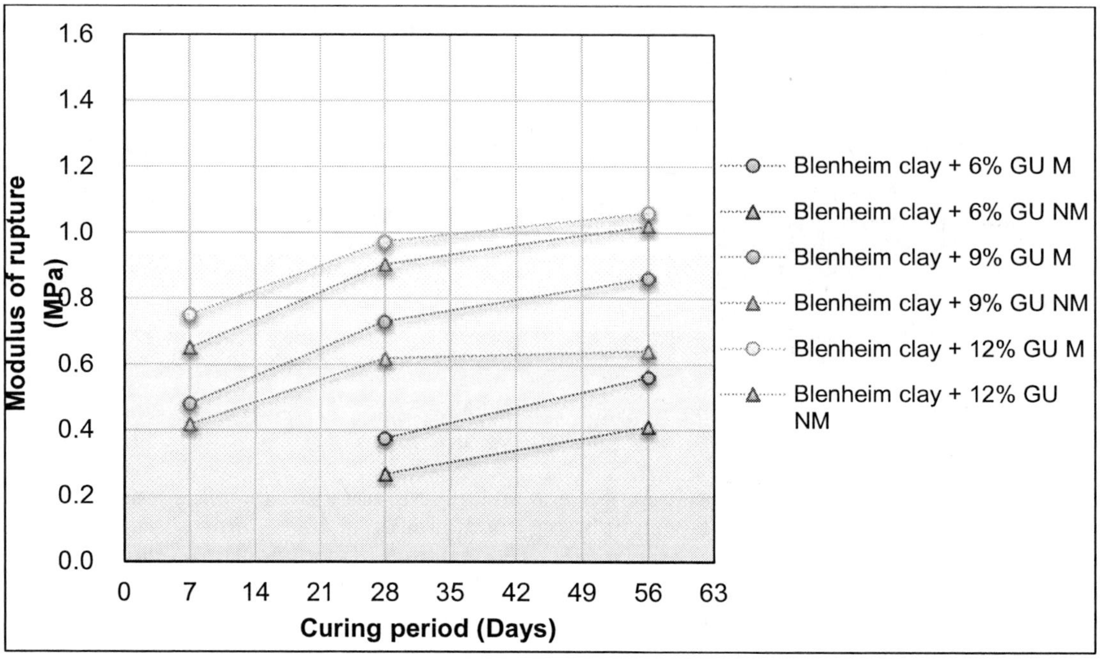

Figure 8: Modulus of rupture of Blenheim clay

As was presented in Figures 7 and 8, with the cement content increasing, both M and NM samples had significant growth of rupture modulus. Furthermore, a longer period of curing will lead to larger modulus values. Between them, stabilized Dresden silt generally had a higher modulus of rupture in all cement ratios and curing ages.

As it was discussed previously, the organic matters reduced the soil pH, and affected the hydration and pozzolanic reactions. Thus the stabilized Blenheim clay exhibited relatively lower strength and stiffness than stabilized Dresden silt. Particularly, Blenheim clay samples with 6% GU at the age of 7d were too weak; and the test results were not able to be acquired. On the contrary, the rupture modulus of 12% GU stabilized Dresden silt with micro-cracking compaction accounted for a high of 1.28 MPa at the curing age of 56d.

The principle for the micro-cracking process is to create intensive, tiny cracks to reduce the stiffness temporally at the early age of curing (Luo 2007). As a consequence, the stiffness and compressive strength will be reduced in the field (Louw and Jones 2015). However, from Figures 7 and 8, there was no obvious decrease of modulus of micro-cracked soils specimens compared to untreated ones. On the contrary, specimens which experienced micro-cracking compaction had higher rupture modulus than their counterparts with the same cement content, even at early ages. From the authors' point of view, the stabilized soils were "secondly compacted" rather than micro-cracked. Several reasons caused this phenomenon: first, 3 days of moist curing did not lead to a stiff soil mass before micro-cracking. In other words, the micro-cracking compaction didn't damage the bonding between soil particles, but creating a more compact and dense soil structure. Second, the soil type and mineral components resulted in a low shrinkage potential and crack propagation property. So the multiple micro-cracks didn't appear after the micro-cracking process. Third, the compaction was conducted on a wet state soil which allows plastic deformation, whereas in the field, inadequate curing will lead to severe drying shrinkage and crack propagation. Overall, soils after stabilization and micro-cracking processes will have an improved rupture modulus, and the values of modulus grew continuously with the curing ongoing. Nevertheless, more research work is needed in order to study the effects of soil types and curing condition on micro-cracking technology.

4 CONCLUSION

From the laboratory tests and analyses of the data, the following conclusions and discussions can be drawn:

1. Cement stabilization effectively improved the durability properties and rupture modulus of the studied soils. Along with the increase of cement content and curing time, the values can be further improved. The lab test results of restrained shrinkage and flexural beam tests, showed that the proper cement content for Dresden silt with the 6~9% range, and the adequate cement content for Blenheim clay should at least be around 9% by dry weight. It was also found that Blenheim clay with 6% GU cement by weight did not have enough durability against wetting and drying cycles. The research provided a preliminary study of cement-soil behaviours, future mix designs will address the physical properties of Dresden silt with cement ratio between 6% to 9%. For Blenheim clay, the researchers intend to investigate the use of other hydraulic road binders including supplementary cementitious materials as this soil type is particularly difficult to stabilize and requires more research.

2. The stabilized Dresden silt had weight fluctuation controlled within ± 5% compared to the initial weight. For the Belenheim clay 6% GU could not maintain the initial shape after the recommended wetting and drying cycles, whereas the 9% and 12% GU stabilized clays were much more stable regarding the wetting and drying durability. Apparent shrinkage cracks only appeared on 12% GU treated Blenheim clay NM at the end of the third wetting and drying cycle. So the micro-cracking process in this test improved the soil's durability and reduced the shrinkage potential.

3. Flexural beam test indicated that all the soil specimens had significant growth of rupture modulus. Furthermore, longer periods of curing will lead to larger values of modulus. The stabilized Dresden silt generally had higher modulus of rupture in all cement ratios and curing ages.

Before micro-cracking process, the stabilized soils were cured for 3 days in an ideal condition, thus the soil was in a moist condition. Furthermore, modulus tests were conducted on specimens within 2 months of curing period, so the recovery of stiffness in a longer period of time was not investigated. Therefore, in the future studies, research is still required to investigate the following aspects of the micro-cracking process: soil types, the time of applying micro-cracking, the moisture control before micro-cracking, and stiffness gaining in longer curing period. Generally, the study presented a preliminary evaluation of restrained soil shrinkage and also attempted to investigate the micro-cracking in laboratory scenarios.

References

Adaska, W. S., and Luhr, D. R. May, 2004. Control of reflective cracking in cement stabilized pavements. In *Proceedings of 5th International RILEM Conference on Cracking in Pavements* (pp. 309-316).

ASTM D1635/D1635M-12 Standard Test Method for Flexural Strength of Soil-Cement Using Simple Beam with Third-Point Loading, ASTM International, West Conshohocken, PA, 2012, https://doi.org/10.1520/D1635_D1635M-12

Das, B. M. 2015. Fundamentals of Geotechnical Engineering. 5th Ed. Cengage Learning, USA.

Ellis, S., and Dudgeon, R. 2004. Pre-cracking as a technique to minimise reflection cracking in semi-rigid pavement structures-long term performance monitoring. In *5th International RILEM Conference on Cracking in Pavements-Mitigation, Risk Assessment and Prevention. Edited by Petit C., Al-Qadi IL et Millien A* (pp. 325-332).

George, K. P. 1973. Mechanism of shrinkage cracking of soil-cement bases. *Highway Research Record*, 442.

Harris, P., Harvey, O., Puppala, A., Sebesta, S., and Rao, S. 2009. Mitigating the Effects of Organics in Stabilized Soils: Technical Report (No. FHWA/TX-09/0-5540-1). Research Report No. FHWA/TX-09/0-5540-1 Texas Transportation Institute, Texas A&M University, College Station, Texas.

Louw, S., and Jones, D. 2015. *Pavement Recycling: Literature Review on Shrinkage Crack Mitigation in Cement-Stabilized Pavement Layers* (No. UCPRC-TM-2015-02).

Luo, R. (2007). *Minimizing longitudinal pavement cracking due to subgrade shrinkage*. The University of Texas at Austin.

Petry, T. M., and Little, D. N. 2002. Review of stabilization of clays and expansive soils in pavements and lightly loaded structures—history, practice, and future. *Journal of Materials in Civil Engineering*, ***14***(6), 447-460.

Portland Cement Association. 1992. *Soil-cement laboratory handbook*. Portland Cement Association.

Prusinski, J., and Bhattacharja, S. 1999. Effectiveness of Portland cement and lime in stabilizing clay soils. *Transportation Research Record: Journal of the Transportation Research Board*, (1652), 215-227.

Sebesta, S., and Scullion, T. 2004. *Effectiveness of minimizing reflective cracking in cement-treated bases by microcracking* (No. FHWA/TX-05/0-4502-1,).

Building Tomorrow's Society

Bâtir la Société de Demain

Fredericton, Canada

June 13 to 16, 2018 / *13 au 16 juin, 2018*

TENSILE PROPERTIES OF SINGLE FLAX FIBRES

Betts, Dillon[1], Sadeghian, Pedram[1,3] and Fam, Amir[2]
[1] Department of Civil and Resource Engineering, Dalhousie University, Canada
[2] Department of Civil Engineering, Queen's University, Canada
[3] pedram.sadeghian@dal.ca

Abstract: The stress-strain response of flax fiber-reinforced polymer (FFRP) composites (and other plant-based FRPs) loaded in tension is nonlinear. It has been determined in previous studies that this nonlinear behaviour is not caused by the matrix or specimen shape. Therefore, there is the potential that this nonlinear behaviour is caused by the individual flax fibres. In this study the mechanical behaviour of flax fibres tested under uniaxial tension is examined. A total of 30 flax fibres from three separate sources were tested and analysed. As a part of this study, the results of eleven tests will be discussed. Two sets of fibres were extracted from manufactured flax fabrics from Europe (one unidirectional fabric and one bidirectional fabric) and the third set of flax fibres were provided by a flax farm in Nova Scotia, Canada. The primary objective of these tests is to determine if the mechanical behaviour of the flax fibre is the cause of the nonlinear stress-strain response exhibited by flax fibre-reinforced polymers. A preliminary set of tests showed that the stress-strain response of these fibres is bilinear, but the accuracy of these tests needs improvement. To accurately measure the tensile behaviour of the fibres a 1 N load cell was built and programmed. The secondary objective of these tests was to determine the effect of the source of the fibre on the mechanical behaviour. The completed tests showed that the fibres from both manufactured fabrics exhibited a bilinear mechanical response. This is strong evidence that the cause of the nonlinearity of FFRPs is the fibres. This research is in progress and more results will be available at the time of the conference.

1 INTRODUCTION

As environmental awareness increases across the globe, it is important that the sustainability of infrastructure is improved. One option to improve sustainability of some structures is the implementation of natural materials. One potential structural application of natural materials is the use of plant fibres, such as flax or hemp, in fibre-reinforced polymer (FRP) composites. FFRPs can be used as sustainable alternatives to synthetic FRPs, such as fibre-glass or carbon-fibre, in structural sandwich panel applications (Mak et al. 2015). These FFRP-foam sandwich panels have been studied under flexure and axial loading in recent past (Codyre et al. 2016; Betts et al. 2017a; Betts et al. 2017b).

To properly model FFRPs and their contribution to structural systems, it is necessary to fully understand their mechanical response. Numerous authors have shown that natural FRPs, including hemp FRPs and FFRPs, exhibit a nonlinear mechanical behaviour (Christian and Billington 2011; Mathura and Cree 2016; Yan et al. 2016; Mak et al. 2015). However, the cause of the nonlinearity is still unknown. The study by Betts et al. (2017c) examines the cause of this nonlinear behaviour by looking at the effect of the FFRP resin, the test specimen shape and the properties of the fibre tows. One conclusion of their study was the hypotheses that the cause of the nonlinear behaviour of FFRPs is the mechanical behaviour of the individual fibres. Therefore, the current study examines this hypothesis by testing a series of flax fibres under uniaxial tension to determine if the fibres have a nonlinear mechanical behaviour.

To complete the fibre tension tests, a load cell was designed, fabricated, and calibrated. Fibre grips were fabricated and attached to the load cell and a hydraulic piston. Eleven fibres were extracted from bidirectional and unidirectional flax fabrics and tested in tension using the proposed test method.

2 PROPOSED TEST METHOD

Elementary flax fibres typically have a diameter of 10 to 20 µm and an ultimate strength of 500 to 900 MPa (Sparnins 2006). Therefore, to measure the tensile load applied to a flax fibre, a small-scale load cell (<1 N) is required. As the smallest available load cell at the university had a capacity of 100 N, a small-scale load cell was designed and fabricated. The load cell design is shown in Figure 1.

Due to the small scale of the tests, the strain measurement was based on the test stroke. As shown in Figure 1, the bottom fibre grip was attached to the hydraulic piston of an Instron 8501 test frame. The tests were performed at a constant displacement rate of 0.01 mm/s and the test specimens had a gauge length of 10 mm. Therefore, the strain could be calculated based on the time, displacement rate and gauge length.

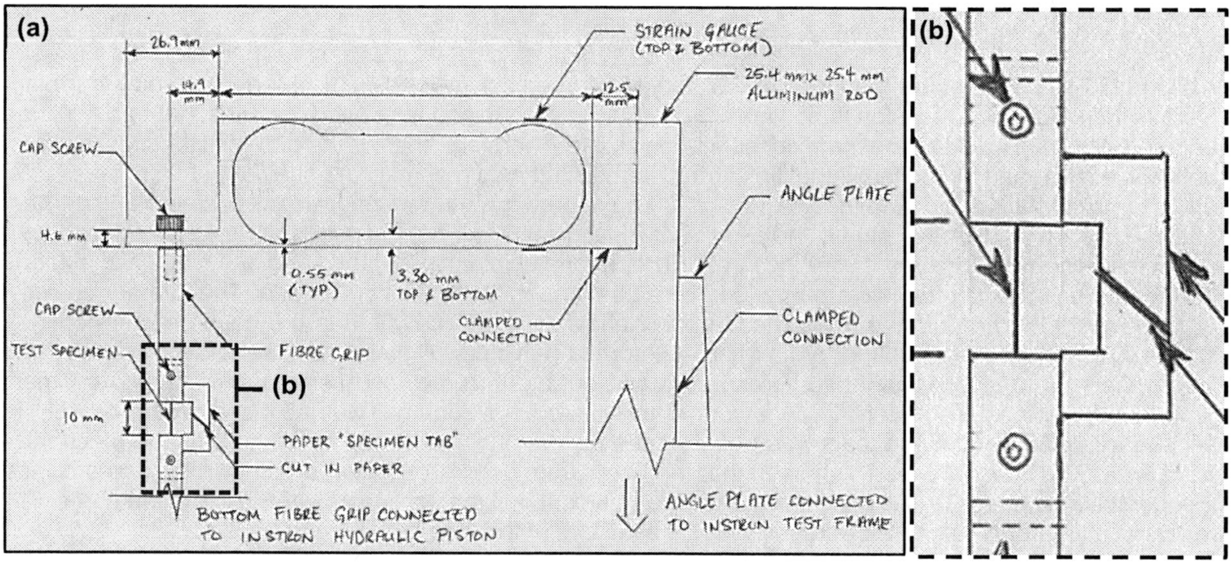

Figure 1: Load cell design: (a) general schematic and (b) detail of fibre grips

The fibre grip was fabricated using an aluminium rod with a diameter of 10 mm. The grip design is shown in Figure 2. Each fibre was glued to a paper tab which had been cut to the correct gauge length. The prepared fibre was then placed into the bottom grip which was closed by tightening the cap screw using an Allen Key. The bottom grip and fibre were then raised into the top grip which was tightened in the same manner. Once both grips were tightened, the bottom grip was raised to ensure that there was no tensile force on the specimen (ie. the distance between the grips was less than the gauge length of the specimen). The paper tab was then cut as shown in Figure 1a. Finally, the bottom grip was lowered until the fiber specimen was almost taut. The test was then ready to start.

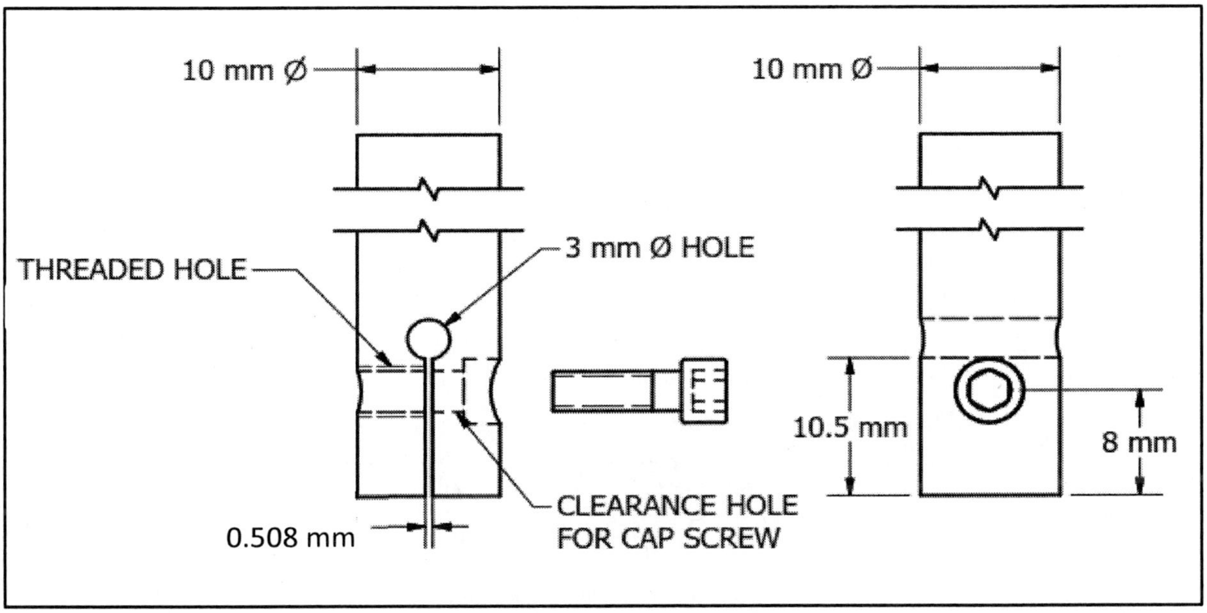

Figure 2: Fibre grip design

The load cell was fabricated using a section of aluminium which was cut using an end mill. The output of the load cell is a strain reading from the strain gauge on the top of the load cell which was placed near the rigid clamped connection to an aluminium rod. Ten known weights were placed on the fixture to calibrate the load cell. The data calibration was completed in the scientific python package, Anaconda, as was the post-processing of the data. The masses of these know weights and the corresponding strain readings from the load cell are provided in Table 1. Based on the values presented in Table 1, the data was calibrated using a linear fit, as shown in Figure 3.

Table 1: Load cell calibration masses

Calibration Mass (g)	Cumulative Mass (g)	Load (N)	Load Cell Output (με)
0	0	0	-545
4.40	4.40	0.04316	-534
4.88	9.28	0.09104	-519
4.88	14.16	0.13891	-506
4.85	19.01	0.18649	-492
4.92	23.93	0.23475	-477
4.91	28.84	0.28292	-465
4.82	33.66	0.33020	-451
4.96	38.62	0.37886	-437
4.30	42.92	0.42105	-425
4.86	47.78	0.46872	-411

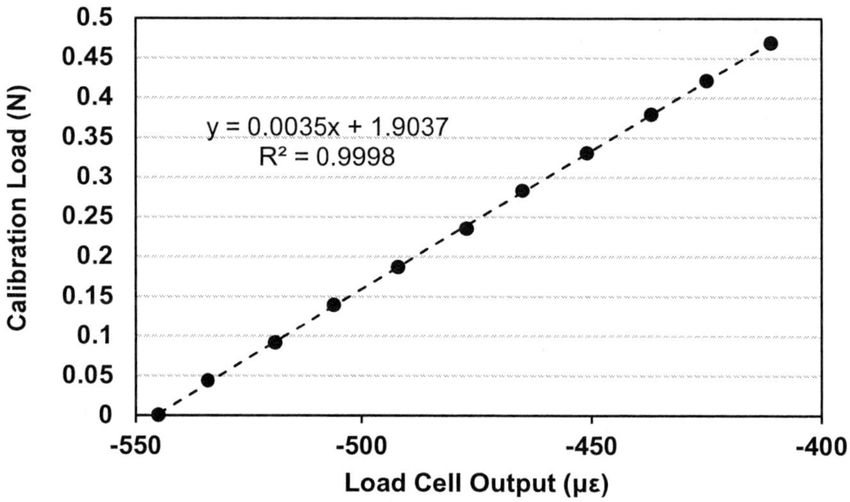

Figure 3: Load cell calibration

The post-processing of the data was performed by a program written in the Python programming language. The program takes the test data, the specimen diameter, the gauge length and the displacement rate as preliminary inputs. It plots the raw data and prompts the user to define the range of the valid test data (i.e. eliminate the data recorded before the testing began and remove the data recorded after the specimen was broken) by providing start and end indices of the valid section of data. The program then implements a moving average filter with an N-value of 100 which improves the resolution of the data from approximately 0.02 N to approximately 0.002 N, which is an improvement by a factor of 10.

Once the data has been averaged, the program converts load to stress using the fibre diameter and time to strain using the gauge length and displacement rate. Figure 4 shows an example of the raw data plotted by the program and the final stress-strain diagram plotted by the program.

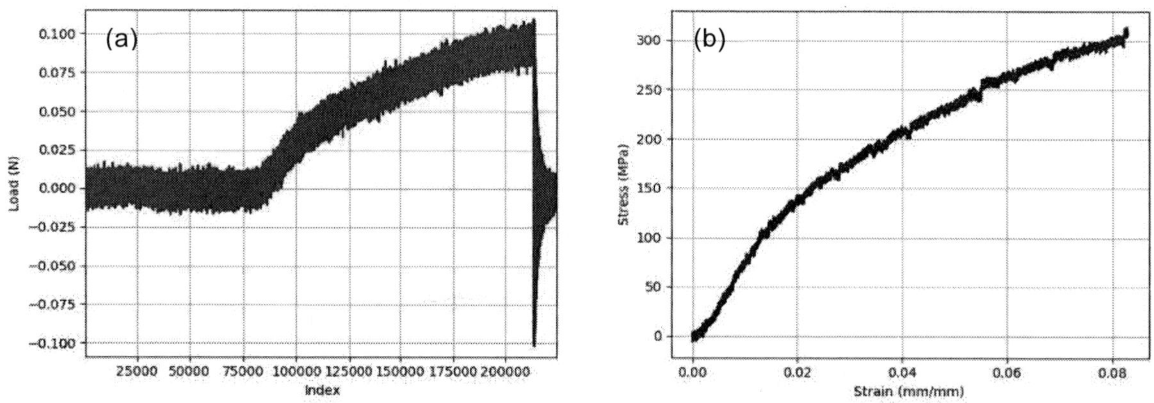

Figure 4: Post-processing program outputs for specimen FF-5 (a) raw load data (b) stress-strain diagram

3 FIBRE TESTS

Thus far, eleven flax fibres have been tested in uniaxial tension as a part of this study. Each specimen had a gauge length of 10 mm and was tested at a rate of 0.01 mm/s using an Instron test frame and the load cell as described in the previous section. Nineteen more fibres will be tested as a part of this study; the full test matrix is presented in Table 2.

Table 2: Test matrix

Specimen Group	Source Location	Source Details	Number of Specimens	Number of Specimens Tested
FF-NS	Nova Scotia	Farmed fibre	10	N/A
FF-BD	United Kingdom	Extracted from 2x2 twill flax fabric	10	5
FF-UD	United Kingdom	Extracted from unidirectional flax fabric	10	6

Figure 5 shows a picture of the tensile test set-up used for the tests of the first eleven specimens. The specimens were carefully extracted from a 2x2 twill flax fabric using tweezers and glued to paper specimen tabs which were cut to the gauge length of 10 mm. The specimen was then placed in the fibre grips and fixed in place by tightening the cap screws. Once the fibre and paper tab were in place the paper tab was cut (as seen in Figure 5b) and the test was ready to start.

4 RESULTS AND DISCUSSIONS

The results of the tested specimens are presented in Table 3. Figure 6 shows the stress-strain plots of the tested specimens. The tested fibres were not observed under a microscope to obtain an accurate measurement of diameter and therefore the stress calculations were based on a nominal fibre diameter of 19.93 μm. This value was determined by taking the average of the fibre diameters from a previous set of preliminary tests. In these previous tests, three photos along the length of each fibre were captured using a microscope. Then the diameter of each fibre was measured at six points along the fibre length using an image processing software. Based on scanning electron microscope (SEM) imagery as shown in Figure 7, it is known that the fibres have a hollow cross section. However, for the calculation of stress in this study, this was neglected; that is, the gross cross-sectional area was used for the stress calculations. Upon examination of the data, it was determined that specimen FF-BD-3 experienced pre-mature failure and therefore, it was not used when calculating the average values shown in Table 3 and Figure 6. This per-mature failure could have been caused from specimen handling or been due to a defect in the fibre prior to extraction from the flax fabric.

Figure 5: Test set-up (a) general test set-up and (b) detail of a tested fibre

Table 3: Fibre test results

Specimen	Primary Modulus (GPa)	Secondary Modulus (GPa)	Secondary to Primary Modulus (%)	Tensile Strength (MPa)	Ultimate Strain (mm/mm)
FF-BD-1	7.67	4.43	58	352	0.0622

Specimen	Primary Modulus (GPa)	Secondary Modulus (GPa)	Secondary to Primary Modulus (%)	Tensile Strength (MPa)	Ultimate Strain (mm/mm)
FF-BD-2	8.16	2.15	26	265	0.0939
FF-BD-3 *	4.69	N/A	N/A	182	0.0389
FF-BD-4	6.67	1.93	29	295	0.0868
FF-BD-5	8.69	2.45	28	314	0.0829
Average	7.80	2.74	35	307	0.0815
St. Dev.	0.86	1.15	15	36	0.0136
FF-UD-1	6.82	2.76	40	316	0.0736
FF-UD-2	10.95	2.47	23	286	0.0646
FF-UD-3	7.71	3.54	46	379	0.0714
FF-UD-4	11.88	2.91	24	333	0.0691
FF-UD-5	8.33	2.99	36	313	0.0607
FF-UD-6	9.70	3.46	36	385	0.0837
Average	9.23	3.02	34	335	0.0705
St. Dev.	1.96	0.41	9	39	0.0080

* Presumed premature failure; not included in the average and standard deviation (St. Dev.) calculations

Table 3 shows that the mechanical properties of the FF-UD and FF-BD fibres are similar and both show that the secondary modulus is approximately one third of the primary fibre modulus. Interestingly, the paper by Betts et al. (2017c) showed that the secondary modulus of FFRPs is approximately two thirds of the primary modulus. Upon examination of Table 3, it seems that the primary modulus, secondary modulus, and ultimate strength of the unidirectional fibres are higher than the bidirectional fibres; however, an analysis of variance (ANOVA) with a confidence level of 95% shows that the difference between the groups is statistically negligible.

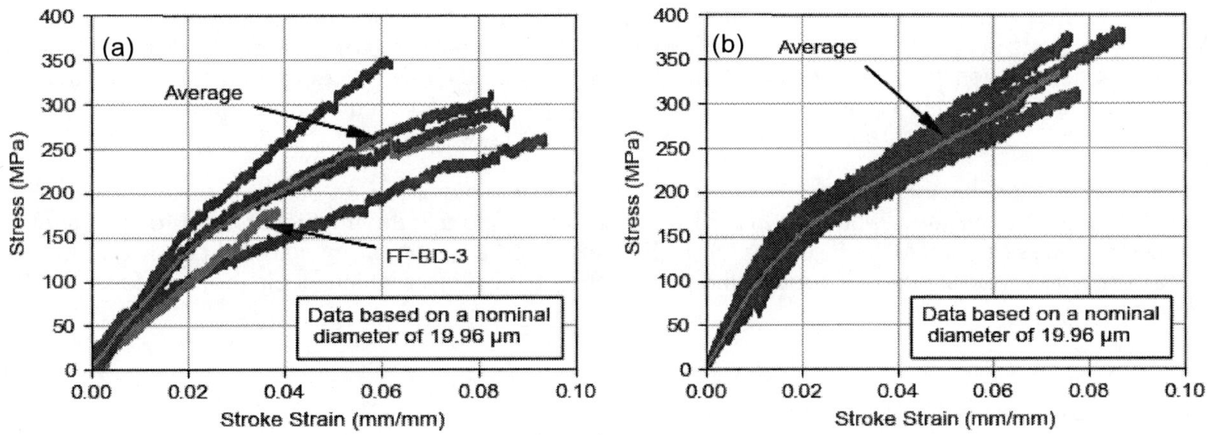

Figure 6: Stress-strain plots of tested flax fibers (a) FF-BD specimens (FF-BD-3 not included in the average) and; (b) FF-UD specimens

4.1 Nonlinear Behaviour of Flax Fibres

As shown in Figure 6, the tested specimens (excluding FF-BD-3) all showed a nonlinear mechanical response to the tensile loading. This result is evidence that the nonlinear behaviour of their composites is potentially caused by the nonlinear nature of the individual fibres. Future studies will include an in-depth analysis of the effect of the fibre nonlinearity on FFRP nonlinearity as well as determining the cause of the fibre nonlinearity.

Per Sparnins (2006), elementary flax fibres are made up of smaller meso fibrils and micro fibrils. This was also observed by the authors as shown in SEM photograph of a flax fibre shown in Figure 7. The current hypothesis of fibre nonlinearity is that until the transition point, the meso and micro fibrils work together and provide the primary fibre stiffness. At that point, the micro fibrils fail and the overall stiffness of the fibre is reduced until ultimate failure of the meso fibrils.

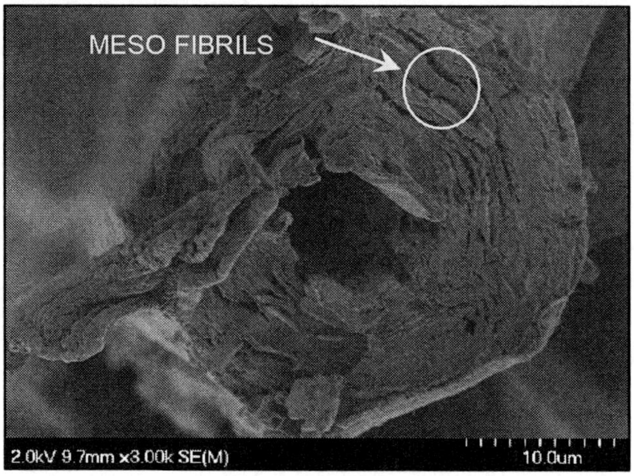

Figure 7: SEM photograph of elementary flax fibre

4.2 Effect of Flax Fibre Behaviour on Composite Behaviour

The unidirectional fibres (i.e. FF-UD specimens) were extracted from the same fabric used in the previous study by (Betts et al. 2017c). Figure 8 presents a plot comparing the results of the fibres, the matrix, and the composite. As discussed later, the strain data provided by the tests is erroneously high. Therefore, to better compare the fibres with the matrix and composite, the fibre strain data was calibrated using data from the study by Betts et al. (2017c). Figure 8a shows the fibre data calibrated using Rule of Mixtures using the simplifying assumption that the composite had no voids. Figure 8b shows the fibre data calibrated such that the rupture strain of the fibre matches the composite rupture strain. To accurately compare the fibre, matrix and composite, a more accurate measurement of strain data is required.

Additionally, based on the data by Betts et al. (2017c), the maximum stress estimation of the fibres according to rule of mixtures would be 273 MPa. As shown in Table 3, the unidirectional fibre tests showed a maximum stress of 335 MPa. Because the FFRPs exhibited a nonlinear mechanical behaviour, the rule of mixtures may not accurately predict the behaviour of the fibres which could be the cause of this discrepancy.

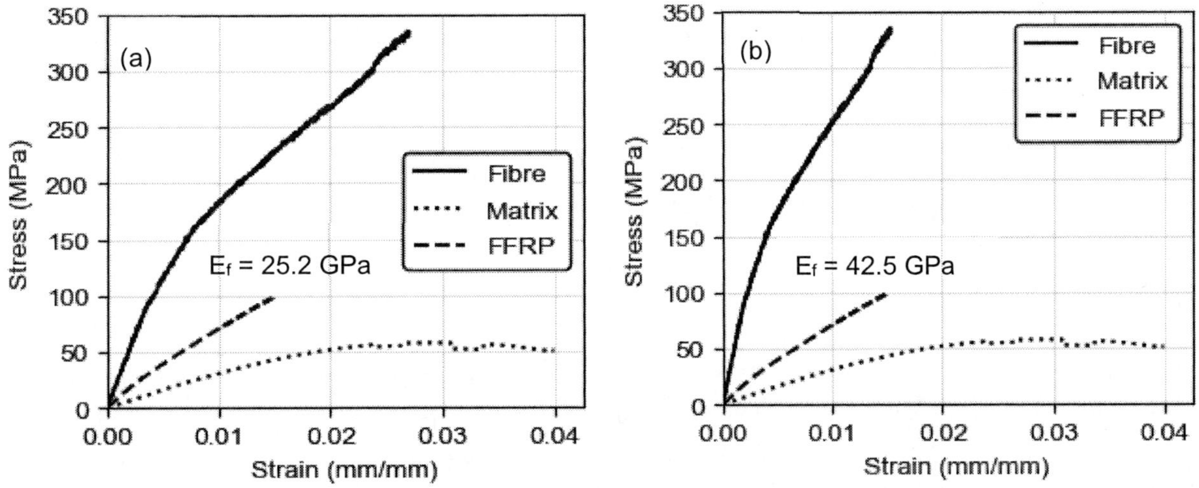

Figure 8: Effect of flax fibres on composite behaviour (a) strain data adjusted to Rule of Mixtures (b) strain data adjusted to composite rupture

4.3 Sources of Error

The young's modulus of flax fibres is typically between 50 GPa and 70 GPa (Sparnins 2006), therefore upon examination the data presented in Table 3, it is apparent that the modulus values determined through testing are low. This is most likely due to erroneously high strain readings, which could be explained by several sources of error, including:

a. Load cell deflection
b. Incorrect gauge length estimation, which is affected by:
   i. fibre twist (or "waviness")
   ii. fibre placement on the paper tabs
c. Fibre slippage
d. The use of a nominal fibre diameter
e. Neglecting the deduction of cross-sectional area due to the fibre hollowness.
f. Fibre fragility (potential to damage fibre before testing)

Some of these sources of error can be mitigated by implementing additional protocols into the test procedure. These protocols include: using an LVDT to measure the deflection of the load cell, using a laser extensometer to measure displacement between fibre grips and measuring the individual fibre diameters before testing.

5 CONCLUSIONS

As a part of this study thirty fibre specimens will be tested in uniaxial tension using a load cell that was designed and fabricated by the authors. Thus far, eleven of the flax fibre specimens have been tested. The results of the tests show that the fibres extracted from the bidirectional fabric have an ultimate tensile strength of 307 MPa and the fibres extracted from the unidirectional fabric have an ultimate tensile strength of 335 MPa. The rule of mixtures predicts an ultimate strength of 273 MPa, however, this discrepancy could be caused by the nonlinear nature of flax fibres and FFRPs. Fibres extracted from both sources exhibit a nonlinear mechanical response, specifically a bilinear response where the secondary modulus is approximately one third of the primary modulus. This supports the hypothesis that the nonlinear behaviour of flax fibre-reinforced polymers is caused by the behaviour of the individual fibres. Currently, the strain estimation is erroneously high which affects the calculation of the young's modulus. The implementation of additional protocols in the test procedure will help mitigate the sources of error. This study is a part of ongoing research and future studies include: testing of more fibres from different sources, determining cause of fibres' nonlinear behaviour and studying the effect of growth location.

Acknowledgements

The authors would like to thank Brian Kennedy and Jesse Keane for their assistance in the lab and acknowledge the use of the fibre grip design by John Noddin and Kevin Borgel. The authors would also like to thank Bioindustrial Innovation Canada (BIC), Queen's University and Dalhousie University for their in-kind financial support.

References

Betts, D, Sadeghian, P, and Fam A. 2017a. Cumulative Impact Energy Absorption of Sandwich Panels with Foam Cores and Flax FRP Faces. *17th International Conference on Non-Conventional Materials and Technologies*, Merida, Yucatan, Mexico, 1–10.

Betts, D, Sadeghian, P, and Fam A. 2017b. Structural Behaviour of Sandwich Panels Constructed of Foam Cores and Flax FRP Facings. *CSCE Annual Conference.* Vancouver, BC, Canada, 1–10.

Betts, D, Sadeghian, P, and Fam A. 2017c. Tensile Properties of Flax FRP Composites. *Asia Pacific Conference on FRP in Structures*. Signapore. 1-4.

Christian, S J, and Billington, S L. 2011. Composites : Part B Mechanical Response of PHB- and Cellulose Acetate Natural Fiber-Reinforced Composites for Construction Applications. *Composites Part B* **42** (7): 1920–1928.

Codyre, L, Mak, K, and Fam, A. 2016. Flexural and Axial Behaviour of Sandwich Panels with Bio-Based Flax Fibre-Reinforced Forced Polymer Skins and Various Foam Core Densities. *Journal of Sandwich Structures and Materials*, 1–22.

Mak, K, Fam, A, MacDougall, C. 2015. Flexural Behavior of Sandwich Panels with Bio-FRP Skins Made of Flax Fibers and Epoxidized Pine-Oil Resin. *Journal of Composites for Construction* **19**(2003): 1–13.

Mathura, N, and Cree, D. 2016. Characterization and Mechanical Property of Trinidad Coir Fibers. *Journal of Applied Polymer Science* **133**(29): 1–9.

Sparnins, E. 2006. Mechanical Properties of Flax Fibers and Their Composites. Lulea University of Technology.

Yan, L, Kasal, B, and Huang, L. 2016. A Review of Recent Research on the Use of Cellulosic Fibres, Their Fibre Fabric Reinforced Cementitious, Geo-Polymer and Polymer Composites in Civil Engineering. *Composites Part B: Engineering,* **92**(April): 94–132.

Building Tomorrow's Society

Bâtir la Société de Demain

Fredericton, Canada

June 13 to 16, 2018 / *13 au 16 juin, 2018*

THE ECONOMIC VERSATILITY OF SELF-CONSOLIDATING CONCRETE – UNDERGROUND AND BEYOND

Maher, Michael L. J.[1,3], Bonin, Grant[2], Dehghan, Alireza[1], Navarra, Michael P.[1]
[1] Golder Associates Ltd., Whitby, ON, Canada
[2] Golder Associates Ltd., Vancouver, B.C., Canada
[3] Michael_Maher@golder.com

Abstract: While originally self-consolidating concrete (SCC) was developed in Japan to address a labour shortage, it is growing in application across North America. The benefits are being recognized mainly in terms of reduced risk to the mechanical and durability properties from improper consolidation. The added product cost can be offset by faster construction, more economical design of reinforcing and concrete section, and reduced construction noise and labour cost. Early in its development, the desired high flowability of SCC was gained by adding excessive amounts of water or superplasticizers to the concrete. Therefore, the mixture had limited application in civil engineering structures when durability requirements were paramount. Establishment of standards and codes of practise for SCC, development of specialized testing equipment, and the introduction of new chemical admixtures, such as high range water reducing (HRWR) and viscosity modifying admixtures, facilitated the production of SCC mixtures with comparable or superior mechanical and durability properties when compared to ordinary Portland cement concrete. Even though SCC has the potential to be used cost-effectively for a broad range of structural concrete, its usage has remained limited mostly to high-end projects. However, this needs to change. This paper will explore some recent project successes where specially formulated SCC mixes were essential components of project feasibility, particularly in underground applications. Some examples are provided of projects where costly remedial measures could have been avoided with the use of SCC. Two case studies are presented that highlight the versatility of SCC and its use in non-conventional underground applications.

1 INTRODUCTION

In the 1980s, the shortage of skilled construction workers in Japan, raised concerns regarding the quality and durability of concrete structures. To solve the issues related to inadequate or improper consolidation, Professor Okamura proposed a concrete that did not require consolidation (Okamara et al. 2000; Vachon 2002). In 1988, the first prototype of SCC, called High-Performance Concrete was introduced to the market (Okamura and Ouchi 1999). The pioneering work done in Japan led to the establishment of multiple Japanese standards and manuals, including "Recommended Practice of High-Fluidity Concrete for Building Construction" in 1997, "Recommendation for Construction of Self-Compacting Concrete" in 1998, and "Manual for Manufacturing of Self-Compacting Concrete" in 1998 (Okamara et al. 2000).

To keep up with the pace of this emerging technology, other organizations and authorities began to explore SCC. RILEM established the Technical Committee TC 174-SCC in 1997 and the first RILEM International Symposium on SCC was held in 1999 with papers presented from 13 countries (RILEM Technical

Committee 2006). In North America, ASTM established Subcommittee C09.47 to explore the appropriate testing procedures for SCC (Vachon, 2002) with its first approved testing method released in 2005. In 2007, the American Concrete Institute (ACI) Committee 237 published the provisional guidelines for SCC. Currently, there are numerous standards and codes of practise for SCC globally.

1.1 Differences between Conventional Concrete and SCC

The primary characteristic of SCC is its high flowability. Flowability is measured in terms of spread using a modified slump test and is defined in ASTM C143 (ASTM, 2015). The spread or slump flow of SCC is typically in the range of 450 to 800 mm (Figure 1).

Figure 1: SCC appearance during Flow Test

The key challenge with the mix design is to achieve this flow without using excess water and preventing instability of the mix and segregation. These attributes are achieved by a combination of adjustments in the mix components and by the use of highly effective chemical additives. Essentially, the high flowability is achieved by utilizing a high-range-water-reducing (HRWR) admixture and the stability of the wet mix is achieved by the use of viscosity modifying admixtures. In addition, SCC mixes generally have smaller maximum aggregate size; less coarse aggregate; higher sand to coarse aggregate ratio and a higher paste volume. A further consideration for construction with SCC is the need to use more substantial formwork since it is required to support full fluid pressure from the wet concrete. For this reason, deep pours need to be placed in lifts so as to give time for the lower lifts to set.

1.2 Advantages of SCC

As can be inferred from its name, when designed and produced properly, a self-consolidating concrete does not require an external source of energy for compaction and it levels under its own mass. Daczko (2013) summarized the benefits that this feature provides as:

- Reduction in the number of workers required for placing and consolidating;
- Reduction in the equipment needed (e.g. vibrators);
- Reduction in noise and nerve and vascular damage to operators;
- Faster construction and higher productivity;
- Easier placement in cases with limited access or congested reinforcement;
- Higher quality of finished surfaces;
- Safer working conditions; and
- Lower construction cost.

Rich et al. (2015) summarized the potential savings due to use of SCC by various authors. In terms of cost, utilization of SCC can lead to up to 21% reduction as compared to the use of conventional concrete. Depending on requirements for fresh and hardened properties, it is estimated that the initial cost of SCC can be up to 15% higher than conventional concrete (Pai 2004), but this cost is offset by labour cost reductions allowing an 8% reduction in overall cost (Cazacu et al. 2015). A Value Engineering study by Huang et al. (2011) on an 8.2 km long underground railway project with an estimated concrete volume of 57,000 m^3 found that usage of SCC can provide a 25% reduction in the overall cost of the project. In the literature, reported reductions in construction time due to usage of SCC vary from 20% (Rich et al. 2015) to 66% (Daczko 2012).

The abovementioned advantages attracted the construction industry, especially precast concrete producers, to use SCC since it allows a high degree of automation in production lines (Elliott and Hamid 2017). Daczko (2012) compared the placement and consolidation cost of precast products made with SCC and conventional concrete and reported a 40% reduction. The suitability for automation has attracted more interest in adopting the SCC technology by the precast concrete industry as compared to ready-mix producers for cast-in-place applications. For ready-mix producers the benefits of using SCC magnify when shorter unloading times for trucks are achieved. However, for the user, the real benefits are in terms of better construction quality and reduced risk of concrete defects.

1.3 Large Projects with SCC Around the World

The first projects that incorporated SCC were the Akashi-Kaikyo Bridge, Japan in 1998 (Dordi 2002) and a liquid natural gas storage tank in Osaka, Japan in 2000 (Nishizaki 2001). During construction of the two anchorages for the Akashi-Kaikyo Bridge, 290,000 m^3 of SCC was used and it reduced the construction period from 2.5 years to 2 years (Han et al. 2017). Successful application of SCC in these early projects accelerated its acceptance globally such that SCC has been successfully used in numerous large projects, including the following examples:

- The Burj Khalifa, UAE: SCC with a slump flow of 600-750 mm was used to build this 160-floor skyscraper which is the tallest free-standing building to date (Baker et al. 2008).

- The International Commerce Centre, Hong Kong: To construct Hong Kong's tallest building (118-floors), a high strength SCC mix with a w/cm ratio of 0.26, a mean 28-day compressive strength of 90 MPa, and an initial slump flow of 720 mm was used (Zhang et al. 2009).

- Mori Tower, Shanghai, China: 37,000 m^3 of SCC was used in the foundation of this 101-floor skyscraper. In one of the pouring phases, 28,000 m^3 of SCC was poured continuously for 40 hours (Phelan 2011).

- Los Angeles Sewer Tunnel Rehabilitation, USA: 37,000 m^3 of SCC with slump flow of 580-610 mm and 28-day compressive strength of 40 MPa was used in this project which, at that time, was considered the largest SCC project in North America to date (Zhang et al. 2009).

- Pearson International Airport Expansion Project, Toronto, Canada: Due to insufficient overhead clearance, 2,000 m^3 of SCC with average slump flow of 600 mm and 28-day compressive strength of 45 MPa was used to fill 180 hollow steel composite columns, each 31 m tall (Lessard et al. 2003).

- Hindaka Electric Power Station, Japan: 22,000 m^3 of SCC with slump flow of 520-620 mm was used to construct the extruded concrete tunnel lining for the Hidaka Power station (Zhang et al. 2009).

- Cleuson Dixence project, Switzerland: 73,000 m^3 of SCC was used to construct 15.8 km of tunnel, 3.9 km of inclined shaft and backfilling between the steel liner and rock. To increase productivity, an underground batching plant was used for producing the SCC. Multiple mix designs were developed including one for a SCC mix that could be pumped over 2.9 km (Botte et al. 1999).

- Sodra Lanken, Sweden: approximately 225,000 m^3 of SCC, with slump flow of 720-770 mm was used in different elements of the 4.5 km long four-lane underground tunnel of the Sodra Lanken link which is the third longest motorway tunnel in Europe. SCC was used for the construction of 16 km of concrete-lined rock tunneling and earth retention walls (Ouchi et al. 2003; SDC-ICT 2011).

2 SLOW ADOPTION OF SCC IN CANADA

Despite the proven benefits of SCC, the broad adoption of SCC on major construction projects in Canada has been slow. The main barriers appear to be the initial cost premium compared to conventional ready-mix and unfamiliarity with the derived benefits. The following are some examples of projects where the use of SCC would have avoided significant additional costs and projects delays.

2.1 Exterior Bulkhead Wall, Transportation Facility

This was a heavily reinforced cast-in-place concrete wall approximately 28 m long, 7 m high and 1.5 m thick. Concerns arose about the integrity of the wall due to the presence of voids that became apparent after formwork removal. Subsequently areas of concrete on one side of the wall were chipped away to reveal the reinforcing steel bars and the condition of the concrete. The inspection revealed frequent defects generally related to zones of segregated and unconsolidated concrete, voids along rebar, cold joints and random voids not associated with rebar extending in 600 mm (Figure 2). A review of the placement records

a. b.

Figure 2: Segregated concrete (a); and voids behind rebars in a large concrete wall using conventional ready-mix concrete (b)

identified a concrete load that took 3 hours to discharge; four rejected loads; pump truck malfunctions; a mix temperature that reached 29°C; ambient temperatures of 33°C; and superplasticizer being added up to four times to the same load to restore slump. Without effective means to define the extent of the problem, or to remediate the defects, the decision was made to demolish and reconstruct the wall, with significant impacts on project costs and schedule (Figure 3).

2.2 Transfer Slab, High Rise Condominium

A 2.3 m thick fourth level large transfer slab in a high rise residential building was poured using self consolidating concrete for the lower 300 mm and then conventional concrete above it. The slab was heavily reinforced with minimum rebar spacing at about 200 mm and also contained a system of post-tensioning cables running through duct work in the slab (Figure 4). When the slab met 75% specified compressive strength, stressing of the post tensioning tendons was initiated. A number of the early tendons tested failed due to the concrete crushing behind the bearing plates.

Figure 3: Demolition of a recently constructed concrete wall due to excessive placement defects

Figure 4: View of reinforcing at time of concrete placement

A GPR survey of the transfer slab was undertaken and while it detected some voids, the extent of the embedded rebar made accurate interpretation unreliable. A coring and probing investigation was undertaken in conjunction with a boroscope camera inspection. The investigation identified voids, cold joints, zones of segregation and honeycombing. One void extended some 700 mm along a post tensioning duct. The initial engineering recommendation was to remove and re-pour the slab. Subsequently, for schedule and commercial reasons, a decision was made to remediate the slab. While this option was still disruptive to the project, it allowed some construction above the fourth level to continue. The remedial plan was complex and prolonged. It involved further exploratory coring and pressure grouting. However, the major component, was to drill in a tight grid a series of small diameter probe holes around the live and dead ends of each of the post tensioning ducts and pressure grout them with epoxy grout (Figure 5).

In addition, some structural redesign was necessary and some additional support columns were added with impact on space utilization. The remedial works took 10 weeks to complete but ultimately allowed all post-tensioning cables to be fully stressed.

The two projects described above are examples of where SCC was the ideal choice and its use could have avoided costly remedial works or replacement as well as schedule disruptions.

a. b.

Figure 5: Extensive voids around post tensioning cable ducts (a) and grout and probe holes around live end of post-tensioning cable ducts prior to grouting (b)

3 SCC IN MINING APPLICATIONS

In urban and transportation infrastructure construction, SCC is supplied as a proprietary product by ready-mix concrete suppliers. In these applications it is usually not practical to customize mix design for specific applications. However, in the mining sector, specialty SCC mixes are gaining acceptance as cost-effective and versatile solutions to underground construction. The use of high fly ash replacement mixes has expanded the application of SCC for large volume concrete pours (Daczko and Phillips 2001.

3.1 SCC Plugs to Reduce Groundwater Inflows, Underground Mine, New York, USA

An active underground mine was managing a flow of ~12,000 L/min into their workings. The mine management was looking for a more sustainable alternative to continuous pumping for removing increasing volumes of water inflows. A hydrogeological investigation devised a plan to construct two, monolithic concrete plugs (or bulkheads) to isolate the inflow while withstanding large hydraulic pressures.

The optimized locations for installation of the plugs were at a depth of ~90 m underground and each opening had an average base width of 8 m, a height of 5 m and length of 5.3 m (Figure 6). From an installation perspective, SCC was essential to streamline installation and to minimize the underground workforce. It was estimated that each plug required about 120 m^3 of SCC to fill the formwork. To reduce the heat of hydration and avoid unacceptable cracking, 65% of the Type I Portland cement was replaced with a Type F fly ash. The design compressive strength at 56-days was 28 MPa and the target slump flow was 650 to 700 mm at the point of discharge.

Some challenges associated with the construction of these two plugs included the long distance between the point of discharge of the concrete above ground and the placement locations, as well as the practical limitations in the construction of the temporary timber formwalls to withstand the lateral loads. With the high fly ash mixes, it was recognized that initial setting would take longer than 24 hours. Ultimately, it was decided to install the plugs in two continuous pours, with the first pour filling the lower 3 m of the plug.

Working with a local ready-mix company, an SCC mix was formulated to meet the project requirements. The mix was delivered to the mine site in conventional ready-mix trucks and then transferred to low profile transmixers. These vehicles have a capacity of 5.6 m^3 and a maximum height of 2.6 m allowing them full access underground.

a. b.

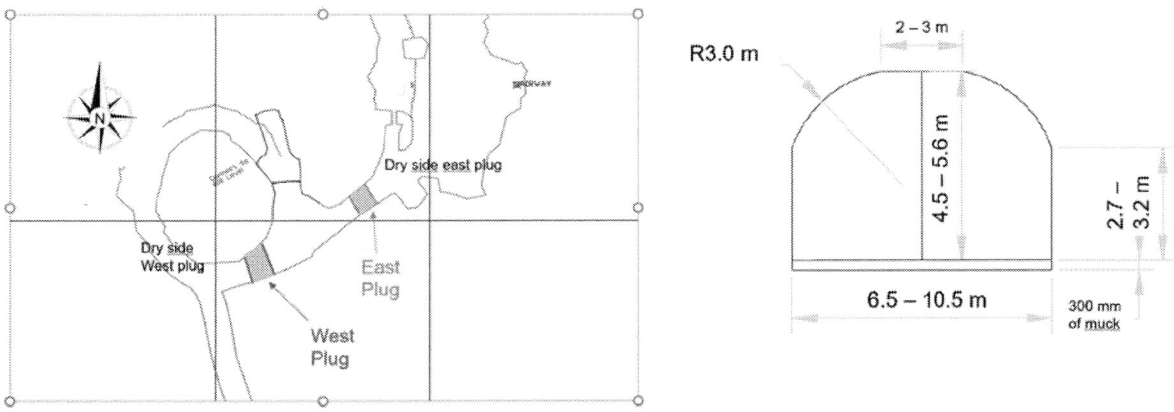

Figure 6: Plug locations (a) and typical plug cross section (b)

The SCC mix was pumped from the transmixer into an opening in the timber bulkhead so as to have a maximum free fall of no greater than 1 m. The transfer of the mix to the transmixer and the placement of SCC in the west plug is shown in Figure 7. The average 56-day compressive strength achieved for the plugs was 55 MPa.

a. b.

Figure 7: Transferring SCC mix from conventional mixer to transmixer (a) and placement of SCC mix underground between formwalls (b)

In the initial mix trials, the final setting time of the SCC was estimated to occur after 10 hours. Three days after completion of the first lift, the aggregate on the top surface was exposed by wire brushing with gentle water jetting removal of the debris. The prepared surface needed to be clean and rough to maintain shear integrity and water tightness. The second lift of concrete was then poured. The option of spraying a top-surface retarding admixture was considered but was found to be unnecessary because of the natural retardation due to the cool ambient temperatures underground.

To ensure water tightness of the plug and to compensate for concrete shrinkage, pressure grouting of the concrete-to-bedrock interface around the plug was required. However, this could not be undertaken until most of the shrinkage had occurred. A thermocouple was installed at the base of the second pour of each plug. The peak temperature at the core of the East plug of 47°C was reached on Day 5. The core temperature of the West plug peaked at 44°C. The target temperature of 12°C for commencement of grouting took over 70 days to achieve (Figure 8).

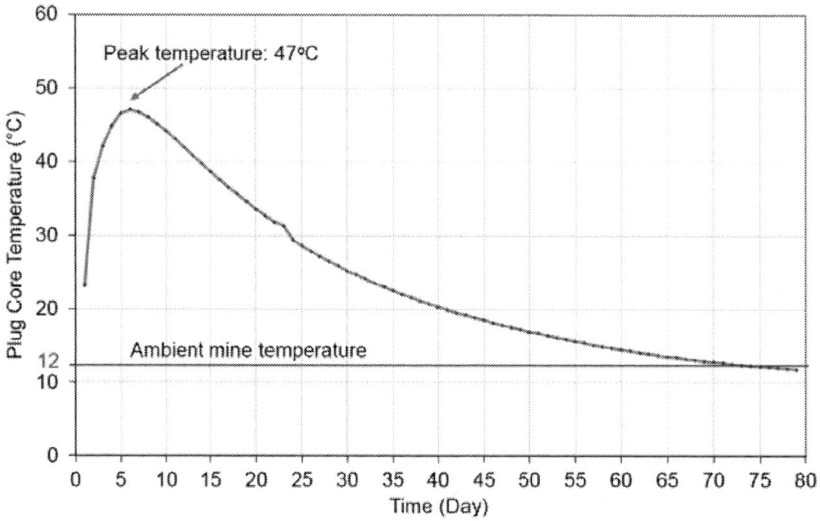

Figure 8: East plug temperature profile

3.2 SCC Plugs for Coal Mine, British Columbia

Two emergency measure, monolithic concrete plugs were designed and installed at an underground coal mine on Vancouver Island, BC. The purpose of the plugs was to manage the risk of a catastrophic in-rush of glaciolacustrine sediments into the mine workings and contain inflowing water. SCC was selected for the plugs for speed and efficiency of placement and to allow the drift to be completely filled without the need for manual consolidation (Figure 9).

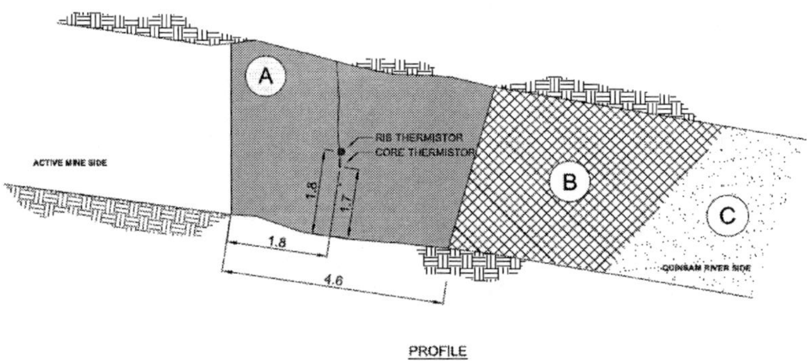

Figure 9: Cross-section showing SCC plug (A) placed against cemented rockfill (B) and waste coal fill (C)

The SCC mix was designed with a 56-day strength of 30 MPa and a design slump flow of 600 to 650 mm. The mix had 55% of the cement replaced by Type F fly ash. The mix also contained HRWR, shrinkage-reducing, and air entraining admixtures. The SCC was batched at a local ready-mix plant approximately a 45 minute drive from the mine.

The SCC was transferred to a pump truck where it was delivered underground through a HDPE pipe inserted through a borehole (Figure 10). This was connected to a 100 mm flexible hose for delivery through a port in the bulkhead. The SCC volumes for the plugs ranged from 120 m^3 to 132 m^3. Test cylinders were cast every 20 m^3 of SCC delivered. The specified 56-day strength of 30 MPa was exceeded at 28-days. The average 56-day strength was 43 MPa.

a. b.

Figure 10: Delivery of SCC to pumper truck connected to vertical delivery borehole underground (a) and completed face of plug being prepared for grouting (b)

The core temperatures within the plugs were monitored to establish when grouting activities, including concrete-to-bedrock contact and consolidation of the surrounding rock mass, could be initiated (i.e. after the majority of the potential shrinkage had occurred and the plug had cooled to the surrounding, ambient rock temperature). A maximum internal concrete temperature of 59°C was reached after 4.5 days following placement. Four and a half months were needed to allow the core temperatures in the plug to drop to 16°C and be suitable for grouting.

4 CONCLUSIONS

SCC is currently an under-utilized technology that can alleviate the risks associated with concrete placement in large and congested structural components in urban and transportation infrastructure. The premium cost can be justified in terms of faster construction, better quality outcomes and a reduced risk of post-construction performance problems. Advances in chemical admixture technology have made SCC mixes extremely versatile and customizable for use in a wide range of complex applications. For specialized applications in the mining sector, SCC mixes provide economic solutions to underground construction where access and work space restrictions impose unique constraints. By using high fly ash ratio mixes, SCC can be used for large volume concrete pours that further enhances the opportunities for use.

SCC mixes provide high strengths when required for high rise buildings and critical infrastructure or they can be designed to generate minimum heat when used in mass concrete. SCC mix designs can be tailored to meet demanding project specifications, in terms of constructability, durability and performance. In mining applications, especially for structural bulkheads, SCC is an effective tool since it can be made with locally available materials and does not require sophisticated materials and equipment.

References

ASTM C143 / C143M-15a, 2015. Standard Test Method for Slump of Hydraulic-Cement Concrete, ASTM International, West Conshohocken, PA, www.astm.org

Baker W. F., Korista S. D., Novak L. C. 2008. Engineering the World's Tallest – Burj Dubai, *Council of Tall Buildings and Urban Habitants 8th World Congress*, March 3- 5, Dubai, UAE.

Botte, J. Burdin, J., Zermatten M. 1999. SCC tunnel applications: Cleuson Dixence project and Loetschberg basis tunnel, Switzerland, *First International RILEM Symposium on Self-Compacting Concrete*, Skarendahl, A. and Petersson, O. (Eds), RILEM Publications SARL: 681-693.

Cazacu N., Bradu A., Florea N. 2015. Technical-Economic Indicators of Self-Compacting Concrete Structures, *Bulletin of the Transilvania University of Braşov*, CIBv 2015, Special Issue No. 1.

Daczko, J.A. and Phillips S.H.E., 2001. Self-Compacting Concrete in Underground and Mining Applications. *Second International Symposium on Self-Compacting Concrete*, Tokyo.

Daczko, J.A. 2012. Self-Consolidating Concrete: Applying What We Know, Spon Press, New York, USA.

Dordi C. M. 2002. Concrete for the millennium, in: Advances in Cement Technology: Chemistry, Manufacture and Testing, Ghosh S. N. (ed), *Tech Books International*, New Delhi, India, 813 p.

Elliott K.S., Hamid Z. 2017. *Modernisation, Mechanisation and Industrialisation of Concrete Structures*, Wiley and Sons Limited, West Sussex, United Kingdom, 483 p.

Han B., Zhang L., Ou J. 2017. *Smart and Multifunctional Concrete Toward Sustainable Infrastructures*, Springer, Singapore, Singapore, 399 p.

Huang C., Lo K., Wu Y., Wang H. 2011. Applying Value Engineering to Evaluate the Use of Concrete in an Underground Railway Construction Project, *IEEE*: 4797-4800.

JSCE 1998. Recommendation for construction of self-compacting concrete, Japan Society of Civil Engineers.

Lessard M., Salazar B., Talbot C. 2003. Self-Consolidating Concrete Solves Challenging Placement Problems, *Concrete International*, December 2003: 80-81.

Nishizaki T., Nakatani M., Miyagawa, K., Kubo, N. 2001. Largest aboveground PC LNG Storage Tank in the World, incorporating the latest technology-construction cost reduction and shortening of work period by employing new construction methods, *Proceedings of the 13th International Conference & Exhibition on Liquefied Natural Gas*, IGU. IIR. GTI, PS6-4.1-4.11.

Okamura, H., Ouchi, M. 1999. Self-compacting concrete. Development, Present Use and Future, *1st International RILEM Symposium on Self-Compacting Concrete*, Skarendahl A., Petersson O. (eds), RILEM Publications S.A.R.L., Cachan Cedex, France, 765 p.

Okamura, H., Ozawa K., Ouchi M. 2000. Self-compacting concrete, *Structural Concrete*, 1: 3-17.

Ouchi M., Nakamura S., Osterberg T., Lwin M. 2003. Applications of Self-Compacting Concrete in Japan, Europe and The United States, *ISHPC*: 1-20.

Phelan W. 2011. Self-Consolidating Concrete (SCC): Today and Tomorrow, *Structure Magazine*, September 2011: 33-35.

Rich D., Glass J., Gibb A. G. F., Goodier C., Sander G. 2015. Optimising construction with self-compacting concrete, *Proceedings of the Institution of Civil Engineers*, **170**(2): 104-114.

RILEM Technical Committee 2006. Final report of RILEM TC 188-CSC, Casting of self compacting concrete, *Materials and Structures*, 39: 937–954.

The Industry Critical Technology Committee on Self-Consolidating Concrete 2011. SCC Contributes to A Project's Success, available at: http://www.concreteconstruction.net/how-to/construction/scc-contributes-to-a-projects-success_o

Vachon, M. 2002. ASTM Puts Self-Consolidating Concrete to the Test, *ASTM Standardization News*, July 2002: 34-37.

Zhang S. Q., Hughes D., Jeknavorian A., Nishimura T., Yang K. 2009. Self-compacting concrete, worldwide experience, *Second International Symposium on Design, Performance and use of Self-Consolidating concrete*, June 5-7, Beijing, China.

Building Tomorrow's Society

Bâtir la Société de Demain

Fredericton, Canada

June 13 to 16, 2018 / *13 au 16 juin, 2018*

THE INFLUENCE OF ASPHALT RHEOLOGICAL BEHAVIOR AND DURABILITY OF RECYCLED ASPHALT CONCRETE UNDER HIGH VISCOSITY RAP AND HIGH ADDITION

Chen, Shih-Huang[1,2], Roro Sulaksitaningrum and Huang, Cheng-Kai[1]

[1] Department of Civil Engineering, National Central University, Taoyuan, Taiwan

[2] hippo601204@gmail.com, shc@ncu.edu.tw

Abstract: This study performs laboratory investigation to discover the rejuvenating effect of Rejuvenating Agent (RA) on the performance of aged asphalt binder and asphalt concrete in order to increase Recycled Asphalt Pavement (RAP) usage. It also distinguishes between asphalt binder employing viscosity blending chart from AI MS-2, and penetration grade from Japanese blending chart, to find the optimum necessary RA in the blends and therefore solve the rutting problem. The material used was 20% and 40% of a 100,000 poises Recycled Asphalt Binder (RAB) with AC 20 as virgin binder and RA. Control samples, short-term aging, and long-term aging samples were used to describe the durability of the mixtures. Penetration blends decreased the RA needed about 40% compared to the viscosity blends and showed better rutting performance, especially for a higher percentage of RAP. Viscosity blending yielded a misleading in the addition of RA and resulted in softer RAB, which is susceptible to rutting. Thus, for the case of rejuvenating severely aged and quite high content of RAP mixtures, relying on only viscosity test and AI MS-2 is not adequate.

Keywords: Recycled Asphalt Pavement (RAP), Rejuvenating agent (RA), AI MS-2、Japanese recycled asphalt pavement handbook

1 INTRODUCTION

Taiwan has been facing limited natural resources due to pavement construction and maintenance projects. In order to reduce the use of natural resources and the big piles of construction waste, Public Works Committee of Executive Yuan promoted the Recycled Asphalt Pavement (RAP) by using Rejuvenating agent (RA). In the first place, RAP waste materials were promoted to reduce harmful environmental impacts during a pavement life cycle, reduce the use of natural resources and reduce their cost. In Taiwan, the optimum rejuvenator agent content is usually determined using the blending chart developed by the Asphalt Institute (AI) . This chart shows a linear relationship between the logarithm of viscosity at 60°C and the percentage of new asphalt or the percentage of rejuvenating agent in the blend. However, when the viscosity of an aged binder exceeds 100 kP, the blending chart might over predict the rejuvenator optimum dose, consequently resulting in rejuvenated RABs that are softer than expected. At RAP binder percentages over 25%, blending charts or equations should be used. NCHRP 9-12. results showed that the predictability of linear blending theory becomes unstable as replaced binder contents approach and exceed the 40%.

The addition of unusually soft binders to compensate high amounts of recycled binder can deviate from the predicted performance expected using blending analysis. Therefore, the Asphalt Institute recommends in no case should the selected virgin binder grade to be more than two grades softer than the binder grade that would be used in a 100 percent virgin (0% RAP) mixture for that location and application. As before, if the project's binder grade is intended to be a modified binder, then the selected virgin binder should also be a modified binder. A modified binder may not be necessary depending on the properties and quantity of the RAP being utilized. Asphaltenes are portions of an asphalt that are insoluble in low molecular weight (MW)-saturated alkanes. Asphaltenes are black, friable solids.

The solvents used to precipitate asphaltenes are usually a straight-chain alkanes ranging in carbon number from 3 (propane) to 10 (n-decane). The branched-chain alkane iso-octane also may be used. In most investigations, n-pentane and n-heptane are the precipitants. Asphaltene contents provide a basis for classification of asphalts into sol-or-gel types. An asphalt high in asphaltene content will have properties characteristic of gel-type asphalts. Among these properties are high degree of complex flow, relatively low change of viscosity with temperature, low ductility, and susceptibility to oxidative age hardening. Asphalts with low asphaltene content will have properties characteristic of sol-type asphalts. Such asphalts are frequently observed to yield "tender" pavements. Sol-type asphalts are ductile, have low degrees of complex flow, and exhibit large changes of viscosity with temperature. Asphalts with intermediate asphaltene content will have properties intermediate between the extremes of sol- and gel-type behavior.

The fractions of saturates, aromatics, resins and asphaltenes directly affects the rheological properties of asphalt. At constant temperature, the viscosity of asphalt increases as the concentration of asphaltenes increases. Likewise, the constant ratio of resins to aromatics and increasing saturates, soften the asphalt. In contrast, addition of resins content, increases the viscosity of asphalt. The viscoelastic behavior of asphalt can be explained by spring and dashpot models The formation of asphaltenes that results from weathering (oxidation and/or volatilization), has long been observed. The increase in viscosity (hardening of asphalt) and the change in colloidal structure (from sol or sol-gel to gel-type materials) or the increase in complex flow that accompanies the increase in asphaltene content have been postulated by many researchers as the cause for asphalt failure by cracking.

Thus the change in asphaltene content was used in this study as an important parameter in asphalt durability evaluation. A refinement of absorption chromatographic separation of asphalts for the purpose of rapid "fingerprinting" is the iatroscan method. This is a thin-layer chromatographic procedure. Solutions of asphalts in carbon disulfide are "spotted" on specially prepared glass rods coated with silica. The spotted rods are then "developed" by partial elution with a series of three solvents, n-heptane, toluene, and THF. Each of the solvents causes different compound types to move along the rods from the original spot. Saturates, naphthene aromatics, and polar aromatics are eluted three solvent, respectively. Iatroscan can provide rapid results similar to those obtained in Corbett analyses. Iatroscan is a method of analysis rather than a physical separation of asphalt fractions, so handling substantial quantities of solvents is not required.

2 STUDY PLAN

According to the RAP limitation to 40% in Taiwan, this study will use 0%, 20%, and 40% RAP in order to compare and investigate the physical, rheological, and chemical properties of the materials. This analysis includes viscosity test, penetration test, lactoscan (TLC-FID), Dynamic Shear Rheometer (DSR) test, Marshall Test, and Hamburg Wheel Tracking test. Moreover, due to the rapid rutting of roads with RAP in Taiwan, this research was carried out using two types of RA blending charts (viscosity index from AI MS-2 and penetration index from Japan) in three conditions (original, short term aging (RTFOT), and long-term aging (PAV)) and compare their performances.

3 MIXTURE DESIGN

RAP with 100,000 poises of viscosity was used in this study. The targeted values for viscosity and penetration are 3,000 poises and 55, respectively, for the additives removal (20% and 40%). Since the Rejuvenating agent will be exposed to a high temperature during both of the mixing and paving processes; then the Rejuvenating agent needs to be characterized by a certain trait of high degrees heat resistance, that should not be lost due to high temperature or rapid aging. Therefore, the RTFOT and PAV tests are used to simulate how would the Rejuvenating agent react to the aging process. Laboratory tests results on the physical properties, chemical properties, rheological properties, and recycled asphalt concrete of asphalt cement would be analyzed in this study.

3.1 Basic physical property test

3.1.1 Rejuvenating Agent

This study used RA-1 to increase the performance and durability of asphalt concrete. RA-1 was subject to properties and chemical analysis to inspect its accordance with Taiwan standards.

3.1.2 Virgin Asphalt Binder

In this study, basic physical tests were carried out using AC-20 grade asphalt mastic and passed the National Association of the People's Republic of China (CNS) related asphalt specifications.

3.1.3 Recycle Asphalt Pavement Properties

RAP material was examined following the standard ASTM D2172 to know its viscosity and asphalt content. The properties are shown in Table 1.

Table 1: RAP properties

Properties	Experimental value	Specification
Viscosity (%)	5.4	>3.8
Penetration (0.1mm)	20	>20
RAP viscosity	100,000	-

3.2 Calculate the Additional amount of Rejuvenating agent

Based on the AI MS-2 graphic method and the Japanese recycled asphalt pavement handbook, the amount of Rejuvenating agent to be added is determined based on the target viscosity of AC-30, which is 3,000 poises, and the target penetration that is of 55. The calculated results are shown in Table 2.

Table 2: Percentage of rejuvenating agent in different materials

Method	0% RAP (Additives %)	20% RAP (Additives %)	40% RAP (Additives %)
AI MS-2	0	6.7	16.2
Japanese recycled asphalt pavement handbook	0	3.6	4.8

3.3 Aging Test

3.3.1 Rolling Thin Film Oven Test (RTFOT)

Following the ASTM-D2872 to simulate short-term aging of asphalt binders, which is preceded by production, and construction of asphalt pavement.

3.3.2 Pressure Aging Vessel (PAV)

By following the ASTM-D6521. Simulating the in-service oxidative aging of asphalt binder was conducted by exposing the asphalt to elevated temperatures in a pressurized environment. The improved PAV prevents the complications from running and documenting asphalt binder aging operations.

3.4 Chemical Analysis Thin Layer Chromatography – Flame Ionization Detection (TLC-FID)

SARA separation results suggested that changes in chemical fractions were responsible for the stiffening effect with aging and the improvement of mechanical properties with the addition of the rejuvenators. Resins mainly affect the asphalt's viscosity by decreasing it and therefore softening the asphalt, unlike the Asphaltene that hardens the asphalt.

3.5 Rheological Analysis Dynamic Shear Rheometer (DSR)

Analysis under different frequencies, temperatures, and torsions are carried out in order to classify and evaluate the bitumen binders according to their performance properties. This has led to a better knowledge of the bitumen behavior that occurs when subjected to different thermal and mechanical conditions, as seen during road construction and service in the field.

In order to simulate the fatigue and rutting behavior of Asphalt cement after repeated loading, the study referred to the experimental configuration of SHRP-A-379 (Specifications, Test Methods and Practices of SUPERPAVE Hybrid Design System Manual), by applying continuous torsion to asphalt cement. Shear force simulates the stoppage of the vehicle, the frequency of the up and down vibrations simulates the rolling of the traffic flow, and adjusts their values to predict the effectiveness of the asphalt cement in different environments.

3.6 Rutting Wheel Track Test

To obtain the rutting resistance of the asphalt concrete; The Public Construction commission Executive Yuan in Taiwan regulated the procedure according to "Chapter 02798 V1.0 Porous asphalt concrete paving" to simulate the field traffic effects on hot mixture asphalt specimens in terms of rutting and moisture-susceptibility. This study used rectangular slabs (30 cm long, 30 cm wide and 5 cm thick) with 6 to 8% of air voids. The test temperature is usually set at 40° C (104° F) or 60° C (140° F) for a time length of approximately 0.5 to 2 hours. Each wheel makes about fifty passes per minute under 686±10N, then by using the equation [1], [2] and , which can evaluate the rutting resistance of the asphalt concrete, are computed.

[1] Dynamic stability, $DS(\text{passes/mm}) = \frac{\Delta n}{\Delta d}$

[2] Rate of Deformation, $RD(\text{mm/min}) = \frac{\Delta d}{\Delta t}$

ΔN = difference in number of times of rolling (number of times).

Δd = difference in deformation (mm).

RD = Deformation rate (in centimeters per minute)

ΔT = rolling time (minutes)

4 RECYCLED ASPHALT PAVEMENT PERFORMANCE RESULTS

4.1 Viscosity and Penetration Physical Test Results

Four different ratios: AI-20%, AI-40%, J-20%, J-40%'s penetration and viscosity all fall within the target AC-30. After long and short-term aging, the viscosity is expected to increase, while the penetration is expected to drop. The Asphalt bitumen acquired a hardening behavior, as shown in Figure 1 and Figure 2, the ratio with a 20% of RAP: AI-20% and J-20%'s performances are the same, but AI-40% and J-40% that used the same additional amount of RAP (40%) showed very different performances. AI-40% has the lowest viscosity and the highest penetration in both short-term aging and long-term aging while the J-40% showed an opposite performance. AI-40% asphalt bitumen is even softer than the one using only 20% of RAP.

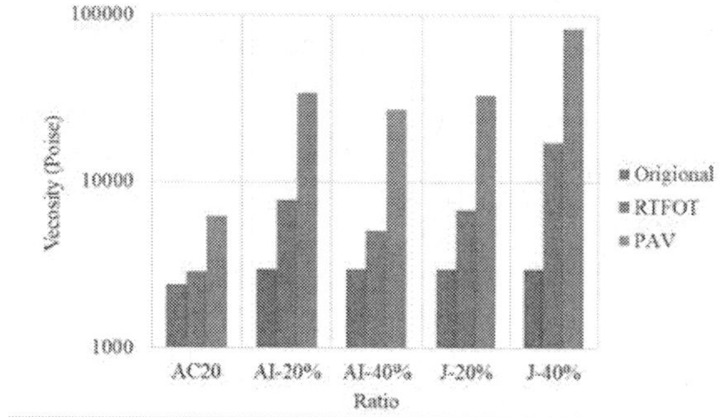

Figure 1: Viscosity of the blend

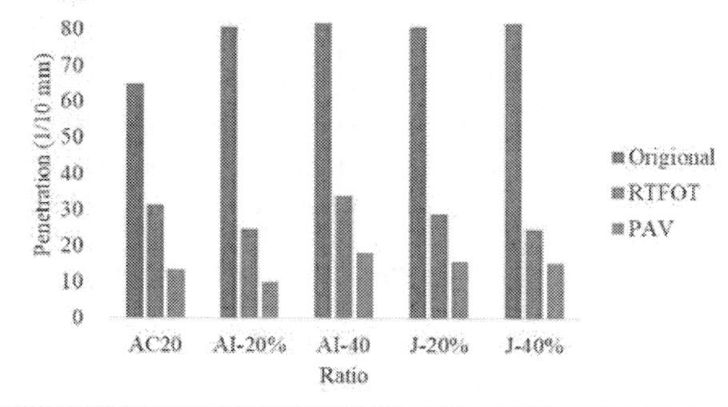

Figure 2: Penetration of the blend

4.2 Chemical Analysis (TLC-FID)

RA are able to change the chemical components of RAB. Figures 3, 4 and 5 show that each RAB Blend behaves differently. However, all of the blends increase the amount of Resin and asphaltene after the aging process, indicating that the asphalt becomes more brittle after aging. In which support our result in Figure 1 and Figure 2.

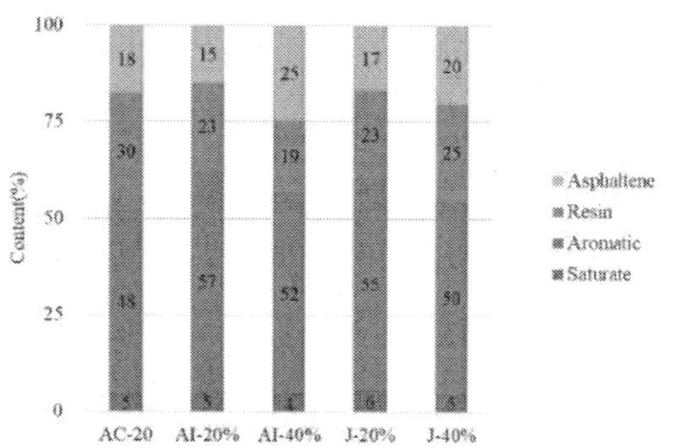

Figure 3: Chemical analysis on non aging

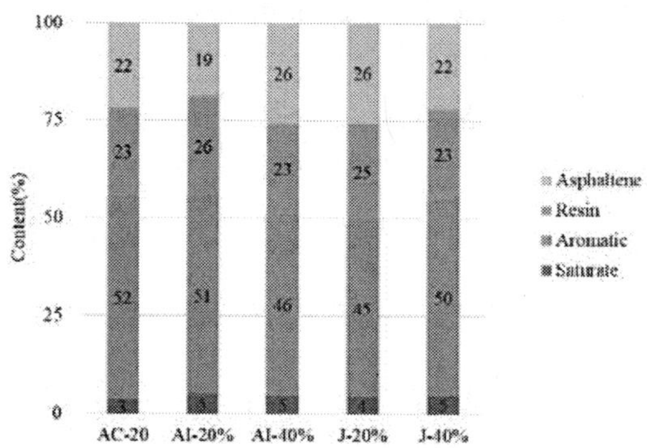

Figure 4: Chemical analysis on short term aging

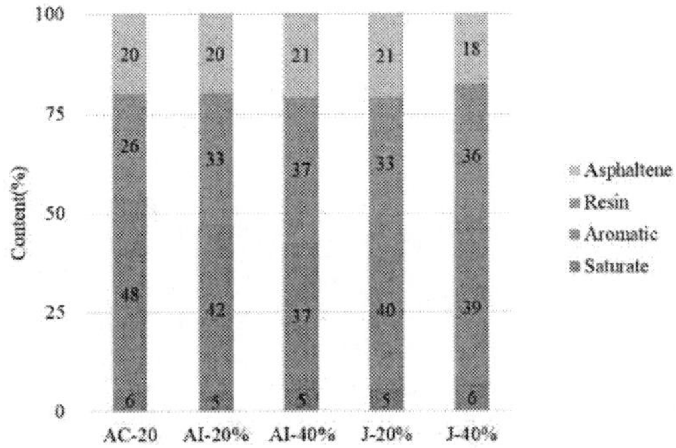

Figure 5: Chemical analysis on long term aging

4.3 Rheological Analysis Dynamic Shear Rheometer (DSR)

4.3.1 G* Value and Δ Angle at Multiple Temperatures for Different Ratio

Figure 6 and 7 show the complex shear modulus (G*) and the phase angle (δ) subjected to the change of temperature. G* decreases with the increasing of temperature, while δ increases. It shows that the RAB softens with the increasing of temperature and leads to rutting. J-40% shows as the strongest blend, and AI-40 shows as the softer blend.

Rheological analysis by G* and δ on figure 8 and 9 show that after subjection to RTFO and PAV. G* gets higher with the asphalt's aging. It indicates that the asphalt binder become stiffer due to the aging treatments. However, δ of all blends slightly decreases. Hence, the results show that asphalt binders will act from viscous to more elastic with the increase of the traffic speed.

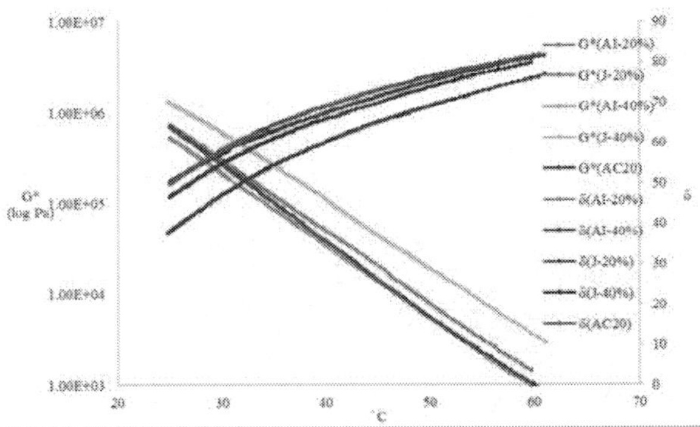

Figure 6: G* and δ for short term aging

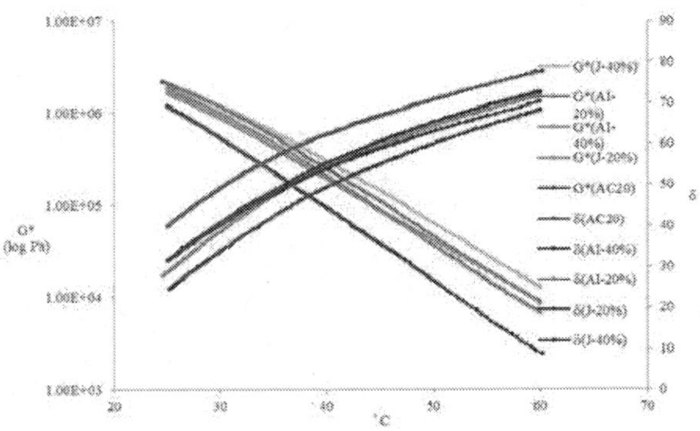

Figure 7: G* and δ for long term aging

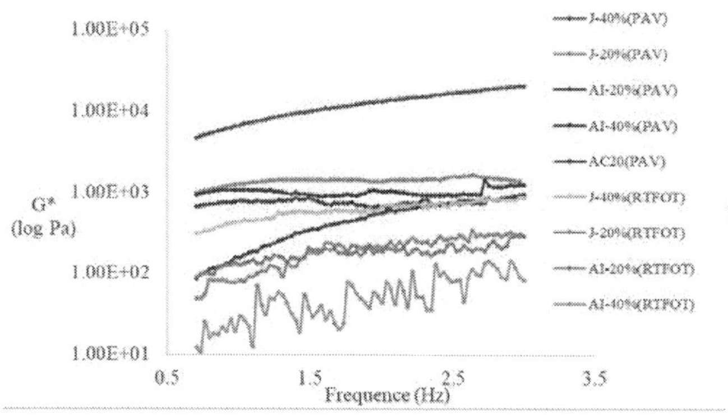

Figure 8: Phase angle for short and long term aging

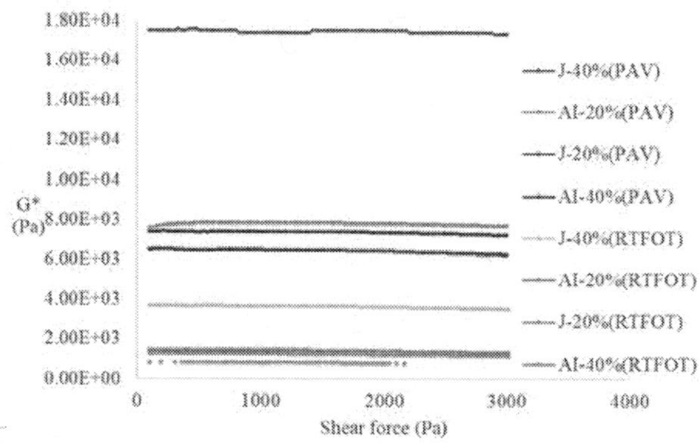

Figure 9: G* for short and long term aging

4.4 Rutting Wheel Track Test

The Rutting wheel track test results from figure 10 show the evaluation of the rut depth with the increase of number of passes. The curve will stop when the wheel reaches 1260 passes. The results of Al-40 shows the highest rutting, while the other blends have not shown any similar behavior. From Table 3 we can notice the RD from Al-40% is almost the double compared to the others, indicating Al-40% is not recommended and still susceptible to rutting. High rejuvenate agent amount in this blend increases its softness than any other blend.

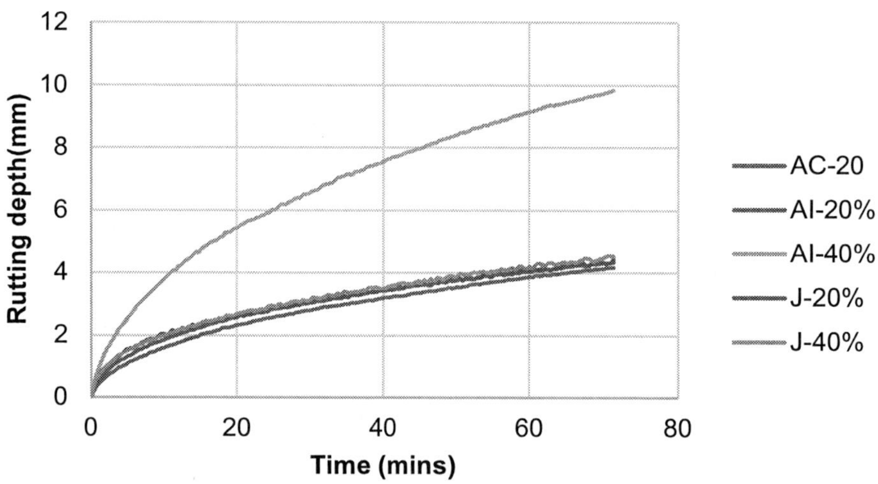

Figure 10: Rutting Hamburg Wheel Track Test for different ratio

Table 3: Dynamic stability and Deformation rate for different ratio

Ratio	DS(passes/mm)	RD(mm/min)
AC-20	1,341.0	0.03
Al-20%	1,074.1	0.04
Al-40%	534.8	0.08
J-20%	1,279.4	0.03
J-40%	1,203.3	0.04

5 CONCLUSION

5.1 Conclusion

1. The addition of a rejuvenating agent can effectively recover the recycled asphalt pavement in most cases and the resulted strength is similar to AC-20.

2. It is known from the Hamburg Wheel Track Test that the AI-40% ratio is significantly softer than the others. The AI MS-2 calculation graph depends on the hand drawing and will lead to errors. The amount of added RA is not accurate, the amount of addition would be overestimated. As a result, AI-40% asphalt concrete is too soft and less resistant to rutting.

5.2 Recommendations

It is recommended that follow-up studies can be considered based on the Japanese recycled asphalt pavement handbook to compare the impact of penetration and viscosity on asphalt in order to increase the amount of addition of the recycled asphalt to be used in the study of pavement and also to improve the problem of stock recycled asphalt pavement.

References

Barnes, H. A. (1989). Shear-thickening ("Dilatancy") in suspensions of nonaggregating solid particles dispersed in Newtonian liquids. Journal of Rheology, 33(2), 329-366.

Brown, S. (1990). The Shell bitumen handbook. Shell Bitumen, UK, 91-95.

Cin, W. C. (2009). The Recycle Agent on promotion the Recycling Asphalt Concrete quality. National Central University, Dept. of Civil Engineering.

Curtis, C. W., Terrel, R. L., Perry, L. M., Al-Swailm, S., & Braanan, C. J. (1991). IMPORTANCE OF ASPHALT-AGGREGATE INTERACTIONS IN ADHESION (WITH DISCUSSION). Journal of the Association of Asphalt Paving Technologists, 60.

Finn, F.N. (1987). Factors Involved in the Density of Asphaltic Pavement Surfaces, NCHRP Report, 39.

Huang, Y., Bird, R. N., & Heidrich, O. (2007). A review of the use of recycled solid waste materials in asphalt pavements. Resources, Conservation and Recycling, 52(1), 58-73.

National Research Council, (1994). The SUPERPAVE Mix Design System Manual of Specifications, Test Methods, and Practices, SHRP-A-379. Strategic Highway Research Program, Washington, DC

Noureldin, A. S., & Wood, L. E. (1989). Variations in molecular size distribution of virgin and recycled asphalt binders associated with aging. Transportation Research Record, (1228), 191-197.

Public Construction commission Executive Yuan. Chapter 02798 V1.0 Porous asphalt concrete paving.

Japan Road Association. (2010). Japanese recycled asphalt pavement handbook"

Roberts, F. L., Kandhal, P. S., Brown, E. R., Lee, D. Y., & Kennedy, T. W. (1991). Hot mix asphalt materials, mixture design and construction.

Vallerga, B. A., Monismith, C. L., & Granthem, K. (1957). A study of some factors influencing the weathering of paving asphalts. In Assoc Asphalt Paving Technol Proc.

Yang, C. Y. (2013). The Study in Feasibility Assessment of Different Additives on the Modified Asphalt in Taiwan Area. National Central University, Dept. of Civil Engineering, Master thesis.

S.-C. Huang, H. Di Benedetto,(2015), "Advances in Asphalt Materials", United Kingdom: Woodhead Publishing.

Building Tomorrow's Society

Bâtir la Société de Demain

Fredericton, Canada

June 13 to 16, 2018 / *13 au 16 juin, 2018*

UNDERSTANDING THE FACTORS AFFECTING EXPANSION DUE TO ALKALI-SILICA REACTION IN CONCRETE

Sinno, Noura[1,2] and Shehata, Medhat[1]
[1] Ryerson University, Canada
[2] noura.sinno@ryerson.ca

Abstract: Alkali-silica reaction is a chemical reaction resulting in expansion and disruption of concrete elements. The use of supplementary cementing materials (SCM) is one of the measures leading to reduction in expansion. The concrete prism test (CPT) is considered the most reliable lab test to evaluate efficacy of SCM. However, late expansions and cracking were observed in field for some aggregate/SCM combinations despite passing the 2-year expansion criterion of the CPT. A number of reasons, including alkali leaching from samples, could be behind this. In this research, the focus is to study the effect of sample geometry and testing temperature on alkali leaching and expansion. Cylinders and cubes were studied in addition to prisms at 38°C and 60°C using two aggregates of different reactivity: Spratt and Sudbury. Larger samples showed less leaching compared to standard prisms for both aggregates. For samples with SCM, cylinders casted with Sudbury showed higher expansion than prisms. However, with Spratt, they had similar expansions. This shows that the effect of sample geometry on expansion varies from one aggregate to another. The dependence of a particular aggregate on certain level of pore solution alkalinity to initiate expansion is an important factor that should be taken into consideration. In addition, results showed more leaching at higher temperature. The ultimate expansions of samples without SCM at 60°C were lower than the ones at 38°C. Also, the prisms two-year expansion values at 38°C were obtained using cylinders at 1 year at 60°C, which is a considerable saving in testing time.

1 INTRODUCTION

Alkali-silica reaction (ASR) is a chemical reaction that occurs between a form of reactive silica present in the aggregate and the alkaline pore solution of the concrete. This reaction will cause the formation of a gel that absorbs water and expands (Swamy 1992). Many preventive measures have been implemented to limit the expansion due to ASR. The addition of supplementary cementing materials (SCM) is one of the common measures that showed enhanced results (Shehata and Thomas 2000, Thomas et al. 2006). The results are based on the concrete prism test (CPT) described in the Canadian Standards, CSA A23.2-14A. This test is used to assess the reactivity of aggregates. An expansion limit of 0.040% at 1 year is specified in the standard above which the aggregate is considered reactive. However, many challenges are facing this test. First is the long testing time needed to obtain the results. To check the effectiveness of the SCM, an expansion below 0.040% should be obtained at 2 years. Many attempts are being made to reduce the testing period of the concrete prism test (Ideker et al. 2008, Liu and Mukhopadhyay 2015). Samples at 60ºC were tested in order to accelerate the reaction. However, it was found that more alkali leaching was happening at 60ºC (Ideker et al. 2008). In addition, some non-reactive aggregates at 38ºC showed expansion at 60ºC. Many correlations were made between the 13-week expansion of prisms at 60ºC compared to the actual prisms at 52 weeks for control samples (Fourier et al. 2004, Ideker et al. 2008). Fournier found that the accelerated test gives same diagnostic results as the standard CPT in 88% of the cases. However, Ideker showed that expansions after 13 weeks tend to be reduced at 60°C compared to the standard prisms at 1 year.

Another challenge that is facing the concrete prism test is the excessive alkali leaching. Due to the small size of the prisms compared to the field, more leaching is occurring in the laboratory samples (Thomas et al. 2006, Ideker et al. 2012). Bérubé found that the larger the diameter of the cylinders or the lower the air/concrete ratio the less leaching will occur leading to higher expansion (Bérubé et al. 2012). Many attempts are being made in order to reduce the leaching of the prisms such as the application of silane-based sealer, plastic sleeve or aluminum foil protection (Rivard et al. 2006, Bérubé et al. 2012). However, none of the methods was able to reduce leaching and increase expansion at the same time. Many researches compared expansions of concrete prisms with exposure blocks from the field (Fournier et al. 2008, Ideker et al. 2012). At 10 years, Ideker found good correlation between the expansions of the CPT at 2 years and the samples in the field. However, at later ages, the leaching of the prisms was more revealed leading to lower expansion compared to the field blocks. Hence, it is important to test blocks in the field under environmental conditions to validate the laboratory results. The alkali leaching as well as the long testing time of the concrete prism test are the main challenges that are being investigated to enhance the results obtained in the lab and to obtain better correlations between field and lab samples.

2 EXPERIMENTAL DETAILS

2.1 Materials

2.1.1 Aggregates

In this study, two coarse aggregates of different reactivity were tested: Sudbury and Spratt. The fine aggregate used is a non-reactive natural sand. The bulk relative density, absorption and expansion of the different aggregates are summarized in Table 1. The expansion is obtained based on results from the concrete prism test described in CSA A23.2-14A. The expansion limit at 1 year is 0.040% below which the aggregate is considered non-reactive.

Table 1: Aggregates Properties

Aggregate	Sudbury	Spratt	Non-Reactive Sand
Bulk relative density	2.552	2.691	2.538
Absorption (%)	0.56	0.52	1.30
Expansion at 1 year (%) (CSA A23.2-14A)	0.170	0.211	-

2.1.2 Cementitious Materials

The cement used is a Type GU cement (GU PC) with 0.99% Na_2O_e. Two supplementary cementing materials were implemented in this research: low calcium fly ash (FA) and slag. The chemical compositions of the cementitious materials are provided in Table 2.

Table 2: Chemical compositions of the cementitious materials

Oxide	SiO_2	Al_2O_3	Fe_2O_3	CaO	MgO	SO_3	Na_2O_e	LOI
GU PC	19.54	5.21	2.16	62.39	2.39	4.03	0.99	2.36
Fly Ash	57.0	23.4	3.5	9.5	1.0	0.1	2.89	0.59
Slag	37.0	8.2	0.5	38.5	10.5	2.7	0.67	2.1

2.2 Experimental Procedure

2.2.1 Sample Preparation

Three different sample shapes were investigated: prisms (75x75x285 mm), cylinders (ø 100 mm by 285 mm) and 150 mm cubes. The coarse aggregate portion is composed of three equal masses of materials size between 19.5 mm to 13.2 mm, 13.2 mm to 9.5 mm and 9.5 mm to 4.75 mm. The coarse to fine aggregate ratio is 60:40 by mass. A water to cement ratio of 0.42 is used. The cement content is 420 kg/m^3. The alkali content is boosted to 1.25% by weight of cement by adding sodium hydroxide. The procedure followed is presented in CSA A23.3-14A. The prism molds used are similar to the ones specified in the standard. Since there are no standard test for cylinders and cubes, their molds were fabricated at Ryerson's lab and are shown in Figure 1.

(a)

(b)

Figure 1: Molds for (a) cylinders and (b) cubes

The cubes were casted in 3 layers where each layer was rodded 25 times similar to the prisms. The cylinders were casted in 4 layers and each layer was rodded with 25 strokes to allow better consolidation along the height. After casting, the samples were put in the curing room for 24 hours and then demolded the next day. Containers were made ready by aligning an absorbent cloth on their walls. The clothes were made wet before putting the samples in. The samples were elevated by around 4 cm above the bottom of the containers and put above water. The pails for prisms and cylinders fit 3 samples at a time as for the cubes one sample is put in each container. The pails and containers used to put the samples in are shown in Figure 2.

(a)

(b)

Figure 2: (a) Pails for the prism and cylinder samples (b) containers for cube samples

Each set of samples is composed of 3 prisms, 3 cylinders and 3 cubes. Two sets of the same mix were prepared. One is put in a heat room at 38ºC and the other set in an oven at 60ºC.

2.2.2 Expansion Measurements

The zero reading was taken before putting the samples at their designated temperatures. Measurements were made after 1, 2, 4, 8, 13, 18, 26, 39, 52 weeks and then once every 3 months until 2 years. The cylinders were measured the same way as the prisms using the length comparator. As for the cubes, a demec style digital strain gauge was used. The pins are put diagonally on two adjacent faces of the cubes and they are 150 mm apart. The average expansion of three samples of the same shape at the same temperature is presented in this paper. The length comparator and the strain gauges used for the different samples are shown in Figure 3.

(a)

(b)

Figure 3: Measurements of (a) cylinders and (b) cubes

2.2.3 Leaching Measurements

Leaching measurements were performed at 1.5 years to obtain the amount of alkalis that leached out from the samples. To do so, the volume of the solution at the bottom of the containers was measured and then a 10 mL sample was collected. The concentrations of Na^+ and K^+ in the solution were obtained by flame photometry. The amount of leached Na^+ and K^+ ions were measured as concentration then converted to Na_2O_e mass taken as a percent of the initial 1.25% boosted alkalis by Portland cement mass.

3 RESULTS AND DISCUSSION

3.1 Testing at 38 ºC

The 1-year expansion of prisms, cylinders and cubes at 38ºC for Sudbury and Spratt without SCM are presented in Figure 4.

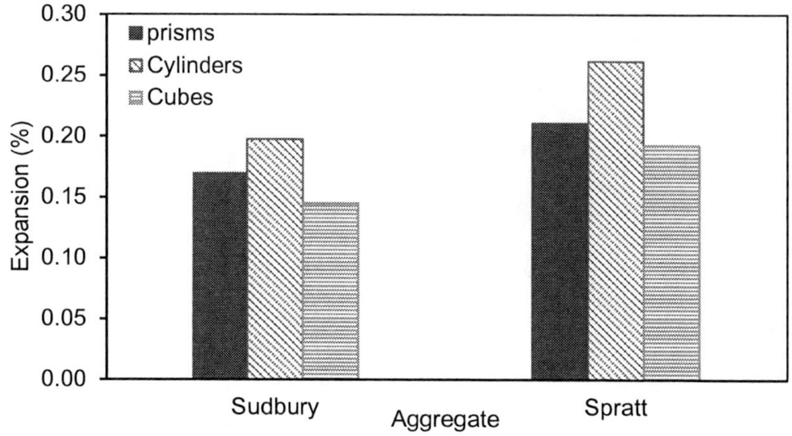

Figure 4: 1-year expansion of Sudbury and Spratt without SCM at 38°C for prisms, cylinders and cubes

With both aggregates, the expansion of the cylinders is higher compared to that of the prisms. This is obtained with the samples without SCM. As for the samples with SCM, the expansion data of the Sudbury samples with 15% FA and 25% slag are shown in Figure 5.

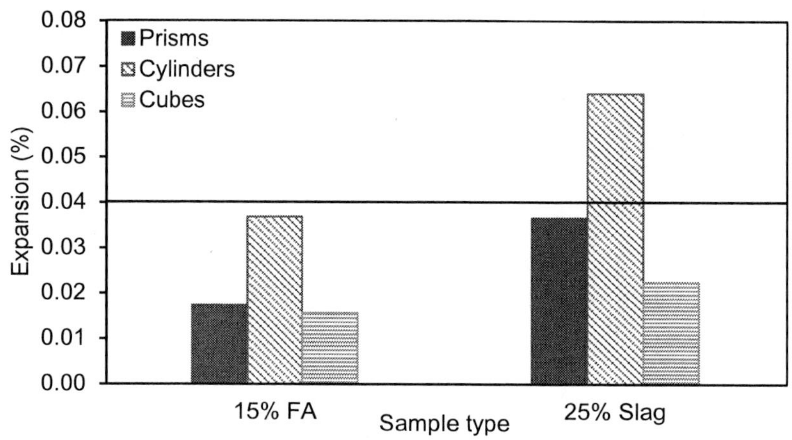

Figure 5: 2-year expansion of Sudbury with SCM at 38°C for prisms, cylinders and cubes

The cylinders showed much higher expansion than the prisms and cubes. For the samples with 15% FA, the cylinders were about to fail with an expansion equal to 0.037% as compared to only 0.017% for the prisms. In addition, the cylinders casted with Sudbury and 25% slag passed the expansion limit of 0.040% as compared to prisms which had an expansion below the limit. At the same age and temperature, cylinders casted with Sudbury aggregate with and without SCM have higher expansion than prisms. As for the Spratt samples with SCM, the expansion at 38°C at 2 years for samples with FA and 1.5 years for samples with slag are shown in Figure 6. The 2-year expansion of samples with slag is yet to be measured.

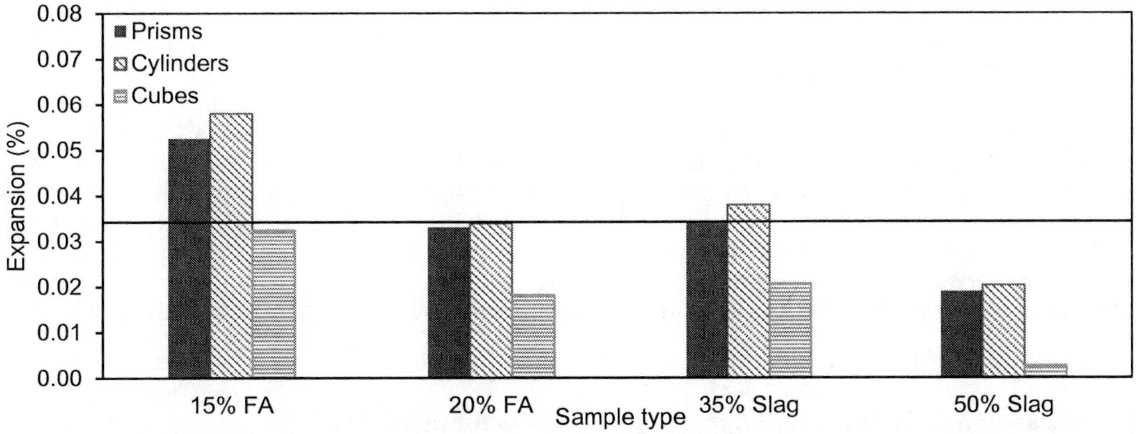

Figure 6: 2-year expansion of samples with FA and 1.5-year expansion of sample with slag at 38°C for prisms, cylinders and cubes casted with Spratt

Prisms and cylinders casted with Spratt at 38°C have the same expansions with all the tested SCM. For the Spratt with 15% FA, prisms and cylinders both passed the expansion limit of 0.040%. At the same age, same expansion was obtained for prisms and cylinders casted with Spratt as opposed to the behavior observed with the Sudbury aggregate. To understand these findings, alkalis leached from the samples to the bottom of the containers were measured for different sample geometries at 38°C. The results are presented in Figure 7 which compares the alkalis leached at 38°C in prisms with that of cylinders after 1.5 years. The shaded markers correspond to the sample without SCM for the specified aggregate.

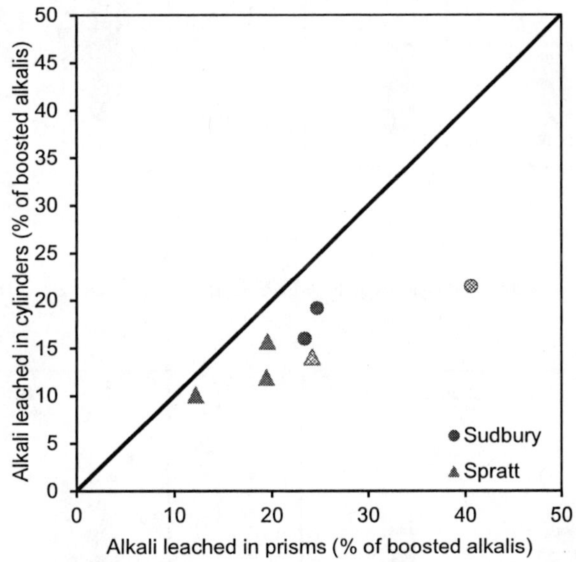

Figure 7: Comparison of alkali leaching at 1.5 years at 38°C between prisms and cylinders for Sudbury and Spratt aggregates

Figure 7 shows that more leaching is occurring with prisms compared to cylinders with both aggregates leading to the conclusion that larger samples reduce the amount of leaching. It is expected that the expansion will be higher for larger samples since more alkalis are available for the formation of ASR gel. For Sudbury aggregate, the expansion of cylinders at 38°C was higher than that of the prisms for the control as well as for the samples with SCM. However, with Spratt, the cylinders and prisms had similar expansions at 38°C for the samples with SCM. Although the leaching was higher with the prisms, however, Spratt aggregate was not affected by the change in the alkali level as compared to Sudbury. It seems that Sudbury aggregate is more sensitive to changes in alkalinity resulting in the observed difference in expansion. This might be due to the fact that Spratt reacts faster and consumes the alkalis before they leach as compared to Sudbury, a slowly reactive aggregate. Sudbury showed higher leaching results compared to Spratt. The prisms and cylinders casted with Sudbury showed a leaching of 41% and 21%, respectively as compared to 24% and 14% for Spratt (shaded markers in Figure 7). In addition to this, it was found that Sudbury aggregate releases more alkalis to the pore solution (Bérubé et al. 2002). Hence the higher alkali leaching in case of Sudbury could be due to: (i) alkalis contributed from the aggregates, and (ii) slow reaction of Sudbury leaving more alkalis in pore solution of the concrete that is "free" to be leached. It should be noted also that although the results showed lower leaching for the cubes compared to cylinders and prims, however the expansion of cubes was still lower than the cylinders at 2 years. The reasons behind this are still not clearly understood but it could be due to the restraints effect that might occur in the cubes.

3.2 Testing at 60°C

3.2.1 Expansion of Sudbury and Spratt Aggregates without SCM

Expansion of prisms, cylinders and cubes were measured for Sudbury and Spratt samples without SCM. The data obtained at 6 months and 1 year are presented in Figures 8 and 9 for Sudbury and Spratt, respectively at 38°C and 60°C.

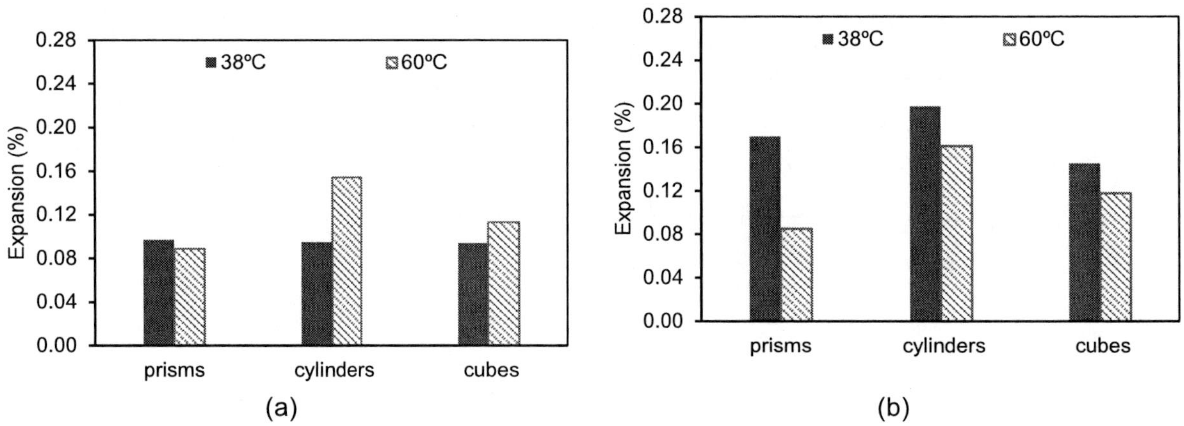

Figure 8: Expansion of Sudbury aggregate at 38°C and 60°C at (a) 6 months and (b) 1 year

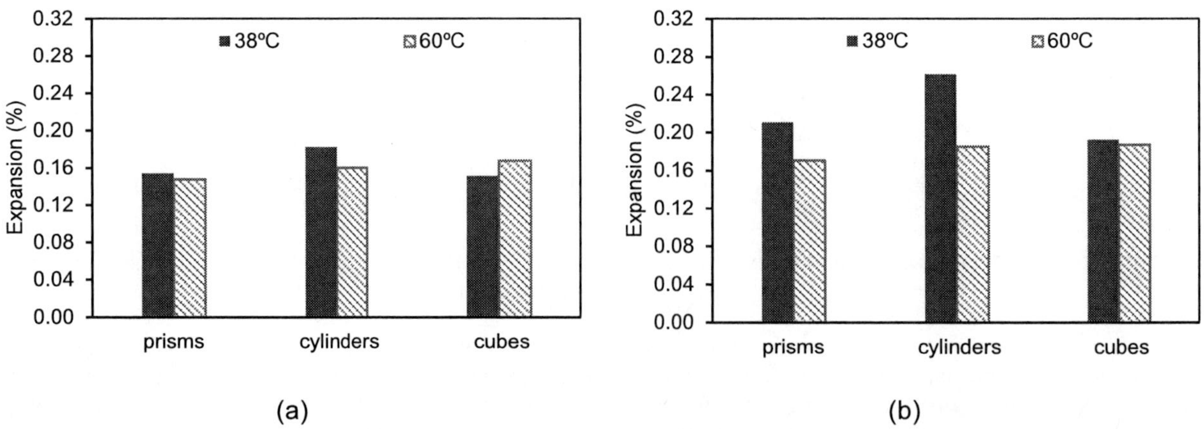

Figure 9: Expansion of Spratt aggregate at 38°C and 60°C at (a) 6 months and (b) 1 year

Expansions of Spratt at 6 months were the same at 38°C and 60°C. However, cylinders containing Sudbury showed higher expansion at 60°C compared to 38°C. In addition, at 1 year, it is clear for both aggregates that the ultimate expansions at 38°C are higher than the expansions at 60°C for all the different sample shapes. Leaching of alkalis from the samples at 60°C was measured at 1.5 years and compared to the leaching of the same samples tested at 38°C. Results are shown in Figure 10.

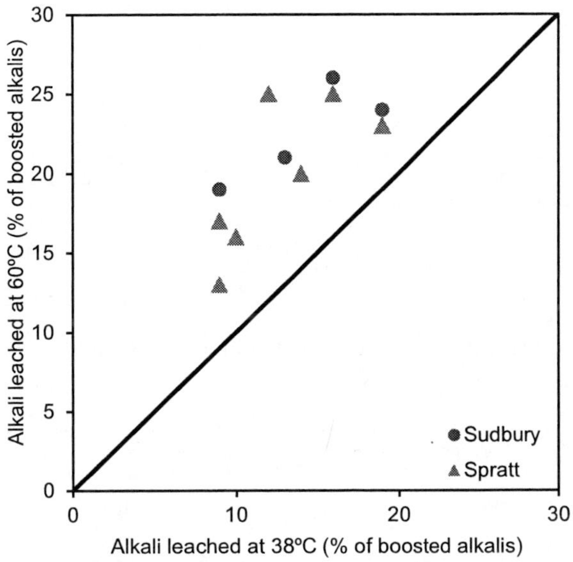

Figure 10: Comparison of alkali leaching at 1.5 years between 38°C and 60°C

The samples at 60°C leached more than the samples at 38°C for both aggregates and for all the different sample geometries. This is due to the higher diffusivity of alkalis with increased temperature (Lindgard et al. 2012). The rate of alkali leached is increased at 60°C lowering the alkalinity of the pore solution faster than the samples at 38°C. This explains partly the fact that Spratt and Sudbury control samples showed lower ultimate expansions at 60°C compared to 38°C. This could also be due to the effect of temperature on the gel viscosity exerting less pressure at higher temperatures (Swamy 1992). In addition, it was found that there is a decrease in pore solution alkalinity at 60°C due to the higher solubility of the ettringite (Fournier et al. 2004).

3.2.2 Expansion of Sudbury and Spratt Aggregates with SCM

Samples with FA and slag at different replacement levels were also measured for expansion. The data obtained at 1 year and 2 years for samples with 15% FA are presented in Figures 11 and 12 for Sudbury and Spratt, respectively.

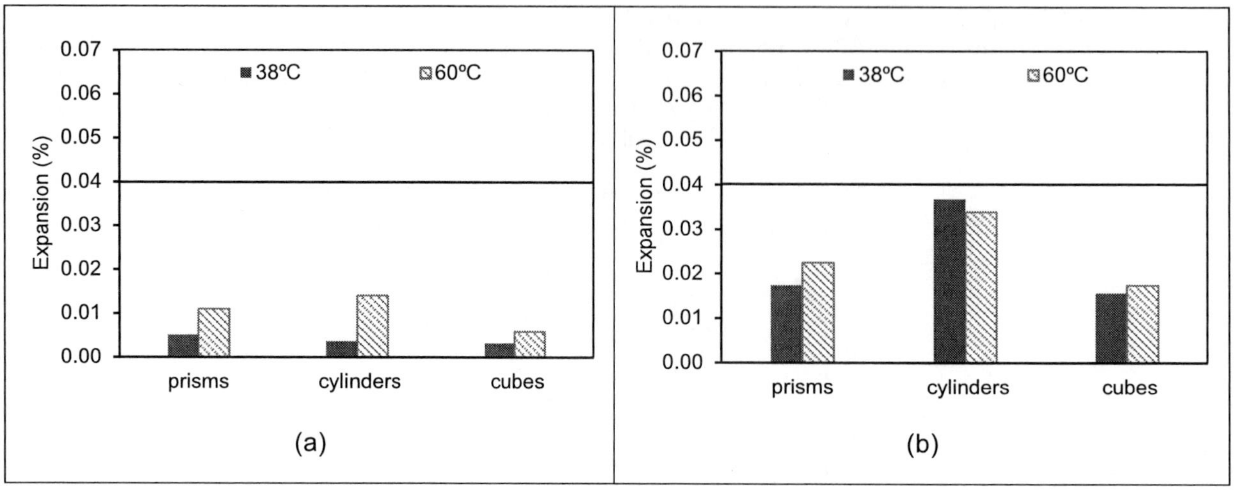

Figure 11: Expansion of Sudbury aggregate with 15% FA at (a) 1 year and (b) 2 years

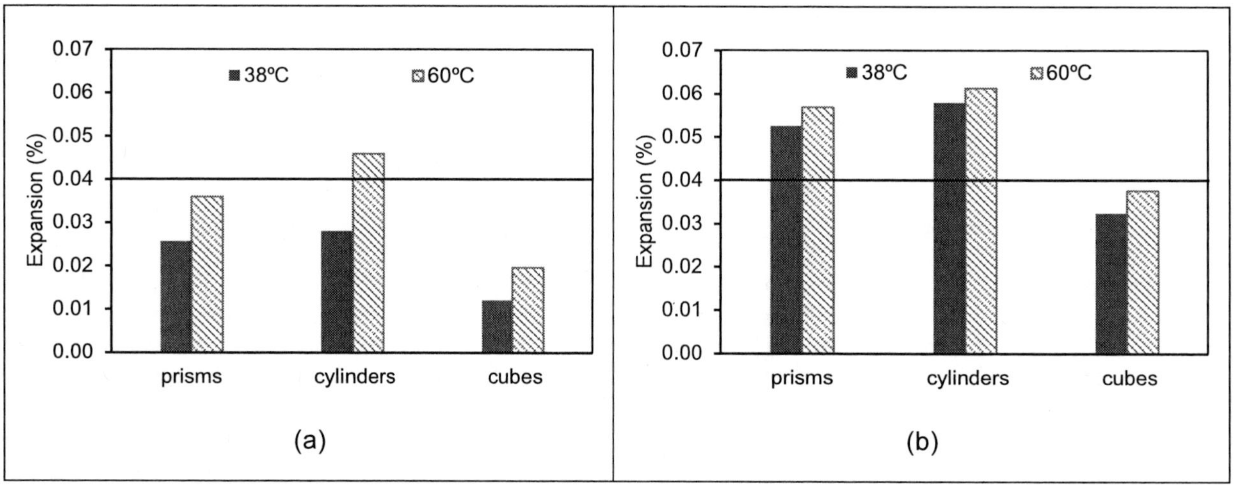

Figure 12: Expansion of Spratt aggregate with 15% FA at (a) 1 year and (b) 2 years

According to CSA A23.2-27A, replacing cement by 15% FA is supposed to reduce the 2-year expansion of concrete prisms to below 0.040% when using the CPT with Sudbury aggregate but not with the highly reactive aggregate, Spratt. This was found to be the case as shown in Figures 11-b and 12-b. Looking at the cylinders expansions at 60°C at 1 year, same conclusions are drawn (Figures 11-a and 12-a). Sudbury cylinders had an expansion lower than 0.040% as compared to the Spratt cylinders which showed an expansion higher than 0.040% at 1 year at 60°C. Table 3 shows the expansion of prisms at 38°C at 1 year and 2 years for the control samples and samples with SCM, respectively. These expansions are compared to the expansions of the cylinders at 60°C at 6 months and 1 year, respectively to see if testing at 60°C can reduce the testing period by 50% when cylinders are used.

Table 3: Comparison of prisms expansion at 38°C and cylinders expansion at 60°C

Sample type		Expansion of cylinders at 60°C*	Expansion of prisms at 38°C**	Difference in expansion (%)	Pass/Fail Criterion
	Control	0.154	0.170	0.016	Fail/Fail
Sudbury	15% FA	0.014	0.017	0.003	Pass/Pass
	25% Slag	0.022	0.037	0.015	Pass/Pass
Spratt	Control	0.160	0.211	0.050	Fail/Fail

Sample type	Expansion of cylinders at 60°C*	Expansion of prisms at 38°C**	Difference in expansion (%)	Pass/Fail Criterion
15% FA	0.046	0.053	0.007	Fail/Fail
20% FA	0.032	0.035	0.001	Pass/Pass
35% Slag	0.041	0.037	0.004	Fail/Pass

*Expansion of cylinders for the control samples is at 6 months and for the samples with SCM at 1 year.

**Expansion of prisms for the control samples is at 1 year and for the samples with SCM at 2 years.

Both expansions obtained from the prisms at 2 years at 38°C and the cylinders at 1 year at 60°C with SCM showed a pass/pass or fail/fail relationship except the Spratt sample with 35% slag. This was due to the fact that the expansion of the prisms was taken at 1 year and 9 months because they did not reach 2 years at the time of writing the paper. However, the expansion of the prisms was very close to fail with a value of 0.037%. In addition, although there was a pass/pass relationship for the Sudbury samples with 25% slag, however, the expansion of the cylinders at 60°C at 1 year was much lower than the prisms at 38°C at 2 years. The reasons for this behavior is still under investigation. In conclusion, the 2-year expansion of prisms with SCM at 38°C can be obtained with the cylinders in a shorter time at 60°C. Although at 60°C, more leaching is occurring, but the cylinders leach less compared to prisms. In addition, the increase in temperature will accelerate the ASR reaction. Hence, by reducing leaching and accelerating the reaction, the same conclusions obtained with the standard prisms at 2 years at 38°C can be reached with the cylinders at 60°C at 1 year. The testing time is reduced significantly with the use of cylinders at 60°C. However, it should be noted that more samples need to be tested to confirm the applicability of the above finding to a wide range of aggregate/cementing blends.

4 CONCLUSIONS

The main conclusions that can be obtained from this study are as follows:

1. Larger samples as well as lower temperatures will lead to reduction in alkali leaching.
2. Higher expansion can be obtained with cylinders compared to prisms at the same age and temperature due to reduced leaching however it depends also on the aggregate type.
3. Sudbury aggregate showed higher expansion with the cylinders compared to prisms. However, this was not the case with the Spratt aggregate. The dependence of the aggregate on a certain level of pore solution alkalinity might also affect expansion.
4. Samples without SCM showed lower ultimate expansions at 60°C as compared to 38°C.
5. The 2-year expansion of the standard prisms casted with SCM at 38°C using the CPT described in CSA A23.2-14A can be obtained in a shorter duration with the cylinders at 60°C at 1 year.

Acknowledgements

Funding of this research project is obtained through a Discovery Grant from the Natural Sciences and Engineering Research Council of Canada (NSERC). The financial support of NSERC is highly appreciated.

References

Bérubé, M., Duchesne, J., Dorion, J. and Rivet, M. 2002. Laboratory Assessment of Alkali Contribution by Aggregate to Concrete and Application to Concrete Structures Affected by Alkali-Silica Reactivity. *Cement and Concrete Research*, **32**(8): 1215-1227.

Bérubé, M., Fournier, B. and Côté T. 2012. Testing Concrete Cores For Residual Expansion Due to AAR-An Attempt to Minimize Alkali Leaching and Consequent Unrealistic Expansion Decrease. *14th International Conference on Alkali-Aggregate Reaction,* Austin, Texas, USA.

Fournier, B., Chevrier, R., Grosbois, M.d., Lisella, R., Folliard, K., Ideker, J., Shehata, M., Thomas, M. and Baxter, S. 2004. The Accelerated Concrete Prism Test (60°C) - Variability of the Test Method and Proposed Expansion Limits. *12th International Conference on Alkali-Aggregate Reaction,* Beijing, China.

Fournier, B., Ideker, J., Folliard, K., Thomas, M. , Nkinamubanzi, P. and Chevrier, R. 2008. Effect of Environmental Conditions on Expansion in Concrete due to Alkali-Silica Reaction. *13th International Conference on Alkali-Aggregate Reaction*, Trondheim, Norway.

Ideker, J., Drimalas, T., Bentivegna, A., Folliard, K., Fournier, B., Thomas, M., Hooton, D. and Rogers, C. 2012. The Importance of Outdoor Exposure Site Testing. *14th International Conference on Alkali-Aggregate Reaction,* Austin, Texas, USA.

Ideker, J., East, B., Folliard, K., Thomas, M. and Fournier, B. 2008. The current state of the accelerated concrete prism test. *Cement and Concrete Research*, **40**(4): 550-555.

Lindgård, J., Andiç-Çakır, Ö., Fernandes, I., Rønning, T. and Thomas, M. 2012. Alkali–silica reactions (ASR): Literature review on parameters influencing laboratory performance testing. *Cement and Concrete Research*, **42**(2): 223-243.

Liu, K. and Mukhopadhyay, A. 2015. Accelerated Concrete-Cylinder Test For Alkali-Silica Reaction. *Journal of Testing and Evaluation*, **44**(3): 1229-1238.

Rivard, P., Bérubé, M., Ollivier, J. and Ballivy, G. 2006. Decrease of pore solution alkalinity in concrete tested for alkali-silica reaction. *Materials and Structures*, **40**(9): 909-921.

Shehata, M. and Thomas, M. 2000. The effect of fly ash composition on the expansion of concrete due to alkali–silica reaction. *Cement and Concrete Research*, **30**(7):1063-1072.

Swamy, R. 1992. *The Alkali-silica reaction in concrete*. Glasgow: Blackie.

Thomas, M., Fournier, B., Folliard, K., Ideker, K. and Shehata, M. 2006. *Test Methods for Evaluating Preventive Measures for Controlling Expansion due to Alkali-Silica Reaction in Concrete*. Austin: International Center for Aggregates Research.

Building Tomorrow's Society
Bâtir la Société de Demain

Fredericton, Canada
June 13 to 16, 2018 / *13 au 16 juin, 2018*

UNSATURATED POLYESTER (UP) RESIN MIXED WITH AGGREGATES AS THE SURFACE COURSE FOR PERMEABLE PAVEMENTS

Chang, Jia-Ruey[1,2], Wang, Song-Jun[1] and Luo, Yi-Ting [1]
[1] Graduate Institute of Architecture and Sustainable Planning, National Ilan University, No. 1, Sec. 1, Shennong Road, Yilan City, Yilan County 26047, Taiwan (R.O.C.)
[2] changjr@niu.edu.tw

Abstract: In this study, unsaturated polyester (UP) resin was used as the adhesive material for blending with alabaster or natural aggregate at ambient temperature for the surface course on pavements. UP resin is non-toxic and has good weather resistance. It is a good material for replacing the commonly seen epoxy resin. The mixture after blending UP resin with alabaster or natural aggregates has a very high strength and its stability can reach two to three times of that of asphalt mixture within a few hours. Moreover, the mixture's coefficient of permeability is nearly 40 times higher than that of porous asphalt mixture. In this study, the effects of temperature and water content of aggregates on the hardening of UP resin mixture were studied. We found that different additions of accelerator and hardener were required at different temperatures to adjust the hardening time of mixture. Higher additions resulted in faster hardening rate, but adding an excessive amount reduced the final strength of the mixture instead because of the residues of the accelerator and hardener. The water content of aggregates caused the hardening reaction to be sluggish and incomplete. Only 1% of water content in aggregates substantially slowed down the hardening. Conditions at higher temperatures reduced the effect of water on hardening.

1 INTRODUCTION

There has been a need to ease the regional drainage burden resulting from instantaneous heavy rainfall and the heat island effect in urban areas in recent years, as well as a need to reduce the thermal energy, carbon emissions, and costs incurred during paving by the asphalt industry. Meanwhile, there has also been a desire to provide landscape environments that are beautiful and culturally innovative. Hence, the promotion of permeable pavement is currently in full swing. In this study, unsaturated polyester (UP) resin was selected as the adhesive material for blending with alabaster or natural aggregates under ambient temperature for a surface course on pavements. The color pigments thus can be added to the mixture to create beautiful and culturally innovative landscape environments. The surface course can be laid out on a permeable pavement for pedestrian pathways and bicycle lanes. The surface water can be permeated, conserved, and slowly drained through the surface course during a certain period. This innovative structure thus enables road permeation and water conservation and obviates the need for building a huge drainage system considering the disappearance of surface vegetation within the urban areas. Its use is in line with the "On-site Water Retention Indicators" for green building, high-performance green building material for water permeability, recyclable green building material, and the eco-city policy that have recently been promoted by the Taiwan government. In this study, the effects of temperature and water content of aggregates on the hardening of UP resin mixture were studied in depth.

2 THE UP RESIN IN THIS STUDY

UP resin is not toxic and has good weather resistance (Lee and Han 1987, Lem and Han 1984). It is a good material for replacing the commonly seen epoxy resin (Daid et al. 1985, Fowler and Kukacka 1981, Manson 1990). As shown by relevant literature (Spiecker et al. 2003), the strength and wear resistance of the mixture obtained by blending UP resin with aggregates after addition of accelerator and hardener performs very well. The blending process can be performed at room temperature, unlike the blending of asphalt mixture, which needs temperatures exceeding 120°C. UP resin has excellent adhesive force and the contact area between coarse aggregates within mixture can be increased without more mineral aggregates. Meanwhile, the strength of UP resin mixture is acceptable despite the presence of many internal voids, so the drainage effect of mixture is extremely good. However, many factors affect the hardening of UP resin mixture, including the composition of the resin, the additions of accelerator and hardener, the hardening temperature, and the hardening time. The UP resin of ETERSET 2519P-S type suitable for resin concrete provided by Eternal Materials Co., Ltd. was used in this study. Its characteristics include medium-to-high viscosity, prepromoted setting, good flexural strength and compressive strength, good surface dryness, and low shrinkage rates (Pusatcioglu et al. 1979) (as shown in Table 1).

Table 1: Basic characteristics of UP resin used in this study

Appearance	Purple-red opaque
Styrene monomer content (%)	38-41
Viscosity - Brookfield at 25°C, spindle #3, 60 rpm (cps)	500-600
Acid value (mg KOH/g resin)	14-22
Specific gravity	1.1-1.16

Pure-resin hardening characteristics: Room-temperature hardening at 25°C, 1.0% Hardener (55% Methyl Ethyl Ketone Peroxide (MEKPO) included)

Gelation time (min)	10-13
Minimum hardening time (min)	15-30
Peak exotherm (°C)	165-200
Physical properties of the hardened injection molding product	Transparent injection molding product, ETERSET 2519P-S
Barcol (934-1) hardness	50
Tensile strength (kgf/mm^2, ASTM D 638)	6.5
Elongation rate (%, ASTM D 638)	3.5
Flexural strength (kg/mm^2, ASTM D 790)	13
Flexural modulus (kg/mm^2, ASTM D 790)	375
Heat distortion temperature (°C, ASTM D 648)	93
Water absorption rate (%, JIS K6919)	0.18

3 UP RESIN BLENDING TEST

3.1 Gradation of Natural Aggregates

The gradation of natural aggregates was adjusted based on the gradation of porous asphalt mixture. We used 20% of 19.0 mm (3/4 in) aggregate, 35% of 9.5 mm (3/8 in) aggregate, 30% of 6.35 mm (1/4 in) aggregate, and 15% of sand. Because the resin is easily absorbed by fine aggregates such as mineral filler to sequentially reduce its adhesive force, the mineral filler was not used in the gradation. The gradation for natural aggregates is shown in Table 2.

Table 2: Gradation for natural aggregates

Aggregates Sieve size (in)	19.0 mm (3/4 in)	9.5 mm (3/8 in)	6.35 mm (1/4 in)	Sand	Mineral filler	Amount of Usage
1.5	100	100	100	100	100	100.00
1.0	100	100	100	100	100	100.00
0.75	79	100	100	100	100	95.80
0.5	1	68	100	100	100	69.00
0.375	0	29	100	100	100	55.15
No. 4	0	1	51	99	100	30.50
No. 8	0	0	9	80	100	14.70
No. 16	0	0	4	51	100	8.85
No. 30	0	0	2	26	100	4.50
No. 50	0	0	2	7	100	1.65
No. 100	0	0	1	7	99	1.35
No. 200	0	0.1	0.9	2.8	88.4	0.73
Ratio of Usage (%)	20	35	30	15	0	

3.2 Compacted UP Mixtures Test

The 3.175 mm (1/8 in) alabaster (single aggregate), 3.81 mm (1.2/8 in) alabaster (single aggregate), and natural aggregate (gradation in Table 2) were used to respectively prepare the 4-in specimens of compacted UP mixtures. Figure 1 shows the dry alabaster. The UP resin content includes 4.5%, 5.0%, and 5.5%. The specimens and test data are listed in Table 3.

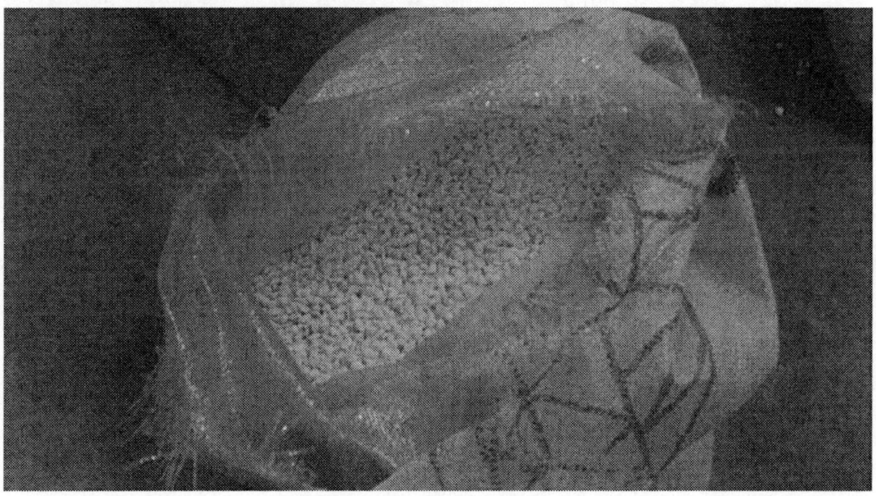

Figure 1: Dry alabaster

In Table 3, the bulk specific gravity of 3.175 mm (1/8 in) alabaster and 3.81 mm (1.2/8 in) alabaster specimens decreased because of the increase in resin content, whereas the change in resin content did not significantly affect the bulk specific gravity of the natural aggregate specimen. Moreover, the water absorption test indicates that the water absorption rate of 3.175 mm (1/8 in) alabaster specimen is directly proportional to the resin content, whereas the natural aggregate specimens had a lower water absorption rate despite the increasing resin content. In addition, the Marshall indoor permeameter was used in the study to measure the specimens' water permeability rate (ml/sec), which was converted into the coefficient of permeability (k, cm/sec). The k values lie between 0.137 cm/sec and 0.365 cm/sec. According to the coefficient of permeability for permeable concrete paving blocks specified in Chinese National Standards (CNS) 14995, the k value should be greater than 1.0 × 10−2 cm/sec. The k values of UP resin mixture specimens were nearly 40 times this standard, with the 3.81 mm (1.2/8 in) alabaster specimens showing the best values and the natural aggregate specimens showing the worst. This is probably because the fewer voids in the gradation of natural aggregates. The content of mineral aggregates in the gradation of natural aggregates can be reduced to increase the water permeation.

Table 3: Specimens and test data

Aggregate type	Specimen No.	UP resin content (%)	Weight in air (g)	Weight in water for saturated surface-dry specimens (g)	Weight in air for saturated surface-dry specimens (g)	Specimen thickness (mm)	Correction factor	Bulk specific gravity	Water absorption rate (%)	Water permeability rate (ml/sec)
3.175 mm (1/8 in) alabaster (single aggregate)	1	4.5	967.4	607.8	993.5	63.6	1	2.51	6.77	77.56
	2	4.5	959.8	602.7	985.3	63.6	1	2.51	6.66	81.66
	3	4.5	968.3	600.1	986.7	64.4	1	2.49	5.25	72.96
	Mean						1	2.50	6.23	77.39
3.175 mm (1/8 in) alabaster (single aggregate)	1	5.0	966.0	604.7	993.9	62.8	1	2.48	7.17	89.87
	2	5.0	965.7	594.8	983.9	62.3	1	2.46	6.16	83.17
	3	5.0	972.6	607.8	1003.3	63.7	1	2.46	7.76	89.93
	Mean						1	2.47	7.03	87.66
3.175 mm (1/8 in) alabaster (single aggregate)	1	5.5	956.8	596.7	984.5	61.6	1	2.47	7.14	75.47
	2	5.5	961.6	599.6	990.4	62.6	1	2.46	7.37	79.41
	3	5.5	959.7	596.0	986.1	63.5	1	2.46	6.77	90.80
	Mean						1	2.46	7.09	81.89
3.81 mm (1.2/8 in) alabaster (single aggregate)	1	4.5	964.8	606.3	987.7	64.3	1	2.53	6.00	96.15
	2	4.5	971.2	610.1	992.7	65.5	1	2.54	5.62	91.05
	3	4.5	960.3	601.5	985.5	65.4	1	2.50	6.56	94.28
	Mean						1	2.52	6.06	93.83
3.81 mm (1.2/8 in) alabaster (single aggregate)	1	5.0	980.0	614.1	1004.3	64.0	1	2.51	6.23	88.37
	2	5.0	966.9	608.2	991.2	63.8	1	2.52	6.34	98.62
	3	5.0	979.7	613.5	1003.2	64.9	1	2.51	6.03	82.87
	Mean						1	2.52	6.20	89.95
3.81 mm (1.2/8 in) alabaster (single aggregate)	1	5.5	974.0	609.7	998.0	63.3	1	2.51	6.18	92.82
	2	5.5	967.0	602.2	989.8	64.4	1	2.49	5.88	91.83
	3	5.5	973.1	604.6	997.3	65.4	1	2.48	6.16	88.50
	Mean						1	2.49	6.08	91.05
Natural aggregate (gradation in Table 2)	1	4.5	1027.7	609.9	1054.5	63.1	1	2.31	6.03	38.66
	2	4.5	1033.8	610.4	1061.0	65.8	1	2.29	6.04	59.74
	3	4.5	1028.0	612.5	1050.0	63.8	1	2.35	5.03	47.34

	Mean						1	2.32	5.70	48.58
Natural aggregate (gradation in Table 2)	1	5.0	1028.6	604.2	1049.3	61.6	1	2.31	4.65	38.95
	2	5.0	1032.6	610.0	1056.0	62.2	1	2.32	5.25	40.15
	3	5.0	1033.6	608.7	1055.4	62.1	1	2.31	4.88	42.31
	Mean						1	2.31	4.93	40.47
Natural aggregate (gradation in Table 2)	1	5.5	1033.9	602.0	1049.2	61.5	1	2.31	3.42	36.05
	2	5.5	1023.6	601.3	1043.0	62.4	1	2.32	4.39	32.78
	3	5.5	1033.4	605.1	1049.8	61.6	1	2.32	3.69	37.74
	Mean						1	2.32	3.83	35.52

3.2.1 Marshall Stability and Flow Tests

The strength of the UP resin mixtures was very high. Figure 2 shows the 4-in specimens of compacted UP mixtures. The specimens were not destroyed even when the Marshall stability had reached 6,000 kgf. If the loading had been continued to 8,000 kgf, the Marshall apparatus would have not been able to bear the load. The stability of asphalt mixtures are usually lower than 3,000 kgf. Evidently, the strength of UP resin mixtures far exceeds the strength of asphalt mixtures. Moreover, the flow was only 8 units (0.25 mm per unit) when the stability of the UP resin specimens exceeded 6,000 kgf, indicating that the flexibility of the UP resin mixtures is very low and that they are not prone to deformation.

Figure 2: The 4-in specimens of compacted UP mixtures

4 UP RESIN HARDENING TEST

4.1 Effect of Temperature on UP Resin Hardening

To understand the effect of temperature on the hardening of UP resin mixture, we conducted tests at lower temperatures (about 11°C to 13°C). Table 4 shows the test design and test results. The test was first attempted by adding 0.3% and 0.4% accelerator, as well as 0.8% and 1.0% hardener (the additions are relative to the UP resin's weight) to be mixed with UP resin (test design (1) in Table 4). Figure 3 shows the resulting mixture. Then, the resulting mixture blended with alabaster, as shown in Figure 4. However, the room temperature was so low that the UP resin mixture still had not hardened six hours after blending. Until the next day, it was barely thickened like jelly. The strength of mixture was still far from expectation after standing for a week. We used different additions of accelerator and hardener at lower temperature and the improvements were observed with additions of 0.5% accelerator and 1.2% hardener (test design (2) in

Table 4). Although the UP resin mixture hardened after one day, the final strength still did not reach the expectation.

Figure 3: UP resin mixed with accelerator and hardener

Figure 4: UP resin mixture blended with alabaster

We continued to conduct the combinations of 0.6% accelerator + 1.3% hardener, 0.8% accelerator + 1.5% hardener, and 1.0% accelerator + 1.7% hardener (test designs (3), (4), and (5) in Table 4). The hardening condition was better. The study was continued by increasing the UP resin content to 4.5% of the total mixture weight mixed with the combinations of 0.6% accelerator + 1.3% hardener and 0.8% accelerator + 1.5% hardener (test designs (6) and (7) in Table 4). The hardening time and the strength of mixtures were found to be good.

As proven by the test results, the UP resin is very sensitive to temperature; high temperature and low temperature drastically accelerated and slowed down the hardening rate. Adjusting the additions of accelerator and hardener reduced the effect of temperature on the hardening rate. Moreover, increasing the UP resin content allowed the mixture to be harder, resulting in a significant decrease in flow. The stability for test designs (3), (5), and (6) were greater than 5,000 kgf (11,000 pound-force). Taking considerations of the cost and required flow, the study selected test design (3) (4.0% UP resin, 0.6% accelerator, and 1.3% hardener) for the subsequent water content of aggregates test.

Table 4: Test design and test results

Test design	(1)	(2)	(3)	(4)	(5)	(6)	(7)
Total weight of mixtures (g)	1000	1000	1000	1000	1000	1000	1000
Alabaster size (mm)	3.81	3.81	3.81	3.81	3.81	3.81	3.81
UP resin content (% of total weight)	4.0	4.0	4.0	4.0	4.0	4.5	4.5
Accelerator content (% of resin's weight)	0.3, 0.4	0.50	0.60	0.80	1.00	0.60	0.80
Hardener content (% of resin's weight)	0.8, 1.0	1.20	1.30	1.50	1.70	1.30	1.50
Moisture content (%)	60-70	60-70	60-70	60-70	60-70	60-70	60-70
Temperature (℃)	11-13	11-13	11-13	11-13	11-13	11-13	11-13
Water content (%)	0	0	0	0	0	0	0
Color pigment	Nil	Nil	Nil	Nil	Nil	Nil	Nil
Number of compaction	25 times (only one side)	25 times (only one side)	25 times (only one side)	25 times (only one side)	25 times (only one side)	25 times (only one side)	25 times (only one side)
Hardening time (hr)	> 24	12-15	2.0-2.5	1.5-2.0	1.5-2.0	2.0-2.5	1.5-2.0
Stability (kgf)	/	/	> 5,833	> 5,918	> 6,030	> 6,042	> 6,185
Flow (mm)	/	/	1.59	> 1.6	> 1.54	> 0.93	> 0.94
Remarks	After blending, no hardening reaction in 4-7 hr	After blending, no hardening reaction in 4-6 hr; slow hardening after 1 day	After blending, specimen can be demolded in 1.5 hr and mixture can be hardened in about 2.0-2.5 hr	After blending, specimen can be demolded in 1 hr and mixture can be hardened in about 1.5-2.0 hr	After blending, specimen can be demolded in 1 hr and mixture can be hardened in about 1.5-2.0 hr	After blending, specimen can be demolded in 1.5 hr and mixture can be hardened in about 2.0-2.5 hr	After blending, specimen can be demolded in 1 hr and mixture can be hardened in about 1.5-2.0 hr

4.2 Water Content of Aggregates Test

To understand the effects of water content of alabaster on the hardening of UP resin mixture, the study used alabasters with 0.0% (dry), 1.0%, and 1.5% of water content, along with test design (3) in Table 4 (4.0% UP resin, 0.6% accelerator, and 1.3% hardener) to measure the hardening time and final strength of UP resin mixture. The test design and the test results are shown in Table 5.

According to the test results, the specimen (UP resin mixed with water-containing alabaster) can be demolded in 1.0 hr to 1.5 hr after blending but the mixture were congealed like jelly and did not have strength. Hardening took about 2.0 hr to 2.5 hr, and the final strength was far less than that of the specimen without water-containing alabaster. With the test design (3), the stability at room temperature (21°C to 23°C) were lower than those at lower temperatures (11°C to 13°C) whether the specimens were lacked water (0% water content of aggregates) or had water content of aggregates. This again shows the high temperature sensitivity of the UP resin. In accordance with the suggestion of the resin supplier, the uses of UP resin are room temperatures of 23°C to 25°C, as well as the addition of 0.3% to 0.5% accelerator and 0.8% to 1.2% hardener. Adding a large amount of accelerator and hardener to the same UP resin content may not necessarily result in high strength in a short period; hence, the additions of accelerator and hardener at different temperatures must be adjusted accordingly.

Table 5: Water content of aggregates test and test results

Test design	(8)	(9)	(10)
Water content (%)	0.0	1.0	1.5
Total weight of mixtures (g)	1000	1000	1000
Alabaster size (mm)	3.81	3.81	3.81
Resin content (% of total weight)	4.0	4.0	4.0
Accelerator content (% of resin's weight)	0.60	0.60	0.60
Hardener content (% of resin's weight)	1.30	1.30	1.30
Moisture content (%)	64	64	64
Temperature (°C)	21-23	21-23	21-23
Color pigment	Nil	Nil	Nil
Number of compaction	25 times (only one side)	25 times (only one side)	25 times (only one side)
Hardening time (hr)	1.0-1.5	> 2	> 2.5
Stability (kgf)	3,797	2,461	2,529
Flow (mm)	1.37	0.54	0.56

5 COST ANALYSIS

Cost analysis was conducted for the mixture of alabaster and UP resin without mixed with color pigment (i.e., uncolored). The total weight of mixture was 1,000 grams (4.5% UP resin, 0.3% accelerator, and 0.8% hardener). Alabaster, UP resin, accelerator, and hardener accounted for 955 g, 44.5 g, 0.135 g, and 0.365 g, respectively. The cost of the mixture is about CAD 0.346/kg to CAD 0.390/kg. The cost of the mixture of natural aggregates and UP resin is about CAD 0.173/kg to CAD 0.195/kg. Color pigment must be stirred in to impart different colors and a beautiful appearance to the mixtures. The coloring effect of alabaster was better. Color pigment (1.5% with respect to resin content) was required to increases the cost by CAD 0.005/kg to CAD 0.013/kg. Coloring of natural aggregates was more difficult. More color pigment (about 25% with respect to resin content) was required, thus increasing the cost by CAD 0.072/kg to CAD 0.234/kg. Table 6 shows the cost comparison for the alabaster and natural aggregates.

Table 6: Cost comparison of alabaster and natural aggregates

	Alabaster	Natural aggregates
Aggregate (CAD/kg)	0.238	0.017
UP resin (CAD/kg)	2.598 - 3.031	
Color pigment (CAD/kg)	6.495 - 21.649 (varies with the color)	
Accelerator and hardener (CAD/kg)	8.660 - 12.990	

6 CONCLUSIONS

UP resin after blending with alabaster or natural aggregates has very high strength. It can immediately present stability two to three times than that of asphalt mixtures in a few hours, and even have coefficient of permeability (k) nearly 40 times than that of porous asphalt mixture. We used alabaster because it can be colored easily and it can be laid out on an environment with landscape requirements, such as pedestrian pathways and bicycle lanes. Because the aggregates adhered through the UP resin, the resin and asphalt are equally subject to the problem of aging. Moreover, the resin price is higher than that of asphalt. We can adopt natural aggregates with less mineral aggregates to reduce the cost and maintain their advantages of high strength and high water permeability in practice. This study obtained the following findings:

1. At different temperatures, different additions of accelerators and hardeners are needed to adjust the hardening rate of resin mixture. Higher addition resulted in faster hardening rate, and vice versa; but adding an excessive amount reduced the final strength due to the residue of accelerator and hardener. According to the resin supplier's suggestion for room temperature (23°C to 25°C), the addition of accelerator should be from 0.3% to 0.5% and that of hardener should be from 0.8% to 1.2%. Excessively low temperature prolonged the hardening time, and excessively high temperature weakened the final strength. Tests were specifically conducted at temperature of 11°C to 13°C in the study, and the appropriate addition of accelerator is 0.6% to 1%, and that of the hardener is 1.3% to 1.7%.

2. Aggregates containing water caused the hardening reaction to be sluggish and incomplete. Only 1% of water content in aggregates had significant effects on the hardening of the mixtures. Therefore, the aggregates should be kept as dry as possible. At higher temperatures, the effect of water on hardening was reduced.

3. The strength of the mixtures originates from the mechanical properties of the UP resin itself. Even the pure UP resin without blended aggregates possessed extremely high strength; therefore, the strength of the resin mixture was less likely to be affected by the compaction process.

4. After hardening, the UP resin became hard and brittle. Adding more UP resin resulted in lower flow. Using larger and single-size aggregates can reduce the addition of resin and increase more voids, resulting in better water permeability of the road.

Acknowledgements

The authors would like to thank the Ministry of Science and Technology of Taiwan for the financial support provided under the project MOST 105-2622-E-197-008-CC3.

References

Daid, H., Dor, B. and Heitener-Wirguin, C. 1985. The Porosity and Compressive Strength of Polymer Cement. Cement and Concrete Research, 15(6): 1061-1067.

Fowler, D.W. and Kukacka, L.E. 1981. Application of Polymer Concrete. Publication SP-69, American Concrete Institute.

Lee, D.S. and Han, C.D. 1987. The Effect of the Chemical Structure of Low-profile Additives on the Curing Behavior and Chemorheology of Unsaturated Polyester Resin. Polymer Engineering and Science, 27(13): 964-975.

Lem, K.W. and Han, C.D. 1984. Thermokinetics of Unsaturated Polyester and Vinyl Ester Resin. Polymer Engineering and Science, 24(3): 175-184.

Manson, J.A. 1990. Applications of Polymer Concrete. Publication SP-69, American Concrete Institute.

Pusatcioglu, S.Y., Fricke, A. L., and Hassler, J. C. 1979. Heats of Reaction and Kinetics of a Thermoset Polyester. Journal of Applied Polymer Science, 24(4): 937-946.

Spiecker, P.M., Gawrys, K.L., Trail, C.B., and Kilpatrick, P.K. 2003. Effects of Petroleum Resins on Asphaltene Aggregation and Water-in-oil Emulsion Formation. Colloids and Surfaces A: Physicochem. Eng. Aspects, 220: 9-27.

Building Tomorrow's Society
Bâtir la Société de Demain

Fredericton, Canada
June 13 to 16, 2018 / *13 au 16 juin, 2018*

USE OF EVA FOAM IN PORTLAND CEMENT CONCRETE

Farhoud, Aleya M.[1], Mansour, Menatallah[1], Shoukry, Radwa M.[1,2], El Bagoury, Omar[1] El Akkad, Shehab[1], Farag, Mayer[1], Hamza, Ahmed[1], El Nahas, Eman[1], Hussam, Ahmed[1], and Abou-Zeid, Mohamed N.[1]
[1] The Department of Construction Engineering, the American University in Cairo, Egypt
[2] radwamsaeed@aucegypt.edu

Abstract: EVA foam is a copolymer, composed of Ethylene and Vinyl Acetate, that is used in several industrial products. It is known to be lightweight, moisture resistant, chemical insulating, heat insulating and energy absorbent. During the manufacturing of EVA products, such as athletic footwear, swimming mats, etc., large quantities of EVA waste is generated. This waste is non-recyclable and poses various health harms when burnt, such as severe irritation to the eyes, skin and mucous membrane. Thus, it is currently being disposed of in landfills, as this deems to be the easiest way for manufacturers to handle the EVA waste. This study aims to assess the incorporation of EVA Foam as a non-recyclable; toxic solid waste in Portland cement concrete. Thirteen different mixes were poured using different sizes of EVA Foam which are 9.5, 19 and 38mm. These sizes were used as replacement for coarse aggregates by the following percentages 5,10,15 and 20%. Concrete with the replaced EVA aggregates demonstrated a decrease in compressive strength, but showed little or no difference in flexural strength. EVA concrete demonstrated a good potential in being considered as lightweight concrete as well as an abrasion resistant one. Although the results of the water penetration test showed a higher water penetration for EVA concrete, EVA concrete seems to be a good candidate for thermal insulation applications. Building on these tests, EVA concrete did not seem to witness any significant change with regards to its fire resistance properties. In addition, the incorporation of EVA in concrete seemed to make it more ductile as opposed to its brittle nature. Moreover, the high replacement percentages showed an initial potential for pervious concrete applications.

1 INTRODUCTION

EVA is short for Ethylene Vinyl Acetate, which is a co-polymer of ethylene and vinyl acetate with the chemical formula, $C_6H_{10}O_2$ (Seedahmed A.I., 332). Ethylene Vinyl Acetate is known for its following properties: moisture resistance in its raw form, light weight, thermal insulation and chemical resistance.

These properties are what make it widely used in various industries including the sports, and healthcare industries. It is especially widely used in the footwear industry both globally and locally. According to a study by UNIDO, there are 17 billion pair of shoes produced globally every year, with each million of which generating approximately 80 tons of waste. Out of this amount, EVA Foam waste represents 14%, summing up to a total of 194,000 tons produced every year (UNIDO). In a more focused context, primary research suggests that the footwear industry in Egypt generates an annual sum of 28,000 tons of EVA waste.

Given that EVA foam is considered to be a thermosetting material, recycling these amounts of waste requires special procedures that are quite expensive. Accordingly, these piles of waste end up being thrown in landfills, occupying large areas (Diana Lopes, 231). Not only does this option take up much space, it also negatively impacts the environment given that it is non-biodegradable and will need more than one thousand years to decompose.

1.1 Objective and Scope

In an attempt to reduce the environmental impact of such waste material and concrete, the creation of greener concrete using EVA waste was investigated. The experimental work was designed to assess the effect of replacing different percentages of coarse aggregates with EVA waste aggregate of various sizes on the concrete's performance. This was done through testing for the concrete's compressive and flexural strength, its permeability and resistance to abrasion. In addition, the concrete's thermal resistance, fire insulation, infiltration rate and modulus of elasticity were tested in light of EVA foam's properties.

2 EXPERIMENTAL PROGRAM

2.1 Materials

Cement: Ordinary Portland cement, Type I, was used with specific gravity of 3.15

EVA Foam: Sizes of interest: 9.5mm, 19mm and 38mm

The EVA waste is originally attained in the form shown in Figure 1. This form was then shredded using fabricated steel plates to acquire the specified sizes, a sample of which is shown in Figure 2.

Coarse Aggregates: Well-graded coarse aggregates with a SSD specific gravity of 2.66 was used

Fine Aggregates: Natural sand with a SSD Specific gravity of 2.5 was used

Water: Normal tap water was used in concrete mixing and curing

Super Plasticizer: 1% by weight of cement of Super plasticizer was used in order to increase the workability of the concrete mix

Figure 1: EVA Waste in its Original Form

Figure 2: EVA Waste After Shredding

2.2 Concrete Mix Design

Table 1: Mix Design

EVA Aggregate Size (mm)	Replacement Percentage (%)	Cement (kg)	Water (kg)	w/c	Fine aggregates (kg)	Coarse Aggregates (kg)	EVA Foam (kg)
-	0	350	175	0.5	650	1170	0
9.5	5	350	175	0.5	650	1111	46
9.5	10	350	175	0.5	650	1053	91
9.5	15	350	175	0.5	650	994	137
9.5	20	350	175	0.5	650	936	182
19	5	350	175	0.5	650	1111	46
19	10	350	175	0.5	650	1053	91
19	15	350	175	0.5	650	994	137
19	20	350	175	0.5	650	936	182
38	5	350	175	0.5	650	1111	46
38	10	350	175	0.5	650	1053	91

2.3 Tests

2.3.1 Fresh Concrete:

Tests were conducted on fresh concrete according to the designated ASTM standards. These tests were: slump, unit weight and air content.

2.3.2 Hardened Concrete

Compressive Strength: This test was conducted according to British Standards [BS 1881] for 150mmx 150mmx150mm cubes to find the 7 and 28-compressive strength. For each mix, three cubes were tested and the average strength was calculated.

Flexural Strength: This test was conducted according to ASTM Standards [ASTM C78] for 150mmx150mmx750mm beams to find the flexural strength after 28 days.

Water Penetration: This test was conducted according to German Standards [DIN 1048] for 150mmx 150mmx150mm cubes to find out the depth of water penetration under pressure.

Abrasion: This test was conducted according to ASTM Standards [ASTM C779] to determine the relative abrasion resistance of horizontal concrete surfaces. Each specimen of size 50mmx50mmx25mm was subjected to abrasive material on a rotating disk for three 2-minute intervals. This procedure was repeated with three samples of the same mix to obtain the average weight loss percentage.

2.3.3 Specialized Tests

Thermal Conductivity: The test was conducted according to ASTM Standards [ASTM C518] to cover the measurement of steady state thermal transmission through flat slab specimens 300mmx300mmx40mm using a heat flow meter apparatus. It measures the thermal conductivity (K) which is the ability of a material to conduct heat.

Fire Resistance: This test was conducted according to ASTM Standards [ASTM E119] to examine the concrete's ability to hinder the progression of fire from one surface of concrete to the other. One side of the specimen of size 300mmx300mmx120mm was subjected to specific temperatures. Temperature sensors were placed on the other side of the concrete specimen, measuring the temperature every 5 minutes for the duration of two hours.

Modulus of Elasticity: This test was done according to ASTM Standards [ASTM C580] to determine the modulus of elasticity by obtaining the stress versus strain curves for the samples. The specimen was subjected to a uniform load until fracture. With the generated load-deflection data from the MTS machine, the stress-strain curve was drawn for each specimen.

Infiltration Test: This test was done according to ASTM Standards [ASTM C1701] to assess the samples' potential as pervious concrete. Time is monitored to determine how long it takes for a known quantity of water to infiltrate through the concrete.

3 RESULTS AND DISCUSSION:

3.1 Fresh Concrete Tests:

3.1.1 Slump Test

In comparing the results obtained for the slump test, shown in Table 2, for the EVA concrete samples with the control sample, the results suggest that the incorporation of EVA in concrete decreases the value of slump. However, in analyzing each mix's workability, the lighter nature of the EVA aggregates compared to the conventional aggregates must be taken into consideration.

Table 2: Slump Test Results for Fresh Concrete Mixes

EVA Aggregate Size (mm)	Replacement Percentage (%)	Slump (mm)
Control	0	45
9.5	5	20
	10	10

EVA Aggregate Size (mm)	Replacement Percentage (%)	Slump (mm)
	15	30
	20	10
19	5	5
	10	20
	15	0
	20	0
38	5	15
	10	25

3.1.2 Air Content

When assessing the effect that the incorporation of EVA aggregates has on the air content of the concrete, the data proposes that there is a level of air entrapment associated with the replacement of the conventional aggregates with the EVA aggregates. In addition, there seems to be an increase in the air content associated with the increase in replacement percentage, as demonstrated in Table 3. The highest air content was that of the 20% replacement mix.

Table 3: The results of the Air Content test

EVA Aggregate Size (mm)	Replacement Percentage (%)	Air Content (%)
Control	0	3.4
9.5	5	4.5
	10	7.0
	15	9.0
	20	9.5
19	5	4.3
	10	7.9
	15	7.0
	20	8.5
38	5	3.5
	10	3.9

3.1.3 Unit Weight:

The results of the unit weight test suggest that the increase of the EVA aggregate replacement percentage results in a decrease in the unit weight of the concrete. Similarly, an increase in the EVA aggregate size seems to contribute to the decrease in unit weight as displayed in Table 4. To conclude, the mix with the lowest unit weight was found to be the one associated with the 19 mm size with a 20% replacement.

Table 4: Unit Weight Test Results for Fresh Concrete Mixes

EVA Aggregate Size (mm)	Mix Replacement Percentage (%)	Unit Weight (Kg/m^3)
Control	0	2344
9.5	5	2136
	10	1933
	15	1706
	20	1504
19	5	2019
	10	1756
	15	1474
	20	1393

EVA Aggregate Size (mm)	Mix Replacement Percentage (%)	Unit Weight (Kg/m^3)
38	5	2020
	10	1673

3.2 Hardened Concrete Tests:

3.2.1 Compressive Strength Test:

The results for the 7-day compressive strength shown in Figure 3 show a decrease in the compressive strength when a higher percentage of the conventional aggregate is replaced for the EVA aggregate. This can be explained by the weaker nature of the EVA aggregate when compared to the conventional one. There also seems to be a decrease associated with the increase in the EVA aggregate size.

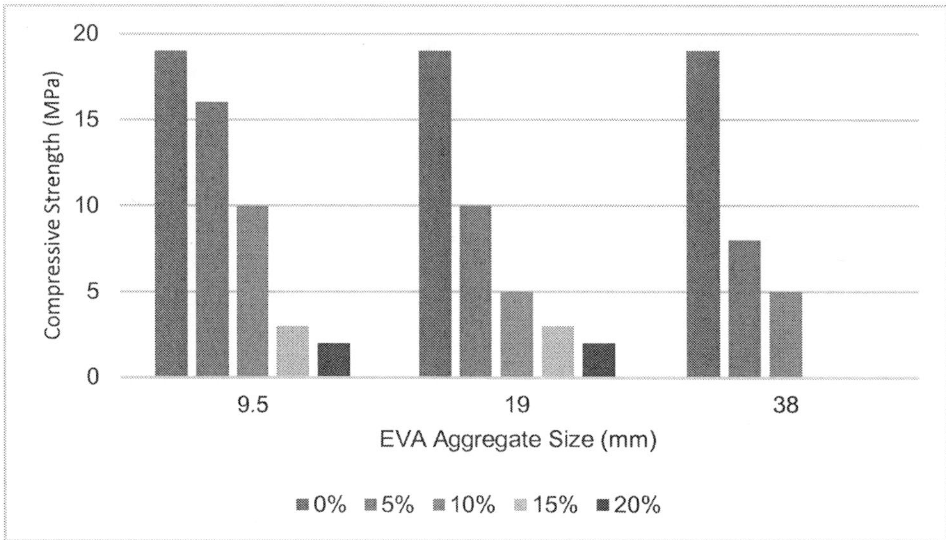

Figure 3: 7-Day Compressive Strength Results

As for the 28-day compressive strength, displayed in Figure 4, the results showed a slight improvement, moving the 9.5 mm 5% replacement mix into the starting range of structural concrete with a strength of 22 MPa. Although the results continue to show a wide gap between the strength of the control mix and the EVA concrete, the ratio of 7/28-day strength for all EVA samples is smaller than that of the control mix. Accordingly, long term testing should be conducted to study EVA concrete's potential of longer term strength gain.

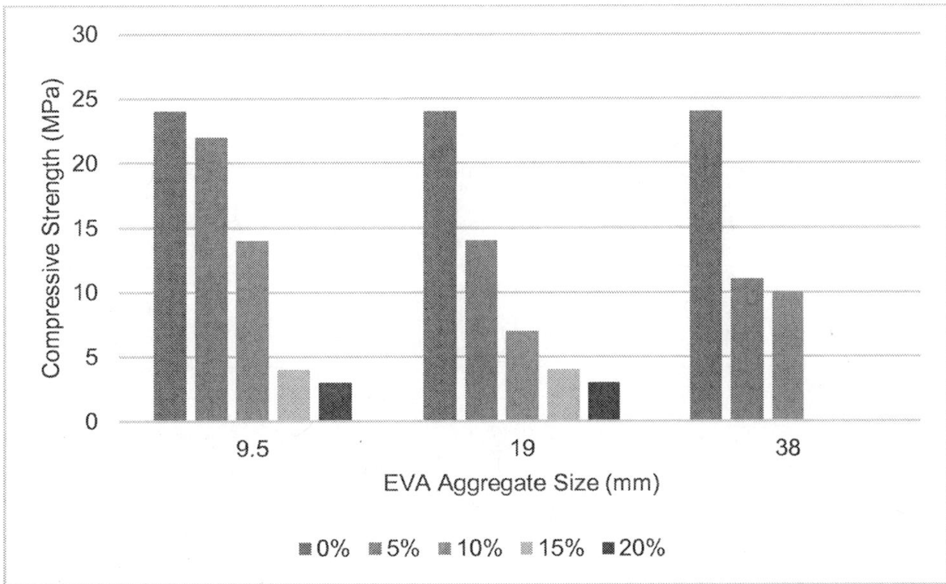

Figure 2: 28-Day Compressive Strength Results

3.2.2 Flexural Strength:

The results of the flexural strength displayed in Figure 5 suggest that the incorporation of EVA aggregates does not negatively impact the flexural strength of the concrete. In fact, there are some mixes such as the 9.5-5% and the 19-20% which yield higher values of flexural strength. The remaining mixes are all close in value to the control mix. This can be attributed to the nature of binding between the EVA aggregate particles. It can also be as a result of the EVA's high absorption, which can enhance the binding between the particles.

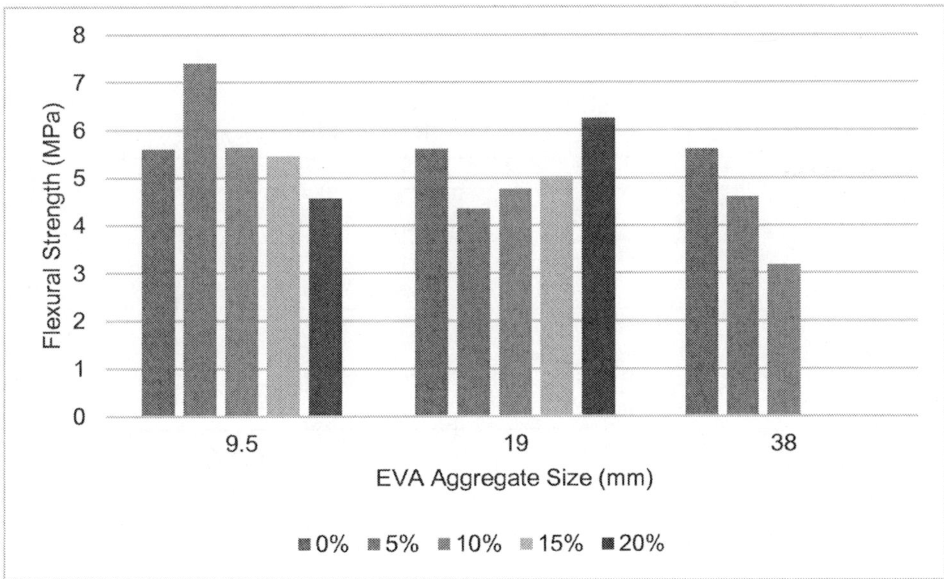

Figure 3: Flexural Strength Results

3.2.3 Abrasion:

The results of the Abrasion test, shown in Table 5, suggest that the EVA concrete has a higher abrasion resistance than the conventional concrete. This is especially true for the mixes with 10% aggregate replacement as these mixes show higher potential in abrasion resistance displayed by the lowest percentage weight loss when compared to the other mixes.

Table 5: Abrasion Test Results for Concrete Mixes

Size of EVA (mm)	Replacement Percentage (%)	Abrasion (% loss)
Control	0	6.78
9.5	5	4.42
	10	2.72
	15	3.56
	20	2.80
19	5	3.39
	10	2.00
	15	2.43
	20	2.29
38	5	3.69
	10	2.39

3.2.4 Water Penetration Test

The water penetration test results shown in Table 6 show an increase in the concrete's permeability with the increase in the EVA replacement percentage. This can be attributed to the presence of more voids in the EVA concrete when compared to the conventional concrete.

Table 6: Water Penetration Test Results for Concrete Mixes

Size of EVA (mm)	Replacement Percentage (%)	Permeability (cm)
Control	0	2.9
9.5	5	3.3
	10	4.7
	15	3.7
	20	5.3
19	5	3.6
	10	5.0
	15	**
	20	**
38	5	3.0
	10	**

** Test could not be conducted due to mixes' pervious nature.

3.3 Non-conventional Testing

3.3.1 Thermal Conductivity

The results in Table 7 demonstrate a potential for the 9.5 mm 10% replacement mix to be a good thermal insulator when compared to its conventional counterpart. This reflects a potential for higher replacement percentages to act as thermal insulators.

Table 7: Thermal Conductivity Test Results

EVA Aggregate Size (mm)	Replacement Percentage (%)	Thermal Conductivity (W m-1K-1)	Thermal Resistivity (mK W-1)
Control	0	1.26	0.79
9.5	10	1.00	1.00

3.3.2 Fire Resistance

After subjecting two samples; one control and the other with 9.5mm-10% replacement, to a specified temperature curve, the results suggest, as shown in Figure 6. that the incorporation of EVA aggregates does not affect the concrete's performance in case of fire.

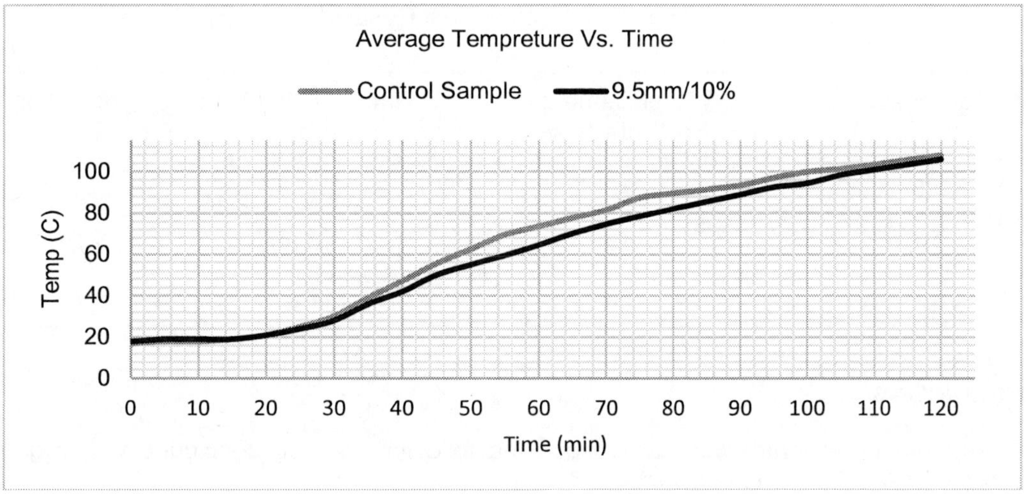

Figure 6: Fire Resistance Results

3.3.3 Modulus of Elasticity

After generating the stress-strain curves, outlined in Figure 7 for the two mixes; control and the 9.5mm-10% replacement, the results suggest that the incorporation of EVA aggregate in the concrete mix makes the concrete more ductile as opposed to its brittle nature. The EVA concrete specimen witnessed a 40% decrease in its modulus of elasticity.

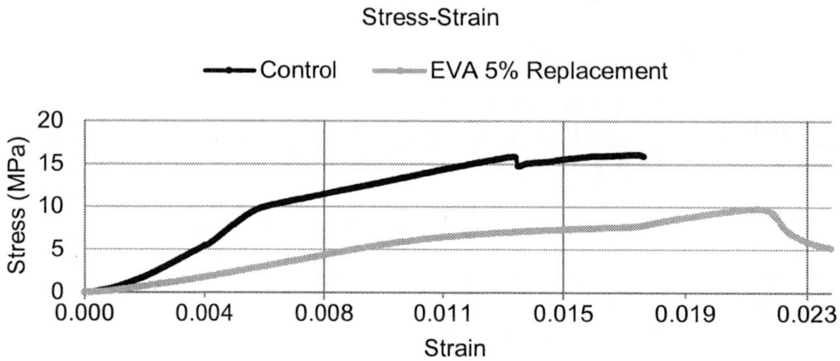

Figure 7: Stress-Strain Curves

3.3.4 Infiltration Test

The results of the infiltration test, shown in Table 8, suggest that the 19-mm size with 15% and 20% replacement show potential to be considered as pervious concrete. This was assessed in comparison to the moderate concrete infiltration rate of 8580 mm/hr. The 9.5mm-15% replacement mix's infiltration rate lies below that, eliminating it from the potential mixes to be used as pervious concrete.

Table 8: Infiltration Test Results

EVA Aggregate Size (mm)	Replacement Percentage (%)	Time to Infiltrate 1L (s)	Infiltration Rate (mm/hr)
9.5	15	170.67	4047
19	15	18.80	35660
19	20	17.47	39828

4 CONCLUSIONS

After the experimental program was carried out, the following can be concluded with regards to the incorporation of EVA foam waste in concrete:

1. EVA Concrete tends to be more abrasion resistant than conventional concrete, especially the sample incorporating 10% replacement.
2. EVA Sample with 9.5mm and 5% replacement yielded better results in terms of compressive strength.
3. Incorporation of EVA Foam in the concrete does not seem to negatively impact flexural strength.
4. EVA Concrete showed potential results in terms of fire resistance and thermal insulation.
5. When subjected to fire, EVA Concrete does not seem to release toxic fumes.
6. EVA Concrete shows encouraging potential to be used as pervious concrete.
7. EVA Concrete can be considered in non-load bearing concrete elements.

5 RECOMMENDATIONS

With these conclusions in mind, EVA concrete can be used in non-load bearing elements that can utilize its advantageous thermal insulation properties such as non-accessible roofs, shading areas, and concrete insulating shells. It can also be used as partitioning or fill concrete. Furthermore, the pervious mixes that were identified can be used in applications such as pervious pavements.
In light of the potential uncovered during the experimental work, there are various tests that should be conducted to get a more in-depth view of the EVA concrete's properties. These tests include the following list:

8. Conducting further long-term testing such as durability, mechanical properties, fatigue and creep.
9. Further test EVA concrete's potential as pervious concrete.
10. Study the interaction of admixtures with EVA in order to enhance concrete properties.
11. Investigate EVA concrete when exposed to aggressive chemical environments.
12. Conduct Modulus of Elasticity Test using strain gauge in the MTS apparatus.
13. Explore the potential of EVA Concrete in internal curing.
14. Explore the potential of incorporating EVA waste in concrete blocks.

Having opened various topics for research and shown potential for a couple of applications, EVA concrete stands as an example of how the use of waste materials in concrete can be advantageous, not necessarily in terms of mechanical properties, but other special ones. Accordingly, further research should be conducted using various types of waste materials. In addition, a mechanism for the testing of waste materials should be adopted in order to attempt to eliminate inconsistencies in the waste materials used.

Acknowledgment

This work is based primarily on the study conducted by Aleya Farhoud, Mennatallah Mansour, Radwa Shoukry, Omar El Bagoury and Shehab El Akkad as their final thesis project in the Department of Construction Engineering at The American University in Cairo.

References

"24937-78-8." National Center for Biotechnology Information. PubChem Compound Database. Accessed December 16, 2017. https://pubchem.ncbi.nlm.nih.gov/compound/175988#section=Top.

By, Giz. "Country Report on the Solid Waste Management in EGYPT." Management in EGYPT, Apr. 2014, www.sweep-net.org/sites/default/files/EGYPT%20RA%20ANG%2014_1.pdf.

Kodolov, V. "Polyethylene foam waste utilization for lightweight concrete production" International Journal of Polymeric Materials, 47, 1(2000), pp. 7-17.

Mounanga, P; Gbongbon, W; Poullain, P; Turcry, P. "Proportioning and characterization of lightweight concrete mixtures made with rigid polyurethane foam wastes", Cement and Concrete Composites, 30, 9(2008), pp. 806- 814.

UNIDO, Wastes Generated in the Leather Footwear Industry, 14thSession of the Leather and Leather Products Industry Panel, Czech Republic. [Online]: https://www.unido.org, 2000.

Seedahmed A.I., Ibrahim. "Using EVA Foam Waste as the Constituent of Lightweight Concrete Mixture for Construction Application." International Journal of Engineering Sciences & Research Technology, December 2016, 332-338. doi:10.5281/zenodo.199557.

Diana Lopes, Maria J. Ferreira, Rui Russo, Joana M. Dias, Natural and synthetic rubber/waste – Ethylene-Vinyl Acetate composites for sustainable application in the footwear industry, In Journal of Cleaner Production, Volume 92, 2015, Pages 230-236, ISSN 0959-6526, https://doi.org/10.1016/j.jclepro.2014.12.063.

Building Tomorrow's Society

Bâtir la Société de Demain

Fredericton, Canada

June 13 to 16, 2018 / *13 au 16 juin, 2018*

THE USE OF GRANITE SLURRY IN MASONRY MANUFACTURING

Attia, Nada[1,4], Moustafa, Omar[1],Abdeldayem, Mohamed[1],Yanni, Rafik[1], Elsobky, Basma[1], Elweleily, Omar[1], Darwish, Mohamed[2], Rashwan, Mohamed[2], Abouzeid, Mohamed[3] and Fahmy, Ezzat[3]
[1]Student, American University in Cairo, Egypt
[2]Adjunct Faculty, American University in Cairo, Egypt
[3]Professor, American University in Cairo, Egypt
[4]Corresponding Author: nadatarattia@aucegypt.edu

Abstract: The waste powder produced during granite manufacturing has been a source of significant pollution and health hazards. The use of this waste powder in concrete has been studied by several researchers. The current study involves partial replacement of the fine aggregates by five different proportions of granite waste powder in the mix of the concrete used for producing concrete bricks. Wet and dry tests on concrete bricks are performed. The effect of varying the percentage of granite waste powder on the various properties of the produced masonry are assessed and compared to a control group with no powder involved. The mixtures having the highest strengths are determined and used to construct masonry prisms subjected to compression. The mix used to construct the prism surviving the highest compression is considered as the one adequate for future masonry production. The initial results indicate that incorporating such waste in masonry walls can be one promising venue for its utilization in the construction industry.

Keywords: Granite Waste; Cement bricks; Granite Slurry; Masonry

1 INTRODUCTION

Egypt has got huge stores of amazing granite and stone. World Stone generation achieved the pinnacle of somewhere in the range of 75 million tons (or 820 million m^2 proportionate), net of quarry waste (El-Sayed, 2016). The official generation figures of Egypt are striking; yet, the genuine creation is extensively higher than the level demonstrated by the official insights and possibly past the levels assessed over the span of the examination featured. The estimations in light of the data recovered through neighborhood appraisal credited to Egypt: a quarry generation of around 3.2 million tons and more than 25 unique sorts of Egyptian stone (El-Sayed, 2016). This shows the nation lies among the main 8 world makers of crude material. The normal yearly rate of increment has achieved 8.8% since 2002(Allam, 2014).

Concerning quarry creation and crude fare, Egypt is positioned the fourth separately with an offer of 4.3% and 6.6% of aggregate world market of Granite and rock. This implies Egypt can be viewed as the seventh exporter on the planet, as far as volume, after China, India, Italy, Spain, Turkey and Brazil. During the process of granite manufacturing, a massive amount of water is used while cutting the granite piles with diamond blades to cool off the process. The result is water absorbing dust and granite slurry powder without any proper way to recycle it or getting rid of it. If the water is disposed on the ground, the slurry powder dries, and it causes air pollution. The main issue that water use in the granite manufacturing process is essential to cool down the blades.

Several researchers have worked on utilizing the use of this slurry powder byproduct in cement bricks. One of these works was done by Hamza et al (2011) who produced cementbricks with partial replacement of the aggregates with the slurry. The partial replacement was done for increments of 10%, 20%, 30% and 40% of the aggregates. The bricks were tested for physical and mechanical properties and the test results revealed that the recycled products have physical and mechanical properties that qualify them for use in the building sector, where all cement brick samples tested in this study comply with the Egyptian code requirement for structural bricks, with granite slurry having a positive effect on cement brick samples that reach its optimum at 10% slurry incorporation (Hamza et al, 2011).

Similarly, other researchers such as (Allam and Bakhoum, 2014), (Kumara, 2015) and (Dhanapandian and Shanthib, 2009) did similar works to that done by Hamza et al (2011). However, none of these researchers narrowed down the window of the study of varying the percentage slurry than increments of 10%. Also, none of them studied the durability of the produced bricks or their performance under axial loads in the walls. These issues are important as they will affect the decision of using such products in various conditions of loading and in various environments.

The current research covers the variation in usage of the percentage of slurry in increments of 5% by producing cement bricks with percentages of 0%, 5%, 10%, 15%, 20% and 25% slurry in replacement of aggregates and tested for compressive strength. Following on that,and in lieu of the results, the increments were reduced to 2.5% in order to get the most accurate possible percentage of slurry. Also, the durability of the bricks was studied by immersion in salt solutions in addition to regular fresh concrete tests. Finally, the performance of brick prisms under compression was studied.

2 EXPERIMENTAL PROGRAM

The experimental work initially involved preparing 6 mixes, by replacing fine aggregates in conventional brick mix with slurry in different increments of 0%, 5%, 10%, 15%, 20% and 25%. The 0% slurry brick and 2 other mixes, that have the highest compressive strength, were used to build 9 masonry prisms, 3 from each mix, to determine the best mix in terms of performance within a masonry wall. The compressive results after 7 day were considered as indication for the strength after 28 days and selecting the mixtures for the prism tests to reduce the duration of the test program and limit the number of the produced bricks. Furthermore, two additional mixtures with 2.5% and 7.5% of slurry were prepared in order to determine a more accurate figure for the percentage slurry corresponding to the maximum strength.

2.1 Material Properties

The materials in the mixtures were as follows:

- **Cement:** Type I Portland Cement Concrete with specific gravity of 3.15.
- **Fine Aggregates:** Well graded sand with specific gravity of 2.55, 0.4% absorption, and a fineness modulus of 2.85.
- **Coarse Aggregates:** Well graded crushed dolomite with specific gravity of 2.57 and 1.8% absorption.
- **Water:** Ordinary municipal tap water was used in producing and curing process of the concrete mixtures.
- **Granite slurry:** A Granite slurry powder from a plant in Alexandria was used.

2.2 Bricks Mix Design

The only variable targeted within the present study is the slurry in partial replacement of fine aggregates while keeping all the rest of the mix constituents constant. These percentage of slurry replacement ranged between 0% to 25 % with 5% initial increments. This range was determined in accordance to the results published in the literature by previous researchers (Allam and Bakhoum, 2014), (Kumara, 2015), (Dhanapandian and Shanthib, 2009) and (Hamza et al, 2011) which indicated that the percentage slurry producing the maximum compressive strength lies between 0% and 20%. However, these researchers performed their studies for increments of 10% of slurry-to-aggregates replacement.

Two additional mixtures with 2.5% and 7.5% of slurry were prepared in order to determine a more accurate figure for the percentage slurry corresponding to the maximum strength Table 1 shows the mix design for all mixes used in the present study.

Table 1: Bricks mix design

Mix	Cement (kg/m3)	Slurry (kg/m3)	Fine Aggregates (kg/m3)	Coarse Aggregates (kg/m3)	Water (kg/m3)
0%	493	-	661	998	185
2.5%	493	17	646	998	185
5%	493	33	628	998	185
7.5%	493	50	612	998	185
10%	493	66	597	998	185
15%	493	99	564	998	185
20%	493	134	531	998	185
25%	493	165	497	998	185

2.3 Mortar Mix Design

The mix design of the mortar used in the construction of the masonry prisms consist of water, cement and sand with a ratio 0.485: 1: 2.75 respectively. 7.5% of the sand was replaced by granite slurry. Accordingly, after this replacement the mix ratio become 0.485: 1: 2.675: 0.206.

2.4 Tests Performed

2.4.1 Fresh tests

- **Slump test:** This test was conducted according to ASTM C143 to evaluate the consistency of the fresh mix.
- **Air content:** The test was conducted according to ASTM C231 to identify the Air content in the fresh concrete mix.
- **Unit weight test:** This test was conducted according to ASTM C138 to determine the density of the mix.

2.4.2 Hardened Concrete/Mortar tests

- **Brick Compressive Strength:** The test is conducted according to ASTM 109 standards to evaluate and determine the strength of each mix.
- **Mortar cubes test:** Compressive strength of each mortar mix was determine by testing 50x50x50 mm cubes according to ASTM 109 standards..
- **Cycles of Immersion in Salt Solution and Drying:** The bricks were weighed, after 28 days of being casted. Then subjected to heat, at 110 °C for 48 ±2 hrs. After that left for 4-5 hrs to cool down and reach room temperature. After weighting the specimens again theywere immersed in a fully saturated salt solution (sodium chloride), for 24 ±2 hrs. These cycles were repeated for 3 times, then the bricks were weighed after each cycle.

2.4.3 Prism Compressive Strength

In this test, five bricks were tested as a unit as they were used to build a prism with a 10 mm layer of mortar between each two successive bricks to form a prism as shown in Figure 1. Each prism had an aspect ratio of 4 between the its height and the thinnest side which was factored according to pre-specified factors by the ASTM. This was done three times for three different mixes having percentage slurry of 0, 5 and 10%. The load was applied to satisfy a deflection rate of 3 mm/minute in a load control loading scenario. The full record of load versus deflection was recorded and used to plot the stress - strain curves for each specimen.

Figure 1: A prism sample during testing it under compression.

3 TEST RESULTS AND DISCUSSION

3.1 Results of Fresh Tests

Table 2 shows the results of the fresh tests. The results shown indicate variations in the unit weight within limited ranges meaning that the powder/slurry had a limited effect on the air content and the unit weight. This limited effect could be attributed to the fact that originally the difference between slurry and sand (that it is partially replacing) in unit weight is minor as previously reported by (Hamza et al, 2011) which caused the impact on the unit weight and air content to be limited. Furthermore, the low water-to-cement ratio caused the slump to be of zero for all of the mixtures as this low water-to-cement ratio was kept constant for all of the mixtures.

Table 2: Results of Fresh Tests

Percentage Slurry	Slump (mm)	Air Content	Unit Weight kg/m^3
0%	0	8%	2090.0
5%	0	6%	1964.3
10%	0	2.8%	2300.0
15%	0	9.0%	2075.7
20%	0	7.8%	1867.1
25%	0	8.5%	1742.9

3.2 Brick Compressive Strength

The compressive strength results after 7 days for the concrete brick units with 0%, 5%, 10%, 15%, and 20% of slurry replacement are shown in Table 3. The results indicated that the highest strength was obtained at 5% slurry replacement. However, to determine more precise percentage mixes with 2.5% and 5% were tested after 3 days which showed that the mix with 7.5% has the highest compressive strength. Table 3 shows the obtained compressive strength for all mixes after 3 and 7 days

Table 3: 3-Days and 7-Days Brick Compressive Strength

Slurry Percentage	Average Compressive Strength (MPs)	
	After 3 days	After 7 days
0%	-	27.4
2.5%	19.5	-
5%	18.3	29.3
7.5%	21.0	38.5
10%	17.3	21.9
20%	-	20.0
25%	-	19.0

According to the results presented in Table 3, the 7.5% granite slurry replacement shows much higher value of compressive strength when compared to its counterparts. This can indicate that granite slurry can have a better interface with cement paste in the mix beyond purely physical micro filling action. In addition, granite samples show high trend in terms of the degree of strength achieved after 7 days. It is worth mentioning that 10% and 20% slurry samples showed nearly the same range of compressive strengthswhich is consistent with the results presented previously by Hamza et al (2011). The brick was not tested at 28 days because in real life they don't wait 28 days for the brick, so the concern was on 3 & 7 days compressive strength.

3.3 Mortar Cubes Test

Figure 2 shows the compressive strength of mortar cubes. The figure shows that the compressive strength of the mortar with 7.5% slurry replacement was significantly higher than the control mortar mixes. Therefore it is concluded that adding 7.5% granite could significantly increase the mortar compressive strength. The 2.5 % increament was obtained at the end of the scoope because it was important to know excatly the highest compressive strength and it was in the range between 5 % & 10 %. So , another samples was made with the increment of 2.5 % and the results is in figure 2.

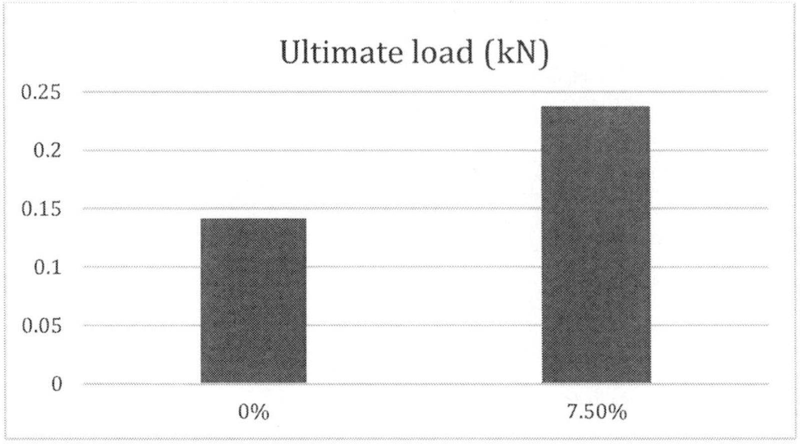

Figure 2: Comparison between the ultimate compressive load carried by the control mortar sample and the mortar sample containing 7.5% slurry

3.4 Cycles of Immersion in Salt Solution and Drying

The results of three cycles of immersion in salt solution followed by oven drying for six mixtures with different percentage slurry are illustrated in Figure 3. As observed, it was found that throughout the cycles the weight increased and this might be because the salt particles that became immersed inside the bricks because of being fully immersed in the solution. However, this variation in weight from one cycle to the other was not major as shown in Figure 3. This may be attributed to the fact that all of the mixes had low water-cement ratios which caused the bricks to have less voids and hence a good level of durability in general.

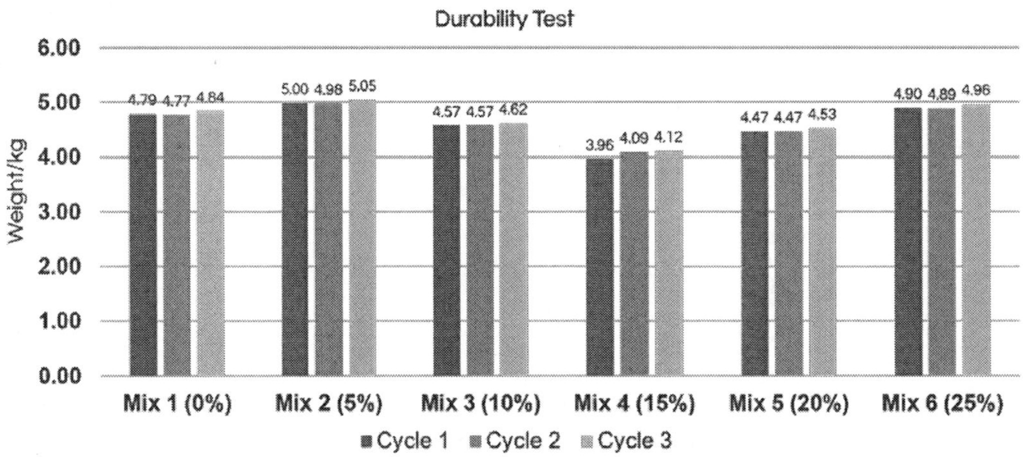

Figure 3: Cycles of immersion in salt solution

3.5 Prisms Test

The maximum compressive load for each prism was recorded and the average value was calculated for each group as shown in Figure 4. It could be obviously seen that the control group (0%) failed at a significantly lower load when compared to its counterparts including a percentage of slurry. It could also be noticed that the 5% and 10% groups got results very near to each other.The obtained stress-strain curves are shown in Figure 5.

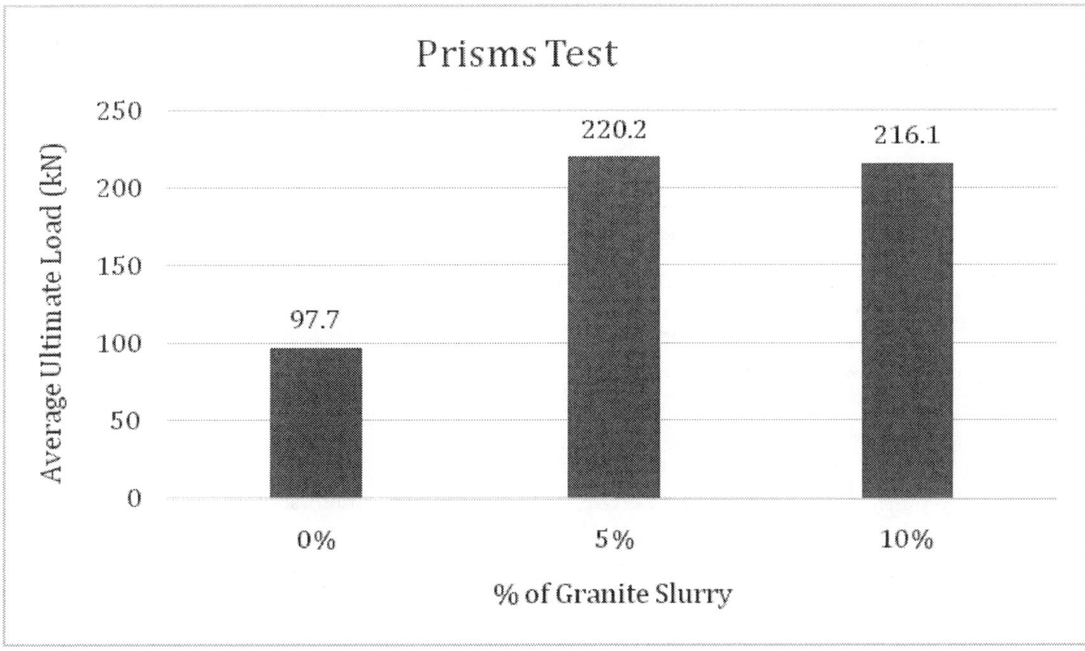

Figure 1: The average ultimate load carried by prisms with different percentages of slurry.

From the three stress-strain diagrams shown in Figure 5 it could be noticed that the three samples had high variability in their results. However, when comparing all of the three control samples to the others shown in Figure 5b and Figure 5c, it could be noticed that all of the control samples failed at stresses lower than these withstood by the two groups that included a percentage of slurry. Meanwhile, the 5% group shown in Figure 5b has some degree of variability that could be attributed to the human error during constructing the prisms as it was very challenging to have exactly the same thickness of the mortar layers between the bricks and having an ideal verticality of the prism. This same variability in construction caused one of the 10% samples shown in Figure 5c to experience early failure as this sample in specific failed at an angle as it was obviously not ideally vertical.

Meanwhile, when comparing the stresses shown in the stress-strain diagrams in Figure 5 to the brick compressive strength shown in Table 3, it could be noticed that the stresses at which the prisms failed were less than what the bricks could actually withstand. This was due to the fact that for all of the prism samples the failure happened in the layer of mortar between the bricks and none of the samples have experienced failure in the bricks themselves.

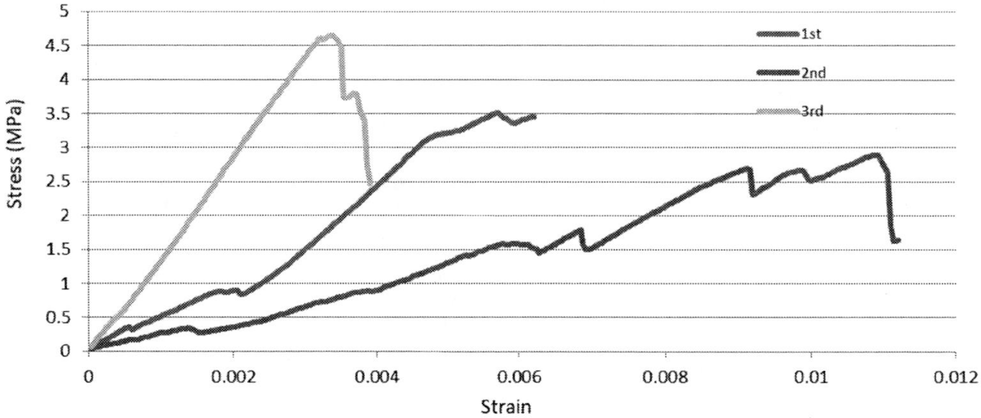

a. Stress-strain curve for the 0% (control) prism sample.

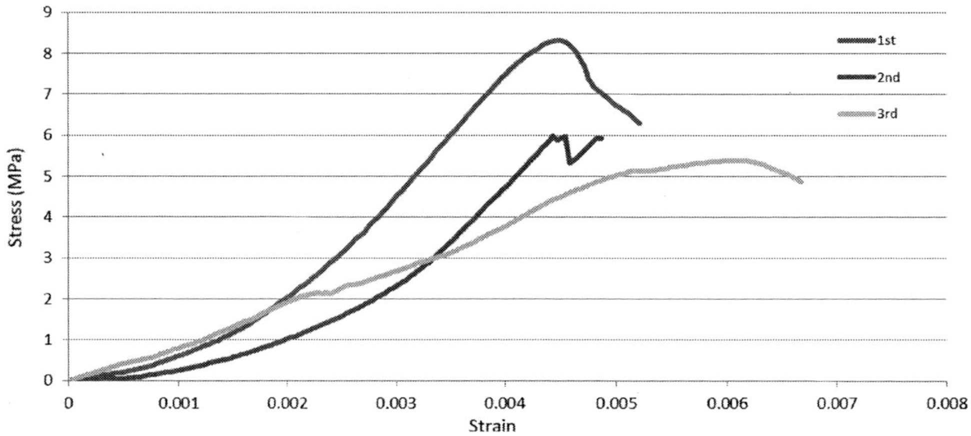

b. Stress-strain curve for the 5% slurry prism sample.

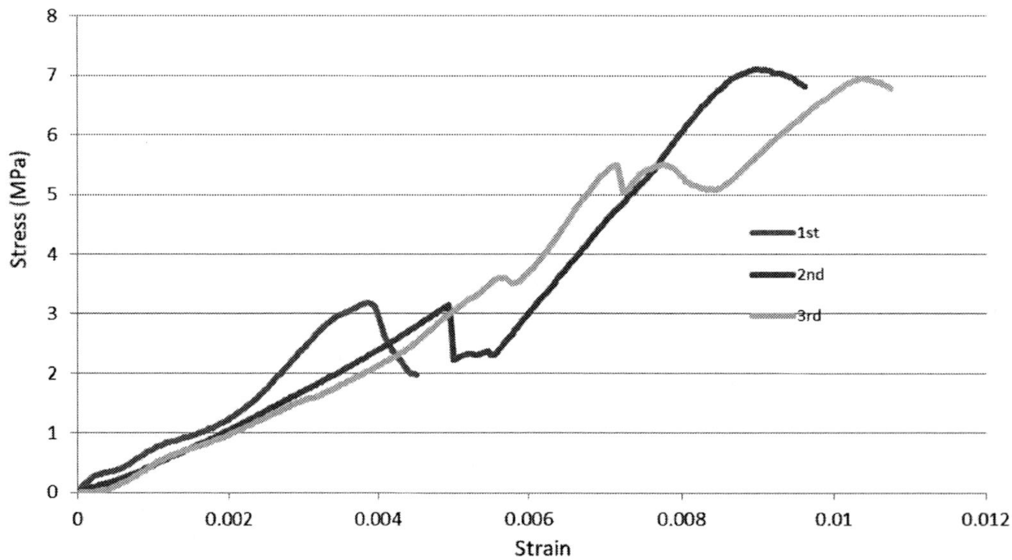

c. Stress-strain curve for the 10% slurry prism sample.

Figure 2: The stress-strain curves for the prism samples.

4 CONCLUSIONS AND RECOMMENDATIONS

4.1 Conclusions

The application of granite slurry into the bricks was investigated. Many tests were conducted to further understand this unique material and enhance the research available on it. Four tests were conducted and their results analyzed. Taking the materials and techniques implements into consideration, the following can be concluded:

1. Granite slurry is a promising innovative material that has potential yet to be explored in the construction industry.

2. Manufacturing of bricks with granite slurry contribute to both environmental as well as technical merits as the incorporation of slurry into the bricks compute to the enhancement of mechanical properties.

3. The fresh test results were quite similar while the slurry had a unique behavior in the slump tests as the result was zero because the w/c ratio was very low.

4. The compressive strengthof the prisms made with brick with granite slurry replacement is more than double the that of the conventional units.

5. It is also important to note that the granite and slurry exhibited superior performance and advantageous qualities in the chemical durability test which means that these bricks can withstand chemical solution at least for a short period of time.

4.2 Recommendations

For the purposes of further research, the following is recommended:

- • Investigating mixes with different cementitious component.
- • Investigating the thermal resistance of Granite Slurry bricks.
- • Further research can be done in exploring the possibility of producing mortar with Granite Slurry and exploring its physical properties and durability.
-
- In the application of usage of granite slurry in the industry the authors would recommend the following:
- • Study the possibility of producing bricks with Granite slurry on large scale in order to replace the conventional cement bricks.
- • As units of brick with mortar in between (prisms) have resulted in more than double the load carrying capacity of conventional units, it is worth to apply this technology in the industry and monitor its performance.

Acknowledgements

The authors would like to acknowledge the efforts of the teaching assistants, lab engineers and lab technicians in the construction engineering labs at the American University in Cairo for their extensive help and assistance during performing the work.

References

Allam, M. E., and E. S. Bakhoum. "Re-use of granite sludge in producing green concrete." *ARPN Journal of Engineering and Applied Sciences,* vol. 9, no. 12, Dec. 2014.

El-Sayed, Hamdy A. “Characteristics of the marble processing powder waste at Shaq El-Thoaban industrial area, Egypt, and its suitability for cement manufacture.” *HBRC Journal*, 13 June 2016.

Kumara, Virendra. “Experimental study on granite sludge blocks.” International Journal of Informative & Futuristic research, vol. 2, no. 10, June 2015, pp. 3773–3781.

Mehta, Jitesh, and JayeshkumarPitroda. “Effective utilization of granite powder as replacement of marble powder in marble brick.” effective utilization of granite powder as replacement of marble powder in marble brick, vol. 2, no. 3, may 2016, doi:10.15438/rr.4.4.22.

Hamza,Rania A., Salah EL-Haggar, and SafwanKhedr, “Marble and Granite waste: Characterization and Utilization in concrete bricks”. International Journal of bioscience, biochemistry and bioinformatics, vol. 1, no. 4, November 2011.

DhanapandianS. and OM. Shanthib., Utilization of marble and granite wastes in brick products”, Jr. of industrial pollution control, volume 25, issue 2, 2009, pp 145-150.

Standard Test Methods for Sampling and Testing Concrete Masonry Units and Related Units, ASTM C140/C140M – 17a.

Standard Test Methods for Conducting Strength Tests of Masonry Wall Panels, ASTM C1717 – 12.

Standard Test Method for Compressive Strength of Masonry Prisms, ASTM C1314 – 16.

Building Tomorrow's Society

Bâtir la Société de Demain

Fredericton, Canada

June 13 to 16, 2018 / *13 au 16 juin, 2018*

UTILIZING THE WASTE OF MARBLE AND GRANITE INDUSTRY IN PORTLAND CEMENT CONCRETE

Emam, Hala[1], Naguib, Merna[1],El Rashidy, Nouran[1], Khalil, Rana[1], Awad-Allah, Yasmein[1], Ghaly, Athnasious[1], Abou-Zeid, Mohamed[1]
[1] Department of Construction Engineering, The American University in Cairo, Egypt
[2] yasmein@aucegypt.edu

Abstract: The waste of the marble and granite industry imposes an imminent environmental and health hazard in many parts of the world creating various problems that are not only limited to the aforementioned hazards but are also related to high economic costs. Such waste could be used to replace both the fine and coarse aggregates in a Portland cement concrete mix enhancing its properties while simultaneously offering burial destination to such waste. Hence, it is vital to explore and research the properties of such waste products before incorporating it into Portland cement concrete. In this work, Portland cement concrete mixtures are prepared using the solid debris waste of marble and granite to partially replace coarse aggregates in three dosages. In Addition, 3 mixes using marble debris and slurry to replace both the fine and coarse aggregates is prepared to test the effect of the slurry. Aggregate, thermal, fresh and hardened concrete testing were conducted in addition to other tests such as the water permeability, chemical resistance, RCPT and abrasion. The main aim of this work was to reach a good quality concrete mix that could be positively compared to the conventional concrete; in addition, a feasibility study exploring the means of incorporating such waste in terms of material and transportation costs as well as the costs of the health and environmental hazards is prepared and analyzed. Conclusions and recommendations are acquired to define the most appropriate dosage range to implement as well as to highlight possible advantages of using such waste.

1 INTRODUCTION

Marble and Granite industry is considered one of the largest industries in the world, where the global production capacity of marble and granite have been rising on average by an 8.8% rate of growth annually, starting the year 2000. Moreover, around 60-80% of the entire production process is attributed to waste in the form of debris; as for the generated sludge, it is estimated to be over 30% of the block's volume. The wastes produced from the Marble and Granite industry have the capacity to highly pollute the surroundings and land ecology due to the stones' high alkalinity, imposing an imminent risk to the environment. The waste production cycle starts with the squaring of the marble blocks using the Monolama; then it is placed in the Gang saw to adjust the slab thickness before polishing in the polishing machine. Finally, the slabs are cut into tiles using the cutting machine. The solid waste parts (debris) are mainly produced from the Monolama, cutting machine and the polishing machine; after accumulation the waste is deposited outside the factory. Moreover, there is also wastewater resulting from the process, which is collected in a well and deposited in a water silo after a powder is added for deposition infection. Then, the marble dust particles go through filter presses for slurry treatment. The slurry, which is in a semi-moist dust form, is then accumulated outside the factories and dumped in public waste deposits using trucks. All of the slurry is then dried and scattered all over.

The harming impact of such waste is not only centered on the environmental damage of reduction in soil fertility and plant productivity, but it also contributes the threatening health diseases such as damaging the cells of the respiratory system, chronic coughing, shortening of breath, vision pollution, lung cancer and skin and eye irritation. Hence, it is of vital importance to explore feasible means of utilizing such waste to limit its negative impacts and highlight means of potential positive attributes.

2 LITERATURE REVIEW

2.1 Re-use of Granite Sludge in Producing Green Concrete:

Allam M. E., Bakhoum E. S. and Garas G. L.

In this paper, the environmental hazards and impacts of waste generated in the production process of granite are discussed. A shocking 60-80% is attributed to the waste generated from mining the granite, up to its final finishing process; as for the generated sludge, it was estimated to be over 30% of the block's volume. The work performed in this paper, discusses the possibility and experimental program of utilizing such waste, namely the sludge, to replace the cement and sand to produce a green concrete mix. The compressive strength results indicated that the highest compressive strength in comparison with the control was obtained with a 10% substitution of sand by the granite sludge, while that the replacement of cement, decreased the compressive strength regardless of any replacement percentage. Hence, indicating that the utilization of such waste material in the replacement of sand can have a positive impact on the compressive strength.

2.2 Marble and Granite Waste: Characterization and Utilization in Concrete Bricks:

Rania A. Hamza, Salah El-Haggar, and Safwan Khedr

This work discusses the impact of the flourishing industry of marble and granite in Egypt, which in turn increased the piles of mined and processed waste. Due to its high alkalinity, the waste generated is immensely polluting, imposing health and environmental hazards on the affiliated surroundings. Moreover, it particularly discusses the large amounts of waste deposited in Shaq Al-Thuban, the largest industrial area in Egypt and its effect on the neighboring areas, their communities and ecology. The work performed in this paper, aimed to utilize the stone waste in the manufacturing process of concrete bricks, fully replacing fine and coarse aggregates with waste debris and slurry reaching 40%. After testing the physical as well as the mechanical properties of the bricks incorporating such waste according to the ASTM and Egyptian Code (EC), it was concluded that such waste was rendered suitable for usage in the building industry as all bricks complied with all the structural bricks requirements in the EC, furthermore, that with 10% granite slurry replacement; the cement bricks reached their optimum properties.

2.3 Impact of Marble Waste as Coarse Aggregate on Properties of Lean Cement Concrete:

Sudarshan D. Kore*, A.K. Vyas

In the following journal entry, the work performed aimed to effectively employ the waste generated from marble, to replace the conventional coarse aggregate in the production of concrete, in order to alleviate the environmental hazards resulting from such waste. In order to conduct multiple experiments to study the validity and feasibility of such an attempt, various concrete mixes were prepared replacing the coarse with marble aggregate by weight through varying percentages (0-100%), with a constant w/c ratio throughout. The results indicated that the mixes containing marble, had a 14% higher workability than that of the control mix. Moreover, that the compressive strength of the mixes utilizing marble waste were observed to increase, on average, by 40% and 18% when tested at 7 days and 28 days, correspondingly.

3 OBJECTIVE AND SCOPE

In this research, the main objective was to explore feasible means to utilize waste resulting from the marble and granite industry not only to be incorporated into Portland cement concrete but also to potentially enhance the concrete industry. This objective was obtained through conducting experimental work for the marble and granite waste in Portland cement concrete.

4 EXPERIMENTAL PROGRAM

4.1 Material Properties

Cement: Type I Ordinary Portland Cement Concrete

Fine Aggregates: Natural sand will be used.

Coarse Aggregates: Well-graded crushed dolomite was used as coarse aggregates. The aggregates had MNA of 38 mm.

***Marble slurry*:** A marble sludge powder was used, which was obtained from the area of Shaq Al Thu'ban, Cairo, Egypt.

Marble debris: Marble waste used was from the area of Shaq Al Thu'ban and crushed into crusher with MNA of 38 mm.

Granite debris: Granite waste used was from the area of Shaq Al Thu'ban and crushed into crusher with MNA of 38 mm.

4.2 Concrete Mix Design

The characteristic concrete mixture that was used in this study has a cement content of 400 kg/m^3 and a water-to-cement ratio of 0.45. based on literature review and material properties other mixtures will be prepared by partially replacing the coarse aggregates with marble debris with dosages of 20%, 40% and 60% and granite debris with dosages of 30%, 60% and 100%. Other mixtures will be made by replacing the fine aggregates by the marble slurry and the coarse aggregate by marble debris. Concrete mixing and casting of specimens were carried out according to ASTM standards.

Table 1 Concrete mix

Mix No.	mix name	Cement (kg/m^3)	Fine aggregate (kg/m^3)	Coarse aggregate (kg/m^3)	Marble debris (kg/m^3)	Granite debris (kg/m^3)	Marble slurry (kg/m^3)	Water (kg/m^3)	w/c ratio
1	Control Mix	400	621	1117	-	-	-	180	0.45
2	20% Marble	400	620	893	223	-	-	180	0.45
3	40% Marble	400	619	669	446	-	-	180	0.45
4	60% Marble	400	619	446	668	-	-	180	0.45
5	30% Granite	400	623	785	-	336	-	180	0.45
6	60% Granite	400	625	450	-	675	-	180	0.45
7	100%Granite	400	619	-	-	1114	-	180	0.45

Mix No.	mix name	Cement (kg/m³)	Fine aggregate (kg/m³)	Coarse aggregate (kg/m³)	Marble debris (kg/m³)	Granite debris (kg/m³)	Marble slurry (kg/m³)	Water (kg/m³)	w/c ratio
8	10% slurry	400	558	893	223	-	62	180	0.45
9	20% slurry	400	496	893	223	-	124	180	0.45
10	30% slurry	400	435	893	223	-	186	180	0.45

4.3 Tests

4.3.1 Aggregate Tests:

- Sieve analysis: Sieve analysis was performed to determine the gradation of aggregates
- Specific gravity: Test was conducted to determining the specific gravity of the aggregates.
- Absorption: Test was conducted to determine the percentage of water absorption by weight of the aggregates.
- LAA (Los Angeles abrasion): Test was conducted to determine the degradation of aggregates by abrasion

4.3.2 Fresh Concrete Tests:

- Slump: This test is conducted to check the workability of freshly made concrete in accordance with [ASTM C232]
- Air content: This test is conducted to determine the air content of concrete in accordance with [ASTM C231]
- Unit weight: This test is conducted to determine the unit weight of concrete.

4.3.3 Hardened Concrete Tests:

- Compressive Strength: The test is conducted to evaluate the strength of concrete using cubes in accordance with British Standards [BS 1881]. This test was performed on cubes of 15cm*15cm*45cm at 7 days and 28 days.
- Flexural Strength: The test is conducted in accordance with the ASTM standards [ASTM C78] to measure the flexural strength of the mixes. Three Point Test was performed on beams of 15cm*15cm*15cm at 3 days, 7 days and 28 days.
- Thermal Expansion: The test is conducted to evaluate the thermal expansion, where samples of 2.5cm*2.5cm*8cm, where a steam generator is used to heat the samples.
- Thermal Conductivity: The test is conducted to evaluate the thermal conductivity of the concrete cubes in accordance with the ASTM Standards [ASTM C518] on a sample of 15cm*15cm*5cm after 28 days of curing.
- Water permeability: This test measures the permeability of mortar or concrete to water by measuring how deep the water penetrates the specimen through three days under pressure. The test is done on specimen after 28 days.
- Chemical resistance:
  - Sulphuric acid (H_2SO_4): cubes were submerged in diluted sulphuric acid with concentration 10% for 1 cycle.
  - Sulphate salts ($MgSO_4$): cubes were submerged in solution of salt with concentration of 60% for 4 cycles.
- RCPT (rapid chloride penetration test): In this test, we were trying to study the performance of concrete in resisting the flow of chloride ions through it. This can help us in determining whether this material can be used in applications that require less permeability of chloride ions.

- Abrasion: This test is based on relative value comparison, specimens of 5cm*5cm*5cm cubes was fixed on a rotating surface and abrasive material and water was added to find the reduction in length.
- Results and Discussion

4.4 Aggregate tests:

4.4.1 Sieve analysis:

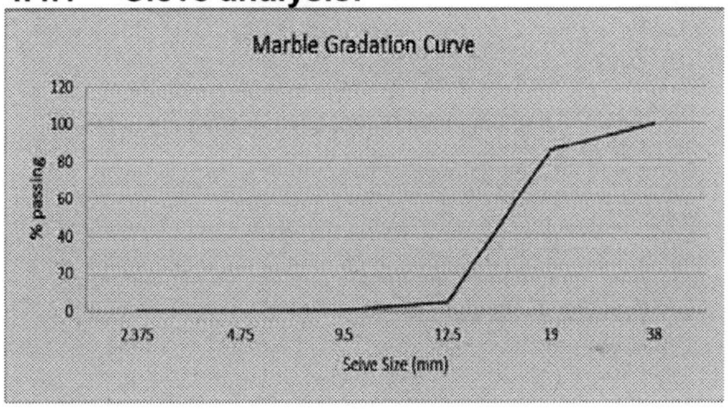

Figure 1

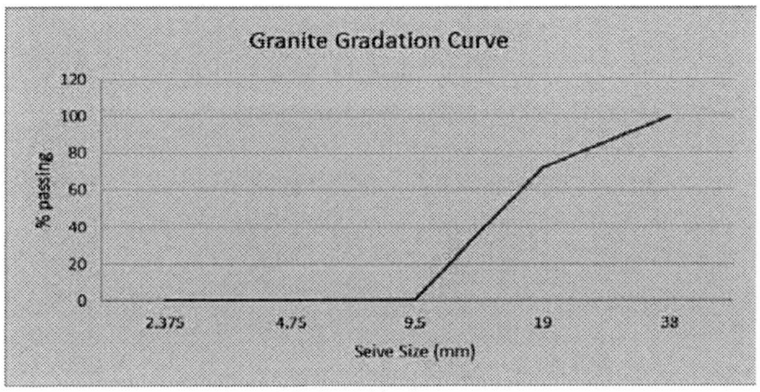

Figure 2

Table 2 Aggregates properties

	Marble	granite	Course aggregates	Fine aggregates
specific gravity	2.58	2.59	2.57	2.55
Absorption	1.55%	0.50%	1.80%	0.40%
LAA	24.50%	48.10%	-	-

The sieve analysis tests for both the marble and granite showed that both samples were poorly graded; however other properties such as the specific gravity and the absorption percentage were aligning with the materials' properties from the research done. The LAA Abrasion test, used to measure the aggregates' resistance to abrasion and impact, produced an adequate result for the marble, but a slightly higher than average percentage loss in the granite.

4.4.2 Fresh concrete results

Table 3 Fresh test results

	Slump (cm)	Unit Weight (kg/m^3)	Air Content
Control Mix	2.5	2417	2.4%
20% Marble	2.5	2399	2.5%
40% Marble	3.5	2393	2.0%
60% Marble	4.0	2379	2.4%
30% Granite	4.0	2376	2.5%
60% Granite	6.5	2376	1.7%
100% Granite	5.0	2390	2.0%
10% Slurry	7.5	2374	2.7%
20% Slurry	11.0	2366	1.5%
20% Slurry	11.0	2340	1.4%

The slump test gave adequate results for the marble and granite mixes; however, it was high in the slurry mixes due to its unique behaviour when it's mixed with the rest of the concrete components. Moreover, the unit weight and the air content tests gave similar results to those obtained from the research.

4.5 Hardened tests:

4.5.1 Compressive Strength

Table 4 compressive strength results

Mix	3 days (MPa)	7 days (MPa)	28 days (MPa)
Control Mix	28.1	34.4	40.5
20% Marble	25.2	34.3	43.5
40% Marble	26.2	29.4	34.7
60% Marble	27	29.8	39.8
30% Granite	27.2	28.1	33
60% Granite	24.1	25.7	33.1
100% Granite	22.6	19.2	31.1
10% Slurry	20.9	28.3	34.9
20% Slurry	21.7	27.5	33.1
30% Slurry	17.5	20.7	27.1

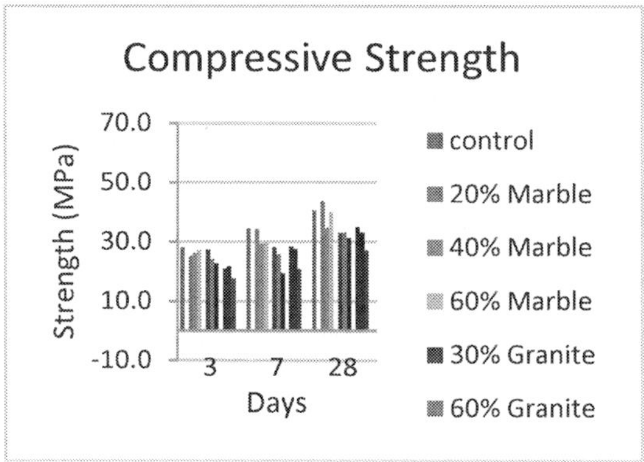

Figure 3 Compressive strength results

The compressive strength at 3 days, 7 days and 28 days was tested for all of the samples and compared with the performance of the control mix. For the marble mix the results show that there is a slightly decrease trend and that all of the three samples at 28 days were of adequate values with respect to the control mix with the 20% marble replacement dosage reaching the maximum compressive strength with a value of 43 Mpa. As for the granite mix, the results exhibit adequate results as well with a minimum of 30 Mpa for the 100% replacement, although they were all less than the value of the control mix. Finally, for the slurry mixes at 28 days, the compressive strength results ranged from 26-35 Mpa which could also be considered adequate. The value of the 40% marble is very low compared with the 20% and 60% as it may be because of an error in compaction.

4.5.2 Flexural Strength (ASTM C-78)

Table 5 Flexural strength results

Mix	7 days (MPa)	28 days (MPa)
Control	3.4	3.9
20% Marble	3.5	3.9
40% Marble	3.7	4.5
60% Marble	4	5.6
30% Granite	3.4	3.6
60% Granite	3.1	3.5
100% Granite	3.1	3.5
10% slurry	2.9	2.8
20% slurry	2.7	2.7
30% slurry	2.4	3

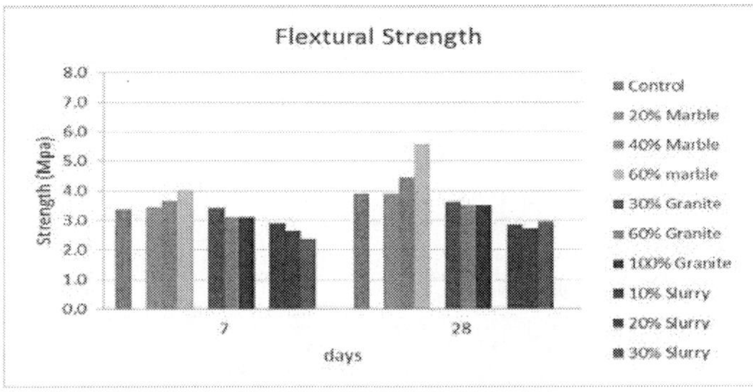

Figure 3 Flexural strength results

The flexural strength results showed an optimum replacement of 60% marble with 5.56 Mpa, which is greater than the control mix. Moreover, the remaining results showed comparable values to that of the control. Moreover, there was an observable trend in the Marble mixes; as we increased the dosage of the marble replacement, the flexural strength increased. The Granite and the slurry mixes show approximately similar results for the three percentages.

4.5.3 RCPT:

Table 6 RCPT test results

	Charge Passed (Coulombs)
Control	3640
20% Marble	4674
40% Marble	4267
60% Marble	4130
30% Granite	3375
60% Granite	4046
100% Granite	7474
10% Slurry	3641
20% Slurry	4838
30% Slurry	4860

Since some mixes resulted in very high results, the EDX (Energy Dispersive X-Ray) test was conducted in order to know and asses the different components in the marble and granite debris. The test resulted in finding some steel in the granite aggregates and hence, interfered with the RCPT results. Moreover, since all of the waste is deposited in the same area, there is a possibility that the marble and the slurry would be affected as well. Hence, the RCPT test results were deemed not reliable to assess the concrete properties.

Contaminated granite (EDX Test):

Standard :
O SiO2 1-Jun-1999 12:00 AM
Na Albite 1-Jun-1999 12:00 AM
Al Al2O3 1-Jun-1999 12:00 AM
Si SiO2 1-Jun-1999 12:00 AM
K MAD-10 Feldspar 1-Jun-1999 12:00 AM
Ca Wollastonite 1-Jun-1999 12:00 AM
Fe Fe 1-Jun-1999 12:00 AM

Element	App Conc.	Intensity Corrn.	Weight%	Weight% Sigma	Atomic%
O K	23.71	0.8878	54.88	0.81	68.80
Na K	0.51	0.8239	1.27	0.19	1.11
Al K	1.80	0.8739	4.23	0.25	3.15
Si K	14.67	0.8927	33.77	0.66	24.12
K K	1.27	0.9746	2.67	0.22	1.37
Ca K	0.99	0.9310	2.18	0.21	1.09
Fe K	0.40	0.8193	0.99	0.26	0.36
Totals			100.00		

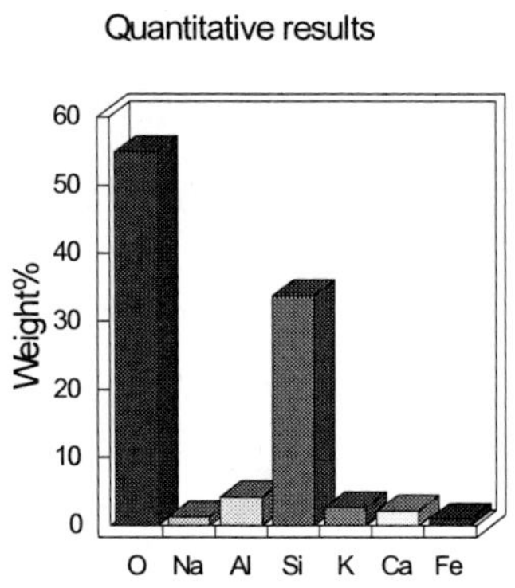

Contaminated Marble (EDX Test):

Standard:
C CaCO3 1-Jun-1999 12:00 AM
O SiO2 1-Jun-1999 12:00 AM
Si SiO2 1-Jun-1999 12:00 AM
Ca Wollastonite 1-Jun-1999 12:00 AM

Element	App Conc.	Intensity Corrn.	Weight%	Weight% Sigma	Atomic%
C K	3.11	0.8646	4.14	0.76	7.25
O K	17.61	0.3788	53.47	0.90	70.29
Si K	0.72	0.8629	0.95	0.11	0.72
Ca K	37.34	1.0368	41.43	0.80	21.74
Totals			100.00		

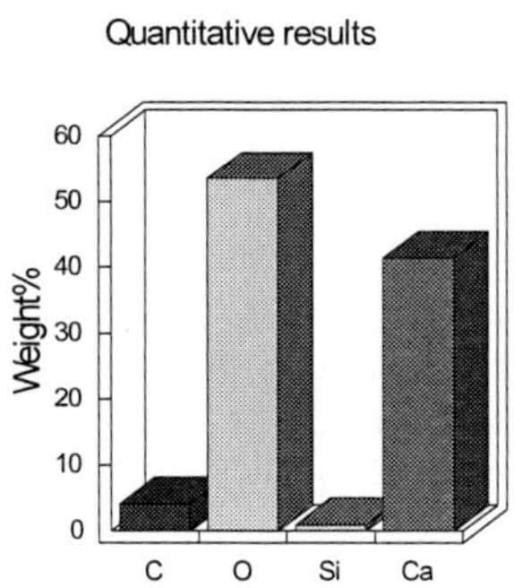

4.5.4 Water permeability:

Table 7 Water permeability results

	Average (cm)
Control	1.0
20% marble	1.2
40% marble	1.4
60% marble	1.5
30% granite	3.3
60% granite	3.5
100% granite	3.7
10% Slurry	2.2
20% Slurry	2.7
30% Slurry	3.1

Figure 6

The water permeability test exhibited a common trend in all of the waste mixes, since the water permeability increased with increasing the percentage replacement of waste. However, the results are not prohibitive to utilize the marble and slurry waste as their values are comparable to that of the control mix.

4.5.5 Abrasion:

Table 8 Abrasion test results

	% loss
Control	1.83%
60% Marble	2.44%
60% Granite	0.50%
20% Slurry	0.90%

The results show that the properties of the Granite and slurry are better in abrasion than those of the control mix; Moreover, even though the percentage loss in the marble mix was about 0.6% more than that in the control mix, it is still a relatively small increase and does not disqualify the usage of marble waste.

4.5.6 Chemical Resistance:

- Sulphuric Acid Resistance:

Table 9 chemical test results

Mix	% Reduction
Control Mix	8.50%
Marble 20%	8.50%
Marble 40%	7.00%
Marble 60%	2.20%
Granite 30%	0.50%
Granite 60%	0.40%
Granite 100%	0.10%
Slurry 10%	1.50%
Slurry 20%	0.50%
Slurry 30%	0.40%

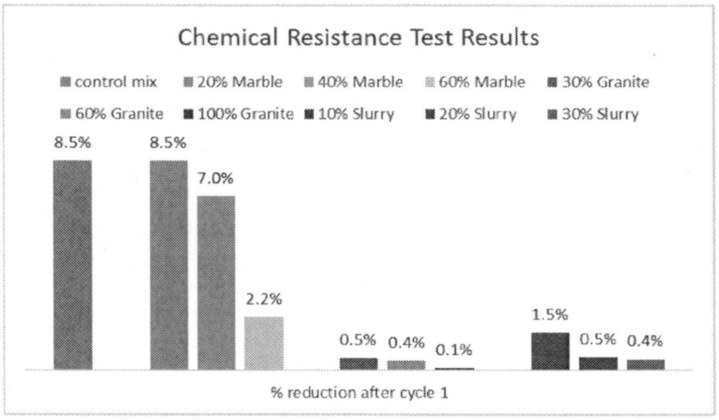

Figure 7 chemical resistance results

The results of the acid chemical tests indicate a common trend amongst all waste materials; as we increase the percentage replacement, the percentage reduction is decreased indicating superior chemical resistance qualities of the marble, granite and slurry.

- Sulphate Salts Resistance:

	% reduction after cycle 1	% reduction after cycle 2	% reduction after cycle 3	% reduction after cycle 4
control mix	5.0%	3.7%	2.2%	3.2%
marble 20%	5.1%	3.7%	2.2%	3.3%
marble 40%	4.7%	0.3%	-0.8%	-0.1%
marble 60%	0.9%	-1.6%	0.6%	1.8%
Granite 30%	-1.6%	-0.1%	-0.9%	-2.5%
Granite 60%	-1.9%	-0.3%	0.3%	-0.8%
Granite 100%	-2.6%	-0.3%	0.3%	-0.5%
Slurry 10%	2.1%	2.4%	-0.7%	1.9%
Slurry 20%	1.6%	2.1%	-0.5%	0.4%
Slurry 30%	1.3%	1.1%	0.1%	-1.5%

The magnesium sulfate test is deemed not reliable in assessing the salts chemical resistance of the mixes due to the fact that the salts and other chemicals significantly precipitate inside the mixes as we add the magnesium sulfate giving negative percentage reduction at cases; and hence, the results of this test will not be deemed reliable in assessing the concrete properties.

4.5.7 Thermal Conductivity:

Table 10 Thermal conductivity test results

Sample Type	Density (Kg/m3)	Thermal Conductivity (W/m.K)
control	2317	1.29
60% marble	2320	1.34
60% Granite	2270	0.92
20% Slurry	2213	0.70

The thermal conductivity test results show that the marble has a slightly higher thermal conductivity coefficient than the control mix, while the granite and slurry mixes had lower values than that of the control indicating superior thermal properties; which means the heat transfer at low rate from one side to another.

4.5.8 Thermal expansion

Table 11 Thermal expansion results

Mix	Thermal Expansion Coefficient (c^{0-1})
Control	$9.6x10^{-6}$
60% Marble	$9.5x10^{-6}$
60% Granite	$9.2x10^{-6}$
20% Slurry	$9.4x10^{-6}$

The thermal expansion test shows that the control has the highest coefficient and the granite showed the lowest thermal coefficient.

5 CONCLUSIONS

While our study was limited in time and scope the following conclusions could be withdrawn:

- The aggregate testing revealed that all of the waste material was poorly graded while the granite in particular was flaky and elongated
- The fresh test results were quite similar while the slurry had a unique behaviour in the slump tests
- The compressive test results were comparable and the variations in strengths don't show a low value not to be used in construction applications while flexural results showed adequate results for the waste materials compared to the control and the reduction in flexural strength was less than the compressive strength for waste materials
- The RCPT test was deemed not reliable due to the inclusion of metallic deposits and hence, the water permeability test was found more suitable in assessing the permeability of waste mixes. It is also important to note that the granite and slurry exhibited better performance in abrasion than the control mix and that the waste materials exhibited a superior performance and advantageous qualities in the chemical durability acid test and in the thermal tests.

6 RECOMMENDATIONS

The recommendations of our study are as follows:

- Conduct a larger scope experimental work to verify the findings of this study as well as to carry out further tests on non-reinforced as well as reinforced concrete
- Conduct the Radon test to test for radioactive materials, as well as conduct tests for long term properties such as creep and fatigue.
- Use this concrete in structures exposed to acids and high temperatures due to their beneficial properties in such conditions.
- Urge the Egyptian Government as well as the construction industry to not only consider investing in utilizing the waste products but to also consider including them in the Egyptian code of practice
- Include this study as one of several steps towards more environmentally friendly and greener construction industry.

Acknowledgements

- Mr. Ashraf Awad, CEO of KiroMarble Factory
- Dr. Mahmoud Abd El Razek, HBRC
- Dr. Medhat Al- Mahalawy, HBRC
- Engineer Mohamed Salah, Physics Department, American University in Cairo
- Engineer Dina, Mechanical Department, American University in Cairo
- Dr. Ahmed Noor, Research Centre, American University in Cairo

References

Abukersh, S.A., and C.A. Farfield. "Recycled aggregate concrete produced with red granite dust as a partial cement replacement." pag. *School of Engineering and the Built Environment, Edinburgh Napier University, 10 Colinton Road, Edinburgh, EH10 5DT, UK*. Web. 20 Jan. 2017.

Allam, M.E., E.S. Bakhoum, and G.L. Garas. “Re-use of granite sludge in producing green concrete.” pag. *Department of Civil Engineering, National Research Centre, Egypt*. Web. 20 Jan. 2017.

Al-Zboon, Kamel, and Jehad Al-Zouby. "Recycling of stone cutting slurry in concrete mixes." *Springer Japan* (2014): Web. 20 Jan. 2017.

ASTM C518-15, Standard Test Method for Steady-State Thermal Transmission Properties by Means of the Heat Flow Meter Apparatus, ASTM International, West Conshohocken, PA, 2015, www.astm.org

Haggag, Farah.” Economic and financial valuation of the marble industry in Egypt”. *Department of Economics, American university in Cairo, Egypt*. Web. May. 2012.

"Health Risks of Marble Dust." *LIVESTRONG.COM*. Leaf Group, 24 Oct. 2015. Web. 11 Feb. 2017.

Khedr, Safwan, Salah El- Haggar, and Rania Hamza. "Marble and Granite Waste: Characterization and Utilization in Concrete Bricks." *International Journal of Bioscience, Biochemistry and Bioinformatics* 1.4 (2011): Web. 20 Jan. 2017.

AUTHOR INDEX

AUTHOR INDEX

Canadian Society for Civil Engineering (CSCE)
521-300 rue St-Sacrement
Montréal, QC
H2Y 1x4

CURRAN ASSOCIATES INC.
proceedings.com

ISBN 978-1-5108-8975-0
9 781510 889750